KU-408-084

ENDURANCE

100 TALES OF SURVIVAL, ADVENTURE AND EXPLORATION

Also in the anthology series

THE ART OF THE GLIMPSE
100 Irish Short Stories
Chosen by Sinéad Gleeson

THE BIG BOOK OF CHRISTMAS MYSTERIES
The Very Best Yuletide Whodunnits
Chosen by Otto Penzler

DEADLIER
100 of the Best Crime Stories Written by Women
Chosen by Sophie Hannah

DESIRE
100 of Literature's Sexiest Stories
Chosen by Mariella Frostrup and the *Erotic Review*

FOUND IN TRANSLATION
100 of the Finest Short Stories Ever Translated
Chosen by Frank Wynne

FUNNY HA, HA
80 of the Funniest Stories Ever Written
Chosen by Paul Merton

GHOST
100 Stories to Read With the Lights On
Chosen by Louise Welsh

HOUSE OF SNOW
An Anthology of the Greatest Writing About Nepal
Chosen by Ed Douglas

JACK THE RIPPER
The Ultimate Compendium of the Legacy and Legend of History's Most Notorious Killer
Chosen by Otto Penzler

LIFE SUPPORT
100 Poems to Reach for on Dark Nights
Chosen by Julia Copus

OF GODS AND MEN
100 Stories from Ancient Greece and Rome
Chosen by Daisy Dunn

QUEER
A Collection of LGBTQ Writing from Ancient Times to Yesterday
Chosen by Frank Wynne

SHERLOCK
Over 80 Stories Featuring the Greatest Detective of all Time
Chosen by Otto Penzler

THAT GLIMPSE OF TRUTH
100 of the Finest Short Stories Ever Written
Chosen by David Miller

THE TIME TRAVELLER'S ALMANAC
100 Stories Brought to You From the Future
Chosen by Jeff and Ann VanderMeer

THE STORY
100 Great Short Stories Written by Women
Chosen by Victoria Hislop

WE, ROBOTS
Artificial Intelligence in 100 Stories
Chosen by Simon Ings

THE WILD ISLES
An Anthology of the Best of British and Irish Nature Writing
Chosen by Patrick Barkham

WILD WOMEN
and their Amazing Adventures Over Land, Sea & Air
Chosen by Mariella Frostrup

ENDURANCE

100 TALES OF SURVIVAL, ADVENTURE AND EXPLORATION

LEVISON WOOD

An Apollo Book

First published in the UK in 2022 by Head of Zeus Ltd,
part of Bloomsbury Publishing Plc

All excerpts have been reproduced according to the styles found in the original works. As a result, some spellings and accents used can vary throughout this anthology.

9 7 5 3 1 2 4 6 8

A catalogue record for this book is available from the British Library.

ISBN (HB) 9781801102568
ISBN (E) 9781801102582

Typeset by PDQ
Printed and bound in Germany by CPI Books GmbH

Head of Zeus Ltd
5–8 Hardwick Street
London EC1R 4RG
WWW.HEADOFZEUS.COM

For all those who never gave up

CONTENTS

Frontiers of Discovery

Pioneers and Mavericks

Pushing the Limits

INTRODUCTION

A remarkable thing happened while I was compiling this anthology. One hundred and seven years after the sinking of the famed ship *Endurance* in cold, Antarctic seas, its wreckage was discovered on none other than the one hundredth anniversary of the death of its captain, Sir Ernest Shackleton. The story of this ship's demise has gone down in history as one of the greatest displays of leadership, courage and perseverance of all time. It has inspired countless people and was one of the influences for me to create this book.

In 1915, Shackleton and his team of twenty-seven men embarked on an expedition to be the first to cross the South Pole. On their way, their ship became trapped in pack ice in the most treacherous and freezing ocean in the world – slowly crushing the hull. It seemed there was no way out. Yet Shackleton and his men persisted in their quest to survive. After a dark and difficult winter, they were forced to abandon their sinking ship and camp on the floe. When the sea ice finally started melting, they began their escape, rowing in lifeboats to Elephant Island. Shackleton and five others continued rowing for a further 800 miles to South Georgia to get help, and returned to rescue their crew. Almost two years after they set off, all of them made it home alive. Against all the odds, Shackleton managed to turn a seemingly catastrophic situation into a miraculous success.

This new, thrilling chapter in the story has been called the most difficult shipwreck discovery of all time. The ship was found perfectly preserved, standing almost upright, with its name still emblazoned in gold on its hull, some 10,000 feet at the bottom of the icy Weddell Sea. Shackleton's tale is so far removed from most people's lives, yet it resonates with many as a blueprint for resilience and leadership. These are qualities it is possible for every human to possess.

This anthology is filled with stories of courage and heroism. I think, even from this one example, it is clear why tales of endurance are so powerful. Stories are a force to educate and inspire, to make you curious about what drives others to do extraordinary things. For as long as I can remember, I've had a burning desire to explore – to learn and to understand – and I can say that much of this desire can be attributed to storytelling. I have always devoured books of all kinds, but especially those about history, exploration, or great feats of survival. As a child, I would sneak into my father's study and select the most heroic or gruesome accounts I could find. My parents were both teachers and introduced me to some of the great works of history. I was instantly enamoured, my eyes wide and open to the incredible tales on the pages in front of me, filled with lessons in survival, determination and exceptional courage. The stories that most fascinated me were those that spoke of the enduring nature of the human spirit.

I came across the *Odyssey* as soon as I could read and was hooked by the tales of great adventure and discovery in verse. As most kids are, I was thrilled by monsters, Greek Gods and heroes – the terrifying Cyclops and the Sirens who lured men to their doom. I have re-read it many times since and what grips me is the sheer perseverance of Odysseus's two-decade-long journey and fight for survival. I think, as humans, we are innately drawn to stories of people defying the odds. It reminds us of what we are capable of. There's no fakery, no façade, and in those stories, deep truths about human nature are revealed.

When I was first asked to create this anthology, I was faced with the daunting task of selecting which books and excerpts to include. I cannot say that I had a method for this process, other than that each of these books has, at some point or another in my life, taught me a valuable lesson about the human spirit. I would be the first to admit that this is not an exhaustive list of the most important, influential, or fascinating stories of human endurance ever written. There are millions of stories out there, happening every day, that could fall under this theme and there are many more books that I wanted to include and simply did not have the space. But I hope that the ones I have chosen will resonate with you as they did with me. For I truly believe that some of the most important lessons in life can be derived from the pages of great literature; the seeds of an idea or the spark of inspiration can spring off of a page and be the catalyst for lifelong passions and we can further our understanding of the world by viewing it through the eyes of another.

I was initially drawn to the theme of endurance by the stories that I read growing up, of explorers and pioneers who faced the odds in order to advance human knowledge and our understanding of the world. In my childhood bedroom in North Staffordshire, I pored over Sir Edmund Hillary and Tenzing Norgay's accounts of summiting Everest and longed for my own mountain adventure. Wilfred Thesiger's exploits in the deserts and marshes of Arabia sparked my life-long fascination with the Middle East and made me dream of the feeling of freedom that he described. Most of these accounts are not glamorous by any stretch of the imagination; logically, it seems ridiculous to crave to suffer as Thesiger did on some of his journeys. In 1946, he lay starving and dehydrated in the vast desert of the Empty Quarter, having walked for days under the relentless sun. He comforted himself with the assurance that there was no place he would rather be: 'I lay with my eyes shut, insisting to myself, "If I were in London, I would give anything to be here",' he wrote in *Arabian Sands*. Despite the comforts of my childhood home, I could think of nothing better than joining Thesiger on his adventures.

Some years later, I did follow in his footsteps. I ventured into the notorious expanse of the Empty Quarter desert. With each laborious step further into this barren, hostile environment, his words rang in my ears: 'the harder the way, the more worthwhile the journey'.

Humans are fascinating creatures. When faced with even the greatest physical and mental challenges, we can dig deep and surprise ourselves with our resilience and strength. The fact is, it's addictive: the rush of adrenalin that athletes get when they reach their goal, the fear and tension pulsing through a mountaineer's veins as he battles the elements to summit a peak, the allure of the unknown. There is no rhyme or reason as to why humans choose to put themselves through trials and challenges. The three most famous words in mountaineering were uttered by George Mallory about summiting Everest. When a journalist asked him the question: 'Why do you want to climb Mount Everest?', he retorted, 'Because it's there'.

In 2013, I walked the length of the Nile River. The expedition took me nine months and it was the hardest thing I have ever done. I remember one low point in particular was when I reached the Sahara Desert in Sudan; it wasn't even halfway, and I had already been walking for six months. I looked ahead and all I could see was endless sand. I was tired, hot and fed up, but I kept going, one foot in front of the other. In that journey, I faced fear, loss, pain, monotony – I experienced dreadful lows and ecstatic highs. It was a rollercoaster ride of emotions and an adventure that I will never forget. When I completed it, I started planning my next expedition.

Neil Armstrong believed in the unseen reality of going to the moon. He wasn't deterred by the risks, the unknown and by countless others telling him it was impossible. By persevering and sharing his story, he inspired generations to dream of endless possibilities.

What struck me the most when doing the research for this book, was that the greatest feats of survival and endurance are as much about mental strength, bravery and grit as they are about physical strength.

Having the power of endurance is necessary in many different situations – it can bring out the best qualities in some and the worst in others – the full range of human emotions and behaviours. Of course, in many situations that people endure, there's no choice in the matter. I felt that it was important to include the stories of people who have endured the worst that humankind has to offer, and who have experienced some of history's most horrific atrocities. Books can give a voice to the suffering of ordinary people and serve as a stark reminder of humanity's brutalities that we cannot ignore.

In that vein, there are examples in this book of great tragedy, of people plunged into life-or-death situations, forced to face horrific suffering at the hands of slave traders or watch their families murdered in horrendous genocides. In my career in the army and as a journalist, I have visited the frontline of war zones and seen with my own eyes the devastating impact of conflict. In my work as a writer and filmmaker, I try to immerse myself in the lives of locals and to challenge the myths and stereotypes of the places that I visit. It is heart-breaking to see the scenes of destruction in places like Ukraine and to witness the poverty brought about by the ongoing civil war in Yemen. Most of the time, in these places of unimaginable hardship, I find resilience, kindness and hospitality in the face of immeasurable evil. I have found that the people with less are often the ones who want to give more.

One of the great challenges I faced when creating this anthology was categorising the excerpts. I chose to organise them into the following themes: Against All Odds, Courage Under Fire, Frontiers of Discovery, Pioneers and Mavericks and Pushing the Limits. Within these sections, the extracts are grouped geographically. Each designation is by no means exclusive; many overlap and could be placed elsewhere. For example, from diving into these books it became starkly clear that all enduring experiences require mental fortitude. My purpose with these sections is to make this anthology something that can be dipped into rather than necessarily read all at once – unless you are so inclined.

My motivation in compiling this anthology is simple: to share portions of those stories that inspired me, in the hope that you might feel compelled to reach for the original texts and delve deeper into the human spirit of endurance that resides in their pages; and to decide for yourself which excerpts you would have chosen. If, after the glimpse I have offered here, you decide you want to learn more about these tales, then I have achieved my ambition with this book.

At the end of the day, endurance is a mindset. It's more than just taking that next gruelling step forward. It's having an incessant belief that, despite the challenges that lie ahead, you have the will to make it. The thread that holds all these stories together is the power of human beings to believe, to cling on tight to a shard of hope, to push our limits and to rely on our intrinsic survival instinct that allows us to defy all the odds. Nelson Mandela once said: 'Do not judge me by my success, judge me by how many times I fell down and got back up again.' Even when faced with the most challenging of shipwrecks to uncover, deserts to cross, mountains to climb, or the torments of conflict, these stories will be a reminder that you too have the strength to get back up again.

In the words of Ernest Shackleton: 'Through endurance we conquer'.

Levison Wood, April 2022

AGAINST ALL ODDS

MIRACLE IN THE ANDES

72 Days on the Mountain and My Long Trek Home

Nando Parrado

Nando Parrado (born 1949) survived the tragedy of Uruguayan Air Force Flight 571 which crashed in the Andes Mountains in October of 1972. Of the forty-five passengers and crew, eleven were killed during the crash, including Parrado's mother, Eugenia. Parrado was in a coma for three days due to a skull fracture and his sister, Susana, was among those who later succumbed to their injuries. Parrado and sixteen others remained trapped near the crash site for two months. They had little warm clothing and resorted to eating the flesh of their dead friends and family to stay alive. Eventually, Parrado and two other members of the group, Robert Canessa and Antonion Vizintín, embarked on a ten-day trek through the mountains to find help. His heroism ensured the survival of the group.

None of us had much to say as we followed the gentle incline of the glacier up to the mountain's lower slopes. We thought we knew what lay ahead, and how dangerous the mountain could be. We had learned that even the mildest storm could kill us if it trapped us in the open. We understood that the heavily corniced snow on the high ridges was unstable, and that the smallest avalanche would whisk us down the mountain like a broom sweeping crumbs. We knew that deep crevasses lay hidden beneath the thin crust of frozen snow, and that rocks the size of television sets often came crashing down from crumbling outcrops high on the mountain. But we knew nothing about the techniques and strategies of mountaineering, and what we didn't know was enough to kill us.

We didn't know, for example, that the Fairchild's altimeter was wrong; the crash site wasn't at 7,000 feet, as we thought, but close to 12,000. Nor did we know that the mountain we were about to challenge was one of the highest in the Andes. soaring to the height of nearly seventeen thousand feet, with slopes so steep and difficult they would test a team of expert climbers.

Experienced mountaineers, in fact, would not have gone anywhere near this mountain without an arsenal of specialised gear, including steel pitons, ice screws, safety lines and other critical gadgets designed to keep them safely anchored to the slopes. They would carry ice axes, weatherproof tents and sturdy thermal boots fitted with crampons – metal spikes that provide traction on the steepest, iciest inclines. They would be in peak physical condition, of course, and they would climb at a time of their own choosing, and carefully plot the safest route to the top. The three of us were climbing in street clothes, with only the crude tools we could fashion out of materials salvaged from the plane. Our bodies were already ravaged from months of exhaustion, starvation and exposure, and our backgrounds had done little to prepare us for the task. Uruguay was a warm and low-lying country. None of us had ever seen real mountains before. Prior to the crash, Roberto and Tintin had never even seen snow. If we had known anything about climbing, we'd have seen we were already doomed. Luckily, we knew nothing, and our ignorance provided our only chance.

Our first task was to choose a path up the slopes. Experienced climbers would have quickly spotted a ridge winding down from the summit to meet the glacier at a point less than a mile south of the crash site. If we had known enough to hike to that ridge and climb its long, narrow spine, we would have found better footing, gentler slopes, and a safer and swifter path to the top. We never even noticed the ridge. For days I had marked with my eye the spot where the sun set behind the ridges, and, thinking that the best path was the shortest path, we used that point to chart a beeline path due west. It was an amateurish mistake that would force us to weave our way up the mountain's steepest and most dangerous slopes.

Our beginning, though, was promising. The snow on the mountain's lower flank was firm and fairly level, and the cleats of my rugby boots bit well into the frozen crust. Driven by an intense adrenaline surge, I moved quickly up the slope, and in no time I had pulled fifty yards ahead of the others. But soon I was forced to slow my pace. The slope had grown much steeper, and it seemed to grow steeper yet with every step, like a treadmill that constantly increases its incline. The effort left me gasping in the thin air, and I had to rest, with my hands on my knees, after every few yards of progress.

Soon the sun was strong enough to warm us as we climbed, but it warmed the snow as well, and the firm surface beneath my feet began to weaken. Now, with every step, my foot was breaking through the thinning crust and I would sink up to my knees in the soft, deep drifts below. Each step required extreme effort. I would lift my knee almost to my chest to clear my boot from the snow. Then I would swing that foot forward, shift my weight onto it, and break through the ice again. In the thin air I had to rest, exhausted, after every step. When I looked behind me I saw the others struggling, too. I glanced at the sun above us, and realised that we had waited too long that morning to start the climb. Logic told us it would be wiser to climb in daylight, so we'd waited for the sun to rise. Experts, on the other hand, know that the best time for climbing is in the pre-dawn hours, before the sun can turn the slopes to mush. The mountain was making us pay for another amateur mistake. I wondered what other blunders lay ahead, and how many of them we'd be able to survive.

Eventually all the crust melted away, and we were forced to wade uphill through heavy drifts that sometimes were as deep as my hips. 'Let's try the snowshoes!' I shouted. The others nodded, and in moments we had slipped Fito's makeshift snowshoes off our backs and strapped them to our feet. They worked well at first, allowing us to climb without sinking into the snow. But the size and bulk of the cushions forced us to bow our legs as we walked, and to swing our feet in unnaturally wide circles to keep the fat cushions from colliding. To make things worse, the stuffing and upholstery quickly became soaked with melted snow. In my exhausted state, I felt as if I were climbing the mountain with manhole covers bolted to my shoes. My spirits were rapidly sinking. We were already on the verge of exhaustion, and the real climbing hadn't even begun.

The incline of the mountain grew steadily sharper and soon we reached slopes that were too steep and windblown to hold deep drifts of snow. With relief, we removed the snowshoes, strapped them to our backs, and kept climbing. By mid-morning we had worked our way to a dizzying altitude. The world around us now was more blue air and sunlight than rock and snow. We had literally climbed into the sky. The sheer altitude and the yawning openness of the vast slopes left me reeling with a sense of dreamlike disbelief. The mountain fell away so steeply behind me now that when I looked down on Tintin and Roberto, I saw only their heads and shoulders outlined against two thousand feet of empty sky. The angle of the slope was as steep as a roofer's ladder, but imagine a ladder you could climb to the moon! The height made my head swim and sent tingling spasms along my hamstrings and spine. Turning to look behind me was like pirouetting on the ledge of a skyscraper.

On steep, open slopes like these, where the incline seems intent on tipping you off the mountain and good handholds are hard to find, experts would use safety lines tethered to steel anchors driven into rock or ice, and they'd count on their crampons to give their footing a secure grip on the mountain-side. We had none of those things, but only the fading strength of our arms, legs, fingertips and freezing toes, to keep us from sailing off into the blue void behind us. I was terrified, of course, but still I could not deny the wild beauty all around me – the flawless sky, the frosted mountains, the glowing landscape of deep virgin snow. It was all so vast, so perfect, so silent and still. But something troubling was hiding behind all that beauty, something ancient and hostile and profound. I looked down the mountain to the crash site. From this altitude it was just a ragged smudge on the pristine snow. I saw how crass and out of place it seemed, how fundamentally *wrong*. Everything about us was wrong here — the violence and racket of our arrival, our garish suffering, the noise and mess of our lurid struggle to survive. None of it fit here. *Life* did not fit here. It was all a violation of the perfect serenity that had reigned here for millions of years. I had sensed it the first time I gazed at this place: we had upset an ancient balance, and balance would have to be restored. It was all around me, in the silence, in the cold. Something wanted all that perfect silence back again; something in the mountain wanted us to be still.

By late morning we had climbed some two thousand feet from the crash site and were probably fourteen thousand feet or more above sea level. I was moving inch by inch now as a vicious headache tightened like an iron ring around my skull. My fingers felt thick and clumsy, and my limbs wen heavy with fatigue. The slightest effort – lifting my head, turning to speak to Roberto – left me sucking for air as if I'd just run a mile, but no matter how forcefully I inhaled, I couldn't fill my lungs. I felt as if I were drawing breath through a piece of felt.

I would not have guessed it at the time, but I was suffering from the effects of high altitude. The physiological stress of climbing in oxygen-depleted air is one of the great dangen mountain climbers face. Altitude sickness, which generally strikes in the zone above 8,000 feet, can cause a range of debilitating symptoms, including headache, intense fatigue and dizziness. Above 12,000 feet, the condition can lead to cerebral and pulmonary oedemas, both of which can cause permanent brain damage and rapid death. At high altitude, it's hard to escape the effects of mild and moderate altitude sickness, but the condition is worsened by rapid climbing. Experts recommend that climbers ascend no more than 1,000 feet per day, a rate that gives the body a chance to acclimatise itself to the thinning air. We had climbed twice that far ln a single morning, and were making matters worse by continuing to climb when our bodies desperately needed time to rest.

In response, my oxygen-starved body was struggling to cope with the thinning air. My heart rate soared and my blood thickened in my veins – the body's way of conserving oxygen in the bloodstream and sending it more rapidly to vital organs and tissues. My respiration rate rose to the brink of hyperventilation, and with all the moisture I lost as I exhaled, I was becoming more severely dehydrated with every breath. To supply themselves with the huge amounts of water needed to stay safely hydrated at high altitude, expert climbers use portable gas stoves to melt pots of snow, and they guzzle gallons of fluids each day. Our only source of fluids was the snow we gulped in handfuls, or melted in the glass bottle we had in one of the packs. It did little good. Dehydration was rapidly sapping our strength, and we climbed with a constant, searing thirst.

After five or six hours of hard climbing, we had probably ascended some two and a half thousand feet, but for all our striving, the summit seemed no closer. My spirits sagged as I gauged the vast distance to the top, and realised that each of my tortured steps brought me no more than fifteen inches closer. I saw with brutal clarity that we had taken on an inhuman task. Overwhelmed with fear and a sense of futility, I felt the urge to sink to my knees and stay there. Then I heard the calm voice in my head, the voice that had steadied me in so many moments of crisis. *You are drowning in distances*, it said. *Cut the mountain down to size*. I knew what I had to do. Ahead of me on the slope was a large rock. I decided I would forget about the summit and make that rock my only goal. I trudged for it, but like the summit it seemed to recede from me as I climbed. I knew then that I was being tricked by the mountain's huge scale of reference. With nothing on those vast empty slopes to give me perspective – no houses, no people, no trees – a rock that seemed ten feet wide and a hundred yards away might really be ten times larger and more than a mile distant. Still, I climbed towards the rock without resting, and when I finally got there I picked another landmark and started all over again.

I climbed that way for hours, focusing my attention completely on some target – a rock, a shadow, an unusual ruffle in the snow – until the distance to that target became all that mattered in the world. The only sounds were my own heavy breathing and the rhythmic crunch of my shoes in the snow. My pace soon became automatic, and I slipped into a trance. Somewhere in my mind I still longed for my father, I still suffered from fatigue, I still worried that our mission was doomed. but now those thoughts seemed muted and secondary, like a voice on a radio playing in another room. *Step-push, step-push*. Nothing else mattered. Sometimes I promised myself I'd rest when the next goal was reached, but I never kept my promise. Time melted away, distances dwindled, the snow seemed to glide beneath my feet. I was a locomotive lumbering up the slope. I was lunacy in slow motion. I kept up that pace until I had pulled far ahead of Roberto and Tintin, who had to shout to make me stop. I waited for them at an outcrop that offered a level place to rest. We ate some meat and melted some snow to drink. None of us had much to say. We all knew the kind of trouble we were in.

'Do you still think we can make it by nightfall?' asked Roberto. He was looking at the summit.

I shrugged. 'We should look for a place to camp.'

I looked down to the crash site. I could still make out the tiny shapes of our friends watching us from seats they'd dragged outside the fuselage. I wondered how things looked from their perspective. Could they tell how desperately we were struggling? Were their hopes beginning to fade? If at some point we stopped moving, how long would they wait for us to start moving again? And what would they do if we didn't? These thoughts occurred to me only as passing observations. I was no longer in the same world as the boys down below. My world had narrowed, and the feelings of compassion or responsibility I had felt for the other survivors were now crowded out by my own terror and my own furious struggle to survive. I knew it was the same for Tintin and Roberto, and while I was certain we would fight side by side as long as possible, I understood that each of us, in his desperation and fear, was already alone. The mountain was teaching me a hard lesson: camaraderie is a noble thing, but in the end death is an opponent each of us would face in solitude.

I looked at Roberto and Tintin, resting sullenly on the ledge of rock. 'What did we do to deserve this?' muttered Roberto. I looked up the mountain, searching for a cliff or a boulder that might give us shelter for the night. I saw nothing but a steep, endless blanket of snow.

As we worked our way up the mountain, that snow-cover gave way to an even more difficult landscape. Now rocks were jutting from the snow, some of them huge and impossible to climb. Massive ridges and outcrops above us blocked my view of the slope ahead, and I was forced to choose my path by instinct. Often I chose wrong, and found myself trapped under an impassable ledge, or at the base of a vertical rock wall. Usually I could backtrack, or inch my way diagonally across the slope to find a new path. Sometimes I had no choice but to press on.

At one point in the early afternoon, I found my way blocked by an extremely steep, snow-covered incline. I could see a level rock shelf at the far upper edge. Unless we could climb the incline diagonally and scramble up onto that narrow shelf, we'd have to backtrack. That could cost us hours, and with sunset growing closer by the minute, I knew that was not an option. I looked back at Tintin and Roberto. They were watching to see what I would do. I studied the incline. The slope was sheer and smooth, there was nothing to grip with my hands. But the snow looked stable enough to support me. I'd have to dig my feet into the snow and keep my weight tilted forward as I climbed. It would all be a matter of balance.

I began to climb the frozen wall, carving the snow with the edges of my shoes and pressing my chest against the slope to keep from toppling backwards. The footing was stable, and with great caution I inched my way to the rock ledge and scrambled up onto level ground. I waved to Tintin and Roberto. 'Follow my steps,' I shouted. 'Be careful, it is very steep.'

I turned away from them and began to climb the slopes above me. Moments later I glanced back to see that Roberto had made it across the incline. Now it was Tintin's turn. I began climbing again, and had ascended thirty yards or so when a terrified shout echoed up the mountain.

'I'm stuck! I can't make it!'

I turned to see Tintin frozen in the middle of the incline. 'Come on, Tintin!' I shouted. 'You can do it!'

He shook his head. 'I can't move.'

'It's the backpack!' said Roberto. 'It's too heavy.' Roberto was right. The weight of Tintin' s backpack, which he carried very high on his back, was pulling him off the face of the mountain. He was struggling to shift his balance forward, but there was nothing to offer him a handhold, and the look on his face told me he could not hold out for long. From my vantage point I could see the dizzying drop behind him. and I knew what would happen if Tintin fell. First he would swim away from us for a long time in thin air, then he would hit the slope, or an outcrop, and tumble down the mountain like a rag doll until some drift or crag eventually brought his broken body to a stop.

'Tintin, hold on!' I shouted. Roberto was at the lip of the rock shelf above the incline, stretching his arm down to Tintin. His reach was short by inches. 'Take off your backpack!' he shouted. 'Give it to me!' Tintin removed the backpack carefully, struggling to keep his balance as he slowly worked the straps off his arms and handed it up to Roberto. Without the weight of the backpack, Tintin was able to find his balance and climb safely up the incline. When he reached the ledge, he slumped to the snow. 'I can't go any farther,' he said. 'I'm too tired. I can't lift my legs.'

Tintin's voice betrayed his exhaustion and fear, but I knew we had to climb until we found a sheltered place to rest for the night, so I kept going, leaving the others no choice but to follow. As I climbed, I scanned the slopes in every direction, but the mountain was so rocky and steep there was no safe place to spread our sleeping bag. It was late afternoon now. The sun had drifted behind the western ridges, and shadows were already stretching down the slopes. The temperature began to fall. At the crash site below, I saw that our friends had retreated into the fuselage to escape the cold. A clot of panic was rising in my throat as I frantically searched the slopes for a safe, level place to spend the night.

At twilight I scaled a tall rock outcropping to get a better view. As I climbed, I wedged my right foot in a small crevice in the rock, then, with my left hand, reached up to grab a horn of boulder jutting from the snow. The boulder seemed solid, but when I pulled myself up, a rock the size of a cannon ball broke free and plummeted past me.

'Watch out! Watch out below!' I shouted. looked down to see Roberto beneath me. There was no time to react. His eyes widened as he waited for the impact of the rock, which missed his head by inches. After a moment of stunned silence, Roberto glared at me. 'You son of a bitch! You son of a bitch!' he shouted. 'Are you trying to kill me? Be careful! Watch what the fuck you are doing!' Then he fell silent and leaned forward, and his shoulders started to heave. I realised he was crying. Hearing his sobs, I felt a pang of hopelessness so sharp I could taste it on my tongue. Then I was overtaken by a sudden, inarticulate rage '*Fuck* this! *Fuck* this!' I muttered. 'I have had *enough!* I have had *enough!*' I just wanted it to be over. I wanted to rest. To sink into the snow. To lie still and quiet. I can't remember any other thoughts, so I don't know what led me to keep going, but once Roberto had gathered himself, we started climbing again in the fading light. Finally I found a shallow depression in the snow beneath a large boulder. The sun had warmed the boulder all day, then the heat radiating from the rock had melted out this compact hollow. It was cramped, and its floor tilted sharply down the slope, but it would shelter us from the night-time cold and wind. We laid the seat cushions on the floor of the hollow to insulate us from the cold, then spread the sleeping bag over the cushions. Our lives depended upon the bag, and the body warmth it would conserve, but it was a fragile thing, sewn together crudely with strands of copper wire, so we handled it with great care. To keep from tearing the seams, we removed our shoes before sliding in.

'Did you pee?' asked Roberto, as I eased myself into the bag 'We can't be getting in and out of this bag all night.'

It reassured me that Roberto was becoming his grumbling self again.

'I peed,' I answered. 'Did you pee? I don't want you peeing in this bag.'

Roberto huffed at me. 'If anyone pees in the bag it will be you. And. be careful with those big feet.'

When the three of us were all inside the sleeping bag, we tried to get comfortable, but the ground beneath us was very hard, and the floor of the hollow was so steep we were almost standing up, with our backs pressed to the mountain and our feet braced against the downhill rim of the hollow.

That small rim of snow was all that kept us from sliding down the slope. We were all exhausted, but I was far too frightened and cold to relax.

'Roberto,' I said, 'you are the medical student. How does one die of exhaustion? Is it painful? Or do you just drift off?'

The question seemed to irk him. 'What does it matter how you die?' he said. 'You'll be dead and that's all that matters.'

We were quiet for a long time. The sky was as black as ink now, and studded with a billion brilliant stars, each of them impossibly clear and blazing like a point of fire. At this altitude I felt I could reach out and touch them. In another time and place I would have been awestruck by all this beauty. But here, and now, it seemed a brutal show of force. The world was showing me how tiny I was, how weak and insignificant. And temporary. I listened to my own breathing, reminding myself that as long as I drew breath I was still alive. I promised myself I would not think of the future. I would live from moment to moment and from breath to breath, until I had used up all the life I had.

The temperature dropped so low that night that the water bottle we carried shattered from the cold. Huddled together in the sleeping bag, we kept ourselves from freezing, but still we suffered terribly. In the morning we placed our frozen shoes in the sun and rested in the bag until they thawed. Then, after eating and packing our things, we began to climb. The sun was bright. It was another perfect day.

We were climbing above fifteen thousand feet now, and with every hundred yards or so the incline of the mountain tilted closer to the vertical. The open slopes were becoming unclimbable, so we began to work our way up the rocky edges of the winding couloirs – the steep plunging ravines that gashed the side of the mountain. Experienced climbers know couloirs can be killing zones – their shape makes them efficient chutes for all the rocks that tumble down the mountain – but the packed snow inside them gave us good footing, and the tall rock walls at their rims gave us something firm to grip.

At times, one edge of a couloir would lead us to an impassable point. Then I would work my way across the snow-covered centre of the couloir to the opposite edge. As we climbed the couloirs, I found myself worrying more and more about the lethal void behind me. Perhaps it was the dizzying altitude, perhaps it was fatigue or a trick of my oxygen-starved brain, but I felt that the emptiness at my back was no longer a passive danger. Now it had presence and intention, very bad intention, and I knew that if I didn't resist it with all my strength, it would lure me off the mountain and toss me down the slope. Death was tapping me on the shoulder, and the thought of it made me slow and tentative. I second-guessed every movement, and lost faith in my balance. I realised with searing clarity that there were no second chances here, there was no margin for error. One slip, one moment of inattention, one bit of bad judgement, would send me headlong down the slope. The tug of the void was constant. It wanted me, and the only thing that could keep me from it was the level of my own performance. My life had collapsed to a simple game – climb well and live, or falter and die – and my consciousness had narrowed until there was no room in my thoughts for anything but a close and careful study of the rock I was reaching for, or the ledge on which I was about to brace my foot. Never had I felt such a sense of concentrated presence. Never had my mind experienced such a pure, uncomplicated sense of purpose.

Put the left foot there. Yes, that edge will hold. Now, with the left hand, reach up for the crack in that boulder. Is it sturdy? Good. Lift yourself. Now, put the right foot on that ledge. Is it safe? Trust your balance. Watch the ice!

I forgot myself in the intensity of my concentration, forgot my fears and fatigue, and for a while I felt as if everything I had ever been had disappeared, and that I was now nothing more than the pure will to climb. It was a moment of pure animal exhilaration.

I had never felt so focused, so driven, so fiercely alive. For those astonishing moments, my suffering was over, my life had become pure flow. But those moments did not last. The fear and exhaustion soon returned, and climbing once again became an ordeal. We were very high on the mountain now, and altitude was making my motions heavy and my thinking slow. The slopes had become almost vertical and were harder than ever to climb, but I told myself that inclines this steep could only mean we were nearing the summit. To steady myself, I imagined the scene I'd see from the summit just as I'd imagined it so many times before – the rolling hills partitioned into green and brown parcels of farmland, the roads leading off to safety, and somewhere a hut or a farmhouse . . .

How we continued to climb, I cannot say. I was shivering uncontrollably from cold and fatigue. My body was on the verge of complete collapse. Only the simplest thoughts could take shape in my mind. Then, in the distance above me, I saw the outline of a sloping ridge in sharp relief against a background of clear blue sky, and no more mountain above it. The summit! 'We made it!' I shouted, and with renewed energy I clawed my way to the ridge. But as I pulled myself over its edge, the ridge gave way to a level shelf several yards wide, and above the shelf the mountain rose again. It was the steep angle of the slope that had fooled me. This was only another trick of the mountain, a false summit. And it wasn't the last. We spent the afternoon struggling towards one false summit after another until, well before sunset, we found a sheltered spot and decided to pitch our camp.

Roberto was sullen that night as we lay in the sleeping bag. 'We will die if we keep climbing,' he said. 'The mountain is too high.'

'What can we do but climb?' I asked.

'Go back,' he said. For a moment I was speechless.

'Go back and wait to die?' I asked.

He shook his head. 'Do you see across there, that dark line on the mountain? I think it's a road.' Roberto pointed across a wide valley to a mountain ridge miles away.

'I don't know,' I said. 'It looks like some sort of fault line in the rock.'

'Nando, you can barely see,' he snapped. 'I tell you it's a road!'

'What are you thinking?' I asked.

'I think we should go back and follow that road. It must lead somewhere.'

That was the last thing I wanted to hear. Since the moment we'd left the fuselage, I had secretly been tormented by doubts and misgivings. *Are we doing the right thing? What if rescuers come while we're in the mountains? What if the farmlands of Chile are not just over the ridge?* Roberto's plan seemed like lunacy, but it forced me to consider other options, and I did not have the heart for that now.

'That mountain must be twenty-five miles away,' I said. 'If we hike there and climb to that black line, and find that it is just a layer of shale, we won't have the strength to return.'

'It's a road, Nando, I'm sure of it!'

'Perhaps it's a road, perhaps it's not,' I replied. 'The only thing we know for sure is that to the west is Chile.'

Roberto scowled. 'You've been saying that for months. but we'll break our necks before we get there.'

Roberto and I argued about the road for hours, but as we settled down to sleep, I knew the matter had not been resolved. I woke the next morning to yet another clear sky.

'We've been lucky with the weather,' said Roberto. He was still inside the sleeping bag.

'What have you decided?' I asked him. 'Are you going back?'

'I'm not sure,' he said. 'I need to think.'

'I'm going to climb,' I said, 'maybe we'll reach the summit soon.'

Roberto nodded. 'Leave your packs here,' he said. 'I'll wait until you return.'

I nodded. The thought of going on without Roberto terrified me, but I had no intention of turning back now. I waited for Tintin to gather his pack, then we turned to the slope and began to climb. After hours of slow progress, we found ourselves trapped at the base of a cliff towering hundreds of feet above us. Its face was almost dead vertical and covered with hard-packed snow.

'How can we climb this?' asked Tintin.

I studied the wall. My mind was sluggish, but soon I remembered the aluminium walking stick strapped to my back.

'We need a stairway,' I said. I drew the stick off my back, and with its sharp tip, I began to carve crude steps into the snow. Using the steps like the rungs of a ladder, we continued to climb. It was excruciating work, but I kept at it with the dull persistence of a farm animal, and we ascended one slow step at a time. Tintin followed behind me. He was frightened, I know, but he never complained. In any case, I was just dimly aware of his presence. My attention was focused on the task at hand: *Dig, climb, dig, climb.* I felt, at times, that we were climbing the sheer sides of a frozen skyscraper, and it was very difficult to keep my balance as I dug, but I no longer worried about the void at my back. I respected it, but I had learned to tolerate its presence. A human being, as I've said before, gets used to anything.

It was an agonising process, inching up the mountain that way, and the hours passed slowly. Sometime in late morning I spotted blue sky above a ridgeline and worked my way towards it. After so many false summits, I had learned to keep my hopes in check, but this time, as I climbed over the ridge's edge, the slope fell away flat and I found myself standing on a gloomy hump of rock and wind-scoured snow. It dawned on me slowly that there was no more mountain above me. I had reached the top.

I don't remember if I felt any joy or sense of achievement in that moment. If I did, it vanished as soon as I glanced around. The summit gave me an unobstructed 360-degree view of creation. From here I could see the horizon circling the world like the rim of a colossal bowl, and in every direction off into the fading blue distance, the bowl was crowded with legions of snow-covered mountains, each as steep and forbidding as the one I had just climbed. I understood immediately that the Fairchild's co-pilot had been badly mistaken. We had not passed Curicó. We were nowhere near the western limits of the Andes. Our plane had fallen somewhere in the middle of the vast cordillera.

I don't know how long I stood there, staring. A minute. Maybe two. I stood motionless until I felt a burning pressure in my lungs, and realised I had forgotten to breathe. I sucked air. My legs went rubbery and I fell to the ground. I cursed God and raged at the mountains.

TOUCHING THE VOID
The True Story of One Man's Miraculous Survival

Joe Simpson

Joe Simpson (born 1960) is a British mountaineer who, in 1985, fell into a crevasse while trying to make the first ascent of the West Face of the Siula Grande in Peru. Despite suffering severe injuries and having been assumed lost by his partner Simon Yates, he managed to crawl his way back to their base camp three days later, just before Yates had intended to leave. His descent is regarded by many as one of the most extraordinary examples of survival against the odds. Simpson later underwent six surgeries to fix his broken right leg and now communicates the lessons learned from this experience in his career as a motivational speaker.

I lolled on the top, scarcely able to hold my head up. An awful weariness washed through me, and with it a fervent hope that this endless hanging would soon be over. There was no need for the torture. I wanted with all my heart for it to finish.

The rope jolted down a few inches. How long will you be, Simon? I thought. How long before you join me? It would be soon. I could feel the rope tremble again; wire-tight, it told me the truth as well as any phone call. So! It ends here. Pity! I hope somebody finds us, and knows we climbed the West Face. I don't want to disappear without trace. They'd never know we did it.

The wind swung me in a gentle circle. I looked at the crevasse beneath me, waiting for me. It was big. Twenty feet wide at least. I guessed that I was hanging fifty feet above it. It stretched along the base of the ice cliff. Below me it was covered with a roof of snow, but to the right it opened out and a dark space yawned there. Bottomless, I thought idly. No. They're never bottomless. I wonder how deep I will go? To the bottom... to the water at the bottom? God! I hope not!

Another jerk. Above me the rope sawed through the cliff edge, dislodging chunks of crusty ice. I stared at it stretching into the darkness above. Cold had long since won its battle. There was no feeling in my arms and legs. Everything slowed and softened. Thought became idle questions, never answered. I accepted that I was to die. There was no alternative. It caused me no dreadful fear. I was numb with cold and felt no pain; so senselessly cold that I craved sleep and cared nothing for the consequences. It would be a dreamless sleep. Reality had become a nightmare, and sleep beckoned insistently; a black hole calling me, pain-free, lost in time, like death.

My torch beam died. The cold had killed the batteries. I saw stars in a dark gap above me. Stars, or lights in my head. The storm was over. The stars were good to see. I was glad to see them again. Old friends come back. They seemed far away; further than I'd ever seen them before. And bright: you'd think them gemstones hanging there, floating in the air above. Some moved, little winking moves, on and off, on and off, floating the brightest sparks of light down to me.

Then, what I had waited for pounced on me. The stars went out, and I fell. Like something come alive, the rope lashed violently against my face and I fell silently, endlessly into nothingness as if dreaming of falling. I fell fast, faster than thought, and my own stomach protested at the swooping speed of it. I swept down, and from far above I saw myself falling and felt nothing. No thought, and all fears gone away. So this is it!

A whoomphing impact on my back broke the dream and the snow engulfed me. I felt cold wetness on my cheeks. I wasn't stopping, and for an instant blinding moment I was frightened. Now, the crevasse! Ahh… NO!

The acceleration took me again, mercifully fast, too fast for the scream which died above me…

The whitest flashes burst in my eyes as a terrible impact whipped me into stillness. The flashes continued, bursting electric flashes in my eyes as I heard, but never felt, the air rush from my body. Snow followed down on to me, and I registered its soft blows from far away, hearing it scrape over me in a distant disembodied way. Something in my head seemed to pulse and fade, and the flashes came less frequently. The shock had stunned me so that for an immeasurable time I lay numb, hardly conscious of what had happened. As in dreams, time had slowed, and I seemed motionless in the air, unsupported, without mass. I lay still, with open mouth, open eyes staring into blackness, thinking they were closed, and noting every sensation, all the pulsing messages in my body, and did nothing.

I couldn't breathe, I retched. Nothing. Pressure pain in my chest. Retching, and gagging, trying hard for the air. Nothing. I felt a familiar dull roaring sound of shingles on a beach, and relaxed. I shut my eyes, and gave in to grey fading shadows. My chest spasmed, then heaved out, and the roaring in my head suddenly cleared as cold air flowed in.

I was alive.

A burning, searing agony reached up from my leg. It was bent beneath me. As the burning increased so the sense of living became fact. Heck! I couldn't be dead and feel that! It kept burning, and I laughed – Alive! Well, fuck me! – and laughed again, a real happy laugh. I laughed through the burning, and kept laughing hard, feeling tears rolling down my face. I couldn't see what was so damned funny, but I laughed anyway. Crying and laughing at high pitch as something uncurled within me, something tight twisted in my guts that laughed itself apart and left me.

I stopped laughing abruptly. My chest tightened, and the tension took hold again.

What stopped me?

I could see nothing. I lay on my side, crumbled strangely. I moved an arm cautiously in an arc. I touched a hard wall. Ice! It was the wall of the crevasse. I continued the search, and suddenly felt my arm drop into space. There was a drop close by me. I stifled the urge to move away from it. It also sloped steeply beneath me. I was on a ledge, or a bridge. I wasn't slipping, but I didn't know which way to move to make myself safe. Face down in the snow I tried to gather my confused ideas into a plan. What should I do now?

Just keep still. That's it... *don't move...* Ah!

I couldn't stop myself. Pain in my knee jolted through me, demanding movement. I had to get my weight off it. I moved, and slipped. Every muscle gripped down at the snow – *DON'T MOVE.*

The movement slowed, then stopped. I gasped, having held my breath for too long. Reaching out again I felt my hand touch the hard ice wall. Then I groped for the ice hammer attached to a lanyard of the thin cord clipped to my harness. Fumbling in the dark, I found the cord running tightly away from me and pulled it, bringing the hammer up out of the drop in front of me. I had to hammer an ice screw into the wall without pushing myself off the ledge I was perched on.

It proved harder than expected. Once I had found the last remaining screw attached to my harness I had to twist round and face the wall. My eyes had adjusted to the darkness. Starlight and the moon glimmering through my entry hole in the roof above gave enough light for me to see the abysses on either side of me. I could see grey-shadowed ice walls and the stark blackness of the drops, too deep for the light to penetrate. As I began to hammer the screw into the ice I tried to ignore the black space beyond my shoulder. The hammer blows echoed around the ice walls, and from deep below me, from the depths of blackness at my shoulder I heard second and third echoes drift up. I shuddered. The black space held untold horrors. I hit the screw, and felt my body slide sideways with each blow. When it was driven in to its hilt I clipped a karabiner through the eye and hurriedly searched for the rope at my waist. The black spaces menaced and my stomach knotted in empty squeezing clenches.

I hauled myself into a half-sitting position close to the wall, facing the drop on my left. My legs kept sliding on the slow so that I had constantly to shuffle back to the wall. Dared not let go of the ice screw for more than a few seconds, but my fingers needed a lot longer to tie the knot. I swore bitterly each time I made a mess of the knot and feverishly tried again. I couldn't see the rope, and although normally I could tie the knot blindfold, I was now hampered by frozen hands. I couldn't feel the rope well enough to thread it back on itself and form the knot. After six attempts I was at the point of tears. I dropped the rope. Reaching for it I slipped forward towards the drop and lunged back scrabbling at the wall for the screw. My mitt slipped across the wall, and I began to fall backwards. I clawed at the ice trying to get my fingers to grip through the mitts, and then felt the screw hit my hand. My fingers locked around it, and the fall stopped. I stayed motionless, staring at the back hole in front of me.

After several abortive attempts suddenly I found that I had tied a knot of sorts. I held it close to my face and looked up through it at the dim light shining through the entry hole in the roof above. I could see the bulge of the knot, and above it the loop I had been struggling to tie. I chuckled excitedly, feeling ridiculously pleased with myself, and clipped it to the ice screw, smiling foolishly into the darkness. I was safe from the black spaces.

I relaxed against the comforting tightness of the rope and looked up at the small hole in the roof, where the sky was cloudless, packed with stars, and moonlight was adding its glow to their bright sparkle. The screwed-up tensions in my stomach flowed away, and for the first time in many hours I began to order my mind into normal thoughts. I'm only, what… fifty feet down this crevasse. It's sheltered. I can get out in the morning if I wait for Simon…

'SIMON!?'

I spoke his name aloud in a startled voice. The word echoed softly back. It hadn't occurred to me that he might be dead, and as I thought about what had happened the enormity of it struck me. Dead? I couldn't conceive of him dead, *not now, not after I survived*. The chill silence of the crevasse came over me; the feel of the tombs, of space for the lifeless, coldly impersonal. No one had ever been here. Simon, dead? Can't be! I'd have heard him, seen him come over the cliff. He would have come on to the rope or down here.

I began to giggle again. Despite my efforts I couldn't prevent it, and the echoes bounced back at me from the ice walls, sounding cracked and manic.

It became so that I couldn't work out whether I was laughing or sobbing. The noises that returned from the darkness were distorted and inhuman, cackling echoes rolling up and around me. I giggled more, listened and giggled again, and for a moment forgot Simon, and the crevasse, and even my leg. I sat, hunched against the ice wall, laughing convulsively, and shivering. It was the cold. Part of me recognised this; a calm rational voice in my head told me it was the cold and the shock. The rest of me went quietly mad while this calm voice told me what was happening and left me feeling as if I were split in two – one half laughing, and the other looking on with unemotional objectivity. After a time I realised it had all stopped, and I was whole again. I had shivered some warmth back, and the adrenalin from the fall had gone.

I searched in my rucksack for the spare torch battery I knew was there. When I had fitted it, I switched on the beam and looked into the black space by my side. The bright new beam cut down through the blackness and lit ice walls that danced away down into depths my torch couldn't reach. The ice caught the light, so that it gleamed in blue, silver and green reflections, and I could see small rocks frozen into the surface dotted the walls at regular intervals. They glistened wetly as I swept the beam down the smooth scalloped dimples. I swallowed nervously. By the light I could see down into 100 feet of space. The walls of the crevasse reared up in a tangle of broken ice blocks and fifty feet above me they arched over to form a roof. The slope to my right fell away steeply for about thirty feet, after which it disappeared. Beyond it lay a drop into darkness. The darkness beyond the light gripped my attention. I could guess what it hid, and I was filled with dread. I felt trapped, and looked quickly around me for some break in the walls. There was none. Ice flashed light back from the hard blank walls, or else the beam was swallowed by the impenetrable blackness of the holes on either side. The roof covered the crevasse to my right and fell down in frozen chaos to my left, blocking the open end of the crevasse from my view. I was in a huge cavern of snow and ice. Only the small black hole above, winking starlights at me, gave any view of another world and unless I climbed the blocks it was as unreachable as the stars.

I turned the torch off to save the batteries. The darkness seemed more oppressive than ever. Discovering what I had fallen into hadn't cleared my mind. I was alone. The silent emptiness, and the dark and the star-filled hole above, mocked my thoughts of escape. I could only think of Simon. He was the only chance of escape, but somehow I was convinced that if he was not dead, then he would think I was. I shouted his name as loud as I could and the sound jumped back at me, and then faded in dying echoes in the holes below me. The sound would never be heard through the walls of snow and ice. The roof was fifty feet above me. On the rope I had hung at least fifty feet above the roof. Simon would see the huge open side of the crevasse, and the cliff, and he would know at once that I was dead. You can't fall that far and survive. That's what he would think. I knew it. I would think the same if I were in his place. He would see the endless black hole and know that I had died in it. The irony of falling 100 feet and surviving unscathed was almost unbearable.

I swore bitterly, and the echoes from the darkness made it a futile gesture. I swore again, and kept swearing, filling the chamber with angry obscenities which cursed me back in echoes. I screamed in frustration and anger until my throat dried, and I could shout no more. When I was silent I tried to think of what would happen. If he looks in he will see me. He might even hear me. Maybe he heard me just then? He won't leave unless he's sure. How do you know he's not dead already? Did he fall with me? Find out... pull the rope!

I tugged on the loose rope. It moved easily. When I turned my torch on I noticed it hanging down from the hole in the roof. It hung in a slack curve. I pulled again and soft snow flurried on to me. I pulled steadily, and as I did so I became excited. This was a chance to escape. I waited for the rope to come tight. I wanted it to come tight. It kept moving easily. It was strange to want the weight of Simon's body to come on to the rope. I had instantly found a way to get out, and it meant only that. When Simon had fallen he would have swept out and clear of the crevasse. So he must have hit the slope and stopped. He would be dead. He must be after that fall.

When the rope comes tight I can Prussik up it. His body will anchor it solidly. Yes. That's it...

I saw the rope flick down, and my hopes sank. I drew the slack rope to me, and stared at the frayed end. Cut! I couldn't take my eyes from it. White and pink nylon filaments sprayed out from the end. I suppose I had known all along. It was a madness. Crazy to have believed in it, but everything was getting that way. I wasn't meant to get out of here. Damn it! I shouldn't even have got this far. He should have left me on the ridge. It would have saved so much... I'll die here after all that. Why bother trying?

I turned off the torch and sobbed quietly in the dark, feeling overwhelmed. I cried in bursts, and between them listened to the childlike sounds fade beneath me, then cried.

It was cold when I awoke. I came up slowly from a long emptiness and wondered where I was. Sleep had taken me unawares, and I was startled. The cold had woken me. That was a good sign. It could easily have taken me. I felt calm. It was going to end in the crevasse. Perhaps I had always known it would end this way. I felt pleased to be able to accept it calmly. All that sobbing and shouting had been too much. Acceptance seemed better. There was no trauma this way. I was certain then that Simon would leave me for dead. It didn't surprise me. Indeed it made things easier. There was one less thing to worry about. I thought it might take me a few days to die. In the end I decided that three days would pass. It was sheltered in the crevasse, and with my sleeping bag I could survive a good few days. I imagined how long it would seem; a long period of twilight, and darkness, drifting from exhausted sleep into half-consciousness. Maybe the last half would be dreamless sleeping, ebbing away quietly. I thought carefully of the end. It wasn't how I had ever imagined it. It seemed pretty sordid. I hadn't expected a blaze of glory when it came, nor had I thought it would be like this slow pathetic fade into nothing. I didn't want it to be like that.

I sat up and turned on the torch. Looking at the wall above the ice screw, I thought it might be possible to climb out. Deep inside I knew it would be impossible, but I urged the faint hope on, deciding that if I fell at least it would be swift. My resolve failed me when I looked at the black void on either side of me. The ice bridge seemed to be desperately precarious. I fastened a Prussik knot to the rope above the screw. I would climb while still attached to the screw. I could let slack rope out through the Prussik but if I fell the Prussik might stop me. I knew it would probably snap but I couldn't summon enough nerve to climb unroped.

An hour later I gave up trying. I had made four attempts to climb the vertical ice wall. Only once had I managed to get myself clear of the ledge. I had planted both axes above me, and hauled myself up. When I had kicked the crampons on my left boot into the wall, I reached up again with an axe. Before I could swing at the ice above, my crampon points broke free, and I slipped heavily on to my ice hammer. It ripped from the ice, and I fell back to the bridge, my injured leg folding agonisingly beneath me. I screamed, and twisted to free it. Then I lay still, waiting for the pain to ease. I would not try again.

I sat on my sack, turned off the torch, and slumped on to the rope which I had retied to the ice screw. I could see my legs in the gloom. There was a delay before I realised the significance of being to see them. I glanced up to the patch of dim light in the roof and checked my watch. It was five o'clock. It would be fully light in an hour, and Simon would be coming down the cliff as soon as it was light. I had been alone in the dark for seven hours, and until then I hadn't realised how demoralising the lack of light had been. I shouted Simon's name loudly. It echoed round me, and I shouted again. I would shout regularly until he heard me, or until I was certain he had gone.

A long time later I stopped shouting. He had gone. I knew he would, and I knew he wouldn't return. I was dead. There would be nothing for him to come back for. I took my mitts and inner gloves off, and examined my fingers. Two blackened fingers on each hand, and one bluish thumb. I curled them into fists and tried to squeeze hard but couldn't feel the pressure. It wasn't as bad as I had thought. Sunlight streamed through the hole in the roof. I glanced at the hole to my left. I could see deeper into it, but there was no sign of it closing up. It just faded into dark shadows a long way down. To my right the slope angled away to the drop I had seen the previous night. Far away to the right of this sunlight sprayed against the back wall of the crevasse.

I picked absently at the frayed end of the rope, trying to come to a decision. I already knew I wasn't prepared to spend another night on the ledge. I wasn't going through that madness again, but I cringed from doing the only thing left to me. I wasn't ready for such a choice. Without deciding I took some coils in one hand, and then threw the rope down to the right. It flew clean out into space, and curled over the drop before falling out of sight. The rope jerked tight. I clipped my figure-of-eight to the rope, and lay on my side.

I hesitated, looking at the ice screw buried in the wall. It wouldn't pull free under my weight. The Prussik knot hung unused just below the ice screw. I thought that I should take it with me. If there was empty space at the end of the rope I would be unable to regain the ledge without it. I let myself slide off the edge and watched the Prussik get smaller as I abseiled down the slope to the drop. If there was nothing there I didn't want to come back.

WHEN I FELL FROM THE SKY
The True Story of One Woman's Miraculous Survival

Juliane Koepcke

Juliane Koepcke (born 1954) is a German scientist who, in 1971, became the sole survivor of the disastrous LANSA Flight 508 crash, which claimed the life of her mother. After surviving a fall of 3,000 meters – almost two miles – Koepcke managed to stay alive in the Amazon rainforest for eleven days on her own. She was eventually rescued by a group of Peruvian fishermen after making her way downriver towards civilization. She has run the Panguana biological research station founded by her parents in Peru since 2002.

The water flows around my feet. Doggedly I set one foot in front of the other. The stream turns into a larger stream, finally almost into a small river. The days are all alike. I try to count them as I go, so I don't lose my sense of time. The intensity of the daylight indicates to me the approximate time of day. In the tropics it gets light at six o'clock in the morning; at six o'clock in the evening, on the dot, it gets dark. The sun itself I rarely see, however, since the canopy of the jungle giants is too dense.

Eventually I've sucked the last sweet. I don't dare to eat anything else. Since it's the rainy season, there's barely any fruit. I don't have a knife with me and cannot hack palm hearts out of the stems. Nor can I catch fish or cook roots. I know that much of what grows in the jungle is poisonous, so I keep my hands off what I don't recognize. But I do drink a great deal of water from the stream, which is brown with floating soil. That might be the reason I don't feel hungry. I don't feel any particular aversion to drinking this water. From living in the jungle, I know that the water of the forest creeks is clean. There isn't much danger of dysentery in uninhabited areas where people don't contaminate the water. Nevertheless, in Panguana we had always boiled the river water before drinking it. That crosses my mind as I take in the muddy water from the stream, but I'm aware that I have no choice. Because I haven't had any food, I must drink a lot to survive.

Despite my counting, the days get mixed up for me. On December 29 or 30, the fifth or sixth day of my trek, I hear a birdcall, and my apathetic mood immediately turns into euphoria. It's the distinct, unmistakeable call of hoatzins, a mixture of buzzing and groaning. At home in Panguana I heard this call often. These birds nest exclusively near open stretches of water, near larger rivers, and that is my hope exactly, because that's also where people settle!

With new impetus I try to make more rapid progress and follow the birdcalls. And indeed I soon find myself at the outlet of "my" stream into a river. But if I was hoping to reach it quickly, I was deceiving myself. The mouth is blocked by a great deal of driftwood and overgrown with thick underbrush. Soon I accept that I will never get through here with nothing but my bare hands. So I decide to leave the streambed and go around the barriers. It costs me hours to fight my way through the jungle. The mouth is densely covered with about fifteen-foot tall reeds, the *caña brava*, and the sharp stalks cut my arms and legs when I'm not careful. But the calls of the hoatzins and the roar of the search planes embolden me.

My mother had studied the hoatzins extensively, observing and describing important details of their breeding behaviour. These interesting animals not only look gorgeous, but they also belong to a very primitive family and are faintly reminiscent of the archaeopteryx. Like that first-known bird, their young also have claws on their wings. Since their parents build their nests not only extremely sloppily but also over the water, the claws are really useful. Frequently a hoatzin chick falls out of the nest, catches itself on the branches with the help of its wing claws and climbs back up. The chicks can also swim really well.

Finally I'm standing on the bank of the large river. I estimate its width at thirty feet, a beautiful stretch of water, but there's not a human soul in sight. Immediately I notice that it can't be navigable, since numerous logs and other driftwood make that impossible. I look up at the sky. After so many days in the half-light of the jungle, I can finally see it open above me again. Where are the search planes? I hear them only in the distance. At one point one more makes a half-hearted sweep over me, and I wave and shout, but in vain. The plane turns away and disappears, just like the others. Silence. *They'll come back,* I tell myself, *for certain.* But time passes, and the engine noise I've heard almost constantly over the past few days doesn't return. Finally I grasp the fact that they've given up the search. Probably all the others have been rescued, except me. *Except me.*

*

A boundless anger overcomes me. I had no idea that I still had the strength for such intense feelings. How can they simply turn around, now that I've finally reached an open stretch of water after all these days! Now that I can make myself noticeable! But as quickly as the anger flares up, it dies out and gives way to a terrible despair. Here I am on the bank of a really large river, feeling utterly alone. Only now that I have a little bit of distance from it do I become distinctly aware of the vastness of the jungle around me. I'm afraid that it's uninhabited for thousands of square miles. I know that there's an extremely slim chance of meeting a person here. I suspect that my odds are virtually zero. But I don't give up.

This is still a real river. And where there's a river, people cannot be far. My father repeated that all the time. *Sooner or later*, I now encourage myself, *I will reach them. There's no reason to despair right now. On the contrary, my rescue lies just ahead.*

I pull myself together and think about how best to proceed. The riverbank is much too densely overgrown for me to carry on hiking along it. I'm also afraid that I might step on a poisonous snake or spider with my bare foot. I begin to wade down river in the shallower water near the bank. But beforehand, I look for a stick, not only to avoid slipping but also to check the ground in front of me. I know that there are dangerous stingrays resting in the mud of the riverbanks or lying in rapids, and they can't be seen. If you step on them, they plunge a poisonous stinger into your foot. Your leg swells up and you get a high fever. Though the stingray's poison isn't deadly, mud often gets into the wound with the stinger, which can lead to blood poisoning. In my situation such an injury could be fatal. I learned all this in Panguana from my parents and our neighbours. I know the dangers in the water, and so I walk carefully and warily.

Making progress is very difficult. There are branches and many logs in the water, and the ground consists either of slippery rocks or deep mire into which I sink. So I soon decide to swim in the middle of the river. In the deep water I'm at least safe from stingrays. Instead there are piranhas, but I've learned that they only become dangerous in standing water. Certainly caimans are to be expected, but they, too, generally don't attack people. So I yield to the current. I still have no fear. Plus, the confidence that I'll somehow make it has returned.

It's good that I don't know but only suspect that they will soon stop searching for survivors. It's also better that I can't imagine that so far no one has been rescued—and that the searchers haven't even been able to find the slightest trace of the plane wreck. But most important of all is that I don't know that other than me some people actually did survive the crash, without being lucky enough to be able to leave the spot where they hit the earth. As I later found out, my mother was among those people. And I think of my mother during each of my nights, in which I barely sleep. Ever since my concussion has abated, I no longer fall into that sleeplike state that is more akin to a stupor. My nights are long, pitch black and without peace.

When the sun descends and I estimate the time at around five o'clock, I search for a reasonably safe spot on the bank where I can spend the night. I always try to find a place where I have protection at my back, either from a slight slope or a large tree. Still, sleep is almost unthinkable. Either mosquitoes or tiny tormenting midges, which are also among the gnats, keep me awake. They seem to want to devour me alive. There's buzzing around my head, and the bothersome pests try to crawl into my ears and nose. Those night hours are unbearable. Dead tired, I fall into a half-sleep and wake up again and again from the burning and biting of new stings. Or, even worse, it rains. Then the mosquitoes leave me alone, but the ice-cold rain pelts down on me mercilessly. I freeze in my thin summer dress, constantly soaked to my skin. As hot as it may be in the daytime, during the rainy season it cools down dramatically at night, and each of the hard drops torments me like an icy pinprick. And then the wind comes and makes me shiver to my core. I search for spots under dense trees or in bushes, collect large leaves and try to protect myself with them. Nothing helps. On those black nights, which seem never-ending, as I cower somewhere soaked to my bones, I cannot protect myself, and I experience a boundless feeling of abandonment. It is as if I were all alone, somewhere out in the universe. Those are the times when I despair.

I think about my mother a lot. How might she be doing? Has she already been rescued? I don't dare to think of the possibility that she could have suffered the same fate as the three people who had been rammed into the ground along with their seat bench. I wonder what my father might be doing right now. How is he? Where is he? Has he heard about the crash yet?

I ruminate a great deal on how it could have happened that I woke up alone in the jungle. I wonder where all the other passengers are, why I couldn't find a swathe cut anywhere through the forest, where in the world the airplane itself has gone. I think about my life up to now, which was so completely unspectacular. At least in my eyes, nothing really exciting happened before. I'm a young girl, like all the rest. I love animals, read avidly, go to the movies with my friends, get good marks, adapt to whatever place I'm dragged to—whether it be the jungle or Lima. I've never worried much about the meaning of my life. Though I was baptized as a Protestant and was just recently confirmed, my parents adhere more to a sort of philosophical nature religion and view the sun as the basis of all life. They have not brought me up to be particularly religious. They believe that I should form my own opinion; they certainly provided the fundamentals of a Christian upbringing, but more was not necessary, in their view.

On those nights I pray. The prayers are mainly about my mother. I've always had a very close relationship with her. She is my mother and a sort of friend. We're more intimately bound to each other than I am to my father, who scarcely lets anyone get close to him other than my mother. I'm aware that it's a miracle I'm still alive, and I wonder why me of all people. I have survived the crash and believe that I now have to get through this too. I pray that I will find people. I pray for my rescue. I want to live. With every fibre of my gradually weakening body, I want to live. And then I wonder what I will do with that life when this is finally over.

I think about that for a long time.

Of course, like all my friends, I've thought about what I will do after school. From an early age I wanted to study biology, like my parents. But I never asked myself why and for what purpose. I was fond of animals, interested in plants, and I liked what my parents did. Up to now, that was reason enough for me. On those rainy nights I now think that it would be great to devote my life to something big, something important, something that would benefit humanity and nature. What that might be, I have no idea. I just feel that my life from now on should have meaning in the fabric of the world. For it has to mean something that I fell out of the airplane and walked away from that with only a few scrapes.

Those "few scrapes," though, do begin to worry me in the days that follow. The cut on my calf swells up and the flesh is whitish. Still I feel no pain. The hidden wound on the back of my right upper arm is a different story. I have to twist my head a long way to see what's going on there. To my horror I discover white maggots, whose bodies peek out of the wound like tiny asparagus heads! Apparently, flies have laid their egg sacs in my wound, and the brood is already half an inch long. That is something else I know all about, and this time my knowledge worries me.

Lobo, my German shepherd cross, was once infested with fly larvae. Unbeknown to us, he had a small cut in his shoulder and the flies laid their eggs in it. Hidden under the skin, the maggots hatched and burrowed ever deeper into the flesh. They do that very skillfully, avoiding blood vessels so that the wound doesn't bleed. Back then, they ate a deep canal for themselves under Lobo's fur, down his leg to his paw. At night Lobo whimpered, and we wondered what was wrong with him. The maggots were still completely invisible. Then the leg eventually swelled and began to smell. By then, it was already so bad that the dog would no longer let anyone touch him. Finally we discovered what had happened. Normally you can get maggots out of the body with alcohol, but my father said we couldn't do that, because the dog would go crazy with pain. So we poured kerosene into the wound, which doesn't burn, until the maggots came crawling out, one after another, and we could patch up poor Lobo. Fortunately, his wound healed without any problems after that.

So I know what I have to do: the maggots have to come out. But I have neither alcohol nor kerosene. Only a silver spiral-shaped ring, and I bend it open and try to fish the maggots out with it. However, as soon as my self-made tweezers get near them, the maggots disappear into my flesh. I try with the buckle on my watchband, but that too gets me nowhere. Then a rather queasy feeling creeps up on me. It's not a pleasant thought, to be devoured alive from inside. Though I know that the maggots themselves would not do anything dangerous to me, since like all good parasites they initially avoid harming the host, the wound can, of course, get infected. After all, I swim all day long in brown, dirty river water. And if that should occur, then it's not out of the question that my arm would ultimately have to be amputated. I've heard of cases like that. I wouldn't be the first person this had happened to.

Since there's nothing I can do about it at the moment, I go on swimming. I noticed a long time ago that the wild animals on the riverbank are extremely trusting. I see martens and brocket deer, which aren't frightened of me at all. I hear howler monkeys, very close, and that makes me think, because usually those animals are extremely shy. I know what that means, but I try to keep the thought at bay: this river and the surrounding forest have not yet come into contact with people, and it will be many miles before that might change.

Meanwhile, I'm getting weaker all the time. Though I don't feel hungry at all, I notice how everything is getting harder. I drink a lot of the river water, which fills my stomach, and I know that I should eat something. How many days have I been on the move already? Seven? Or eight? I count on my fingers, and I realize that the new year of 1972 has possibly already begun. My mother really wanted to celebrate the turn of the year with my father. That was the reason she didn't want to wait any longer for a flight. And now I wonder where my father might be.

Only recently did I find among my aunt's papers letters that my father wrote during those days. On December 31, 1971, he wrote:

Now a week has already passed, and still the plane has not been found. The weather is mostly good, so that search operations could be launched in all directions. I'm on Herr Wyrwich's hacienda, which has an airfield and is therefore equipped with transmitters and receivers. We can inquire in Pucallpa and are then informed of the status of the search operation.

This paragraph is followed by an enumeration of the various theories and statements of witnesses, all of whom claim to have heard the airplane itself or an explosion. But it turns out that in the nearby Sira Mountain Range, due to the constant rainfall, a landslide had occurred that might have caused this sound. As I read this letter for the first time in my father's usual neat handwriting, I tried to imagine what might have been going on inside him. Only the second part of the letter, written after an interruption, bears witness to his emotions. In the meantime an American missionary named Clyde Peters had landed his plane on Herr Wyrwich' s airfield and had given him encouraging news. There was some support for the theory that the LANSA plane must have made an emergency landing somewhere. I could tell by my father's handwriting how hope was rekindled in him.

Of course, I don't have the slightest inkling of all this during my odyssey. I have only one thing in my mind: I have to find some people. During the day I swim or let myself drift, and at night I now have a few encounters with larger animals. At one point, while I'm trying to sleep in the middle of some bushes, I hear hissing and pawing right next to me. I know it's most likely not a jaguar or an ocelot. Probably what's making the noise next to me is a *majás*, known as a paca in English, a rodent as large as a medium-sized dog, with brown fur and white spots arranged in rows. I clear my throat, which gives the animal a terrible scare. It runs away in wild bounds, loudly grunting.

The next morning I feel a sharp pain in my upper back. When I touch it with my hand, it's bloody. While I've been swimming in the water, the sun has burned my skin, which is already peeling. They are second-degree burns, I will later learn. I can't do anything about that either and continue to let myself drift in the water. Luckily, the current is getting stronger. As weak as I am, I only have to be careful not to collide with a log floating in the river or injure myself on another obstacle.

My bad eyes repeatedly fool me. Often I'm convinced that I see the roof of a house on the riverbank. My ears deceive me too, and I'm completely sure that I hear chickens clucking. But of course, it's not chickens; it's the call of a particular bird. Even though I know this call well, I'm taken in by these sounds, again and again. Then I'm annoyed and scold myself: *How can you be so stupid, you know that those aren't chickens*. And yet it happens to me again and again. The hope of finally, finally finding people is stronger. And ultimately I fall into an apathy unlike anything I've ever experienced before.

I'm tired. So horribly tired. During the nights I fantasize about food. About elaborate feasts and really simple meals. Each morning it gets harder for me to stand up from my uncomfortable spot and get into the cold water. Is there any sense in going on? *Yes*, I tell myself, gathering all my strength, *I have to keep going. Keep going. Here I will perish.*

At one point I sink in the middle of the day onto a sandbank in the river under the glaring sun. It seems to me an ideal place to rest for a bit. I've almost dozed off, hardly noticing any more the ubiquitous blackflies on the riverbank that are constantly pestering me. Suddenly I hear a squawk near me that I know; young crocodiles make those noises. When I open my eyes, I see baby caimans, only eight inches large, very close to me. I jump up. I know that I'm in danger. As soon as the mother of these babies notices my presence, she will attack me. And there she is already, very close. She rises on her legs and comes towards me threateningly.

And me? I slide back into the water and drift on. I've already had encounters with spectacled caimans that were dozing on the riverbank. When they noticed me they were frightened and jumped into the water and came towards me. If I didn't know this jungle so well, I undoubtedly would have gone ashore full of panic and run into the forest, where I would probably have died. But instead I trust that what I've learned in Panguana is true: caimans always flee into the water, no matter what direction they suspect danger is coming from, and they will swim past me or under me, but definitely won't attack me. However, the very presence of so many caimans is a sign for me that there are no people living on this river. Later I will learn that at that time the entire river was uninhabited. If I had simply lain down somewhere and stayed there, I would never have been found.

So I keep going.

I'm getting weaker, can scarcely struggle to my feet any more. I know that I have to eat something if I don't want to die. But what?

It's the rainy season, and frogs are jumping around everywhere. And I'm seized by the idea that I have to catch one of these animals and eat it, even though I know they are poison arrow frogs and will not agree with me. The Indians use certain species to poison their arrows, but the effect of these frogs here is too weak to kill an adult. Still, I'm not sure how well I will stomach them in my weakened state. Nonetheless, I try again and again to catch one of the frogs, but I don't manage to do it. At one point one of them is sitting less than six inches from my mouth. The moment I grab it, it's gone again. That depresses me more than anything else.

Again I hear the false chickens clucking, and again I'm fooled by them. At one point I'm close to tears when 1 realize that I've been deceived again.

I spend the tenth day drifting in the water. I'm constantly bumping into logs, and it costs me a great deal of strength to climb over them and to be careful not to break any bones in these collisions. In the evening I find a gravel bank that looks like a good place to sleep. I settle down on it, doze a little, blink; then I see something that doesn't belong here. I think I'm dreaming, but I open my eyes wide and it's really true. There on the riverbank is a boat. A quite large one, actually, of the sort the natives use. I tell myself that it's not possible, that I'm hallucinating. I rub my eyes, look three times, and still it's there. A boat.

I swim over to it and touch it. Only then can I really believe it. It's new and in full working order. Now I notice a beaten trail leading from the river fifteen to twenty feet up the slope of the bank. There are even visible footsteps in the ground. Why didn't I see that before? I have to get up there. Here I will definitely find people! But I'm so weak. It takes me hours to cover those few yards.

And then I'm finally there. I see a tambo, a simple shelter, poles with a palm leaf roof, a floor made out of the bark of the pona palm, about ten feet by fifteen. The boat's outboard motor is stored here—forty horsepower, I notice, as if that were important right now—and a drum of petrol. I can't see anyone around, but a path leads into the forest, and I'm certain that the owner of the boat will emerge at any moment. As I look at the petrol, I remember my maggots, which sometimes hurt horribly and have already got a bit bigger. I will trickle some of the petrol into my wound, and then they will come out as they did with Lobo. It takes an endlessly long time for me to get the drum's screw cap open. With a little piece of hose I find next to it, I suck up the petrol and let it drip into my wound. At first that hurts excruciatingly, as the maggots inside my arm try to escape downwards and bite their way still deeper into my flesh. But finally they come to the surface. I take thirty of them out of the wound with the bent-open ring, and then I'm exhausted. Later it will turn out that that was far from all of them, but for the time being I'm pretty proud of my achievement.

Still no one has come. It gets dark, and I decide to spend the night here. At first I try the floor of the hut, but the pona bark is so hard that I'd rather find a spot on the sandy riverbank. I borrow a tarpaulin that is also lying in the hut, cover myself up with it and, thus protected from the gnats, sleep divinely that night, better than in any five-star hotel.

The next morning I wake up, and still no one has shown up. I wonder what I should do. Perhaps no one will come here for the next few weeks. I know that there are shelters like this in the jungle that trappers or woodcutters use only sporadically. Maybe I really should keep going? Only briefly do I consider taking the boat and heading down river, but it doesn't seem right to me. Who knows? Perhaps the owner is somewhere in the forest here, I think, and when he comes back, he'll need his boat. I can't possibly save my own life and jeopardize someone else's. Besides, I'm not sure whether in my weakened state I'm even capable of manoeuvring the boat down this river. While I'm thinking about this and unable to decide whether to get back into the river, midday comes. And then it begins to pour with rain. I crawl into the *tambo*, wrap the tarpaulin around my shoulders and feel nothing. Now and then I try to catch a frog, in vain.

In the afternoon the rain stops, and my mind tells me that I have to keep going. Against all common sense I simply remain seated. I don't have the strength any longer to struggle to my feet. I will rest for one more day, I think, and tomorrow I'll carry on. Despair alternates with hope; powerlessness with new resolve.

I think that all the others must have been found a long time ago, and I'm the only one still out here. The thought occurs to me how strange it is that a person can disappear just like that, and no one knows about it. It's a peculiar feeling, which fills my chest and goes to my depths. I worry that I might die here, and no one will ever know what became of me. No one will ever find out what an arduous journey I undertook, how far I have come. I'm aware that I am slowly but surely starving. I have gone too long without eating anything. I always thought that when you starve, it hurts terribly. But I have no pain. I don't even feel hungry. I am only so horribly exhausted and weak. Again I try to catch one of the frogs. Again and again. Thus the day passes.

It's already twilight when I suddenly hear voices. I can't believe it! After all this time in solitude, it's inconceivable to me. *I'm imagining it*, I think, *as I have so many other things already*. But they really are human voices. They're approaching. And then three men come out of the forest and stop in shock. They even recoil involuntarily. I begin to speak to them in Spanish.

"I'm a girl who was in the LANSA crash," I say. "My name is Juliane."

They come closer and stare at me in astonishment.

Translated by Ross Benjamin

LOST IN THE JUNGLE

Yossi Ghinsberg

Yossi Ghinsberg (born 1959) is known for surviving for three weeks stranded in a remote and uncharted part of the Bolivian jungle. Twenty-two-year-old Ghinsberg, his two friends and a guide, ventured into an unexplored area of the Amazon in search of adventure, but quickly got separated. Alone, Ghinsberg almost drowned in a flood, sank in bogs, nearly starved to death and had hallucinations of a woman who kept him company. Upon his rescue he was reunited with one of his friends but his other companion and their guide, who they subsequently found out was a wanted criminal, were never found.

It was the seventeenth morning of my solitude. The storm was over. I was in sad shape. I was far from my destination and doubted that I would be able to go on. My feet were infected. From now on walking would be torture. How could my body take any more? I was weak with hunger. I had eaten nothing for the past two days. Now how would I find eggs or fruit? The storm had washed everything away. Was I going to die of hunger or injury? Morbid thoughts filled my mind; there was no chance of my escaping into fantasy. I was distressed to the point of despondency. All my hopes of reaching San José died away. I hadn't yesterday. I apparently I wasn't going to today. Who knew if I ever would?

What an idiot I was. I should never have left Curiplaya. I could have waited there in my hut. I could have survived there for at least a month, and by then surely someone would have come looking for me. Someone would have done something.

Now what would I do? Where should I go? I no longer believed there was much chance of my reaching San José. I doubted that I would be able to cross the river. Though the storm had died away, the whole jungle was submerged. I was bitter and on the edge of absolute despair, almost ready to give up. I started back to Curiplaya.

Overcome by self-pity, I hobbled painfully on until I came to a *trestepita* tree. The tree was bent low, almost touching the ground. It still had fruit on it, and I eagerly sucked the sweet sour pulp from the pits. The small quantity of nourishment tormented my aching belly, but it helped restore my hopes.

Someone is still watching over me. Uncle Nissim's book will protect me. I won't die as long as I have it in my pocket. I shouldn't underestimate its powers. I mustn't lose hope. I am stronger than I think I am. If I have been able to survive this far, I can go on.

I gave myself a good talking-to and turned toward San José once again. I was going on, no matter what. I trod through flood waters, swam across streams, climbed up wadi walls. I don't know where I got the energy. While I was wading through the mud, I believe that I was one of the Zionist pioneers, draining the swamps. A long black snake passing near my foot startled me. I there my walking stick at it but missed.

'Wait a minute,' I called, chasing after it. 'Wait a minute. I want to eat you.'

My shirt caught on a branch and tore. The sharp branch slashed my upper arm down to the elbow. Blood spurted from the wound. I fought back tears of desperation.

It doesn't matter. I'll get over it. I'm going on.

I could neither see nor hear the river but followed the streams that cut in front of me. I knew that they would lead me to the Tuichi. It wasn't raining, but the wind was blowing, and it was very cold. The humidity formed a heavy mist.

Suddenly I heard a sputter, a drone the sound of an engine…an aeroplane.

Don't be a fool, Yossi. It's only your imagination.

But the sound grew louder. It was an aeroplane!

They're looking for me! Hooray! I'm saved!

The sound grew louder, and I ran like a lunatic, ignoring my tattered feet. I had to get to the Tuichi. I had to signal the plane. The sound was right overhead. I stopped, panting, and looked up. Between the treetops I saw a few grey clouds, and amid them, at a moderate altitude, a small white place glided past.

'Hello, here I am! Help! I'm down here.' I waved my arms frantically. 'Don't go. Don't leave me here. Here I am.'

The plane vanished from the sky, its drone fading away.

Now I became aware of my feet. The frantic running had torn the flesh from them, and I felt as if they were on fire. I collapsed to the earth, my face buried in the mud. I lay sprawled there and wanted to cry, but the tears wouldn't come.

I can't take any more. I can't budge another inch. That's it.

From the bottom of my heart I prayed, not for rescue, not even for survival. I prayed for death. *Please, good God, stop this suffering. Let me die.*

And then she appeared. I knew it was all in my mind, but there she lay, next to me. I didn't know who she was. I didn't know her name. I knew we'd never met, and yet I knew that we were in love. She was weeping despondently. Her fragile body trembled.

'There, there, stop crying,' I tried to comfort her.

Take is easy. It's all right. Get up, Yossi, I urged myself, you have to lead the way, keep her spirits up.

I plucked myself up out of the mud and very gently helped her up. Tears still poured down her cheeks.

'The plane didn't see us. It just went by,' she wailed.

'Don't worry, my love. It will surely be back this way. It didn't see us through the jungle trees. We can't be seen from the air. If we could get a fire going, the smoke might be spotted.'

But everything was soaking wet.

When I heard the drone of the plane's engine once again, I knew we had no hope of being found that day.

I had made it back to the Tuichi, but there was no bank. I stood on the bluff, about twenty feet above the river, its rapids tumbling beneath me. I took out the poncho and waved it frantically, but I knew there was no chance of being spotted through the trees. The plane was flying too high and too fast. I watched it go past with longing eyes.

She looked up at me forlornly.

'Don't worry. They'll be back tomorrow,' I promised. 'Look, we were almost saved today. I'm sure that that's Kevin up there. It has to be Kevin. I just know it is. He must have gone to my embassy for help.'

I still did not recognise her: where she was from, why she was here. I just kept comforting her.

'They knew they'd have a hard time finding us today since the weather is so cloudy, but I'm sure they'll come back tomorrow and won't give up until they find us.

'You know, once in a while some guy gets lost in the Judeaen desert, and they call out the army and volunteers and trackers. Sometimes they have to keep looking for a whole week before they find the guy, dead or alive. They never just stop looking.

'What we have to do is help them find us. We have to find a shore to stand on, so they'll be able to see us.'

I remembered the beach where the jaguars had been. I had better head back there.

'Yes, that's a great idea. It's a huge beach.'

I had marked it clearly, and while I assumed that the markings had all been washed away, the beach itself must still be there. It was so wide. I quickly figured the distance. I had first arrived at Jaguar Beach on the afternoon of the fourteenth. I had wasted the rest of that day trying unsuccessfully to ford the river. On the fifteenth, as well, I had stopped walking relatively early. That meant that a day's walk was between me and the beach. I could still get in a good few hours' walk today. Tomorrow I would start walking at dawn and perhaps make Jaguar Beach in the morning hours.

I explained my plan to her.

'Come on, love. Another day's walk, maybe less, and we'll be there,' I said encouragingly. 'There they will spot us easily. First the plane will go over and see us. The pilot will signal us with a tilt of the wings and go back to base. Within a few hours a helicopter will arrive, land on the beach, and pick us up. We'll be saved. It'll all happen tomorrow. We have to stick it out one more day. Come on, let's get going.'

I changed direction for the third time that day. This time without hesitation. I knew that I was doing the right thing.

My feet barely obeyed my will, almost refusing to carry my weight. They couldn't stand much pressure. Every time I stepped on a rock or root, terrible pains pierced through me. When I had to climb a hill and descend the other side, it seemed an impossible effort. I had to get down and crawl, drag myself along with my elbows. But I kept my suffering to myself She was with me. She was also injured, weak, and hungry. It was harder for her than for me. If I wasn't strong, she would break.

I have to push myself harder, hide my own feelings, and keep her morale up.

When we were climbing upward, I would bite my lip and plead with her, Just a little farther, my love. Yes, I know how much it hurts. Here, I'll give you a hand. One more little push. That's all. You see? We made it. We're at the top. Now we have to get down. Sit like this and slide. Slowly, take it easy. Watch out. Be careful you don't slip.'

Rocks and thorns sliced into my buttocks. I noticed with concern that the rash had spread to other parts of my body. Red dots had broken out under my armpits and around my elbows. The cut on my arm hadn't formed a scab. The edges were white. I had been constantly wet for several days.

My body is rotting.

We walked until late evening. I didn't stop talking for a minute, chattering all day long, trying to keep her spirits up, trying to keep her from losing hope. When she stumbled or slowed down, I offered her my hand, caressing her sad face. I was so anxious to cover as much ground as possible that I didn't even notice that the sun had almost set. I had to hurry and find a place to rest our heads before darkness fell.

I tore off some palm fronds and spread them over some muddy tree roots. I didn't bother trying to get comfortable; my body was inured to discomfort. I covered myself with the wet nets and the poncho. Taking my shoes off had been agony. I didn't remove my socks. They would have to remain wet and dirty with mud, blood and pus. I pulled the sack over my feet very carefully, knowing how tormenting the slightest contact would be. I didn't change position all night long in order to give my feet a rest.

I believed with all my heart that tomorrow would be my last day of hardship. Tomorrow a plane would find me.

'Thank you, my love. Thank you for being here. Tomorrow you'll get the kid-glove treatment. Don't cry. Try to shut your eyes, to get some sleep. Tomorrow we still have a few more hours to walk. We have to get there early, before the plane comes.

'Good night, my love.'

At the break of dawn a heavy rain began pouring down. My prayers and please were to no avail. She was awakened by the first drops.

'Today is the big day, the last day,' I told her. 'We aren't going to let a little rain stop us. Don't let it get you down. It's not so bad. When we get to Jaguar Beach, I'll build you a strong shelter. You'll be able to rest, to sleep, until the helicopter comes.

'You're hungry? Yes, I know you're hungry, but we don't have anything left for breakfast. Don't worry, I'll find something to eat in the jungle. You can count on me.'

I couldn't stand. My were soft and mushy, as if a skinless mass of raw, bloody flesh had been poured into my shoes. I couldn't take a single step, but I knew that only chance for survival was to walk. I had to get to the shore. If I stayed in the jungle, no one would ever find me. I stumbled forward like a zombie. I discovered traces of the path, but it vanished after a while.

Walking through the dense growth was like marching through hell. I tried to stay as much as possible on soft, muddy ground, to ease the pain of every step I took. I tried to keep my weight on my trusty walking stick and often pulled myself forward by clutching at the brushes and branches. When I came to an incline, even a gentle slope, I got down on all fours and crawled, my face caked with mud, my clothes torn and weighing me down. I was weak and afraid of losing consciousness. All I had was water. Water had become the enemy. Other than water nothing had passed my lips. The girl was my only consolation.

We walked on together for a few hours, but Jaguar Beach was nowhere to be seen. I tried to locate it by looking for the four islands that had been strung across the river. I remembered them being very close to the beach, but I saw only one solitary island in the river. I feared that the current had swept the islands away but found that hard to believe. The islands had been large and well forested. They couldn't have vanished without a trace.

I trudged on and on through the mud and finally came upon a fruit tree.

It was tall, a species of palm. At the top were large clusters of dates. A family of monkeys were up there having a noisy feast. A few pieces of fruit were strewn on the ground. They were squashed into the mud and rotting. My body quivered, twitched with craving, an age-old primordial instinct. I was hungry like a wild beast. I pounced upon the dates in the mud. I didn't even care if they were rotten. The worms did not disgust me. I put the fruit into my mouth, rolled it around with my tongue, cleaned it off with my saliva, spat it out into the palm of my had, and then spat out the residue of mud in my mouth before putting the fruit back. Soon, however, I lost patience and swallowed the fruit together with the mud. I didn't leave a single piece on the ground. Even the worms were a source of protein. The monkeys started throwing half-eaten dates down at me. They laughed at me and tossed pits down on my head. I was grateful to them, for the monkeys didn't take more than one bite out of each piece and discarded a thick layer of edible pulp. I could see their teeth marks on the dates before I ate them.

I went on for several hours without stopping to rest. It required effort, a supreme and painful effort. Jaguar Beach was nowhere to be seen. I began to worry, though I didn't think that I could have missed it. It had been the widest strip of shore along the entire length of river. I must be moving more slowly than I thought. I was injured, and walking through the mud was slow and laborious. I can't give up. *I have to make it back there before the plane passes overhead again.*

Then I lost my head for a moment. It wasn't her fault. The hill was just too steep. I knew that I wouldn't make it to the top without a great deal of suffering and pain. Here I collapsed. She burst into tears and refused to go on. I was sick of speaking to her kindly and lovingly.

What the hell does she think? I wondered, enraged. *That I'm having a picnic?*

'Stop coddling yourself,' I shouted. 'I'm sick of you and your whining, do you hear? Who needs you anyway? I don't have enough problems without schlepping a cry-baby along? You don't help with anything. All you do is cry. Would you like to trade places with me for a while and carry this lousy pack on your back? I've had it with your bawling. You can cry your eyes out, for all I care, but you'd better not stop walking, because I'm not going to wait for you anymore.'

I behaved cruelly but felt relieved to have let off steam.

Afterward I felt ashamed of myself. I went over to her, gave her a hug, stroked her hair gently, and told her that I was sorry for having lost my temper and hadn't really meant any of it. I told her that I loved her, that I would protect her and bring her back to safety, but she had to make the effort and walk.

I had by now grown faint and dizzy, become weaker and weaker. When

I came across a fallen tree that blocked my path, I had to walk around it. I couldn't lift my legs over it.

I have to make it to Jaguar Beach. Have to, have to, have to!

I could hear the plane's engine in the distance. I waited as it drew near. I knew that the plane wouldn't be able to see me, but I at least wanted to see it. The sound was dull and distant, then faded altogether. Had I imagined it? Maybe they were looking for me somewhere else. But Kevin was there, I was sure of that, and he knew where I was.

Toward evening I came to an area where a puddle of water floated on the mud. I walked on, oblivious, and before I had a chance to comprehend what was happening, the earth swallowed me up. I sank swiftly. Shocked and in a panic. I found myself up to my waist in bog. I went into a frenzy, like a trapped animal, screaming, trying to get out, but the mud was thick and sticky, and I couldn't move. My walking stick cut through it like a hot knife through butter and was of no help at all. I reached out to some reeds and bushes, stretching my body and arms in their direction. I tried pulling myself out by them, but they came loose in my hands. I continued sinking slowly.

I came out of my convulsive throes and calmed down. I tried to act rationally. I stuck my hands down deep into the mud, wrapped them around one knee and tried to force one leg up out of the mud. I pulled with all my might, but to no avail. It was as if I had been set in concrete. I couldn't budge. I wanted to cry again but felt only a thick lump in my throat.

So this is it, death. I end my life in this bog.

I was resigned. I knew that I didn't have the strength to get myself out, and no power in the world would reach down and pluck me out of the swamp.

It would be a slow, horrible death. The mud was already up to my belly button. The pack rested on the mud, and I was relieved of its weight. Suddenly I had a brilliant idea. I would commit suicide. I took the pack from my shoulders and rummaged through it hurriedly until I found the first-aid kit. There were about twenty amphetamines and perhaps thirty other, unidentified pills. That was it. I would take all of them. I was sure they would kill me or at least make everything good and hazy before I drowned. First I opened the tin of speed. I held a few of them in the palm of my hand.

You' re being selfish, Yossi, really egotistical. It's easy enough for you to die, just swallow the pills, and you' re off to paradise. But what about your parents? Your mother: what will this do to her?

You can't die like this. Not after all you've already been through. It wouldn't be so bad it you had died on the first day in a sudden accident. But now, after all this suffering? It isn't fair to just give up now.

I put the pills back in the tin. I strained forward, leaning my torso out across the mud and moved my arms forward as if I were swimming. I moved my arms back and forth, pulling and wriggling in the mud. I kicked my legs in fluttering movements. I fought with every ounce of strength. Fought for my life.

It took about half an hour, maybe more. As soon as I got my legs free of the mud, I crept forward without sinking. I left neither the pack not the walking stick behind. After I'd advanced another six feet, I was out of the quagmire.

My entire body was caked with a thick layer of black, sticky mud. I cleared it out of my nostrils, wiped my eyes, and spat it out of my mouth.

To live. I want to live. I'll suffer any torment, but I'll go on. I'll make it to Jaguar Beach, no matter what.

ROBINSON CRUSOE

Daniel Defoe

Robinson Crusoe is the fictional lead character of **Daniel Defoe's** novel of the same name. The book opens with Crusoe marooned on a deserted island, forced to survive alone and live off the land. It is believed that Crusoe is based on the real story of Alexander Selkirk, a Scottish privateer who was stranded on a South Pacific island and survived several years before being rescued. The two stories mirror each other in several ways, and both provide insights into the psychological and physiological challenges of being left for dead on a deserted island. Daniel Defoe was a prolific pamphleteer and journalist, and *Robinson Crusoe* is seen as a contender for the first English novel.

And now our Case was very dismal indeed; for we all saw plainly, that the Sea went so high, that the Boat could not live, and that we should be inevitably drowned. As to making Sail, we had none, nor, if we had, could we ha' done any thing with it; so we work'd at the Oar towards the Land, tho' with heavy Hearts, like Men going Execution; for we all knew, that when the Boat came nearer the Shore, she would be dash'd in a Thousand Pieces by the Breach of the Sea. However, we committed our Souls to God in the most earnest Manner, and the Wind driving us towards the Shore, we hasten'd our Destruction with our own Hands, pulling as well as we could towards Land.

What the Shore was, whether Rock or Sand, whether Steep or Shoal, we knew not; the only Hope that could rationally give us the least Shadow of Expectation, was, if we might happen into some Bay or Gulph, or the Mouth of some River, where by great Chance we might have run our Boat in, or got under the Lee of the Land, and perhaps made smooth Water. But there was nothing of this appeared; but as we made nearer and nearer the Shore, the Land look'd more frightful than the Sea.

After we had row'd, or rather driven about a League and a Half, as we reckon'd it, a raging Wave, Mountain-like, came rowling a-stern of us and plainly bad us expect the *Coup de Grace*. In a word, it took us with such a Fury, that it overset the Boat at once; and separating us as well from the Boat, as from one another, gave us not time hardly to say, O God! for we were all swallowed up in a Moment.

Nothing can describe the Confusion of Thought which I felt when I sunk into the Water; for tho' I swam very well, yet I could not deliver my self from the Waves so as to draw Breath, till that Wave having driven me, or rather carried me a vast Way on towards the Shore, and having spent it self, went back, and left me upon the Land almost dry, but half-dead with the Water I took in. I had so much Presence of Mind as well as Breath left, that seeing my self nearer the main Land than I expected, I got upon my Feet, and endeavoured to make on towards the Land as fast as I could, before another Wave should return, and take me up again. But I soon found it was impossible to avoid it; for I saw the Sea come after me as high as a great Hill, and as furious as an Enemy which I had no Means or Strength to contend with; my Business was to hold my Breath, and raise my self upon the Water, if I could; and so by swimming to preserve my Breathing, and Pilot my self towards the Shore, if possible; my greatest Concern now being, that the Sea, as it would carry me a great Way towards the Shore when it came on, might not carry me back again with it when it gave back towards the Sea.

The Wave that came upon me again, buried me at once 20 or 30 Foot deep in its own Body; and I could feel my self carried with a mighty Force and Swiftness towards the Shore a very great Way; but I held my Breath, and assisted my self to swim still forward with all my Might. I was ready to burst with holding my Breath when as I felt my self rising up, so to my immediate Relief, I found my Head and Hands shoot out above the Surface of the Water; and tho' it was not two Seconds of Time that I could keep my self so, yet it reliev'd me greatly, gave me Breath and new Courage. I was covered again with Water a good while, but not so long but I held it out; and finding the Water had spent it self, and began to return I strook forward against the Return of the Waves, and felt Ground again with my Feet. I stood still a few Moments to recover Breath, and till the Water went from me, and then took to my Heels, and run with what Strength I had farther towards the Shore. But neither would this deliver me from the Fury of the Sea, which came pouring in after me again, and twice more I was lifted up by the Waves, and carried forwards as before, the Shore being very flat.

The last Time of these two had well near been fatal to me for the Sea having hurried me along as before, landed me, or rather dash'd me against a Piece of a Rock, and that with such Force as it left me senseless, and indeed helpless, as to my own Deliverance for the Blow taking my Side and Breast, beat the Breath as it were quite out of my Body; and had it returned again immediately, I must have been strangled in the Water; but I recover'd a little before the return of the Waves, and seeing I should be cover'd again with the Water I resolv'd to hold fast by a Piece of the Rock, and so to hold my Breath, if possible, till the Wave went back; now as the Waves were not so high as at first, being nearer Land, I held my Hold till the Wave abated, and then fetch'd another Run, which brought me so near the Shore, that the next Wave, tho' it went over me, yet did not so swallow me up as to carry me away, and the next run I took, I got to the main Land, where, to my great Comfort I clamber'd up the Clifts of the Shore, and sat me down upon the Grass, free from Danger, and quite out of the Reach of the Water.

I was now landed, and safe on Shore, and began to look up and thank God that my Life was sav'd in a Case wherein there was some Minutes before scarce any room to hope. I believe it is impossible to express to the Life what the Extasies and Transports of the Soul are, when it is so sav'd, as I may say, out of the very Grave, and I do not wonder now at that Custom, *viz.* That when a Malefactor who has the Halter about his Neck, is tyed up, and just going to be turn'd off, and has a Reprieve brought to him: I say, I do not wonder that they bring a Surgeon with it, to let him Blood that very Moment they tell him of it, that the Surprise may not drive the Animal Spirits from the Heart, and overwhelm him:

For sudden Joys, like Griefs, confound at first.

I walk'd about on the Shore, lifting up my Hands, and my whole Being, as I may say, wrapt up in the Contemplation of my Deliverance making a Thousand Gestures and Motions which I cannot describe, reflecting upon all my Comerades that were drown'd, and that there should not be one Soul sav'd but my self; for, as for them, I never saw them afterwards, or any Sign of them, except three of their Hats, one Cap, and two Shoes that were not Fellows.

I cast my Eyes to the stranded Vessel, when the Breach and Froth of the Sea being so big, I could hardly see it, it lay so far off, and considered, Lord! how was it possible I could get on Shore?

After I had solac'd my Mind with the comfortable Part of my Condition, I began to look round me to see what kind of Place I was in and what was next to be done, and I soon found my Comforts abate and that in a word I had a dreadful Deliverance: For I was wet, had no Clothes to shift me, nor any thing either to eat or drink to comfort me neither did I see any Prospect before me, but that of perishing with Hunger, or being devour'd by wild Beasts; and that which particularly afflicting to me, was, that I had no Weapon either to hunt and kill any Creature for my Sustenance, or to defend my self against any other Creature that might desire to kill me for theirs: In a Word I had nothing about me but a Knife, a Tobaccopipe, and a little

Tobacco in a Box, this was all my Provision, and this threw me into terrible Agonies of Mind, that for a while I run about like a Mad-man; Night coming upon me, I began with a heavy Heart to consider what would be my Lot if there were any ravenous Beasts in that Country seeing at Night they always come abroad for their Prey.

All the Remedy that offer'd to my Thoughts at that Time, was, to get up into a thick bushy Tree like a Firr, but thorny, which grew near me, and where I resolv'd to set all Night, and consider the next Day what Death I should dye, for as yet I saw no Prospect of Life; I walk'd about a Furlong from the Shore, to see if l could find any fresh Water to drink, which I did, to my great Joy; and having drank and put a little Tobacco in my Mouth to prevent Hunger, I went to the Tree, and getting up into it, endeavour'd to place my self so, as that if I should sleep I might not fall; and having cut me a short Stick, like a Truncheon, for my Defence, I took up my Lodging, and having been excessively fatigu'd, I fell fast asleep, and slept as comfortably as, I believe, few could have done in my Condition, and found my self the most refresh'd with it, that I think I ever was on such an Occasion.

When I wak'd it was broad Day, the Weather clear, and the Storm abated, so that the Sea did not rage and swell as before: But that which surpris'd me most, was, that the Ship was lifted off in the Night from the Sand where she lay, by the Swelling of the Tyde, and was driven up almost as far as the Rock which I first mention'd, where I had been so bruis'd by the dashing me against it; this being within about a Mile from the Shore where I was, and the Ship seeming to stand upright still, I wish'd my self on board, that, at least, I might save some necessary things for my use.

When I came down from my Appartment in the Tree, I look'd about me again, and the first thing I found was the Boat, which lay as the Wind and the Sea had toss'd her up upon the Land, about two Miles on my right Hand. I walk'd as far as I could upon the Shore to have got to her, but found a Neck or Inlet of Water between me and the Boat, which was about half a Mile broad, so I came back for the present, being more intent upon getting at the Ship, where I hop'd to find something for my present Subsistence.

A little after Noon I found the Sea very calm, and the Tyde ebb'd so far out, that I could come within a Quarter of a Mile of the Ship; and here I found a fresh renewing of my Grief, for I saw evidently, that if we had kept on board, we had been all safe, that is to say, we had all got safe on Shore, and I had not been so miserable as to be left entirely destitute of all Comfort and Company, as I now was; this forc'd Tears from my Eyes again, but as there was little Relief in that, I resolv'd, if possible, to get to the Ship, so I pull'd off my Clothes, for the Weather was hot to Extremity, and took the Water, but when I came to the Ship, my Difficulty was still greater to know how to get on board, for as she lay a ground, and high out of the Water, there was nothing within my Reach to lay hold of; I swam round her twice, and the second Time I spy' d a small Piece of a Rope, which I wonder'd I did not see at first, hang down by the Fore-Chains so low, as that with great Difficulty I got hold of it, and by the help of that Rope, got up into the Forecastle of the Ship; here I found that the Ship was bulg'd, and had a great deal of Water in her Hold, but that she lay so on the Side of a Bank of hard Sand, or rather Earth, that her Stern lay lifted up upon the Bank, and her Head low almost to the Water; by this Means all her Quarter was free, and all that was in that Part was dry; for you may be sure my first Work was to search and to see what was spoil'd and what was free; and first I found that all the Ship's Provisions were dry and untouch'd by the Water, and being very well dispos'd to eat, I went to the Bread-room and fill'd my Pockets with Bisket, and eat it as I went about other things, for I had no time to lose; I also found some Rum in the great Cabbin, of which I took a large Dram, and which I had indeed need enough of to spirit me for what was before me: Now I wanted nothing but a Boat to furnish my self with many things which I forsaw would be very necessary to me.

It was in vain to sit still and wish for what was not to be had, and this Extremity rouz'd my Application; we had several spare Yards, and two or three large sparrs of Wood, and a spare Top-mast or two in the Ship; I resolv'd to fall to work with these, and I flung as many of them over board as I could manage for their Weight, tying every one with a Rope that they might not drive away; when this was done I went down the Ship's Side, and pulling them to me, I ty'd four of them fast together at both Ends as well as I could, in the Form of a Raft, and laying two or three short Pieces of Plank upon them crossways, I found I could walk upon it very well, but that it was not able to bear any great Weight; the Pieces being too light; so I went to work, and with the Carpenter's Saw I cut a spare Top-mast into three Lengths, and added them to my Raft, with a great deal of Labour and Pains, but hope of furnishing my self with Necessaries, encourag'd me to go beyond what I should have been able to have done upon another Occasion.

My Raft was now strong enough to bear any reasonable Weight; my next Care was what to load it with, and how to preserve what I laid upon it from the Surf of the Sea; But I was not long considering this, I first laid all the Plank or Boards upon it that I could get, and having consider'd well what I most wanted, I first got three of the Seamens Chests, which I had broken open and empty'd, and lower'd them down upon my Raft; the first of these I fill'd with Provision, *viz.* Bread, Rice, three Dutch Cheeses, five Pieces of dry'd Goat's Flesh, which we liv'd much upon, and a little Remainder of *European* Corn which had been laid by for some Fowls which we brought to Sea with us, but the Fowls were kill'd; there had been some Barly and Wheat together, but, to my great Disappointment, I found afterwards that the Rats had eaten or spoil'd it all; as for Liquors, I found several Cases of Bottles belonging to our Skipper, in which were some Cordial Waters, and in all about five or six Gallons of Rack, these I stow'd by themselves, there being no need to put them into the Chest, nor no room for them. While I was doing this, I found the Tyde began to flow, tho' very calm, and I had the Mortification to see my Coat, Shirt, and Wast-coat which I had left on Shore upon the Sand, swim away; as for my Breeches which were only Linnen and open knee'd, I swam on board in them and my Stockings: However this put me upon rummaging for Clothes, of which I found enough, but took no more than I wanted for present use, for I had other things which my Eye was more upon, as first Tools to work with on Shore, and it was after long searching that I found out the Carpenter's Chest, which was indeed a very useful Prize to me, and much more valuable than a Ship Loading of Gold would have been at that time; I got it down to my Raft, even whole as it was, without losing time to look into it, for I knew in general what it contain'd.

My next Care was for some Ammunition and Arms; there were two very good Fowling-pieces in the great Cabbin, and two Pistols, these I secur'd first, with some Powder-horns, and a small Bag of Shot, and two old rusty Swords; I knew there were three Barrels of Powder in the Ship, but knew not where our Gunner had stow'd them, but with much search I found them, two of them dry and good, the third had taken Water, those two I got to my Raft, with the Arms, and now I thought my self pretty well freighted, and began to think how I should get to Shore with them, having neither Sail, Oar, or Rudder, and the least Cap full of Wind would have overset all my Navigation.

I had three Encouragements, 1. A smooth calm Sea, 2. The Tide rising and setting in to the Shore, 3. What little Wind there was blew me towards the Land; and thus, having found two or three broken Oars belonging to the Boat, and besides the Tools which were in the Chest, I found two Saws, an Axe, and a Hammer, and with this Cargo I put to Sea; For a Mile, or thereabouts, my Raft went very well, only that I found it drive a little distant from the Place where I had landed before, by which I perceiv'd that there was some Indraft of the Water, and consequently I hop'd to find some Creek or River there, which I might make use of as a Port to get to Land with my Cargo.

As I imagin'd, so it was, there appear'd before me a little opening of the Land, and I found a strong Current of the Tide set into it, so I guided my Raft as well as I could to keep in the Middle of the Stream: But here I had like to have suffer'd a second Shipwreck, which, if I had, I think verily would have broke my Heart, for knowing nothing of the Coast, my Raft run a-ground at one End of it upon a Shoal, and not being a-ground at the other End, it wanted but a little that all my Cargo had slip'd off towards that End that was a-float, and so fall'n into the Water: I did my utmost by setting my Back against the Chests, to keep them in their Places, but could not thrust off the Raft with all my Strength, neither durst I stir from the Posture I was in, but holding up the Chests with all my Might, stood in that Manner near half an Hour, in which time the rising of the Water brought me a little more upon a Level, and a little after, the Water still rising, my Raft floated again, and I thrust her off with the Oar I had, into the Channel, and then driving up higher, I at length found my self in the Mouth of a little River, with Land on both Sides, and a strong Current or Tide running up, I look'd on both Sides for a proper Place to get to Shore, for I was not willing to be driven too high up the River, hoping in time to see some Ship at Sea, and therefore resolv'd to place my self as near the Coast as I could.

At length I spy'd a little Cove on the right Shore of the Creek, to which with great Pain and Difficulty I guided my Raft, and at last got so near, as that, reaching Ground with my Oar, I could thrust her directly in, but here I had like to have dipt all my Cargo in the Sea again; for that Shore lying pretty steep, that is to say sloping, there was no Place to land, but where one End of my Float, if it run on Shore, would lie so high, and the other sink lower as before, that it would endanger my Cargo again: All that I could do, was to wait 'till the Tide was at highest, keeping the Raft with my Oar like an Anchor to hold the Side of it fast to the Shore, near a flat Piece of Ground, which I expected the Water would flow over; and so it did: As soon as I found Water enough, for my Raft drew about a Foot of Water, I thrust her on upon that flat Piece of Ground, and there fasten'd or mor'd her by sticking my two broken Oars into the Ground; one on one Side near one End, and one on the other Side near the other End; and thus I lay 'till the Water ebb'd away, and left my Raft and all my Cargoe safe on Shore.

My next Work was to view the Country, and seek a proper Place for my Habitation, and where to stow my Goods to secure them from whatever might happen; where I was I yet knew not whether on the Continent or on an Island, whether inhabited or not inhabited whether in Danger of wild Beasts or not: There was a Hill not above a Mile from me, which rose up very steep and high, and which seem'd to over-top some other Hills which lay as in a Ridge from it northward; I took out one of the fowling Pieces, and one of the Pistols, and an Horn of Powder, and thus arm'd I travell'd for Discovery up to the Top of that Hill, where after I had with great Labour and Difficulty got to the Top, I saw my Fate to my great Affliction, (*viz*) that I was in an Island environ'd every Way with the Sea, no Land to be seen, except some Rocks which lay a great Way off, and two small Islands less than this, which lay about three Leagues to the West.

I found also that the Island I was in was barren, and, as I saw good Reason to believe, un-inhabited, except by wild Beasts, of whom however I saw none, yet I saw Abundance of Fowls but knew not their Kinds, neither when I kill'd them could I tell what was fit for Food, and what not; at my coming back, I shot at a great Bird which I saw sitting upon a Tree on the Side of a great Wood, I believe it was the first Gun that had been fir'd there since the Creation of the World; I had no sooner fir'd, but from all the Parts of the Wood there arose an innumerable Number of Fowls of many Sorts making a confus'd Screaming, and crying every one according to his usual Note; but not one of them of any Kind that I knew: As for the Creature I kill'd, I took it to be a Kind of a Hawk, its Colour and Beak resembling it, but had no Talons or Claws more than common, its Flesh was Carrion, and fit for nothing.

Contented with this Discovery, I came back to my Raft, and fell to Work to bring my Cargoe on Shore, which took me up the rest of that Day, and what to do with my self at Night I knew not, nor indeed where to rest; for I was afraid to lie down on the Ground, not knowing but some wild Beast might devour me, tho', as I afterwards found, there was really no Need for those Fears.

However, as well as I could, I barricado'd my self round with the Chests and Boards that I had brought on Shore, and made a Kind of a Hut for that Night's Lodging; as for Food, I yet saw not which Way to supply my self, except that I had seen two or three Creatures like Hares run out of the Wood where I shot the Fowl.

I now began to consider, that I might yet get a great many Things out of the Ship, which would be useful to me, and particularly some of the Rigging, and Sails, and such other Things as might come to Land, and I resolv'd to make another Voyage on Board the Vessel, if possible; and as I knew that the first Storm that blew must necessarily break her all in Pieces, I resolv'd to set all other Things apart, 'till I got every Thing out of the Ship that I could get; then I call'd a Council, that is to say, in my Thoughts, whether I should take back the Raft, but this appear'd impracticable; so I resolv'd to go as before, when the Tide was down, and I did so, only that I stripp'd before I went from my Hut, having nothing on but a Chequer'd Shirt, and a Pair of Linnen Drawers, and a Pair of Pumps on my Feet.

I got on Board the Ship, as before, and prepar'd a second Raft, and having had Experience of the first, I neither made this so unweildy, nor loaded it so hard, but yet I brought away several Things very useful to me; as first, in the Carpenter's Stores I found two or three Bags full of Nails and Spikes, a great Skrew-Jack, a Dozen or two of Hatchets, and above all, that most useful Thing call'd a Grindstone; all these I secur'd together, with several Things belonging to the Gunner, particularly two or three Iron Crows, and two

Barrels of Musquet Bullets, seven Musquets, and another fowling Piece, with some small Quantity of Powder more; a large Bag full of small Shot, and a great Roll of Sheet Lead: But this last was so heavy, I could not hoise it up to get it over the Ship's Side.

Besides these Things, I took all the Mens Cloths that I could find, and a spare Fore-top-sail, a Hammock, and some Bedding; and with this I loaded my second Raft, and brought them all safe on Shore to my very great Comfort.

I was under some Apprehensions during my Absence from the Land, that at least my Provisions might be devour'd on Shore; but when I came back, I found no Sign of any Visitor, only there sat a Creature like a wild Cat upon one of the Chests, which when I came towards it, ran away a little Distance, and then stood still; she sat very compos'd, and unconcern'd, and look'd full in my Face, as if she had a Mind to be acquainted with me, I presented my Gun at her, but as she did not understand it, she was perfectly unconcern'd at it, nor did she offer to stir away; upon which I toss'd her a Bit of Bisket, tho' by the Way I was not very free of it, for my Store was not great: However, I spar'd her a Bit, I say, and she went to it, smell'd of it, and ate it, and look'd (as pleas'd) for more, but I thank'd her, and could spare no more; so she march'd off.

A HERO FOR THE AMERICAS

Robert Calder

Gonzalo Guerrero (1470–1536) was a Spanish soldier who was shipwrecked on the coast of the Yucatán Peninsula in 1512 and taken captive by the Maya. Guerrero went on to embrace Mayan culture and married Princess Zazil Há, daughter of the Nachán Can; their children are the first known European – Native American children on record in the mainland Americas. Guerrero led the opposition to the Spanish conquest of the Yucatán, and whilst he was reviled in Spain as a traitor, he is remembered today in Mexico as a hero and the father of Mexican *mestizos*.

Robert Calder is a Canadian writer and professor. *A Hero for the Americas* is the first comprehensive investigation of the life and influence of Gonzalo Guerrero.

When confronted by crisis, extreme hardship, physical pain, psychological threat, and death, people often collapse and perish. Some, however, display remarkable and almost inexplicable resilience, and they survive. A few not only survive but even thrive in the hostile environment. In 1512, a party of Spaniards was shipwrecked off the coast of Jamaica, was blown across the Caribbean, and fell into the hands of Mayan tribes in the Yucatán Peninsula. Only two of the twenty castaways survived the days on the water in the relentless sun and the years of captivity by the hostile natives of a strange land. One was a cleric and the other a common sailor.

The Spanish chroniclers, notably Bernal Díaz del Castillo and Francisco Cervantes de Salazar, have recorded Jerónimo de Aguilar's account of his survival. Aguilar endured, he told Hernán Cortés and his rescue party, because he served his Mayan captors obediently and cheerfully, and he won their admiration by resisting the temptations of the beautiful young women offered to him. Most importantly, he overcame the despair, depression, and sense of dislocation that sapped the will to live of so many of his fellows by strict adherence to his Christian faith. He remained convinced that God, for whom he was an instrument of some higher purpose, would guide him safely back to his own people.

But what of Gonzalo Guerrero? What gave him the strength to withstand the rigours of his shipwreck and capture, and what gave him the will to survive and triumph? Unlike Aguilar, he left no account of his experience, but we can speculate about what might have given a young sailor from Palos, toughened by the ordeal of Darién, the traits necessary to endure. And if, amid the alien deities of his Mayan captors, he retained a belief in his Catholic God, then for what could he have thought God had preserved him?

Unlike Aguilar and almost all the others in the shipwrecked party, Gonzalo was accustomed to hard labour. Anyone who had grown up in the working class of an Andalusian seaport, and who had laboured aboard the ships plying their trade along the Iberian and African coasts, would have been physically strong and resilient. He was hardly a middle-class fortune seeker accustomed to comfort, luxury, and servants to do his heavy labour. If he needed any further hardening, Gonzalo found it scrambling on the rigging or hauling sails on the heaving sea between Cadiz and the Spanish Indies. To have survived the severe hardships of the Darién settlements, when so many others perished, required a strong constitution and determined will to live. Thus, when he staggered ashore along the Yucatán coast, Gonzalo had already been harshly tested, and few men were better suited to survive the long ordeal that he was about to face.

Such strengths would soon have become apparent to his Mayan captors, and they would have made Gonzalo a valuable commodity. Like most early cultures – notably the Egyptians, Greeks, and Romans – the Mayan communities depended heavily on slavery to provide essential labour. Slaves became an important economic commodity, being sold and traded up and down the Yucatán coast and into the Gulf of Mexico. They were most commonly acquired in battle, after which the victors would immediately sacrifice the high-ranking captives – hence the early deaths of Juan de Valdivia and three others of the shipwrecked Spaniards – and keep those of lower rank as slaves. Despite being condemned to a life of servitude, prisoners who worked industriously were treated relatively well, though many were ultimately sacrificed on the deaths of their masters or to propitiate the gods during times of drought, famine, or disease.

Gonzalo undoubtedly had a strong back for hauling wood and water and tilling and planting the maize fields, but he possessed two things that made him unusually valuable to the Maya: knowledge of the sea and sailing, and knowledge of the strange bearded men beginning to intrude on their lands. He was one of seven survivors who had escaped from their original captors, and, after falling into the hands of another chiefdom, he was sold to Nachan Ka'an, the *cacique* of Chactemal, located at the southeasternmost corner of the Yucatán Peninsula. As such, Gonzalo had become part of the Maya's most seagoing community, a maritime commercial society not unlike that in which he had grown up in Palos.

The Chactemal community was practically surrounded by water. It was flanked on the east by the Bay of Chetumal, a large, shallow body of water separated from the Caribbean by a long finger of land, the Xcalak Peninsula, and extending into what is now Belize. Through the southern opening of this nearly inland sea, the Maya could paddle their eighty-foot cedar canoes down the coast to what are now Honduras, Nicaragua, and Panama. Through the Boca Bacalar Chico, a narrow channel possibly dug by the Maya, they could cut across the peninsula directly to the Caribbean and go north to Tulum, Xel Ha, and Cozumel.

The mainland on all sides of the Chactemal settlement contained more bodies of water – lakes and rivers – than any other territory in Yucatán. Connected to the Bay of Chetumal are a long sliver of a lagoon now called Laguna Guerrero and a lake named Lago Guerrero. Farther inland is Laguna Bacalar, a lagoon fifty kilometres long and ten kilometres wide, connected to the bay by a network of shallow channels and a stream draining into the Hondo River. Since the Hondo stretches 150 kilometres into the interior and itself drains into the Bay of Chetumal, there was a natural marine route linking the mainland tribes to the communities along the Caribbean coast.

The Chactemal Maya became prosperous traders, moving seashells, cotton, feathers, salt harvested from the shallow lagoons, and cacao, grown nowhere else in Yucatán, inland and up the northern coast. They exchanged these and other goods such as honey, wax, obsidian, and jade in Honduras and Nicaragua for the metals not found anywhere in the Yucatán. Indeed, it was the gold and silver jewellery imported from the south that fired the early Spanish dreams of finding vast amounts of treasure on the peninsula.

Such maritime commerce led one historian of the Maya to call the Chactemal Maya "the Phoenicians of Middle America," and Guerrero, coming from a similarly rich seagoing tradition, must have felt some kinship with them. The seafarers of Chactemal, however, were also called "the guardians of the sands, the guardians of the seas," because they were the first line of defence of the region against predation by hostile tribes from the Mosquito Coast of Nicaragua and elsewhere. The big trading canoes served equally well as war craft patrolling the mangrove-lined lagoons and open water of the Caribbean shoreline. A practised seaman with experience in combat was thus an invaluable asset to the Chactemal Maya, and Gonzalo soon came to be trusted to lead their maritime defences. His greatest role, however, ultimately became that of captain of the Mayan forces repelling the invasion of the Yucatán by enemies much more familiar to him: his own countrymen, the Spanish.

The survival of Gonzalo, like that of Aguilar, was remarkable, but his rise from slave to respected military captain is much more astonishing. At the beginning, Gonzalo was truly an alien among the Maya, a bearded white man of the sort that they had never seen before, a being who washed up on their shore from a land and a culture whose existence they had never known. If stories of the Spanish presence in the Indies or Darién – particularly its brutal treatment of the natives – had filtered into the Yucatán, he would have begun his enslavement as a deeply suspicious figure, seen to be capable, given the opportunity, of betraying his captors to his compatriots. To become accepted as a tribesman who would marry the *cacique*'s daughter required exceptional tact, resolution, courage, and discipline. Above all, it required great adaptability, and this is what made the experience of Gonzalo so fundamentally different from that of Aguilar.

Because Gonzalo was an exotic captive so unlike the indigenous prisoners taken from other tribes, he was a prize claimed by the *cacique* of the Chactemal community, Nachan Ka'an. Attached to this eminent household, he would first have been employed in heavy labour such as hauling wood and water and tilling fields, simple tasks that, in the absence of a common language, he could be assigned by gestures. Most slaves spent their lives – which for some, unfortunately, were brief – in such service, but Gonzalo still burned with some optimism and ambition, and this made him learn the language of his captors, a difficult task since it bore no resemblance to written or spoken Spanish. As he became more proficient in it, he was given more responsibility and more complex tasks, and he gained the confidence of Nachan Ka'an by fulfilling them capably. Having found his tongue among these strangers, he likely began to tell them of his earlier life and his own people, and one can only begin to imagine what the Chactemal Maya made of his talk of Andalusia, Catholicism, armour and gunpowder, ocean-going sailing ships, horses, wheeled vehicles, and so much else. Most importantly, he gave his captors an essential truth that would remain hidden from the other great Middle American civilisation, the Aztecs, until it was too late: the Spanish were not gods but men like themselves.

As Nachan Ka'an's respect for, and trust in, his slave grew, Gonzalo must have been aware that his rise from the lowest social rank to a position of prominence would never have been possible in his native Spain. Coming from a family of sailors in Palos, he might have been able to make some fortune in the New World, but it would not have raised him much above the working class at home. Living with the Maya, he found what centuries later would be called the "American Dream," a meritocracy in which character, skill, and ambition outweighed birthright and class and could raise one from slave to captain. Listening to Aguilar plead with him to join Cortés' men, Gonzalo must have known that he would, at the least, be going from Mayan captain back to Spanish foot soldier.

Among the Maya, Gonzalo earned the respect of far more than the *cacique*; he won the admiration of Nachan Ka'an's daughter, Zazil Há, and nothing marked his assimilation into Mayan life as indelibly as their marriage and the children, the first *mestizos* in Mexican history, that came of it. It is this union, the first true marriage in American history of a European and a native, that is celebrated in statues and paintings all ove the Yucatán Peninsula.

The story of Gonzalo Guerrero and the Mayan "princess," Zazil Há, like that of John Smith and Pocahontas, or countless Hollywood films such as *Dances with Wolves*, has been romanticized in numerous accounts, and there is an understandable appeal in imagining a growing attraction between the slave and the chief's daughter. Mayan slaves, though not treated as members of the family, were part of the household, and there would have been ample opportunities for the pair to be together. Perhaps, too, an enraptured Zazil Há sat with her father when Gonzalo told his stories of Castile, his youth, and his experiences in Darién. As the exotic stranger revealed more of his character, and perhaps bared something of his soul, a love might have sprung up between him and the beautiful young woman listening attentively in the background. Zazil Há might have pleaded with her father to be allowed to marry the slave rather than some young Mayan warrior.

The marriage of Zazil Há and Gonzalo, however, was likely much more prosaic. Romantic attraction was not an important part of Mayan marriages, most of which were arranged by parents or professional village matchmakers. Indeed, seeking a mate by oneself without the help of experts was considered improper. Fathers in particular were very careful to find appropriate wives, women from the same social class, for their sons, and the family backgrounds of both bride and groom were important. Family lineage ran through the males, with children being given the surname of the father; in addition to this lineage name, however, they adopted a secondary or "house" name taken from the family name of the mother.

Gonzalo, of course, could bring no Mayan lineage to name to the marriage. In this sense, he was classless, and any name that he might be given by the tribe would become that by which his *mestizo* children and their descendants would be known. This was a serious impediment to the marriage of anyone enslaved from a foreign tribe, but Nachan Ka'an overlooked it in giving his blessing to the union of Gonzalo and Zazil Há. As the Spaniard rose in his estimation as a worker, sailor and warrior, he must have looked increasingly like a man with whom the *cacique* could entrust his daughter. If Gonzalo were of average stature for a Spaniard, then he would have been an imposing figure among Mayan men, whose average height was five feet, one inch, and average weight about 120 pounds. As such, he would have looked like someone with whom a woman could be secure and with whom she could raise a protected family. In any case, Gonzalo rose remarkably from slave to man worthy of marrying into the tribe's highest-ranking family.

Following Mayan tradition, the wedding of Zazil Há and Gonzalo took place at the house of Nachan Ka'an, where a feast had been prepared. Guests – including all prominent figures in the community – met the couple and mingled with the bride's relatives. When the priest was assured that all parties had been consulted and were satisfied with the arrangements, he conducted a simple ceremony, essentially the giving of Zazil Há to Gonzalo. Following this was the nearly universal ritual of a feast, and the couple were then allowed to consummate their marriage.

Tradition required that a newly married couple live in the bride's father's house for five or six years, during which the husband worked for his father-in-law. Thereafter, if the groom performed his duties satisfactorily, he was allowed to set up his own household; if not, he was driven from the family. Since Gonzalo had already performed much of that service as a slave, and had proven to be an exemplary worker, his probation was relatively brief. Then, with the help of the community, he built his *ná*, the traditional Mayan house.

12 YEARS A SLAVE

Solomon Northup

Solomon Northup (*c.*1807–1863) was a freeborn African American abolitionist from New York who was kidnapped and sold to a slave plantation in Louisiana. For twelve years, he lived as a slave and bore witness to the horrific conditions of slave life on the plantations. In his memoir *12 Years a Slave*, he recalled numerous instances of beatings, death threats and sexual abuse by white slaveholders towards their slaves. He was eventually set free after his family discovered his location. His kidnappers were charged, but the charges were dropped, as it was illegal at the time for a white man to be charged by an African American man.

"Well, my boy, how do you feel now?" said Burch, as he entered through the open door. I replied that I was sick, and inquired the cause of my imprisonment. He answered that I was his slave—that he had bought me, and that he was about to send me to New-Orleans. I asserted, aloud and boldly, that I was a free man—a resident of Saratoga, where I had a wife and children, who were also free, and that my name was Northup. I complained bitterly of the strange treatment I had received, and threatened, upon my liberation, to have satisfaction for the wrong. He denied that I was free, and with an emphatic oath, declared that I came from Georgia. Again and again I asserted I was no man's slave, and insisted upon his taking off my chains at once. He endeavored to hush me, as if he feared my voice would be overheard. But I would not be silent, and denounced the authors of my imprisonment, whoever they might be, as unmitigated villains. Finding he could not quiet me, he flew into a towering passion. With blasphemous oaths, he called me a black liar, a runaway from Georgia, and every other profane and vulgar epithet that the most indecent fancy could conceive.

During this time Radburn was standing silently by. His business was, to oversee this human, or rather inhuman stable, receiving slaves, feeding and whipping them, at the rate of two shillings a head per day. Turning to him, Burch ordered the paddle and cat-o'-ninetails to be brought in. He disappeared, and in a few moments returned with these instruments of torture.

The paddle, as it is termed in slave-beating parlance, or at least the one with which I first became acquainted, and of which I now speak, was a piece of hard-wood board, eighteen or twenty inches long, moulded to the shape of an old-fashioned pudding stick, or ordinary oar. The flattened portion, which was about the size in circumference of two open hands, was bored with a small auger in numerous places. The cat was a large rope of many strands—the strands unraveled, and a knot tied at the extremity of each.

As soon as these formidable whips appeared, I was seized by both of them, and roughly divested of my clothing. My feet, as has been stated, were fastened to the floor. Drawing me over the bench, face downwards, Radburn placed his heavy foot upon the fetters, between my wrists, holding them painfully to the floor. With the paddle, Burch commenced beating me. Blow after blow was inflicted upon my naked body. When his unrelenting arm grew tired, he stopped and asked if I still insisted I was a free man. I did insist upon it, and then the blows were renewed, faster and more energetically, if possible, than before. When again tired, he would repeat the same question, and receiving the same answer, continue his cruel labor. All this time, the incarnate devil was uttering most fiendish oaths. At length the paddle broke, leaving the useless handle in his hand. Still I would not yield. All his brutal blows could not force from my lips the foul lie that I was a slave. Casting madly on the floor the handle of the broken paddle, he seized the rope. This was far more painful than the other. I struggled with all my power, but it was in vain. I prayed for mercy, but my prayer was only answered with imprecations and with stripes. I thought I must die beneath the lashes of the accursed brute. Even now the flesh crawls upon my bones, as I recall the scene. I was all on fire. My sufferings I can compare to nothing else than the burning agonies of hell!

At last I became silent to his repeated questions. I would make no reply. In fact, I was becoming almost unable to speak. Still he plied the lash without stint upon my poor body, until it seemed that the lacerated flesh was stripped from my bones at every stroke. A man with a particle of mercy in his soul would not have beaten even a dog so cruelly. At length Radburn said that it was useless to whip me any more—that I would be sore enough. Thereupon, Burch desisted, saying, with an admonitory shake of his fist in my face, and hissing the words through his firm-set teeth, that if ever I dared to utter again that I was entitled to my freedom, that I had been kidnapped, or any thing whatever of the kind, the castigation I had just received was nothing in comparison with what would follow. He swore that he would either conquer or kill me. With these consolatory words, the fetters were taken from my wrists, my feet still remaining fastened to the ring; the shutter of the little barred window, which had been opened, was again closed, and going out, locking the great door behind them, I was left in darkness as before.

In an hour, perhaps two, my heart leaped to my throat, as the key rattled in the door again. I, who had been so lonely, and who had longed so ardently to see some one, I cared not who, now shuddered at the thought of man's approach. A human face was fearful to me, especially a white one. Radburn entered, bringing with him, on a tin plate, a piece of shriveled fried pork, a slice of bread and a cup of water. He asked me how I felt, and remarked that I had received a pretty severe flogging. He remonstrated with me against the propriety of asserting my freedom. In rather a patronizing and confidential manner, he gave it to me as his advice, that the less I said on that subject the better it would be for me. The man evidently endeavored to appear kind—whether touched at the sight of my sad condition, or with the view of silencing, on my part, any further expression of my rights, it is not necessary now to conjecture. He unlocked the fetters from my ankles, opened the shutters of the little window, and departed, leaving me again alone.

By this time I had become stiff and sore; my body was covered with blisters, and it was with great pain and difficulty that I could move. From the window I could observe nothing but the roof resting on the adjacent wall. At night I laid down upon the damp, hard floor, without any pillow or covering whatever. Punctually, twice a day, Radburn came in, with his pork, and bread, and water. I had but little appetite, though I was tormented with continual thirst. My wounds would not permit me to remain but a few minutes in any one position; so, sitting, or standing, or moving slowly round, I passed the days and nights. I was heart sick and discouraged. Thoughts of my family, of my wife and children, continually occupied my mind. When sleep overpowered me I dreamed of them—dreamed I was again in Saratoga—that I could see their faces, and hear their voices calling me. Awakening from the pleasant phantasms of sleep to the bitter realities around me, I could but groan and weep. Still my spirit was not broken. I indulged the anticipation of escape, and that speedily. It was impossible, I reasoned, that men could be so unjust as to detain me as a slave, when the truth of my case was known. Burch, ascertaining I was no runaway from Georgia, would certainly let me go. Though suspicions of Brown and Hamilton were not unfrequent, I could not reconcile myself to the idea that they were instrumental to my imprisonment. Surely they would seek me out—they would deliver me from thraldom. Alas! I had not then learned the measure of "man's inhumanity to man," nor to what limitless extent of wickedness he will go for the love of gain.

HERE LIES HUGH GLASS
A Mountain Man, a Bear, and the Rise of the American Nation

Jon T. Coleman

Hugh Glass (*c.*1783–1833) was an American trapper and frontiersman whose near-fatal encounter with a large grizzly bear and subsequent journey to safety secured his status as an American folk hero. Although the details of his story remain disputed by historians, it has served as the inspiration for a variety of books and films, including *The Revenant*, starring Leonardo DiCaprio, who won his first Oscar for his portrayal of Glass.

Jon T. Coleman (born 1970) is a professor of American history at the University of Notre Dame. He is the author of *Vicious: Wolves and Men in America,* which won the W. Turrentine Jackson Prize and the John H. Dunning Prize.

Doctor Willard listened, but he didn't believe, until the old man lifted his kilt. In the glow of the campfire, Hugh Glass stripped, and the sight of his body set his audience abuzz. What a heap! The crumbling physique drew gasps even before its owner disrobed. Glass looked much older than his "middle aged" hunting companions. "Probably 75," Willard guessed. Wearing highland garb to signal his nativity, the mountain man told the doctor that he immigrated to the United States from Scotland as a youth and drifted west to seek his fortune. Glass may have earned a substantial income from beaver pelts, but he amassed nothing. The old man, the doctor reported, bobbed along on a current of alcohol. Booze lured him to the settlements, sucked him dry, and flushed him back into the mountains so that he could accumulate more hides and start the cycle anew.

Willard bumped into the hunter on the Santa Fe Trail in 1825. Glass and his partners, Stone, Andrews, and March, had lost two of their three mules and a majority of their traps in a river on their way to the headwaters of the Rio del Norte (the upper Rio Grande). The remaining mule wobbled under the entire crew's baggage while the hunters ambled to Taos to reequip. Willard's commercial outfit presented the four men a quick money-making opportunity. They were hired on to kill bison and antelope for the greenhorns.

Glass fed Willard's mess and the two struck up a friendship. In time, the old man told the doctor about the bear. Up north, along the upper Missouri, he and a trio of hunters (the same three he traveled with still) broke off from a larger American expedition to try their luck on a side stream. One morning, while his partners checked their traps, a female grizzly bear jumped him while he prepared breakfast for the crew. The animal burst out of the underbrush and bit him. She threw him down and "began her work of destruction." The commotion alerted Stone, who was wading in a nearby creek. He ran to the amp, saw his cook pinned, "the monster mounted upon him," and levelled his gun. He killed the bear with one shot. She fell and smothered Glass.

The bear's death cued the storyteller to hop out of this clothes. He ditched his tartan and let the doctor "examine" him. Willard traced the scars with his eyes and fingers. He noted "the large chasms upon the right arm and shoulder blade and the crest of which was wanting, also the upper portion of the right thigh." The folds and ridges attested to the veracity of the story, though Willard supplemented this "ocular proof" with testimony from Stone. Glass, the doctor felt certain, had nearly been destroyed by a white ear.

The strip-down boosted Willard's confidence, and the doctor lapped up the rest of the tale:

> It sounds from what he told me, that about the time the young men left, his wounds began to supperate [*sic*] as the inflammation gave way, followed by healing action. The weather was warm, and their provisions being meat only, soon spoiled, hence starvation seemed to stare the old man in the face, but fortunately he was left by a spring of water, to which he could crawl upon his hands and knees, and in a few days move. [He] came to the conclusion to work his way up the trail of the company, and gleaning as he went such roots as he could masticate for food, and I think he had not gone far, before he could bear his weight upon his legs and finally succeeded in overtaking the company which had commenced trapping again.

Many American hunters tussled with white bears, but only one scrambled back from the dead, traveling alone over such a vast distance. His incredible journey elevated his story out of the realm of entertaining mishap and into the rarefied air of national origin myth. The crawl turned Glass into an epic environmental American.

In 1825, Rowland Willard watched a mountain man named Glass perform a campfire story about a bear attack and a miraculous recovery. Glass took off his clothes and let the doctor inspect the residue of his trauma. Willard wrote down his observations, and his family collected them for posterity. Eventually, Yale University's Beinecke Library acquired the documents. The memoir, unknown to earlier Glass biographers, qualified as a coup for me: A doctor had examined Hugh Glass's wounds and written about them. A literate, trained medical professional had ogled and poked him. I needed a pinch myself; it felt too good to be true.

And it was.

Willard accompanied a caravan of Missouri traders headed to Santa Fe. Mexican officials had recently opened the trail, and the country's northern frontier intrigued Americans in search of commercial opportunities and close encounters with Spanish exotica. Willard sensed this curiosity and kept a daily journey to fortify his memoir for later publications. Once in Mexico, he wrote articles for the St. Louis newspapers and reported for Timothy Flint, who included the doctor's observations in his *Western Monthly Review* and at the tail end of his edited volume of James O. Pattie's mountain-man narrative. The story of Glass stripping down and enacting the white bear story appeared in none of Willard's contemporary accounts (though the doctor did scribble Glass's name multiple times in the journal). Willard wrote about the bear and the crawl years later in a handwritten memoir he composed for his family.

The nearness of Glass, therefore, was an illusion. Willard zoomed in on the old man from the distance of his own advanced age. (He composed the autobiography in his seventies and eighties.) He could reach out and touch Glass because he actually stood far away from him. To the memoir's further discredit, Willard fumbled verified elements of Glass's biography. Instead of moving north to their winter fort, Willard thought Henry's expedition travelled back to St. Louis after abandoning Glass. He claimed that Southwest Indians clubbed the old man to death along the Colorado River a couple of years after they met. (The Arikaras killed Glass near the Missouri River in 1833.) He abbreviated Glass's first name with a poorly executed W or a sloppy M rather than an H. Time, memory lapses, and fabrications crippled the doctor's account.

Distance both tarnished the doctor's report and made it possible. Had the doctor's report emerged from the past lucid and impeccable, it might never have mentioned the crawl. A more authentic and trustworthy account, one closer to Hugh Glass in space and time, would have cut to the white bear in an effort to comprehend the nature of the attack: In the 1820s, white bears, not their mutilated victims, held the meaning of the wilderness. The imperfections in Willard's vision opened a sight line into the blindest alley in the Glass saga.

After his companions left him for dead, Glass disappeared into a total eclipse of information. He vanished from culture like the "Indianhater *par excellence*." The frontier and nature swallowed him; his crawl existed on the far side of news, biography, and history. No one but Glass could speak about it. He reported his own death struggle, which should have set off alarm bells; Glass wasn't a reputable guy.

Willard acknowledged as much. He connected Glass's wasted body to his wasteful habits. The cycle of debauchery and dissipation that drove Glass back and forth from towns to beaver streams made him a prize storyteller and a social pariah. His cultural authority—the power to attract, to command, and to buffalo an audience—grew from his decaying form. The environment starved, scarred, and aged him; the trade added to his disintegration by facilitating cycles of epic bingeing and purging. Glass consumed Falstaffian proportions of liquor, food, and conviviality during summer rendezvous and settlement sojourns. Back in the hills, he abstained like a monk. The mountain men swelled and shrank with the wild swings in their access to food, markets, and luxuries. The radical changes in their appearance underwrote their environmental Americanism. Through bingeing and purging, nature and markets reworked them into new species and races.

Yet, as Willard noted, these cycles of excessive celebration and extreme privation violated middle-class ideals of frugality and moderation. The men's ethics seemed as unsound and repulsive as their bodies. Written decades after their encounter, the memoir gave Willard the space he needed to perform a contortionist's maneuver: it allowed him to embrace and reject Hugh Glass simultaneously. Willard could admire Glass's capacity for withstanding nature's torments while denouncing the old hunter's self-abuse. His ruined body proved that an epic confrontation between man and nature did indeed occur; his wasteful habits proved that he wasn't fit for civilization. Both dissipations worked in favor of environmental Americanism.

The story of the crawl secured Glass's legend even as it strained belief. Who could buy the proposition that a horribly injured man not only survived for weeks alone in a hostile environment but also dragged his mangled limbs hundreds of miles to the nearest outpost? Who could trust the words of a ne'er-do-well hunter with a drinking problem? Glass's body nourished confidence and cynicism. Something dreadful happened to him; the scars proved that much. But what? The doctor who probed his injuries and listened to his tale offered several reasons for his destruction. His work, his habits, and his immorality wrecked him. Others suggested that the bear, the West, or the wilderness ravaged him. Nationalists thought America got him. Whatever the cause, Glass's body demanded a narrative that explained its impoverished state. Glass furnished a story that fit his experiences as a fur hunter. He oriented his survival tale around food, markets, and freedom. The crawl signaled Glass's final initiation into his western environment, a matriculation process that began with the Arikara attack and continued with the bear mauling. The survival tale detailed the economic whipsaw that flung the mountain men between abundance and famine. The bear and his friends sent Glass to the outer limits of privation. He came back to fill his belly, declare his free agency, and show off the damage he earned at the far reaches of capitalism.

BETWEEN A ROCK AND A HARD PLACE

Aron Lee Ralston

Aron Lee Ralston (born 1975) is an American mountaineer. In 2003 he became famous for amputating his arm with a dull pocketknife, after being trapped by a dislodged boulder in Bluejohn Canyon, Utah, during a solo descent. His miraculous story served as the inspiration for his own memoir *Between a Rock and a Hard Place* and was adapted into the 2010 film *127 Hours*. A few years after the accident, Ralston climbed all of Colorado's 'Fourteeners' – peaks over 14,000 feet – solo and during winter.

Miserable, I watch another empty hour pass by. At least I don't have to fight to stay warm. The cold bite of the outer atmosphere no longer sucks off my body heat as it did throughout the night. But by removing the need to reconfigure the ropes around my legs and the cloth and plastic wraps around my arms, daytime has removed the last bustle from my experience in the canyon. Without even that minimal distraction, I have nothing whatsoever to do. I have no life. Only in action does my life approximate anything more than existence. Without any other task or stimulus, I'm no longer living, no longer surviving. I'm just waiting.

Since the recoiling blows of the hammer rock tenderized my left hand, all I've had left to do is wait. For what, though? Rescue. . . or death? It doesn't matter to me. The two endings represent the same thing—salvation and deliverance from my suffering. I can't stand the inactivity that breeds such apathy. At this point, the waiting itself is the worst part of my entrapment. And when I'm done waiting, all there is, is more waiting. I can touch the face of infinity in these doldrums. Nothing gives even a slight hint that the stillness will break.

But I can make it break. I can ignore the pain in my left hand and resume smashing the chockstone with the handheld wrecking ball. I can continue hacking away at the rock with my knife, despite its in-utility. I can do everything I've done in the past five days for the sake of motion. I reach for the rounded hammer rock, then realize I'm going to want my left sock for a pad. Off with the shoe, off with the sock, and I have the cushioning for my battered palm. The bruises on the meaty pad of my thumb are the most sensitive to the impact, and they scream for reprieve from the first blow through the fifth, when I pause. Adrenaline channels into anger, and I raise the hammer again, this time in retribution for what this wretched piece of geology has done to my left hand. Bonk! Again I strike the boulder, the pain in my hand flaring. Thwock! And again. Screeaatch! The rage blooms purple in my mind, amid a small mushroom cloud of pulverized grit and the burning smell of the sock that comes between the rock and the chockstone, melting with the friction heat of each strike. I bring the rock down again. Carrunch! With animalistic fury, I growl, "Unnngaaarrrrgh!" in response to the throbs pulsing in my left hand.

I force myself to stop, and can't release my grip on the hammer rock. My fingers have been paralyzed in their clench.

Whoa, Aron. You might have taken that too far.

Gradually, my shocked nerves relax, and my digits extend until I can let go of the rock, which I set on the chockstone. I've created a mess once again. I want to brush the collected dirt off my arm, away from the open wound. I take my knife and begin clearing particles from my trapped hand, using the dulled blade like a brush. Sweeping the grit off my thumb, I accidentally gouge myself and rip away a thin piece of decayed flesh. It peels back like a skin of boiled milk before I catch what is going on. I already knew my hand had to be decomposing. Without circulation, it has been dying since I became entrapped. Whenever I considered amputation, it had always been under the premise that the hand was dead and would have to be amputated once I was freed. But I hadn't known how fast the putrefaction had advanced since Saturday afternoon. Now I understand the increase in the interest of the indigenous insect population. They could already smell their next meal, their breeding ground, their larvae's new home.

Out of curiosity, I poke my thumb with the knife blade twice. On the second prodding, the blade punctures the epidermis as if it is dipping into a stick of room-temperature butter, and releases a telltale hissing. Escaping gases are not good; the rot has advanced more quickly than I had guessed. Though the smell is faint to my desensitized nose, it is abjectly unpleasant, the stench of a far-off carcass.

On the heels of the odor, a realization hits my brain—whatever has started in my hand will shortly pass into my forearm, if it hasn't already. I don't know and furthermore don't care if it's gangrene or some other insidious attack, but I know it is poisoning my body. I lash out in fury, trying to yank my forearm straight out from the sandstone handcuff, never wanting more than I do now to simply rid myself of any connection to this decomposing appendage.

I don't want it.

It's not a part of me.

It's garbage.

Throw it away, Aron. Be rid of it.

I thrash myself forward and back, side to side, up and down, down and up. I scream out in pure hate, shrieking as I batter my body to and fro against the canyon walls, losing every bit of composure that I've struggled so intensely to maintain. Then I feel my arm bend unnaturally in the unbudging grip of the chockstone. An epiphany strikes me with the magnificent glory of a holy intervention and instantly brings my seizure to a halt:

If l torque my arm far enough, I can break my forearm bones.

Like bending a two-by-four held in a table vise, I can bow my entire goddamn arm until it snaps in two!

Holy Christ, Aron, that's it, that's it. THAT'S FUCKING IT!

I scramble to clear my stuff off the rock, trying to keep my head on straight. There is no hesitation. Under the power of this divine interaction, I barely realize what I'm about to do. I slip into some kind of autopilot; I'm not.at the controls anymore. Within a minute, I orient my body in a crouch under the boulder, but I can't get low enough to bend my arm before I feel a tugging at my waist. I unclip my daisy chain from the anchor webbing and drop my weight as far down as I can, almost making my buttocks reach the stones on the canyon floor. I put my left hand under the boulder and push hard, harder, HARDER!, to exert a maximum downward force on my radius bone. As I slowly bend my arm down and to the left, a Pow! reverberates like a muted cap-gun shot up and down Blue John Canyon. I don't say a word, but I reach to feel my forearm. There is an abnormal lump on the top of my wrist. I pull my body away from the chockstone and down again, simulating the position I was just in, and feel a gap between the serrated edges of my cleanly broken arm bone.

Without further pause and again in silence, I hump my body up over the chockstone, with a single clear purpose in my mind. Smearing my shoes against the canyon walls, I push with my legs and grab the back of the chockstone with my left hand, pulling with every bit of ferocity I can muster, hard, harder, HARDER!, and a second capgun shot ends my ulna's anticipation. Sweating and euphoric, I again touch my right arm two inches below my wrist, and pull my right shoulder away from the boulder. Both bones have splintered in the same place, the ulna perhaps a half inch closer to my elbow than my radius. Rotating my forearm like a shaft inside its housing, I have an axis of motion freshly independent of my wrist's servitude to the rock vise.

I am overcome with the excitement of having solved the riddle of my imprisonment. Hustling to deploy the shorter and sharper of my multi-tool's two blades, I skip the tourniquet procedure I have rehearsed and place the cutting tip between two blue veins. I push the knife into my wrist, watching my skin stretch inwardly, until the point pierces and sinks to its hilt. In a blaze of pain, I know the job is just starting. With a glance at my watch—it is 10:32 A.M. —1 motivate myself "OK, Aron, here we go. You're in it now."

I leave behind my prior declarations that severing my arm is nothing but a slow act of suicide and move forward on a cresting wave of emotion. Knowing the alternative is to wait for a progressively more certain but assuredly slow demise, I choose to meet the risk of death in action. As surreal as it looks for my arm to disappear into a glove of sandstone, it feels gloriously perfect to have figured out how to amputate it.

My first act is to sever, with a downward sawing motion, as much of the skin on the inside surface of my forearm as I can, without tearing any of the noodle-like veins so close to the skin. Once I've opened a large enough hole in my arm, about four inches below my wrist, I momentarily stow the knife, holding its handle in my teeth, and poke first my left forefinger and then my left thumb inside my arm and feel around. Sorting through the bizarre and unfamiliar textures, I make a mental map of my arm's inner features. I feel bundles of muscle fibers and, working my fingers behind them, find two pairs of cleanly fractured but jagged bone ends. Twisting my right forearm as if to turn my trapped palm down, I feel the proximal bone ends rotate freely around their fixed partners. It's a painful movement, but at the same time, it's a motion I haven't made since Saturday, and it excites me to know that soon I will be free of the rest of my crushed dead hand. It's just a matter of time.

Prodding and pinching, I can distinguish between the hard tendons and ligaments, and the soft, rubbery feel of the more pliable arteries. I should avoid cutting the arteries until the end if! can help it at all, I decide.

Withdrawing my bloody fingers to the edge of my incision point, I isolate a strand of muscle between the knife and my thumb, and using the blade like a paring knife, I slice through a pinky-fingersized filament. I repeat the action a dozen times, slipping the knife through string after string of muscle without hesitation or sound.

Sort, pinch, rotate, slice.

Sort, pinch, rotate, slice.

Patterns; process.

Whatever blood-slimy mass I fit between the cutting edge and my left thumb falls victim to the rocking motion of the multi-tool, back and forth. I'm like a pipe cutter scoring through the outer circumference of a piece of soft tubing. As each muscle bundle yields to the metal, I probe for any of the pencil-thick arteries. When I find one, I tug it a little and remove it from the strand about to be severed. Finally, about a third of the way through the assorted soft tissues of my forearm, I cut a vein. I haven't put on my tourniquet yet, but I'm like a five-year-old unleashed on his Christmas presents—now that I've started, there's no putting the brakes on. The desire to keep cutting, to get myself free, is so powerful that I rationalize I haven't lost that much blood yet, only a few drops, because my crushed hand has been acting like an isolation valve on my circulation.

Another ten, fifteen, or maybe twenty minutes slip past me. I am engrossed in making the surgical work go as fast as possible. Stymied by the half-inch-wide yellowish tendon in the middle of my forearm, I stop the operation to don my improvised tourniquet. By this time, I've cut a second artery, and several ounces of blood, maybe a third of a cup, have dripped onto the canyon wall below my arm. Perhaps because I've removed most of the connecting tissues in the medial half of my forearm, and allowed the vessels to open up, the blood loss has accelerated in the last few minutes. The surgery is slowing down now that I've come to the stubbornly durable tendon, and I don't want to lose blood unnecessarily while I'm still trapped. I'll need every bit of it for the hike to my truck and the drive to Hanksville or Green River.

I still haven't decided which will be the fastest way to medical attention. The closest phone is at Hanksville, an hour's drive to the west, if I'm fast on the left-handed reach-across shifting. But I can't remember if there's a medical clinic there; all that comes to my mind is a gas station and a hamburger place. Green River is two hours of driving to the north, but there is a medical clinic. I'm hoping to find someone at the trailhead who will drive for me, but I think back to when I left there on Saturday—there were only two other vehicles in the three-acre lot. That was a weekend, this is midweek. I have to accept the risk that when I get to the trailhead, there won't be anyone there. I have to pace myself for a six-to-seven-hour effort before I get to definitive medical care.

Setting the knife down on the chockstone, I pick up the neoprene tubing of my CamelBak, which has been sitting off to the top left of the chockstone, unused, for the past two days. I cinch the black insulation tube in a double loop around my forearm, three inches below my elbow. Tying the black stretchy fabric into a doubled overhand knot with one end in my teeth, I tug the other end with my free left hand. Next, I quickly attach a carabiner into the tourniquet and twist it six times, as I did when I first experimented with the tourniquet an eon ago, on Tuesday, or was it Monday?

"Why didn't I figure out how to break my bones then?" I wonder. "Why did I have to suffer all this extra time?" God, I must be the dumbest guy to ever have his hand trapped by a boulder. It took me six days to figure out how I could cut off my arm. Self-disgust catches in my throat until I can clear my head.

Aron, that's all just distraction. It doesn't matter. Get back to work.

I clip the tightly wound carabiner to a second loop of webbing around my biceps to keep the neoprene from untwisting, and reach for my bloody knife again.

Continuing with the surgery, I dear out the last muscles surrounding the tendon and cut a third artery. I still haven't uttered even an "Ow!" I don't think to verbalize the pain; it's a part of this experience, no more important to the procedure than the color of my tourniquet.

I now have relatively open access to the tendon. Sawing aggressively with the blade, as before, I can't put a dent in the amazingly strong fiber. I pull at it with my fingers and realize it has the durability of a flat-wound cable; it's like a double-thick strip of fiber-reinforced box-packaging tape, creased over itself in quarter-inch folds. I can't cut it, so I decide to reconfigure my multi-tool for the pliers. Unfolding the blood-slippery implement, I shove the backside of the blade against my stomach to push the knife back into its storage slot and then expose the pliers. Using them to bite into the edge of the tendon, I squeeze and twist, tearing away a fragment. Yes, this will work just fine. I tackle the most brutish task.

Grip, squeeze, twist, tear.

Grip, squeeze, twist, tear.

Patterns; process.

"This is gonna make one hell of a story to tell my friends-," I think. "They'll never believe how I had to cut off my arm. Hell, I can barely believe it, and I'm watching myself do it."

Little by little, I rip through the tendon until I totally sever the twine-like filament, then switch the tool back to the knife, using my teeth to extract the blade. It's 11:16 A.M.; I've been cutting for over forty minutes. With my fingers, I take an inventory of what I have left: two small clusters of muscle, another artery, and a quarter circumference of skin nearest the wall. There is also a pale white nerve strand, as thick as a swollen piece of angel-hair pasta. Getting through that is going to be unavoidably painful. I purposefully don't get anywhere close to the main nerve with my fingers; I think it's best not to know fully what I'm in for. The smaller elastic nerve branches are so sensitive that even nudging them sends Taser shocks up to my shoulder, momentarily stunning me. All these have to be severed. I put the knife's edge under the nerve and pluck it, like lifting a guitar string two inches off its frets, until it snaps, releasing a flood of pain. It recalibrates my personal scale of what it feels like to be hurt—it's as though I thrust my entire arm into a cauldron of magma.

Minutes later, I recover enough to continue. The last step is stretching the skin of my outer wrist tight and sawing the blade into the wall, as if

I'm slicing a piece of gristle on a cutting board. As I approach that precise moment of liberation, the adrenaline surges through me, as though it is not blood coursing in my arteries but the raw potential of my future. I am drawing power from every memory of my life, and all the possibilities for the future that those memories represent.

It is 11:32 A.M., Thursday, May 1, 2003. For the second time in my life, I am being born. This time I am being delivered from the canyon's pink womb, where I have been incubating. This time I am a grown adult, and I understand the significance and power of this birth as none of us can when it happens the first time. The value of my family, my friends, and my passions well up a heaving rush of energy that is like the burst I get approaching a hard-earned summit, multiplied by ten thousand. Pulling tight the remaining connective tissues of my arm, I rock the knife against the wall, and the final thin strand of flesh tears loose; tensile force rips the skin apart more than the blade cuts it.

A crystalline moment shatters, and the world is a different place. Where there was confinement, now there is release. Recoiling from my sudden liberation, my left arm flings downcanyon, opening my shoulders to the south, and I fall back against the northern wall of the canyon, my mind surfing on euphoria. As I stare at the wall where not twelve hours ago I etched "RIP OCT 75 ARON APR 03," a voice shouts in my head:

I AM FREE!

This is the most intense feeling of my life. I fear I might explode from the exhilarating shock and ecstasy that paralyze my body for a long moment as I lean against the wall. No longer confined to the physical space that I occupied for nearly a week, I feel drugged and off balance but buoyed by my freedom. My head bobs to my right shoulder and dips to my chest before I right it and steady myself against the wall. I stumble as I catch my left foot around the rocks on the canyon floor, but I get my legs under me in time to prevent a hard fall onto the southern wall. It is beautiful to me that I could actually fall over right now. I glance at the bloody afterbirth smeared on the chockstone and the northern canyon wall. The spattering on the chockstone hides the dark mass of my amputated hand and wrist, but the white bone ends of my abandoned ulna and radius protrude visibly from the gory muddle. My glance lingers and becomes a stare. My head whirls, but I am fascinated, looking into the cross section of my forearm.

OK, that's enough. You've got thing; to do. The clock is running, Aron. Get out of here.

ADA BLACKJACK
A True Story of Survival in the Arctic

Jennifer Niven

Ada Blackjack (1898–1983) was a native Alaskan woman who survived alone for eight months on the uninhabited Wrangel Island after being abandoned by her expedition. Employed by the Canadian explorer Allan Crawford as a seamstress and cook, Blackjack accompanied the expedition to the island in the East Siberian Sea under significant misgivings, which proved portentous. The team eventually ran out of rations and three members attempted to cross the 90-mile frozen Sea of Chukotsk, leaving Ada behind with an ailing fourth explorer. They were never seen again, and the fourth, Lorne Knight, succumbed to scurvy, leaving Blackjack alone to fend for herself until her rescue eight months later. She was hailed in American and Canadian newspapers as a courageous and intrepid hero but died in penury after her story was exploited by others for financial gain.

Jennifer Niven is an Emmy award-winning and *New York Times* bestselling author from America. She primarily writes young adult novels but one of her first books was her life of Ada Blackjack.

Each day that lonely summer, before setting out to hunt, Ada sat down at Galle's typewriter and wrote him a note about where she was going. She left a sheet of paper in the machine so that she could add to it daily because she did not want to risk the ship coming and not finding her in camp. She wanted to make certain her rescuers did not leave her there after they had come all that way to fetch her.

It was also a way for her to have a conversation with someone, now that she was alone. It was a way of talking to Galle, just as she had talked to him before he left for Siberia. She began lengthening the entries, confiding and sharing more about herself and her work, and reporting in again at the end of the day, telling him what happened and what she did, how the hunt went and how she felt about it. It was a comfort to be able to write those words, even if he wasn't able to listen and respond.

June 24th. I'm going to the other side of the harbar mouth do some duck hunding.

She brought four eiders home that day and then took pictures of her tent and herself, even though she hadn't quite figured out how to work the camera. She had seen Crawford set the contraption on a box or log and pull the string that snapped the shutter. When she tried to imitate his methods, her photographs only turned out blurred and dark. But she would learn to use that camera, she promised herself, so that she could leave a photographic record of her time as well.

June 25. Going same as yest rday. I got seven eiders.

When she got back to the tent, she plucked the birds and then hung the legs and the breasts to dry.

June 26th. I'm going to take a walk to the smale Island. I saw two Polar bears going in shore from the ice way over west of the camp. It's four oclock now. I write down when I saw them. I don't know what I'm going to do if they come to the camp. Well, God knows.

She collected three seagull eggs that morning and cooked them for her lunch. She drank them down with tea and saccharine and "had a nice picknick all by myself."

The very next day, Ada was able to kill her first seal. There were two of them, basking in the sun, and she fired one shot. The other seal slipped into the water, but Ada's seal lay still. She was filled with pride as she bent over the animal to examine it.

One week later, she shot another seal with Knight's rifle. She had fashioned a stretcher out of skins to haul the animals home, and she was sitting in her tent, cleaning the second seal, when she heard a noise outside her door. It sounded like a dog to her, and for one brief moment she believed it was Crawford, Galle, and Maurer come back to find her. But when she raised the tent flap, her heart stopped. Two bears, a large one and a younger one, stood fifteen feet from the tent, gazing down at her. In seconds, Ada was clutching her rifle. She knew that if she hit them in the shoulder or the foot and only injured them slightly, they would be angry and come after her, so she must not even try to kill them because she might miss. She raised the gun, as she had before, and fired over their heads until they turned and began to run away. When the gunfire stopped, the bears paused and looked back toward camp, considering. Ada raised her gun once more and fired five consecutive shots until they disappeared over the horizon.

June 28th. I clean the seal skin today and lat this afternoon Polar bear and one Cub was very close to the Camp and I didn't take any chances. I was afraid if I didn't hit it right I'd be in danger. I just shot over them and they wend away. I was glad thank the living God.

The bears they had waited so long for, that Knight had prayed for, had returned, but it suited her much better when they stayed away. Now, they haunted her. She found fresh polar bear tracks one morning outside the door of the tent. She investigated, but there was no bear in sight. One of her tins of oil, though, was licked clean, and she knew she would have to be more careful with her stores.

July 1st. I stay home today and I fix the shovel handle that I brack this spring and I saw Polar bear out on the ice and this evening I went to the end of the sand spit shot a eidar duck I shot him right in the head thank God keep me a live till now.

On July 4, Ada crawled on her stomach across the beach after a seal. She remembered what Knight had said about waiting until she was at least 150 yards from the animal before shooting, and so she tried to creep as close as possible. She cocked her gun and was just about to pull the trigger when an enormous cake of ice rose up between Ada and the seal, blocking her view. She moved quickly before the seal could disappear and took aim again, but the gun exploded with a deafening bang before she meant to shoot, and the seal slipped away, unharmed, into the water. Ada jumped to her feet and yelled, "Fourth of July," shattering the vast, impenetrable silence of the landscape. She may have come home empty-handed, but she had enjoyed a celebration, fireworks and all, she would note that evening.

She killed her third seal the following day, just a few yards from the back of her tent. The beach ran behind her little house, and she could see a seal some two hundred yards out on the ice. Once again, she fell to her stomach and wriggled, pretending she was another seal. When she was within range, she shot it once through the head, killing it instantly. She was ecstatic and thanked the Lord Jesus for giving her such a gift. The seal did not slip off the ice and into the water this time. It was hers and it was everything – skins for clothing, oil for lamps, and meat for food. She cut up the seal, hung the meat, and pulled the skin taut over the stretcher. If only Knight were there to share it with her, the victory would have seemed sweeter.

On July 6, she shot a seal at the harbor mouth. Because the seal lay a fair distance from camp and because the animals typically weighed well over six hundred pounds, Ada would need something to help her bring this one back to the tent. She had just fetched a poling line and started back toward her seal when she glimpsed something against the white of the sky that looked like a gigantic yellow ball coming toward her. A polar bear. Ada was four hundred yards from her tent, and now she ran back as fast as she could until she was safe inside.

She was afraid she might faint, but instead she climbed onto the platform she had built at the back of the tent and peered through her field glasses. She could see the outline of the bear and its cub as they bent over her seal – the seal she had labored for – and tore it to pieces. She was helpless to do anything as they devoured it before her eyes, but at least, she thought, "I am glad it is not me polar bear eats." She fired her gun toward them in anger, and watched as they scattered west and then east and then as they seemed to cross the harbor mouth, their noses pointed toward camp. She fired one more shot and waited.

Ada stood up on her platform until the sky grew dark and the fog billowed in like the soft, cold breath of the landscape. She decided it was better just to leave the seal to the bears. The next morning, she walked out to where her seal had fallen and found only smears of blood on the ice.

She purposely did not mention much about the polar bears in her diary. She knew that if the spirits were to take their revenge on her and she should die, that her diary would fall into the hands of others. And if her mother read mention of Nanook, she would always believe that Ada had been eaten by a polar bear and was living in his stomach, no matter how she had really died.

The mother bear and cub returned on July 7, circling the spot where the seal had fallen. Ada fired her gun at them, but missed, and they eventually wandered back the way they had come. She hunted for ducks then, but was unable to retrieve the one she shot because the ice on the harbor mouth was thin and would have given way beneath her. She returned to camp, discouraged.

She sewed a new flap onto the tent door, where the wind blew in through the holes in the canvas, and she added canvas to the tent frame to bolster her house. Ada did not know if a ship would come for her. Mr. Stefansson had not been able to make the island last summer, and there was a good chance Ada might face another winter without relief. Last September, she and the men had been unprepared for the ship's failure to reach them, and now she would ready her equipment and her house before the snow and the storms arrived – so that she would be ready for the worst.

She chewed up the sealskins to make them more pliable to mold into soles for her boots. She added new soles to her short boots. She practiced shooting at her target, and opened a new box of shotgun cartridges. She knitted fingers into her gloves to replace the ones that had worn away from all the work she was doing. And she began making a parka out of fancy reindeer skin and wolf trimming. It would be a beautiful parka and she felt she deserved it. It made her happy to work on it, and she indulged herself for days, picking it up whenever she had time to add a new hook or to work on the hood. When she finished it, she looked upon it with pride, with its fancy trim down the front and around the hood. "It look like a parky alright," she boasted in her diary.

On July 10, Ada rested in bed because she was frazzled and worn from the exertion of her work, the constant, unending work without a break. Vic lay against her, warm and breathing, and Ada opened a new can of tea and thanked her heavenly father for letting her live another day.

She was able to bring home two birds on July 16, gray ones she didn't recognize. And on July 18 she shot two squaws and a duck. She was storing her meat now, in case the ship didn't come or in case Galle and Crawford and Maurer were unable to reach her by sled. She would save the meat for winter because she would need it if the game disappeared again. She filled up her food box with the seal meat she had dried in the sun, and then she gently removed the seal skins from the stretcher.

Ada often stood on the beach and looked out to sea, in the direction of Nome – or where she thought Nome might be. She watched for Crawford and Galle and Maurer or for a ship. It was time for a ship to come. The water offshore was clear and calm and open. Ice still hugged the shore in spots, but the space beyond was free of it. There would be nothing to hold a ship back, to keep it from reaching her. But on the morning of July 20, when she turned her gaze to the sea, she saw the pack of ice to the west side of the harbor mouth and felt a chill in her blood.

July 23. I thank God for living.

On July 24, there was a muffled roaring in the distance that let Ada know the walrus had returned. She could not see them, but she heard their foghorn cries for several days. She loaded her gun with brass shells and went out after them, but got two old ducks instead and was content. The game had improved. And when she went to sleep that night, she dreamed she was singing cheers for the red, white, and blue.

She began another pair of boots, these out of reindeer skins and a set of old slippers, and she stitched herself some deer leggings. She threw out the molded hard biscuits and transferred the ones that were still edible to her food box. And she cleaned some seal flippers and stored them away in case the ship was able to find her, so that she could take them home and share them with her sisters "if the lord let me have it."

One night while Ada slept, the wind swept in and blew her skin boat out to sea. She had only used it twice, but she cried all day when she discovered it was gone. Then she cried all the next day and the next. It was too much. Ada was tired and she was weak and there was no one to help her. If she did not get up out of bed to forage for food, she would go hungry. If she did not light the fire, it would remain unlit. If she did not bring in the snow, there would be no water to drink. Eventually she grew tired of crying and pitying herself. She must get up and make another boat, and so she did.

This one was constructed out of canvas, although skins were preferable, but Ada did not have enough skins for a boat. She gathered driftwood for the frame and placed the wood, piece by piece, into position because the bottom must be built first. The boat should be more canvas than wood because the umiak must sit light in the water and she must be able to drag it easily across the ice and land. With the canvas stretched across the skeletal frame, Ada used her needle to sew the canvas into place until gradually it took the shape of a boat. The little vessel was nothing to look at, but it was sturdy and seaworthy, and she carved a set of driftwood oars to go with it. Every time she finished using it, she tied it up so that the wind could not take it away.

Her world was lonely and silent. Now that there was open water, there was no longer the crash of the ice pack, the long, low grind of the floes churning against one another, the deep and sudden splash of water as masses of the pack broke off and plunged into the sea, or the staccato burst like rifle shots that echoed across the island as the ice expanded. There was only the sound of her own voice as she spoke to Vic. She fussed over the cat like a mother and picked her up and held her in her arms and talked to her like she had talked to Crawford and the others. Vic was a warm, breathing creature, who responded with purrs and rubs and an occasional meow. Ada thought she would go insane without her.

VAGRANT VIKING
My Life and Adventures

Peter Freuchen

Peter Freuchen (1886–1957) was a renowned Danish explorer and adventurer notable for his participation in the Thule Expeditions of Knud Rasmussen. On one such expedition, he was trapped in an avalanche, and claimed to have forged a crude knife from his own faeces in order to extricate himself. During the Second World War, Freuchen played an active role in the Danish resistance to Nazi occupation and was imprisoned by the Germans and sentenced to death. He managed to escape to Sweden, from where he continued to oppose Nazi tyranny.

At last I was really worried. My friends would soon begin to search for me, of course, but the question was whether I could survive until they found me. Perhaps I could dig my way out. But the snow surrounding me was now ice, and it was impossible to make the smallest dent in the surface with my gloved hands. I had left my snow knife outside on the sled with all my other tools. I decided to try digging with my bare hands. My hand would freeze but it would be better to lose one hand than to lose my life. I pulled off my right glove and began scratching with my nails. I got off some tiny pieces of ice, but after a few minutes my fingers lost all feeling, and it was impossible to keep them straight. My hand simply could not be used for digging so I decided to thaw it before it was too late.

I had to pull the arm out of the sleeve and put the icy hand on my chest—a complicated procedure in a space so confined I could not sit up. The ice roof was only a few inches above my face. As I put my hand on my chest I felt the two watches I always carried in a string around my neck, and I felt the time with my fingers. It was the middle of the day, but it was pitch black in my ice house. Strangely enough I never thought of using my watches for digging—they might have been useful.

By now I was really scared. I was buried alive and so far all my efforts had failed. As I moved a little I felt the pillow under my head—the skin of the bear's head. I got a new idea. By an endless moving with my head I managed to get hold of the skin. It had one sharply torn edge which I could use. I put it in my mouth and chewed on it until the edge was saturated with spit. A few minutes after I removed it from my mouth the edge was frozen stiff, and I could do a little digging with it before it got too soft. Over and over again I put it back in my mouth, let the spit freeze and dug some more, and I made some progress. As I got the ice crumbs loose, they fell into my bed and worked their way under my fur jacket and down to my bare stomach. It was most uncomfortable and cold, but I had no choice and kept on digging, spitting, freezing and digging.

My lips and tongue were soon a burning torture, but I kept on as long as I had any spit left—and I succeeded. Gradually the hole grew larger and at last I could see daylight! Disregarding the pain in my mouth and ignoring the growing piles of snow on my bare stomach, I continued frantically to enlarge the hole.

In my hurry to get out and save my frozen legs I got careless. I misjudged the size of the hole through which I could get out. My hand had, naturally, been able to move only above my chest and stomach, and to get my head in the right position seemed impossible. But I suddenly made the right movement and got my head in the right position.

I pushed with all my strength, but the hole was much too small. I got out far enough to expose my face to the drifting snow. My long beard was moist from my breathing and from the spit which had drooled from my bear skin. The moment my face got through the hole, my beard came in contact with the runners of the sled and instantly froze to them. I was trapped. The hole was too small to let me get through, my beard would not let me retire into my grave again. I could see no way out. But what a way to die—my body twisted in an unnatural position, my beard frozen to the sled above, and the storm beating my face without mercy. My eyes and nose were soon filled with snow and I had no way of getting my hands out to wipe my face. The intense cold was penetrating my head, my face was beginning to freeze and would soon lose all feeling.

Full of self-pity I thought of all the things in life I would have to miss, all my unfilled ambitions. And I thought of Magdalene, my friend, whose letters through the years had always been a little melancholy. It was the thought of Magdalene which made me want to go on living.

With all my strength I pulled my head back. At first the beard would not come free, but I went on pulling and my whiskers and some of my skin were torn off, and finally I got loose. I withdrew into my hole and stretched out once more. For a moment I was insanely grateful to be back in my grave, away from the cold and the tortuous position. But after a few seconds I was ready to laugh at my own stupidity. I was even worse off than before! While I had moved about more snow had made its way into the hole and I could hardly move, and the bear skin had settled under my back where I could not possibly get at it.

I gave up once more and let the hours pass without making another move. But I recovered some of my strength while I rested and my morale improved. I was alive after all. I had not eaten for hours, but my digestion felt all right. I got a new idea!

I had often seen dog's dung in the sled track and had noticed that it would freeze as solid as a rock. Would not the cold have the same effect on human discharge? Repulsive as the thought was, I decided to try the experiment. I moved my bowels and from the excrement I managed to fashion a chisellike instrument which I left to freeze. This time I was patient, I did not want to risk breaking my new tool by using it too soon. While I waited, the hole I had made filled up with fresh snow. It was soft and easy to remove, but I had to pull it down into my grave which was slowly filling up. At last I decided to try my chisel and it worked! Very gently and very slowly I worked at the hole. As I dug I could feel the blood trickling down my face from the scars where the beard had been torn away.

Finally I thought the hole was large enough. But if it was still too small that would be the end. I wiggled my way into the hole once more. I got my head out and finally squeezed out my right arm before I was stuck again. My chest was too large.

The heavy sled, weighing more than two hundred pounds, was on the snow above my chest. Normally I could have pushed it and turned it over, but now I had not strength enough. I exhaled all the air in my lungs to make my chest as small as possible, and I moved another inch ahead. If my lungs could move the sled I was safe. And I filled my lungs, I sucked up air, I expanded my chest to the limit—and it worked. The air did the trick. Miraculously the sled moved a fraction of an inch. Once it was moved from its frozen position, it would be only a question of time before I could get out. I continued using my ribs as levers until I had both arms free and could crawl out.

It was dark again outside. The whole day and most of another night had passed. The dogs were out of sight, but their snug little hole by the boulder was completely covered by snow, and I knew they must be asleep under it. As soon as I had rested enough, I got to my feet to get the dogs up. I fell at once and laughed at my weakness. Once more I got to my feet and once more I fell flat on my face. I tried out my legs and discovered the left one was useless and without feeling. I had no control over it any more. I knew it was frozen, but at first I did not think about it. I had to concentrate on moving. I could not stay where I was.

I could only crawl, but I got my knife from the sled, pulled the dogs out of their cave and cut them loose from the harness. I planned to hold to the reins and let the dogs pull me on the snow, but they did not understand. I used the whip with what little strength I had left, and suddenly they set off so fast my weak hands could not hold the reins! The dogs did not go far, but they managed to keep out of my reach as I crawled after them, I crawled for three hours before I reached the camp.

Fortunately I then did not know the ordeal was to cost me my foot.

As soon as I had been inside our igloo for a while and began to warm up, feeling returned to my frozen foot and with it the most agonizing pains. It swelled up so quickly it was impossible to take off my kamik. Patloq, our Canadian Eskimo companion who had had a great deal of experience with such accidents, carefully cut off the kamik, and the sight he revealed was not pleasant. As the foot thawed, it had swollen to the size of a football, and my toes had disappeared completely in the balloon of blue skin. The pain was concentrated above the frozen part of my foot which was still without feeling. Patloq put a needle into the flesh as far as it would go, and I never noticed it.

The only thing to do was to keep the foot frozen, Patloq insisted. Once it really thawed, the pain would make it impossible for me to go on. It was obvious that we could not stay where we were and that we had to give up the whole expedition to Baffin Land. And with my foot bare to keep it frozen, we returned slowly to Danish Island, where Knud Rasmussen was completing all preparations for his long journey to Alaska.

He was horrified when he saw what had happened to me, and he wanted to give up his trip. But I insisted I could take care of myself with the aid of our Eskimo friends, and I persuaded my companions to carry out their plans according to schedule. And after a few days Knud set off to the north with two of the Eskimos, Mathiassen to Ponds Inlet at the northeastern tip of Baffin Land, and Birket-Smith south through Canada.

I was left with Bangsted and the two Eskimo couples from Thule, who refused to leave me.

I was nursed by Patloq's wife Apa and I was in constant discomfort. It felt as if my foot had been tied off very tightly. The leg above was all right but the flesh below turned blue and then black. I had to lie quietly on my back while my nurse entertained me by recounting her experiences with frozen limbs. She knew a number of people who had lost both legs, others their arms or hands, but many had been killed because they were far too much trouble to take care of. And as the flesh began falling away from my foot, she tried out her special treatment. She captured lemmings—small mice—skinned them and put the warm skin on my rotting foot with the bloody side down. Every time she changed this peculiar kind of dressing, some of my decayed flesh peeled off with it but she insisted on this treatment until there was no more flesh left.

Gangrene is actually less painful than it is smelly. As long as I kept my foot inside the warm house the odor was unbearable, so we arranged to keep the foot outside. We made a hole in the wall at the end of my bunk, and I put my foot out into the freezing temperature whenever the odor became too overpowering. As the flesh fell away from the bones I could not bear having anything touch the foot, and at night when I could not sleep I stared with horrible fascination at the bare bones of my toes. The sight gave me nightmares and turned my nerves raw. I felt the old man with the scythe coming closer, and sometimes we seemed to have switched roles and my bare bones to have become part of him.

One day Apa told me that I needed a woman to take my mind off my pains. She brought along a young girl, Siksik, whose husband had kindly put her at my disposal while he went off on a trip with Captain Berthie. I felt like King David who was given young girls to keep him warm at night, but I told Siksik that I was in no condition to take advantage of the kind offer.

In the meantime it seemed as if Apa's cure was having some effect. The gangrene did not spread beyond the toes. Once the decay had bared all five toes to the roots, it did not go farther, and one day I decided to do something about it. I got hold of a pair of pincers, fitted the jaws around one of my toes and hit the handle with a heavy hammer.

The excruciating pain cut into every nerve of my body, an agony I cannot describe. Siksik had watched me and was deeply impressed. She offered to bite off the rest of the toes, and if her teeth hurt as much as the pincer, she said that I could beat her up. Ignoring her offer, I fitted the pincers around the next toe, and this time it did not hurt so much. Perhaps one could get used to cutting off toes, but there were not enough of them to get sufficient practice.

I admit that I cried when I was through with them—partly from pain, partly from self-pity. But it was a great relief to have the toe stumps off since they had kept me from walking and putting on my kamiks. Now I could at least get on my boots and hobble around.

THE ODYSSEY

Homer

Odysseus was the legendary Greek king of Ithaca during the decade-long Trojan War between the Mycenaean Greeks and the Bronze Age city of Troy. He is the subject of **Homer**'s epic poem, the *Odyssey,* which chronicles his harrowing return home following the events of the *Iliad*. The journey back to his Ithacan homeland, which lasted ten years, was fraught with a multitude of dangers that have served classicists and historians alike as metaphors for various psychological and interpersonal afflictions. Odysseus's travails secured him a place in the pantheon of Western heroes, inspiring generations of poets, writers and philosophers from Plato and Pausanius to Dante Aligheri and Alfred Lord Tennyson. His story remains the archetypal heroic journey and a poignant metaphor for perseverance, fate and determination.

Then, being much troubled in mind, I said to my men, 'My friends, it is not right that one or two of us alone should know the prophecies that Circe has made me, I will therefore tell you about them, so that whether we live or die we may do so with our eyes open. First she said we were to keep clear of the Sirens, who sit and sing most beautifully in a field of flowers; but she said I might hear them myself so long as no one else did. Therefore, take me and bind me to the crosspiece half way up the mast; bind me as I stand upright, with a bond so fast that I cannot possibly break away, and lash the rope's ends to the mast itself. If I beg and pray you to set me free, then bind me more tightly still.'

I had hardly finished telling everything to the men before we reached the island of the two Sirens, for the wind had been very favourable. Then all of a sudden it fell dead calm; there was not a breath of wind nor a ripple upon the water, so the men furled the sails and stowed them; then taking to their oars they whitened the water with the foam they raised in rowing. Meanwhile I look a large wheel of wax and cut it up small with my sword. Then I kneaded the wax in my strong hands till it became soft, which it soon did between the kneading and the rays of the sun-god son of Hyperion. Then I stopped the ears of all my men, and they bound me hands and feet to the mast as I stood upright on the cross piece; but they went on rowing themselves. When we had got within earshot of the land, and the ship was going at a good rate, the Sirens saw that we were getting in shore and began with their singing.

'Come here,' they sang, 'renowned Odysseus, honour to the Achaean name, and listen to our two voices. No one ever sailed past us without staying to hear the enchanting sweetness of our song—and he who listens will go on his way not only charmed, but wiser, for we know all the ills that the gods laid upon the Argives and Trojans before Troy, and can tell you everything that is going to happen over the whole world.'

They sang these words most musically, and as I longed to hear them further I made signs by frowning to my men that they should set me free; but they quickened their stroke, and Eurylochus and Perimedes bound me with still stronger bonds till we had got out of hearing of the Sirens' voices. Then my men took the wax from their ears and unbound me.

Immediately after we had got past the island I saw a great wave from which spray was rising, and I heard a loud roaring sound. The men were so frightened that they loosed hold of their oars, for the whole sea resounded with the rushing of the waters, but the ship stayed where it was, for the men had left off rowing. I went round, therefore, and exhorted them man by man not to lose heart.

'My friends,' said I, 'this is not the first time that we have been in danger, and we are in nothing like so bad a case as when the Cyclops shut us up in his cave; nevertheless, my courage and wise counsel saved us then, and we shall live to look back on all this as well. Now, therefore, let us all do as I say, trust in Jove and row on with might and main. As for you, coxswain, these are your orders; attend to them, for the ship is in your hands; turn her head away from these steaming rapids and hug the rock, or she will give you the slip and be over yonder before you know where you are, and you will be the death of us.'

So they did as I told them; but I said nothing about the awful monster Scylla, for I knew the men would not go on rowing if I did, but would huddle together in the hold. In one thing only did I disobey Circe's strict instructions—I put on my armour. Then seizing two strong spears I took my stand on the ship's bows, for it was there that I expected first to see the monster of the rock, who was to do my men so much harm; but I could not make her out anywhere, though I strained my eyes with looking the gloomy rock all over and over.

THE DIVE

Stephen McGinty

Roger Chapman (1945–2020) and **Roger Mallinson** (born 1938) were submariners who became trapped nearly 500 metres below the ocean surface in their submersible *Pisces III*. The resulting multinational rescue effort took more than three days and resulted in history's deepest submarine rescue. When the two men were brought to the surface, a nailbiting attempt to remove the hatch was undertaken, eventually succeeding half an hour later. It was determined that there were approximately twelve minutes of oxygen remaining in the tank.

Stephen McGinty (born 1972) is an author, journalist and documentary producer. His previous books include *Fire in the Night: The Piper Alpha Disaster* and he co-produced the BAFTA-winning documentary based on the book.

There's a blackness so thick you can almost reach out and touch it, and from the blackness comes the keen whine of leaking air.

Mallinson and Chapman are disorientated. The ache of limbs clattered against sharp-angled equipment, even through the padding of a cushion, is a reminder that they are both alive, if in the gravest of danger. Mallinson thinks about the last reading on the depth gauge before the lights went out and knows they are so deep as to be surely beyond rescue, but any further dark thoughts are knocked out by the noise of the leak and the need to shut it off.

He recognises the sound of air escaping from one of the oxygen cylinders, but where is it? The exact configuration of *Pisces III* is not yet clear to her pilots. They remain in the dark, both literally and figuratively, and illumination isn't going to come quickly, not with the lights out and the batteries now in who knows what kind of condition.

But the batteries can wait. The renegade oxygen supply cannot.

Mallinson begins to feel around for the oxygen tank, recognising different pieces of equipment by their contours, and begins plotting their jumbled new location. It takes between two and three minutes before his hand lands on the cold steel case of the oxygen cylinder, finds the round control valve and shuts it down. The whine is reduced to a whisper, then silence. Both men can now hear themselves worry.

Chapman, meanwhile, has got his hands on the torch. He switches it on, summoning a cool white cone of light that acts as a guide to the disarray. Their small spherical world has been turned upside down. *Pisces III* seems to be standing on her tail, with the stern wedged deep into the Atlantic's sandy bottom, leaving the accommodation sphere standing up at right angles to its usual position. The two 'beds' on which they had laid while working on burying the cable are now like a pair of stretchers propped against a wall. The 'floor' they're standing on was, prior to the crash, the rear wall on which the sonar equipment, oxygen regulator, valves and underwater transducer were fixed. Each torch sweep reveals another scene of chaos from the collision as well as glimpses of Mallinson's bearded, tense and worried face.

Mallinson is thinking about the oxygen bottle and how much has flowed into the atmosphere around them. While he knows that it's not 'lost', that they will breathe it in over the next few hours, he's also aware that it must be carefully rationed, eked out into this constricted new world in which they will have to live. How much longer will it last? This leads him to the battery and the power supply, as whatever oxygen supplies they do have will be rendered useless if they can't operate the scrubber and continually clear the carbon dioxide out of the atmosphere.

To power it they have to switch the batteries back on, and Mallinson is more worried about this task than he feared their initial plummet. His greatest dread is a fractured cable igniting an electrical fire. A fire inside the oxygen-rich sealed steel sphere of *Pisces III* would be unstoppable, the underwater equivalent of the flash fire inside the command module of Apollo I that burned three NASA astronauts to death six years ago. Those men had a ground crew struggling to save them. At almost 1,600 feet down, the only witnesses here will be fish, attracted to the light of a raging fire behind the porthole glass, a fatal brief flicker in the darkness of the deep.

Mallinson discusses the dangers with Chapman, but both know there's no choice. They have to turn on the batteries in ascending levels of voltage, starting with the 12-volt, then the 24-volt and finally the 120-volt, potentially the most lethal. Chapman shines the torch on the central control panel and Mallinson reaches out to flick the first switch, steeling himself as he does so.

The first switch operates smoothly.

Then the second.

If a deadly fire burns in their immediate future, it will come from switch three, when 120 volts surges through the cable. Mallinson flicks the switch. There's an audible click, and suddenly the sphere is bathed in light.

The batteries have survived the crash. The two of them have electrical power, at least for now, and a limited supply of air. The next crucial item on their checklist is the efficacy of the carbon dioxide scrubber. Exhaled with every breath, CO_2 is lethal if allowed to build up in a confined space. The operating procedure for *Pisces III* is that with two men breathing inside a steel sphere only 80 inches in diameter, the CO_2 level will quickly rise and so must be monitored by careful consultation of the Ringrose CO_2 indicator, a small thermometer-style device held inside a steel panel the size of a paperback novel, which is bolted on to the sphere's wall. The 'thermometer' indicates the level of CO_2 in the atmosphere, and regulations state that when it passes 0.5 per cent, which usually happens every 30 minutes or so, then the scrubber is activated.

The scrubber looks like a tapered steel plant pot with a tin of paint on top. The 'paint tin' is a canister of lithium hydroxide. The tapered steel plant pot into which it's inserted holds an electric fan, and once activated the fan draws the sphere's stale air into the lithium hydroxide, which binds with the CO_2 molecules, trapping them and thereby purifying the air. This takes between 10 and 15 minutes, and is monitored via the Ringrose indicator. The removed CO_2 must then be replaced with oxygen from one of two cylinders that, once activated, slowly 'bleeds' oxygen back into the atmosphere. The pilots know when to switch off the oxygen – when the atmospheric pressure returns to the initial reading taken and fixed when the sub's hatch is first closed at the beginning of the dive.

If switching on the batteries held the grim possibility of swift and violent immolation, switching on the scrubber provokes its own anxiety. If the scrubber doesn't work, they will still die this day.

The angle at which *Pisces III* has come to rest means the scrubber and its canister are lying horizontal, fixed by a bracket to the wall. Chapman reaches for the scrubber, turns it upright and looks over to Mallinson. Both men know exactly what's at stake.

'Try it,' says Mallinson.

Chapman presses the switch. The click is comforting, but not as much as the gentle purr as the motor and fan are roused back to life. A wave of relief rolls over both men, followed by a second wave of exhaustion that barrels into Chapman. Later he'd write:

> *We had been up most of the previous day and night, and just a short while ago had been so near to a bath, breakfast and welcome rest. Then this. It was still almost impossible to understand our predicament, but my body gave the game away. I was shaking like a leaf and very hot.*

The predicament is that they're trapped on the seabed and any attempt at self-rescue will be suicidal. If they were at a depth of 100 feet, they could hold their breath, allow the submarine to fill up with water until the sphere was entirely submerged and the pressure inside matched the pressure outside. This would then enable them to open the hatch. Even if this were possible now, at a depth of almost 1,600 feet, the weight of water pushing down on them would be the equivalent of 50 tons, and even if their organs didn't collapse and they could begin to swim up and away through the darkness, how long would it take to glide through water up the side of the Empire State Building and beyond? Ten minutes? Fifteen minutes? Twenty?

Oxygen consumption is a matter of imprecise arithmetic. A *Pisces* crew will, on average, consume one litre of oxygen per minute per man, recorded as '1 litre/min/man' and consumed via two oxygen cylinders, each carrying 63 cubic feet of oxygen and capable of taking a pressure of 3,000 psi. Under normal circumstances, two oxygen cylinders can expect to last 30 hours at the standard rate of consumption of 1 litre/min/man. In an emergency scenario, the crew are expected to lie down, reduce all physical action to the absolute minimum and restrict conversation to the bare essentials, all the time breathing in a slow and steady manner. The inhalation rate is then reduced to a target of 0.5 litre/min/man, the goal being to take a little over one day's oxygen supply and stretch it to two, maybe three days. In an extreme scenario they need to extend the oxygen supply from 30 hours to 72 hours, the outer edge of any rescue scenario.

The light is kept on for a few minutes as the men take stock of the disarray and begin to rearrange their new living quarters. The 90° adjustment means equipment previously directly in front of them is now right above them. The portholes that were once three little windows have become 'skylights' out onto the black water above. The hatch is no longer in the ceiling, but at their elbows, like a door to another room. The sonar units are at their feet, while the depth sounder, machine control gauge and switches, as well as the 1,000-watt quartz iodine exterior light, are all above their heads.

The first thing to do is reposition the two beds and foam mattresses, which they move from the vertical to horizontal. Previously there was a 12-inch gap between the bunks to allow movement, but instead they squeeze them together, like two singles into a double. The benches are now resting on the penetrators at one end and the hydraulic valves at the other. The aim is to have everything within easy reach in the dark.

There are two clockwork timers screwed to the sphere. They have a plastic dial, like an egg timer, and are used to trigger an alarm every 30 minutes to prompt the activation of the scrubber. If both men fall asleep and fail to switch it on, there's a good chance they will never wake up, as an excess of CO_2 will tip sleep into a fatal unconsciousness. The timers have a ridged, raised surface, allowing the men to 'feel' the time without switching on the light. Chapman unscrews both timers and hands one to Mallinson, keeping the other for himself. They will now be able to feel time passing through their fingertips.

The coiled lead on the microphone of the underwater telephone is long enough to dangle down in front of them, while the control box is an arm's length away. The main oxygen supply is no longer easily accessible, as it's wedged beneath Mallinson's bench, so they decide to switch to using the reserve bottle to bleed oxygen to replace the CO_2 as it's three inches from Chapman's head and the gauge is, quite literally, staring him in the face. The most vital system is now the most inconveniently positioned. The CO_2 scrubber is positioned above their feet at the other end of the sphere; this will require one of them to move around every 30 to 45 minutes, at a cost of effort and, as a result, of oxygen.

Moving around the sphere, putting in place what limited practical solutions he can, Mallinson is thinking about their plight. They are so deep and their potential rescuers so very far from the mainland that, surely, any help is impossible. He decides not to translate thoughts into words and share his fears with Chapman. Instead he lets them slowly roll around like black marbles. Mallinson is also thinking about the batteries. He can tell that the 120-volt battery has been drained down to 100 volts and is silently berating himself for spending too long on the cable burial, which has clearly taken its toll on the supply.

Chapman makes an inventory of their supplies, which are far from lavish – they have a single and very soggy cheese and chutney sandwich and a can of Corona lemonade. Mallinson reminds Chapman that he doesn't like cheese and chutney, and regrets having eaten his own strawberry jam sandwiches. There's one half flask of black coffee, a tin of powdered milk, a packet of sugar, two apples and some paper cups. There are also three soggy biscuits and the standard lifeboat ration of glucose tablets and biscuits. There's no water.

It's about now that they become aware of the incessant din inside the sphere. This has been a constant since they crashlanded, but a mixture of anxiety, adrenaline and the distraction of emergency tasks has served to render it inaudible. Now, it's all they can hear. Almost every second a loud tone reverberates through the sphere. It's the pinger on the CANTAT cable, dropped as a guide for the next shift at the exact spot where they had finished working. The repetitive piercing ping is deeply uncomfortable, but the only way to silence it is to turn down the underwater telephone (UQC), which would cut them off from all surface communication.

In this state of semi-order, Chapman reaches for the communication breaker, which suddenly bursts into life.

'*Pisces*, *Pisces*, this is *Voyager* ... Do you read? Over.'

Chapman pushes the transducer down, which in the new disorientated world means pushing it up: 'This is *Pisces*. Yes, we read you loud and clear.'

Chapman details their position and condition. He explains that the sub is propped at a 90° angle to the horizontal, that they are uninjured and maintaining morale, and that *Pisces III*'s life-support systems are intact. He then gives as detailed as possible a reading of their oxygen supply. Yet what both men don't know is that their figures and the estimated rescue deadline given by Barrow – who have been handed control of the operation by *Voyager* – is based on an error. As the main oxygen tank is buried and therefore out of sight, the reading Chapman passes on to the surface is the last one taken, which was on the surface and seconds before the sudden sinking. The oxygen content read as 2,400 litres, but the cold fact, unknown to both men and to their rescuers, is that the actual reading is 1,900 litres – a fifth less. On the way to the seabed they have consumed the equivalent of 16 man hours of oxygen at the restricted rate of inhalation. The figures they are passing on are incorrect and any deadline Barrow sets will be out by hours.

They have less oxygen and less time than either they or their rescuers believe.

The temperature inside the sphere is 50°F (10°C), but combined with humidity running at 95 per cent it feels much colder. Mallinson is in a thick red sweater; Chapman is wearing only overalls but is beginning to feel the chill, so he strips down his overalls to his waist, puts on a lifejacket, binds his arms with a few woollen rags left over from when the sub was recently cleaned and then buttons himself back up. Both men are still shaking from a combination of cold and adrenaline.

Another concern both men have is that the batteries might fail, rendering them unable to communicate with the surface, light the sphere or operate the scrubber. The batteries are all on their side and the acid in which they sit could be draining to the bottom, thereby reducing connectivity. There could also be an unseen acid leak, but each time one of them rises from their bed to check the voltage on all three, he lies back reassured. For now.

They are three hours into their ordeal and already they recognise an issue with the scrubber. On a normal dive it takes just ten minutes to reduce the CO_2 level below 0.5 per cent, but the time taken to achieve this is rising. The scrubber has not been changed since the previous nine-hour dive and condensation is seeping into the pellets, reducing their efficacy. Chapman discusses the options with Mallinson. They have two fresh canisters but neither man is anxious to make a switch just yet. Even when they do, condensation will remain a problem. Mallinson suggests that they take turns shaking the canister while the scrubber is running in the hope that this will help, and that they let the CO_2 levels rise higher than normal to make the canister last a little longer.

In a notebook later transcribed into his pilot's log, Chapman records:

> *General state of health: No injuries after impact. Breathing fast because of work and initial fear of situation. Thirsty ... Advice given from surface to maintain atmospheric pressure and no physical exertion.*

On the surface, Ralph Henderson is scribbling down a detailed message in the radio room on *Voyager*. There's a round porthole covered by an orange nylon curtain, and whenever he looks out of the window all he can see is the grey-green expanse of the Atlantic. It's 12.47 pm and the men's morale is, he writes, 'fantastic'. The notes continue:

> *02 1 bottle 3400, other bottle 2600, 2 Lth cans scrubber works. Battery volts 110. Attitud[e] 90 degrees stern down. A Tanks possibly flooded. Air bottles 2x2000 PSI 2(*1) 3300 PSI will call every ½ hour.*

Henderson can also feel the frustration building. Captain Len Edwards, master of the ship, and usually a man of a preternatural calm and confidence that's surprising for a man only in his early 30s, is anxious to leave the scene and return to Cork. He knows the sooner they can leave, the sooner they can return – and the quicker Chapman and Mallinson can be recovered. But Edwards and *Voyager* can't depart the scene until they have replacement cover on the accident site, and cover has yet to be found. As he has handed over operational command to the base back in Barrow, he is now awaiting instructions. His priority at the moment is using all his skill to maintain *Voyager* over the exact spot beneath which he believes *Pisces III* will lie. The wind is freshening up and making this increasingly difficult.

For Edwards it's also proving a long, frustrating afternoon. While a number of ships have offered assistance, each is too far distant. The Irish Navy would have been his best hope, but when a Captain Kavanna called Barrow it was to explain, apologetically, that their nearest ship is 16 hours away. Edwards is now resting his hopes on the arrival of either *British Kiwi* – a commercial BP tanker – or a Royal Navy vessel, *Sir Tristram*. Edwards knows *Voyager* is lying 150 miles from Cork, which is 13½ hours' hard sailing if he can hit top speed, which is 27 hours as a round trip. When you factor in the loading time of rescue equipment and submersibles, he will be lucky to be back on this spot within 30 to 32 hours. Every hour he spends waiting is one more hour struck off his men's oxygen supply.

Henderson, meanwhile, is focusing on keeping both men on the bottom as calm as possible. He's already made a decision to restrict all bad or unpleasant news and spin any mishaps on the grounds that what they need is a sense of steady control, mixed in with a few crude jokes on how two men trapped in a sunken submarine can both keep warm and while away the hours. In the early afternoon he advises Mallinson and Chapman that if they are cold they should unzip the seat covers wrapped around the day beds and use them as an impromptu sleeping bag. Good advice, except the coverings had recently been changed; now they're fixed and so no longer unzip. The suggestion does prompt the pair to find the rubber cover that usually sits over the video equipment and use it as a blanket.

At 2.55 pm many of the crew on *Voyager* head up on deck to watch for the arrival of the RAF Nimrod. Edwards had hoped that the communications plane would be able to stay on the accident site and so relieve him and his crew, but Barrow has informed him that the RAF are refusing to accept responsibility and that *Voyager* is ordered to remain on station until a surface relief vessel arrives later in the afternoon. Upon arrival, the pale grey plane arcs overhead and circles at a radius of two miles, slowly dropping in altitude until it's around 300 feet above the surface of the sea. A bay door then opens and a sonar buoy tumbles through the air and splashes down into the Atlantic, sinking at first, before bobbing back up to the surface. Inside the Nimrod, RAF officers, whose principal task is detecting Soviet submarines, are now scanning their sonar screens for signs of *Pisces III*. Their persistent question, '*Pisces III* do you copy?' will remain unanswered.

At 5 pm Henderson tells the men to keep listening as he is about to try another transducer, but in *Pisces III* all Chapman and Mallinson can hear is the persistent 'ping' of the pinger, once every one and a half seconds. They have turned up the underwater telephone's loudspeaker to maximum and the ping is enough to induce a headache. But they can hear nothing else, and after 45 minutes of silence from the surface are beginning to be concerned.

Shortly before 6 pm a ship is spotted on the horizon. It's HMS *Sir Tristram*, a multi-purpose 6,390-ton fast troop and heavy vehicle carrier, with a crew of 18 officers, 50 ratings and, conveniently, a pair of Wessex helicopters positioned on the fore and aft decks. As the captain positions the ship 50 metres out from the buoys marking the spot where *Pisces III* sank, Henderson and diver David Mayo prepare to transfer across. A message is sent to *Pisces III* – which they do hear – that there will be a break in communications, deliberate this time, as the pair swap ships, but that they 'will be in touch again soon'. A 16-foot Gemini inflatable is prepared by Voyager and Henderson loads it up with a heavy 10-inch reel-to-reel tape recorder, as well as the underwater telephone system and his log books. By 6.15 pm Henderson and Mayo are on the bridge of *Sir Tristram*, being warmly welcomed by the captain and crew, and are setting up the equipment and watching through the naval vessel's vast windows. The engines of *Voyager* start up and her bow slowly begins to turn as the propellers send a flotilla of sonic waves down through the light-filled fathoms and into the darkness.

Lying in the dark of the sphere, with the underwater telephone turned up, Chapman hears a background thrum rippling underneath the repetitive din of the pinger. It takes a second or two for him to recognise the noise as the thrum becomes more consistent, picks up pace and rises into a louder thrashing churn. The sound of *Voyager*'s propellers is reaching them. It's a moment of profound emotion that neither man is willing, or perhaps even capable, of expressing to the other. Their support ship, their friends and colleagues, are, for the moment, leaving them at the bottom of the Atlantic. While each man's cool, rational intellect is aware that the thrum of engines and churn of propellers are the sonic signatures of progress, the first line in a chain of events that, God willing, should see them hauled to the surface and safety, it's very hard to think good thoughts when buried so deep and so much in the dark. The primal emotions on which each man bites down are of fear and abandonment.

TRAVELS IN CANCERLAND

Katherine Page

Katherine Page is an Army Officer and Medical Doctor. She has accompanied multiple operational tours in Afghanistan and Iraq, and supported exercises in Canada, Africa and the Middle East. As an expedition doctor, she travelled to some of the most inhospitable places in the world and became the only woman to complete the world's first crossing on foot of Madagascar in 2012. At the age of 36, she was diagnosed with an aggressive form of breast cancer, which spread despite surgery, chemotherapy and radiotherapy. Her disease is not curable, and she has been told she has only a few years to live. Despite her diagnosis and ongoing drug treatments, she continues to serve as a full time Army GP in London, and is an Army Advocate for the Chronic Conditions and Disabilities in Defence (CANDID) Network, using the benefit of her experiences to support others to live as full and positive a life as possible.

Fear is a visceral, hot sensation. My nerves scald as impulses surge through ancient pathways in my body like uncontrollable fireworks: vessels dilate, sweating starts, muscles tense, skin burns. This is not a momentary electric shock, a single surprise that makes you jump out of your skin, but the unrelenting activation of a primal fight or flight response. Terror comes when the fear of death halts all non-essential physical and mental functions. The fear is both real and imagined. I can feel it coming like the inexorable flow of a river, knowing that once the hormonal and emotional cascade takes over, normal activities and thoughts are impossible until I've swam its length. I become completely submerged. When nightmares come, I am rotting in a grave, but still wracked by physical pain.

I have voluntarily taken poison. Unable to sleep, I have been losing my grip on reality over several days. I become paranoid and agitated as steroids push me closer to psychosis. My fingertips are peeled raw and bleeding, the inside of my mouth is fleshy with ulcers and my hair has shed in clumps. I am hunched over with pain. I do not know it yet, but the injections I have been taking to boost my white blood cells have overstimulated my bone marrow. My immature blast cells are crammed – the very insides of my bones bulge to accommodate them – causing a deep, unrelenting ache. My inflamed liver is tight in its capsule and my bloated gut makes deep breathing difficult. With bloodshot, dry, lash-less eyes and patchy long hair on only one side of my head, I am zombie-like. I grip the edge of a sink, trying to stand, shaking as nausea clouds any thought that might improve my situation.

Three months earlier I had been told I had breast cancer, and tonight, in the grips of a drug-induced crisis, all is chaos. Biochemical storms rage in every organ. Chemotherapy or cancer: one of them was going to kill me.

This was not the only time during my cancer treatment that I knew with certainty how close I was to dying. I had reacted badly to one of my drugs, and tingling lips had suddenly become a closed throat. A crash team responded by flattening my chair ready to help me breathe. The familiar flash of limbic fireworks ensued, and my last sensation was of falling – falling asleep, falling from the last step. Resurrected within minutes, trembling with injected adrenaline, Hobson's Choice was laid in front of me: we would have to try again. The same thing might happen – there were no guarantees – but this time the team would stand around me. And so, as syringes were reattached, and chemical storm clouds gathered, I prepared to fall, and on woefully inadequate encouragement, went to my happy place.

In my happy place, I see a rising sun above an African landscape. Everything is quiet and the temperature is soothingly cool – an antidote to my hot flushes. I close my eyes and am taken back to an early morning many years ago. The air is crisp, and the camp is stirring. I enjoy the sounds of tents unzipping and boots being pulled on, everyone eager for breakfast around the campfire. Some people look forward to the end of the day, the marshmallow-like sensation of slipping into your sleeping bag, but I have always been a morning person. I like the anticipation of what is to come.

Army doctors like me have a mixture of skills that make us exactly the type of all-rounder you want with you in exotic far-flung locations. Be it dermatology, obstetrics, psychiatry or pre-hospital emergency care, our knowledge is multifaceted. We're found in situations where you need to know as much about how to manage a peculiar-looking rash, as you do how to deal with a head injury several hours from the nearest hospital. Often the hardest part of our job is to decide when someone has had enough, when they need to be removed from a situation, either for their own benefit or for the benefit of those around them. It is also our job to know when someone still has enough fuel left in the tank, even when they think they don't. The brain will give up long before the body is truly exhausted and recognising this situation might be key to keeping someone alive. Managing uncertainty is risky; it makes you and your patients vulnerable, and extreme stress in these scenarios can cloud judgement. These skills are both an art and a science.

In my leave time from the army, I would volunteer as a doctor on commercial expeditions with a travel company, which enabled intrepid civilians to experience their first taste of adventure. I travelled to some of the most remote and inaccessible places in the world, from the deserts of Jordan to the wilderness bush of South Sudan. These were no ordinary sightseeing tours; these would-be explorers had signed up to be part of an arduous expedition, executed with military precision. There was an expectation that everyone would muck in and contribute as much to the journey's success, as to its possible failure.

There were no guarantees, of course. Sometimes it got too much. Towards the end of one expedition, hiking coast-to-coast across the northern region of Madagascar, we huddled, exhausted, around a tattered old map that had been made when the French Foreign Legion used to train there. Someone asked, 'Will we cross any rivers today?' The day before, following the examples of local guides, we had waded through rivers in our underwear, later dislodging leeches on the backs of each other's legs with dried tobacco leaves. The day before that, we had done the same, and the day before that too. We were pushing towards the sea, having walked some 400 kilometres – a route never passed before, through dense forest – and the rivers had become a useful way of navigating. We either had to cross more rivers, or hack with machetes through the jungle to the top of the valley. The difference between a paying client and a trained soldier is this type of question. My military experience taught me that they were not really asking about the river crossings – they already knew the answer. What they were really asking, or hoping, was, will this get any easier?

I thought a lot about those rivers during my cancer treatment. 'Will there be rivers?' had become a private joke between myself and the ex-military expedition leader with whom I'd gone on to share many adventures. With our backgrounds, it had always amused us that clients – who had paid thousands of pounds to be deliberately challenged, intentionally spending weeks in the jungle with no route out but the way they had come – were clutching at signs that the experience would somehow come to an end early, or that there might be shortcuts.

We did not think less of them. It was brave of these people, who had little experience outside of their club or charity trekking trips, to be pulled out of their comfort zones during their precious time off work. We knew how hard it can be when you reach your physical limits, when you come up against your wall. Fatigue affects your ability to look after yourself, and in a remote location, this can be life-threatening. The first step of torture is sleep deprivation, because it clouds judgement, and makes even the simplest of activities difficult.

The most effective military tool for training against this type of defeatist thinking is a lesson that I have fallen back on repeatedly. Often, during a physical training session or field exercise, you might find yourself with a view of the end point in sight: a finishing line, breakfast, perhaps, and you allow yourself to imagine that with one last push it will finally be over. You give it all you've got and empty your tanks. A glorious finish! Home for tea and medals. Then, 'No' someone bawls, 'don't put your kit down, take a minute, take a breath, turn around, go again'. This moment is designed to separate your brain from your body, to challenge by force the evolutionary protective tricks of the mind, designed to halt the body before injury occurs, or water, or salts, run out. It taps instead into the stimulus held in reserve in our peripheral muscles, to help us escape from predators. To the uninitiated, a surge is palpable in the pack, and it pushes them onwards. Fatigue is an emotion before it is a function. Every recruit knows the adage 'it's all in your head'.

I realised that the key to enduring my cancer treatment was to stop expecting it to get easier. Almost every major philosophy or conventional spiritual practice observes that one of the primary ingredients in the alchemy of happiness is to exist only in the current moment. So, I learned to meditate, I learned acceptance, and I learned to practice gratitude. Incense triggered my nausea, but I learned to sit with it, not to wish it away. Above all else, I learned to surrender to the chaos inside me. If I could centre myself in the eye of the storm while my fears raged then subsided, my body rotted then regenerated, my plans for the future decimated then settled onto something new; if I could overcome the temptation to tell my body that I had had enough; if I could stop looking for an easier route out, then I would thrive.

Only three months after finishing treatment, I celebrated by running the London Marathon. After being cut off from the reassuring routine of hospital visits, this gave me a focus and a goal. I never begrudged a single early morning training run. I ran every mile as a meditation, focusing on an aspect of my cancer care: first mile for my parents, who had lived through their worst fears; second mile for the other person sitting in my chemotherapy chair while I ran; mile sixteen for losing my hair, and for it growing back; mile twenty for the night terrors and for fearing I would die; mile twenty-five because I was grateful to be running again and it was a privilege to be able to do so. Pushing through the wall, I quietened the part of my mind overwhelmed by the distance yet to cover, pushing away the siren-like defeatist whispers to stop, to rest. I sobbed during the last mile, filled with gratitude for my sore muscles, for the aches and pains, because this time they were a sign of strength rebuilding, not fibres wasting or cells dying. My last mile was for the future, and for moving on.

It was only four weeks later when an almost imperceptible fullness met my fingers just above my collar bone and a familiar rising panic distracted me mid-conversation with one of my patients. A supraclavicular lymph node, this grape-sized lump, hidden to anyone not trained to notice it, told me I had cancer again, that it had spread, and that I could not be cured. This was confirmed when, a few short weeks later, I sat in front of my own doctor, and was told that I would start chemotherapy again, but that one day it would stop working, and then I would die. I would have to ready myself to cross some more rivers, but this time, there would be no finishing line.

I do not recall the following weeks, or how I told my family. Anxiety, sleeplessness and trauma cloud my memories. But there was relief too, that my long-standing fears were finally being reconciled, and there is something immensely solid in certainty, even if that certainty is catastrophic. I had a plan again.

I had written a will before, the first time I had deployed to war in Iraq. Before I deployed my father had given me a pocket mirror engraved with the words 'Stay Safe', although put more bluntly it could have said 'Don't Die'. Soldiers deal with death with shocking black humour. We had all promised to leave money so that our fellow officers could build swimming pools with our faces mosaicked on the bottom. But no one really thought they would need it, and despite deaths, to my knowledge, no swimming pools were ever installed. Despite our bravado, we learned the lesson that you must be careful what you wish for, as extraordinary feats of mental and physical resilience from this time are now only spoken about in doctors' offices. Nothing is ever as you expect it to be, and a plan, as they say in the army, never survives first contact with the enemy. With similar squaddie humour I now joke that I will not lose my battle to cancer, that it will at best be a draw.

Being a doctor with cancer means that I know how I am going to die. When I worked in palliative care I would ask my patients: 'Do you want to know what it feels like?' and no one ever said no. Like those early mornings huddled around a map, I used my limited experience to give my patients a glimpse of what to expect, perhaps helping them to understand the lay of the land they would be travelling. To relieve someone of fear of the unknown is to take from them an enormously heavy burden. I hope for the type of death I would describe, a good death, like a good birth, a calm transition from one state to the other into a new world. This now has become my goal. Cancerland has different physics to it. Certainty doesn't exist, hard work doesn't always pay off, sleep doesn't always restore energy, time is marked in treatment cycles, and the future does not exist. It is disorientating and confusing. Time can expand and contract, expectations and reality merge and then converge within a few hours depending on test results, or drug side effects. The landscape is ever-changing.

To live with a terminal disease is to truly understand that the journey is always more interesting than the destination. There is something immensely freeing in gaining this perspective. The trick is to look up and let go of where you thought you were going, and revel in where you are. To face your own mortality is the definition of Awe. It is being stood on the edge of a mountain with a drop beneath you into the clouds. If they sold tickets to experience this rushing *l'appel du vide*, then I would recommend it, if only for the life-affirming *wow* of being pulled from the brink.

I have found that to live every day as if it is your last is ludicrous. It is expensive and stressful. It is exhausting. To live every day as if you are going to live forever is far more hopeful. Embrace the chaos, let go of the military order of things, travel without a plan, take risks and have adventures. Do not look forward to the end of the day, but to the beginning of the next one. Perhaps things won't turn out as you planned. Perhaps they will be better. Perhaps they will be more challenging. But one thing is likely to be certain: there will be more rivers.

LEFT TO TELL
Discovering God Amidst the Rwandan Holocaust

Immaculée Ilibagiza

Immaculée Ilibagiza (born 1972) is an author and a survivor of the Rwandan Genocide. After hiding in a twelve-square-foot bathroom stall with three other women for an agonizing ninety-one days, she escaped and eventually made her way to the United States. The genocide claimed the lives of most of her immediate family members, leaving only her brother and herself alive. Her story is a remarkable example of the triumph of faith and will in the face of calamity and terror. Ilibagiza is now a motivational speaker and holds an honorary doctorate from the University of Notre Dame.

I was deep in prayer when the killers came to search the house a second time. It was past noon, and I'd been praying the rosary since dawn for God to give His love and forgiveness to all the sinners in the world. But try as I might, I couldn't bring myself to pray for the killers. That was a problem for me because I knew that God expected us to pray for *everyone*, and more than anything, I wanted God on my side.

As a compromise, I prayed the rosary multiple times, as intensely as I could, every day. Working through all those Hail Marys and Our Fathers took 12 or 13 hours – and whenever I reached the part of the Lord's Prayer that calls us to "forgive those who trespass against us," I tried not to think of the killers, because I knew that I couldn't forgive them.

During that second search, the killers' racket reached the edge of my prayers like an angry voice waking me from a dream. Then I heard four or five loud bangs next to my head, and they had my full attention. I realized that they were right there in the pastor's bedroom! They were rummaging through his belongings, ripping things from the wall, lifting up the bed, and overturning chairs.

"Look in that!" one of them yelled. "Now look under here. Move that chest! *Search everything!*"

I covered my mouth with my hands, fearing that they'd hear me breathing. They were only inches from my head ... they were in front of the wardrobe – *the wardrobe!* I thanked God again for it, but my heart still thumped against my chest.

I could hear them *laughing*. They were having fun while going about killing people! I cursed them, wishing that they'd burn in hell.

The wardrobe banged against the door. I covered my ears and prayed: *God, please. You put the wardrobe there ... now keep it there! Don't let them move it. Save us, Lord!*

My scalp was burning, and the ugly whispering slithered in my head again: *Why are you calling on God? Don't you have as much hatred in your heart as the killers do? Aren't you as guilty of hatred as they are? You've wished them dead; in fact, you wished that you could kill them yourself! You even pray God would make them suffer and make them burn in hell.*

I could hear the killers on the other side of the door, and entreated, *God, make them go away ... save us from –*

Don't call on God, Immaculée, the voice broke in. *He knows that you're a liar. You lie every time you pray to Him to say that you love Him. Didn't God create us all in His image? How can you love God but hate so many of His creations?*

My thoughts were paralyzed. I knew that the demon in my head was right – I *was* lying to God every time I prayed to Him. I was so overwhelmed with hatred for the people responsible for the genocide that I had a hard time breathing.

At least 40 or 50 men were in the pastor's bedroom by this time, and they were shouting and jeering. They sounded drunk and mean, and their chanting was more vicious than usual: "Kill the Tutsis big and small ... kill them one and kill them all. *Kill them!*"

I began praying, asking God to keep them away from the wardrobe and out of the house altogether.

Beneath the raucous singing, the dark voice taunted me: *It's no use ... don't call on God. Who do you think sent the killers here for you? He did! Nothing can save you. God doesn't save liars.*

I began to pray for the killers and then stopped. I desperately wanted God's protection, but I believed in my heart that they deserved to die. I couldn't pretend that they hadn't slaughtered and raped thousands of people – I couldn't ignore the awful, evil things that they'd done to so many innocent souls.

Why do You expect the impossible from me? I asked God. How can I forgive people who are trying to kill me, people who may have already slaughtered my family and friends? It isn't logical for me to forgive these killers. Let me pray for their victims instead, for those who've been raped and murdered and mutilated. Let me pray for the orphans and widows ... let me pray for justice. God, I will ask You to punish those wicked men, but I cannot forgive them – I just can't.

Finally, I heard the killers leaving. First they left the bedroom, then the house, and soon they were walking away down the road, their singing fading in the distance.

I resumed my prayers. I thanked God for saving us and for giving me the idea to put the wardrobe in front of the bathroom door. *That was so smart of You, God. You are very smart*, I said mentally, and thanked Him again. I wondered where the killers were off to, then I started praying for my friends and family: *Please look over my mother, God; she worries so much about us. Watch over my father; he can be so stubborn...*

It was no use – my prayers felt hollow. A war had started in my soul, and I could no longer pray to a God of love with a heart full of hatred.

I tried again, praying for Him to forgive the killers, but deep down I couldn't believe that they deserved it at all. It tormented me ... I tried to pray for them myself, but I felt like I was praying for the devil. *Please open my heart, Lord, and show me how to forgive. I'm not strong enough to squash my hatred – they've wronged us all so much ... my hatred is so heavy that it could crush me. Touch my heart, Lord, and show me how to forgive.*

I struggled with the dilemma for hours on end. I prayed late into the night, all through the next day, and the day after that, and the day after that. I prayed all week, scarcely taking food or water. I couldn't remember when or for how long I'd slept, and was only vaguely aware of time passing.

One night I heard screaming not far from the house, and then a baby crying. The killers must have slain the mother and left her infant to die in the road. The child wailed all night; by morning, its cries were feeble and sporadic, and by nightfall, it was silent. I heard dogs snarling nearby and shivered as I thought about how that baby's life had ended. I prayed for God to receive the child's innocent soul, and then asked Him, *How can I forgive people who would do such a thing to an infant?*

I heard His answer as clearly as if we'd been sitting in the same room chatting: *You are <u>all</u> My children ... and the baby is with Me now.*

It was such a simple sentence, but it was the answer to the prayers I'd been lost in for days.

The killers were like children. Yes, they were barbaric creatures who would have to be punished severely for their actions, but they were still children. They were cruel, vicious, and dangerous, as kids sometimes can be, but nevertheless, they were children. They saw, but didn't understand, the terrible harm they'd inflicted. They'd blindly hurt others without thinking, they'd hurt their Tutsi brothers and sisters, they'd hurt God – and they didn't understand how badly they were hurting themselves. Their minds had been infected with the evil that had spread across the country, but their *souls* weren't evil. Despite their atrocities, they were children of God, and I could forgive a child, although it would not be easy ... especially when that child was trying to kill me.

In God's eyes, the killers were part of His family, deserving of love and forgiveness. I knew that I couldn't ask God to love me if I were unwilling to love His children. At that moment, I prayed for the killers, for their sins to be forgiven. I prayed that God would lead them to recognize the horrific error of their ways before their life on Earth ended – before they were called to account for their mortal sins.

I held on to my father's rosary and asked God to help me, and again I heard His voice: *Forgive them; they know not what they do.*

I took a crucial step toward forgiving the killers that day. My anger was draining from me – I'd opened my heart to God, and He'd touched it with His infinite love. For the first time, I pitied the killers. I asked God to forgive their sins and turn their souls toward His beautiful light.

That night I prayed with a clear conscience and a clean heart. For the first time since I entered the bathroom, I slept in peace.

SUFFERINGS IN AFRICA

James Reilly

Captain James Reilly (1777–1840) was an American merchant captain of the trading vessel *Commerce* who, along with his crew, was shipwrecked in 1815 off Western Sahara. He led his crew on an arduous journey through the desert, where they were captured and enslaved by Sahrawi tribesmen. The crew experienced great hardship from the harsh conditions of the Sahara and poor treatment at the hands of their captors. Reilly lived to tell the tale of his ordeal and dedicated the remainder of his life to the cause of abolition in the United States.

THE Arabs had been much amused in observing our difficulty in ascending the height, and kept up a laugh while they were whipping us forward. Their women and children were on foot as well as themselves, and went up without the smallest difficulty or inconvenience, though it was extremely hard for the camels to mount; and before they got to the top they were covered with sweat and froth. Having now selected five camels for the purpose, one for each of us, they put us on behind the humps, to which we were obliged to cling by grasping its long hair with both hands. The back bone of the one I was set on was only covered with skin, and as sharp as the edge of an oar's blade; his belly, distended with water, made him perfectly smooth, leaving no projection of the hips to keep me from sliding off behind; and his back or rump being as steep as the roof of a house, and so broad across as to keep my legs extended to their utmost stretch, I was in this manner slipping down to his tail every moment. I was forced however to keep on, while the camel rendered extremely restive at the sight of his strange rider, was all the time running about among the drove, and making a most woful bellowing; and as they have neither bridle, halter, or any other thing whereby to guide or govern them, all I had to do was stick on as well as I could.

The Arabs, both men and women, were very anxious to know where we had been thrown on shore, whether to the eastward or westward; and being satisfied by me on that point, so soon as they had placed us on the camels, and given the women directions how to steer, they mounted each his camel, seated themselves on the small round saddle, and then crossing their legs on the animal's shoulders, set off to the westward at a great trot, leaving us under the care of the women, some of whom were on foot, and urged the camels forward as fast as they could run. The heavy motions of the camel, not unlike that of a small vessel in a heavy head-beat sea, were so violent, aided by the sharp back bone, as soon to excoriate certain parts of my naked body; the inside of my thighs and legs were also dreadfully chafed, so that the blood dripped from my heels, while the intense heat of the sun had scorched and blistered our bodies and the outside of our legs, so that we were covered with sores, and without any thing to administer relief. Thus bleeding and smarting under the most excruciating pain, we continued to advance in a S. E. direction on a plain flat hard surface of sand, gravel, and rock, covered with small sharp stones. It seemed as if our bones would be dislocated at every step. Hungry and thirsty, the night came on, and no indication of stopping; the cold night wind began to blow, chilling our blood, which ceased to trickle down our lacerated legs; but although it saved our blood, yet acting on our blistered skins, it increased our pains beyond description. We begged to be permitted to get off, but the women paid no attention to our distress nor intreaties, intent only on getting forward. We designedly slipped off the camels when going at a full trot, risking to break our necks by the fall, and tried to excite their compassion and get a drink of water, (which they call sherub,) but they paid no attention to our prayers, and kept the camels running faster than before.

This was the first time I had attempted to walk barefooted since I was a schoolboy: we were obliged to keep up with the camels, running over the stones, which were nearly as sharp as gun flints, and cutting our feet to the bone at every step. It was here that my fortitude and reason failed to support me; I cursed my fate aloud, and wished I had rushed into the sea before I gave myself up to these merciless beings in human forms – it was now too late. I would have put an immediate end to my existence, but had neither knife nor any weapon with which to perform the deed. I searched for a stone, intending if I could find a loose one sufficiently large, to knock out my own brains with it; but searched in vain. This paroxysm passed off in a minute or two, when reason returned, and I recollected that my life was in the hand of the power that gave it, and that "the Judge of all the earth would do right." Then running with all my remaining might, I soon came up with the camels, regardless of my feet and of pain, and felt perfectly resigned and willing to submit to the will of Providence and the fate that awaited me.

From that time forward, through all my succeeding trials and sufferings, I never once murmured in my heart, but at all times kept my spirits up, doing the utmost to obey and please those whom fortune, fate, or an overruling

Providence had placed over me, and to persuade, both by precept and practice, my unhappy comrades to do the same. I had, with my companions, cried aloud with pain, and begged our savage drivers for mercy, and when we had ceased to make a noise, fearing, as it were, to lose us in the dark, they stopped the camels, and again placing us on them as before, drove them on at full speed until about midnight, when we entered a small dell or valley, excavated by the hand of nature, a little below the surface of the desart, about from fifteen to twenty feet deep. Here they stopped the camels, and made them lie down, bidding us to do the same. I judge we must have travelled forty miles this day to the S. E.: the place was hard and rocky, not even sand to lie on, nor any covering to shelter us or keep off the cold damp wind that blew strong from the sea.

They soon set about milking, and then gave us each about a pint of pure milk, warm from the camels, taking great care to divide it for us; it warmed our stomachs, quenched our thirst in some measure, and allayed in a small degree the cravings of hunger. Mr. Savage had been separated from us, and I learned from him afterwards that he fared better than we did, having had a larger allowance of milk. Clark, Horace, and Dick the cook were still with me. We lay down on the ground as close to each other as we could on the sharp stones, without any lee to fend off the wind from us; our bodies all over blistered and mangled, the stones piercing through the sore naked flesh to the ribs and other bones. These distresses, and our sad and desponding reflections, rendered this one of the longest and most dismal nights ever passed by any human beings. We kept shifting births, striving to keep off some of the cold during the night, while sleep, that had hitherto relieved our distresses and fatigues, fled from us in spite of all our efforts and solicitude to embrace it; nor were we able to close our eyes.

The morning of the 11th came on at last, and our industrious mistresses having milked a little from the camels, and allowed the young ones to suck, gave us about half a pint of milk among four of us, being just enough to wet our mouths, and then made us go forward on foot and drive the camels. The situation of our feet was horrible beyond description, and the very recollection of it, even at this moment makes my nerves thrill and quiver. We proceeded forward, having gained the level desart for a considerable time, when entering a small valley, we discovered three or four tents made of coarse cloth near which we were met by our masters and a number of men whom we had not before seen, all armed with either a double barrelled musket, a scimitar, or dagger. They were all of the same nation and tribe, for they shook hands at meeting, and seemed very friendly to each other, though they stopped and examined us; as if disposed to question the right of property.

It now appeared there was still some difficulty in deciding to whom each one of us belonged; for seizing hold of us, some dragged one way and some another, disputing very loudly and frequently drawing their weapons. It was however decided at last, after making us go different ways for the space of two or three hours with different men, that myself and the cook should remain, for the present, in the hands of our first master. They gave Clark to another, and Horace to a third. We had come near a couple of tents, and were certainly disgusting objects, being naked and almost skinless; this was sometime about noon, when three women came out who had not before seen us, and having satisfied their curiosity by gazing at us, they expressed their disgust and contempt by spitting at us as we went along, making their faces still more horrid by every possible contortion of their frightful features; this we afterwards found to be their constant practice wherever we went until after we got off the desart.

Towards evening a great number of the men having collected in a little valley, we were made to stop, and as our bodies were blistered and burnt to such a degree as to excite pity in the breasts of some of the men, they used means to have a tent cleared out for us to sit under. They then allowed all those of our crew present to sit under it, and, as may well be supposed, we were glad to meet one another again, miserable as we all were. Porter and Burns, who had been separated from me shortly after our capture, were still absent. A council was now held by the natives near the tent; they were about one hundred and fifty men, some very old, some middle aged, and some quite young. I soon found they were Mohamedans, and the proper names by which they frequently called each other were *Mohamed, Hamet, Seid, Sideullah, Abdallah,* &c. so that by these and the female names *Fatima, Ezimah, Sarah,* &c. I knew them to be Arabs or Moors.

The council were deliberating about us; and having talked the matter over a long time, seated on the ground, with their legs crossed under them in circles of from ten to twenty each, they afterwards arose and came to us. One of the old men then addressed me; he seemed to be very intelligent, and though he spoke a language I was unacquainted with, yet he explained himself in such a plain and distinct manner, sounding every letter full like the Spaniards, that with the help of signs I was able to understand his meaning. He wanted to know what country we belonged to; I told him we were English; and as I perceived the Spanish language was in sound more like that which they spoke than any other I knew, I used the phrase *Inglesis*; this seemed to please him, and he said "*O Fransah, O Spaniah;*" meaning "or Frenchmen or Spaniards;" I repeated we were English. He next wanted to know which point of the horizon we came from, and I pointed to the North.

They had seen our boat, which they called *Zooerga*, and wanted to know if we had come all the way in that boat: I told them no, and making a kind of coast, by heaping up sand, and forming the shape of a vessel, into which I stuck sticks for masts and bowsprit, &c. I gave him to understand that we had been in a large vessel, and wrecked on the coast by a strong wind; then by tearing down the masts and covering the vessel's form with sand, I signified to him that she was totally lost. Thirty or forty of the other Arabs were sitting around us, paying the strictest attention to every one of my words and gestures, and assisting the old man to comprehend me. He wished to know where we were going, and what cargo the vessel (which I now found they called *Sfenah*) had on board. I satisfied them in the best way I could, on this point, telling them that I had on board among other things, dollars: they wanted to know how many, and gave me a bowl to imitate the measure of them; this I did by filling it with stones and emptying it three times. They were much surprised at the quantity, and seemed to be dissatisfied that they had not got a share of them. They then wanted to know which way the vessel lay from us, and if we had seen any of the natives, whom they called Moslemin.

This I took to be what we call Mussulmen, or followers of the Mahommedan doctrine, and in this I was not mistaken. I then explained to them in what manner we had been treated by the inhabitants; that they had got all our clothing, except what we had on when they found us; all our money and provisions; massacred one of our number, and drove us out to sea. They then told me that they heard of the shipwreck of a vessel a great way North, and of the money &c. but that the crew were drowned in the *el M Bahar*; this was so near the Spanish (La Mar) for the sea, that I could not misunderstand it. Thus having obtained what information they wanted on those points, they next desired to know if I knew any thing about Marocksh; this sounded something like *Morocco*: I answered yes; next of the *Sooltaan* (the Sultan) to which instead of saying yes, I made signs of assent, for I found they did no more themselves, except by a cluck with the tongue.

They wanted me to tell his name, *Soo Mook*, but I could not understand them until they mentioned *Moolay Solimaan*; this I remembered to be the name of the present emperor of Morocco, as pronounced in Spanish, nearly. I gave them to understand that I knew him; had seen him with my eyes, and that he was a friend to me and to my nation. They next made me point out the direction towards his dominions, and having satisfied them that I knew which way his dominions lay from us, I tried to intimate to them, that if they would carry me there, I should be able to pay them for my ransom, and that of my crew. They shook their heads – it was a great distance, and nothing for camels to eat or drink on the way. My shipmates, who were with me, could not understand one syllable of what they said, or of their signs, and did not believe that I was able to communicate at all with them. Having finished their council, and talked the matter over among themselves, they separated, and our masters, taking each his slave, made off, every one his own way. Although from the conference I derived hopes of our getting ransomed, and imparted the same to my mates and crew, yet they all seemed to think I was deluding them with false expectations; nor could I convince them of the contrary. We took another leave of each other, when we parted for the night, having travelled this day, I should guess, about fifteen miles S. E.

I had been so fully occupied since noon, that no thought of victuals or drink had occurred to my mind. We had none of us ate or drank any thing this day, except about half a gill of milk each in the morning at daylight, and about half a pint of black beach water near the middle of the day. I was delivered over to an Arab named *Bickri*, and went with him near his tent, where he made me lie down on the ground like a camel. Near midnight he brought me a bowl containing about a quart of milk and water; its taste was delicious, and as my stomach had become contracted by long hunger and thirst, I considered it quite a plentiful draught. I had been shivering with cold for a long time, as I had no covering no skreen, and not even one of my shipmates to lie near me to keep one side warm at a time. I was so far exhausted by fatigues, privations, &c. that my misery could no longer keep me awake. I sank into a deep sleep, and during this sleep I was troubled in the first place with the most frightful dreams.

I thought I was naked and a slave, and dreamed over the principle incidents which had already actually passed. I then thought I was driven by Arabs with red hot iron spears pointed at me on every side, through the most dreadful fire I had ever imagined, for near a mile, naked and barefoot; the flames up to my eyes, scorched every part of my skin off, and wasted away my flesh by roasting, burning, and drying it off to the bones; my torments were inconceivable – I now thought I looked up towards heaven, and prayed to the Almighty to receive my spirit, and end my sufferings; I was still in the midst of the flames; a bright spot like an eye with rays around it, appeared above me in the firmament, with a point below it, reaching towards the N. E. – I thought if I went that way I should go right, and turned from the south to the N. E.; the fire soon subsided and I went on, still urged by them about me, with their spears pricking me from time to time over high sand hills and rocky steeps, my flesh dropping off in pieces as I went, – then descending a deep valley, I thought I saw green trees – flowering shrubs in blossom – cows feeding on green grass, with horses; sheep, and asses near me, and as I moved on, I discovered a brook of clear running water; my thirst being excessive, I dragged my mangled limbs to the brook, threw myself down, and drank my fill of the most delicious water. When my thirst was quenched, I rolled in the brook to cool my body, which seemed still consuming with heat; then thanked my God in my heart for his mercies.

My masters in the meantime kept hurrying me on in the way pointed out by the All-seeing eye, which was still visible in the heavens above my head, through crooked, thorny, and narrow paths, over high mountains and deep valleys – past hosts of armed men on horseback and on foot, and walled cities, until we met a tall young man dressed in the European and American manner, by the side of a brook, riding on a stately horse, who upon seeing me alighted, and rushing forward, wild with joy, caught me in his arms, and pressed me to his breast, calling me by the endearing name of brother, in my own language – I thought I fainted in his arms from excess of joy, and when I revived, found myself in a neat room, with a table set in the best manner before me, covered with the choicest meats, fruits, and wines, and my deliverer pressing me to eat and drink; but finding me too much overcome to partake of this refreshment, he said, "take courage, my dear friend, God has decreed that you shall again embrace your beloved wife and children." At this instant I was called by my master – I awoke, and found it was a dream.

Being daylight, (Sept. 12th) he ordered me to drive forward the camels; this I did for about an hour, but my feet were so much swelled, being lacerated by the cutting of the stones, which seemed as if they would penetrate to my heart at every step – I could not help stooping and crouching down nearly to the ground. In this situation, my first master Hamet observed me; he was going on the same course, S. E. riding on his camel; he came near my present master, and after talking with him a good while, he took off the blanket from his back and gave it to Bickri – then coming close to me, made signs for me to stop. He next made his camel lie down; then fixing a piece of skin over his back behind the saddle, and making its two ends fast to the girths to keep it from slipping off, he bade me mount on it, while he got on his saddle and steadied me with his hand until the camel rose. He then went on the same course as before, in company with three or four other men, well armed and mounted. The sun beat dreadfully hot upon my bare head and body, and it appeared to me that my head must soon split to pieces, as it was racking and cracking with excruciating pain. Though in this horrible distress, yet I still thought of my dream of the last night – "a drowning man will catch at a straw," says the proverb, and I can verily add, that the very faintest gleam of hope will keep alive the declining spirits of a man in the deepest distress and misery; for from the moment I began to reflect on what had passed through my mind when sleeping, I felt convinced that though this was nothing more than a dream, yet still remembering how narrowly and often I had escaped immediate apparent death, and believing it was through the peculiar interposition of divine Providence, I could not but believe that the

All-seeing eye was watching over my steps, and would in due time conduct me by his unerring wisdom, into paths that would lead to my deliverance, and restoration to my family.

I was never superstitious, nor ever did I believe in dreams or visions, as they are termed, or even remembered them, so as to relate any I may have had; but this dream made such an impression on my mind, that it was not possible for me to remove it from my memory – being now as fresh as at the moment I awoke after dreaming it, and I must add that when I afterwards saw Mr. *Willshire*, I knew him to be the same man I had seen in my sleep. He had a particular mark on his chin – wore a light coloured frock coat, had on a white hat, and rode the same horse. From that time I thought if I could once get to the empire of Morocco, I should be sure to find a friend to relieve me and my companions whose heart was already prepared for it by superior Power. My mind was thus employed until we came to a little valley where half a dozen tents were pitched; as soon as we saw them, Hamet made his camel kneel down, and me to dismount – he was met by several women and children, who seemed very glad to see him, and I soon found that they were his relations. He beckoned me to come towards his tent, for he lived there apparently with his mother, and brothers and sisters, but the woman and girls would not suffer me to approach them, driving me off with sticks, and throwing stones at me; but Hamet brought me a little sour milk and water in a bowl, which refreshed me considerably.

It was about two o'clock in the day, and I was forced to remain broiling in the sun without either tree, shrub, or any other shade to shield me from its scorching rays, until night, when Dick (the cook) came in with the camels. Hamet had kept Dick from the beginning, and made him drive the camels, but allowed him to sleep in one corner of the tent, and gave him for the few first days, as much milk as he could drink, once a day; and as he was a domestic slave, he managed to steal water, and sometimes sour milk when he was dry.

In the evening of this day I was joined by Hogan, and now found that he and myself had been purchased by Hamet that day, and that Horace belonged to an ill-looking old man, whose tent was pitched in company. This old villain came near me, and saluted me by the name of *Rais*, asking me the name of his boy; (Horace) I told him it was Horace, which after repeating a few times he learned so perfectly, that at every instant he was yelling out "*Hoh Rais*" for something or other. Hamet was of a much lighter colour than the other Arabs we were with, and I thought he was less cruel, but in this respect I found I was mistaken, for he made myself and Hogan lie on the ground in a place he chose, where the stones were very thick and baked into the ground so tight that we could not pull them out with our fingers, and we were forced to lie on their sharp points, though at a small distance, not more than fifty yards, was a spot of sand. This I made him understand, (pointing at the same time to my skinless flesh) but he signified to us that if we did not remain where he had ordered, we should get no milk when he milked the camels. I calculate we travelled this day about thirty miles.

Here then we staid, but not to sleep, until about the midnight hour, when Hamet came to us with our milk – It was pure and warm from the camels; and about a pint for each. The wind blew as is usual in the night, and on that part of the desart the air was extremely cold and damp; but its moisture on our bodies was as salt as the ocean. Having received our share of milk, when all was still in the tent, we stole to the sandy place, where we got a little sleep during the remaining part of the night. Horace's master would not permit him to come near me, nor me to approach him, making use of a stick, as well to enforce his commands in this particular, as to teach us to understand him in other respects.

At daylight (Sept. 13th) we were called on to proceed. The families struck their tents, and packed them on camels, together with all their stuff. They made us walk and keep up with the camels, though we were so stiff and sore all over that we could scarcely refrain from crying out at every step: such was our agony: – still pursuing our route to the S. E. In the course of the morning, I saw Mr. Williams; he was mounted on a camel, as we had all been the first day, and had been riding with the drove about three hours – I hobbled along towards him; his camel stopped, and I was enabled to take him by the hand – he was still entirely naked; his skin had been burned off; his whole body was so excessively inflamed and swelled, as well as his face, that I only knew him by his voice, which was very feeble. He told me he had been obliged to sleep naked in the open air every night; that his life was fast wasting away amidst the most dreadful torments; that he could not live one day more in such misery; that his mistress had taken pity on him; and anointed his body that morning with butter or grease, but said he, "I cannot live;" should you ever get clear from this dreadful place, and be restored to your country, tell my dear wife that my last breath was spent in prayers for her happiness: he could say no more; tears and sobs choked his utterance.

His master arrived at this time, and drove on his camel and I could only say to him, "God Almighty bless you," as I look a last look at him, and forgot, for a moment, while contemplating his extreme distress, my own misery. His camel was large, and moved forward with very heavy motions; as he went from me, I could see the inside of his legs and thighs – they hung in strings of torn and chafed flesh – the blood was trickling down the sides of the camel, and off his feet – "my God!" I cried, "suffer us not to live longer in such tortures."

I had stopped about fifteen minutes, and my master's camels had gained a great distance from me, so that I was obliged to run that I might come up with them. My mind was so shocked with the distresses of Mr. Williams, that I thought it would be impious for me to complain, though the sharp stones continued to enter my sore feet at every step. My master saw me and stopped the drove for me to come up; when I got near him he threatened me, shaking his stick over my head, to let me know what I had to expect if I dared to commit another fault. He then rode off, ordering me and Hogan to drive the camels on as fast as we could. About an hour afterwards he came near us, and beckoned to me to come to him, which I did. A tall old man, nearly as black as a negro, one of the most ill-looking and disgusting I had yet seen, soon joined my master, with two young men, whom I found afterwards were his sons – they were also joined by a number more on camels and well armed.

After some time bartering about me, I was given to the old man, whose features showed every sign of the deepest rooted malignity in his disposition. And is this my master, thought I? Great God! defend me from his cruelty! He began to go on – he was on foot; so were his two sons; but they walked faster than camels, and the old man kept snarling at me in the most surly manner, to make me keep up. I tried my very best, as I was extremely anxious to please him, if such a thing was possible, knowing the old adage of "the devil is good when he is pleased," was correct, when applied to human beings; but I could not go fast enough for him; so after he had growled and kept on a considerable time, finding I could not keep up with him, he came behind me and thrust me forward with hard blows repeatedly applied to my exposed back, with a stout stick he had in his hand. Smarting and staggering under my wound, I made the greatest efforts to get on, but one of his still more inhuman sons, (as I then thought him) gave me a double barrelled gun to carry, with his powder horn and other accoutrements: they felt very heavy, yet after I had taken them, the old man did not again strike me but went on towards the place where he meant to pitch his tent, leaving me to follow on as well as I could.

The face of the desart now appeared as smooth as the surface of the ocean, when unruffled by winds or tempests. Camels could be seen on every direction, as soon as they came above the horizon, so that there was no difficulty in knowing which way to go, and I took care to keep sight of my new master's drove, until I reached the valley, in which he had pitched his tent. I was broiling under the sun and tugging along, with my load, which weighed me down to the earth, and should have lain down despairing, had I not seen Mr. Williams in a still worse plight than myself.

Having come near the tent about four P. M. they took the load from me, and bid me lie down in the shade of the tent. I then begged for water, but could get none. The time now came on for prayers, and after the old man and his sons had performed this ceremony very devoutly, they went away. I was in so much pain, I could scarcely contain myself, and my thirst was more painful than it had yet been. I tried to soften the hearts of the women to get me a little water, but they only laughed and spit at me; and to increase my distresses as much as they could, drove me away from the shade of the tent, so that I was forced to remain in the scorching sun for the remainder of this long day.

THE GREAT CAVE RESCUE

James Massola

James Massola is an Australian author and journalist. In *The Great Cave Rescue* he tells the story of the Wild Boars football team trapped in the Tham Luang cave in Thailand in 2018, and the race against time to rescue them. The twelve boys – aged eleven to sixteen – and their coach were trapped after monsoon rains raised the water level in the cave system and blocked their exit. They survived in an air pocket for eighteen days whilst one of the most complex and risky rescue operations the world had ever seen was launched to get them out – and which very nearly went wrong on a number of occasions.

While the rescue operation outside the cave was taking its first, cautious steps towards the trapped boys, the Wild Boars were alone in the dark. On Saturday afternoon they had entered the cave brimming with confidence and full of adrenaline at the prospect of a new adventure into a hidden place most of them had never visited.

Some of the younger boys would later admit they were initially nervous, too, scared about the trek into uncharted territory but spurred on by their team mates, their lust to explore, and the promise of a new adventure. But in the deep dark of Tham Luang cave, the thrill of their expedition was soon replaced by the terrifying realisation that they were trapped, marooned on a sandy slope of Nern Norn Sao chamber, water lapping at their feet. This small sanctuary beneath hundreds of metres of rock was now their prison cell, from which there was no obvious escape.

It was almost always dark inside the cave, too; they did have torches, some of the few useful items they had brought with them, but they were careful to save the batteries and mostly left them switched off.

Although the boys kept track of time via Tee's wristwatch, the difference between day and night began to lose its meaning after a while—especially as they didn't have anything to eat, or much to do. To distract themselves from the growing hunger pains, the Boars busied themselves playing chequers, carefully drawing a board in the sand and using rocks as tokens. They did their best to not think about food, trying to distract themselves from the gnawing sensations in their stomachs, but it was hard.

Sometimes the boys would simply give in and cry.

Tee would later remember that while Titan was the youngest, at just 11 years old, it was 13-year-old Mark—one of the four stateless members trapped in the cave—who cried the most while they were lost. But 13-year-old Mick would attempt to buck up his team mates, telling them not to be discouraged or sad and imploring them to keep fighting—the sadness and the hopelessness would soon pass.

And time and again, Ek would lead the boys in meditation sessions. The former Buddhist monk guided them in the prayers he had spent years learning and reciting after the death of his parents. Fifteen years earlier, at the tender age of only 10, he had watched a fatal illness sweep through his family.

First, his 7-year-old brother had fallen ill and died, then his mother succumbed, and finally his father. Experiences like these fundamentally shape a person's character and help determine the course their life will take. Ek turned to his religion for comfort and guidance, entering a Buddhist seminary in Lamphun Province. Still grieving, the young boy found sanctuary as a novice monk in the monastery and took to his studies with aplomb, achieving Them Ek, the highest level of Dharma education for a Buddhist who is not a monk.

The prayer and meditation sessions with coach Ek were familiar to the boys—whenever they slept the night at his place, for example, they would pray together before bedtime. These sessions had two dividends. First, the meditation helped conserve the boys' energy as each day passed and their hungry bodies fed on the little fat they had left on their frames. Second, it helped keep the group calm and focused, tamping down their feelings of panic and dread as each hour passed with no sign of rescue, or a way out.

Tee was one of the most devoted prayers. His family are followers of the highly respected local monk, Kruba Boonchum, a revered holy man. At the invitation of Ek's grandmother, later in the eighteen-day ordeal, Boonchum would visit the cave to pray with the boys' families at the entrance to the Cave of the Sleeping Lady. He had an unusual connection to the cave, and his visit was rapturously received by locals.

Each night, ever more exhausted, the boys slept near the top of the gentle slope at Nern Norn Sao. The water temperature in the cave, as measured by the divers, hovered between 20 and 23°C, while the air temperature was estimated at around 20°C. The water temperature was higher than it would have been in a cave in Northern Europe or southern Australia, for example, but while it might sound like a pleasant temperature for a short swim or a dip—roughly what you might want the temperature in a swimming pool to be—it was still cause for concern. After hours of exposure to the water the human body begins to lose body heat. Even the divers in their wetsuits lost valuable body heat during the hours they were in the cave. Over time, both the temperature and the dampness in the cave started to gnaw away at the boys, who were clad only in T-shirts and shorts, chilling them to the bone and putting them at risk of hypothermia, which was also a real concern for the rescuers.

Kiang Kamluang, Tee's mother, would later share some of the details of an intimate discussion she had with her son soon after he had been rescued. Kiang says one of his abiding memories of his time in the cave was the bitter cold.

'I talked to Tee about his time in the cave, too, not much though. He said it was dark, and cold inside, as cold as a refrigerator's temperature. He said that he cried once for fear of being stuck in the cave for good. But deep down, he said that he believed there would be someone coming to rescue them. Ek told him to pray and meditate and he did so.'

Outside Tham Luang cave, the rain was still hammering down, hampering the rescue efforts of an ever-expanding team of rescuers, who could not even get 200 metres past the entrance without diving. And the water levels kept on rising.

THE LIFE AND ADVENTURES OF ALEXANDER SELKIRK

The Real Robinson Crusoe – A Narrative Founded on Facts

John Howell

Alexander Selkirk (1676–1721) was the real-life inspiration for Daniel Defoe's *Robinson Crusoe* (see page 41). A Scottish privateer, he was marooned on a deserted island in the South Pacific in 1704. He survived by hunting game and domesticating feral cats that kept him safe from nightly rat attacks. His improbable rescue four years later came when the twin privateering ships *Duke* and *Duchess*, under the command of William Dampier, anchored off the island's coast. Selkirk was overjoyed at the prospect of his rescue, and quickly returned to his former life of privateering.

John Howell (1788–1863) was a Scottish author, editor, bookbinder and inventor. Howell worked as a bookbinder and shopkeeper in Edinburgh. He also built – albeit with mixed results – a flying machine and a submarine.

For many days after being left alone, Selkirk was under such great dejection of mind, that he never tasted food until urged by extreme hunger, nor did he go to sleep until he could watch no longer, but sat with his eyes fixed in the direction where he had seen his shipmates depart, fondly hoping that they would return and free him from his misery. Thus he remained seated upon his chest, until darkness shut out every object from his sight. Then did he close his weary eyes, but not in sleep; for morning found him still anxiously hoping the return of the vessel.

When urged by hunger he fed upon seals, and such shell-fish as he could pick up along the shore. The reason of this was the aversion he felt to leave the beach, and the care he took to save his powder. Though seals and shellfish were but sorry fare, his greatest inconvenience was the want of salt and bread, which made him loathe his food until reconciled to it by long use.

It was now the beginning of October (1704,) which in those southern latitudes is the middle of spring, when nature appears in a thousand varieties of form and fragrance, quite unknown in northern climates; but the agitation of his mind, and the forlorn situation in which he was now placed, caused all its charms to be unregarded. There was present no one to partake of its sweets, – no companion to whom he could communicate the feelings of his mind. He had to contend for life in a mode quite strange to him, and it was with much difficulty that he sustained the horror of being alone in such a desolate place. If we think for a moment how disagreeable it is to most men to be left by themselves for a few days, we may form a faint idea of his situation, and how painful it must have been to him, a sailor, accustomed to enjoy and perform all the offices of life in the midst of bustle and fellowship. What greatly added to the horrors of his condition was the noise of the seals during the night, and the crashing made by falling trees and rocks among the heights; which last often broke the stillness of the scene with horrid sounds that were echoed from valley to valley.

So heart-sinking was his situation, that nothing but Divine Providence could have sustained him from falling into utter despair. Indeed, when we reflect upon the society Alexander Selkirk had for some time been associated with, and the habits he must have either acquired or become accustomed to, we cannot think it strange, that he often thought of putting a period to his sufferings by a violent death; so feeble is all the boasted firmness of the most daring courage when left for a length of time to solitude and its own unassisted resources.

It was in this trying situation, when his mind, deprived of all outward occupation, was turned back upon itself, that the whole advantages of that inestimable blessing, a religious education in his youth, was felt in its consoling influence when every other hope and comfort had fled. When misery had subdued the pride of his hard and stubborn heart, it was then he turned to that Divine Being of whom he had thought so little at an earlier period. Then the uninhabited wilderness of Juan Fernandez was turned into a smiling garden, and the darkness of that despair that had nearly overwhelmed him began to clear away. By slow degrees he became reconciled to his fate; and as winter approached, he saw the necessity of procuring some kind of shelter from the weather; for, even in that genial clime, frost is common during the night, and snow is sometimes found upon the ground in the morning.

The building of a hut was the first object that roused him to exertion; and his necessary absence from the shore gradually weaned his heart from that aim which had alone absorbed all his thoughts, and proved a secondary means of his obtaining that serenity of mind he afterwards enjoyed; but it was eighteen months before he became fully composed, or could be for one whole day absent from the beach, and from his usual hopeless watch for some vessel to relieve from his melancholy situation.

During his stay he built himself two huts with the wood of the pimento-tree, and thatched them with a species of grass that grows to the height of seven or eight feet upon the plains and smaller hills, and produces straw resembling that of oats. The one was much larger than the other, and situated near a spacious wood. This he made his sleeping-room, spreading the bedclothes he had brought on shore with him upon a frame of his own construction; and as these wore out, or were used for other purposes, he supplied their place with goats' skins. His pimento bed-room he used also as his chapel; for here he kept up that simple but beautiful form of family-worship which he had been accustomed to in his father's house. Soon after he left his bed, and before he commenced the duties of the day, he sung a psalm or part of one, then he read a portion of Scripture, and finished with devout prayer. In the evening, before he retired to rest, the same duties were performed. His devotions he repeated aloud to retain the use of speech, and for the satisfaction man feels in hearing the human voice even when it is only his own. The greater part of his days was spent in devotion; for he afterwards said, with tears in his eyes, that "He was a better Christian while in his solitude than ever he was before, and feared he would ever be again." To distinguish the Sabbath, he kept an exact account of the days of every week and month during the time he remained upon the island, although the method he adopted is not mentioned in any document we have procured.

The smaller hut, which Selkirk had erected at some distance from the other, was used by him as a kitchen, in which he dressed his victuals. The furniture was very scanty; but consisted of every convenience his island could afford. His most valuable article was the pot or kettle he had brought from the ship to boil his meat in; the spit was his own handiwork, made of such wood as grew upon the island; the rest was suitable to his rudely-constructed habitation. Around his dwelling browsed a parcel of goats, remarkably tame, which he had taken when young, and lamed, but so as not to injure their health, while he diminished their speed. These he kept as a store, in the event of sickness or any accident befalling him that might prevent him from catching others: his sole method of doing which was running them down by speed of foot. The pimento wood, which burns very bright and clear, served him both for fuel and candle. It gives out an agreeable perfume while burning.

He obtained fire, after the Indian method, by rubbing two pieces of pimento wood together until they ignited. This he did, as being ill able to spare any of his linen for tinder, time being of no value to him, and the labour rather an amusement. Having recovered his peace of mind, he began likewise to enjoy greater variety in his food, and was continually gaining some new acquisition to his store. The crawfish, many of which weighed eight or nine pounds, he broiled or boiled as his fancy led, seasoning it with pimento, (Jamaica pepper,) and at length came to relish his food without salt.

As a substitute for bread, he used the cabbage-palm, which abounded in the island, turnips, or their tops, and likewise a species of parsnip of good taste and flavour. He had also Sicilian radishes and watercresses, which he found in the neighbouring brooks, as well as many other vegetables peculiar to the country, which he ate with his fish or goats' flesh.

Having food in abundance, and the climate being healthy and pleasant, in about eighteen months he became reconciled to his situation. The time hung no longer heavy upon his hands. His devotions and frequent study of the Scriptures soothed and elevated his mind; and this, coupled with the vigour of his health, and a constantly serene sky and temperate air, rendered his life one continual feast: His feelings were now as joyful as they had before been melancholy. He took delight in every thing around him; ornamented the hut in which he lay with fragrant branches, cut from a spacious wood on the side of which it was situated, and thereby formed a delicious bower, fanned with continual breezes soft and balmy as poets describe which made his repose, after the fatigues of the chase, equal to the most exquisite sensual pleasures.

Yet happy and contented as he became, there were minor cares that broke in upon his pleasing solitude, as it were to place his situation on a level with that of other human beings; for man is doomed to care while he inhabits this mortal tenement. During the early part of his residence he was much annoyed by multitudes of rats, which gnawed his feet and other parts of his body as he slept during the night. To remedy this disagreeable annoyance, he caught and tamed, after much exertion and patient perseverance, some of the cats that ran wild on the island. These new friends soon put the rats to flight, and became themselves the companions of his leisure hours. He amused himself by teaching them to dance, and do a number of antic feats. They bred so fast, too, under his fostering hand, that they lay upon his bed and upon the floor in great numbers; and, although thus freed from his former troublesome visitors, yet, so strangely are we formed, that when one care is removed another takes its place. These very protectors became a source of great uneasiness to him; for the idea haunted his mind and made him at times melancholy, that, after his death, as there would be no one to bury his remains, or to supply the cats with food, his body must be devoured by the very animals which he at present nourished for his convenience.

COURAGE UNDER FIRE

ON THE FRONTLINE

Marie Colvin

Marie Colvin (1956–2012) was an American war correspondent who worked for *The Sunday Times* for nearly thirty years. She covered conflicts across the world, including the Middle East, Kosovo, Chechnya and East Timor – where she is credited with saving the lives of 1,500 women and children from a besieged compound. For the last decade of her life, she sported a signature eyepatch after she was hit by a rocket-propelled grenade fired by a Sri Lankan soldier and lost the sight in her eye. Colvin was killed in 2012 during a military confrontation between the Syrian government and a major rebel stronghold in the city of Homs. Following a civil action lawsuit from her family, a judge later found the Syrian government guilty of facilitating Colvin's assassination.

'The shot hit me. Blood poured from my eye – I felt a profound sadness that I was going to die'

22 April 2001

Last Monday, award-winning reporter Marie Colvin was attacked as she returned from a rare interview with Tamil Tiger leaders in Sri Lanka. From her hospital bed in New York she writes about her escape.

It was the most difficult decision of my life. I was lying in an open field with a clump of tall weeds on a slight rise for cover. The moon had not yet risen and the night was pitch black. Every five minutes or so a flare, fired from the nearby Sri Lankan army base, seemed to expose every blade of grass. Advancing soldiers intermittently raked the field with automatic weapons fire. They had to be as scared as I was.

I just wanted to lie still and wait for it all to go away. I thought I would not mind lying here for hours. I noticed little things. One of my trouser legs had come up to my knee and that meant my white calf might draw attention in the dark.

There were three options. I could crawl away. But if one soldier had night-vision goggles—didn't even the poorest armies these days? I would be the only moving object on the field and the would jungle be with shot.

If I was not spotted, I would still be alone in the jungle with no shoes. If I lay here until the soldiers stumbled on me, they would shoot first. If I shouted and identified myself as a journalist they might shoot anyway. There was no fourth option.

It was 10pm, on the forward defence line of the Sri Lankan army at Parayanlankulam, about 3½ miles from the Madhu road junction. I thought of how I came to be here. There didn't seem to be any one moment when it all went wrong.

A week earlier I had secretly entered the Vanni, a 2,000-mile area of northern Sri Lanka that has been the refuge of the rebel Tamil Tigers since the government captured the Jaffna peninsula in 1995. The Sri Lankan government bans journalists from travelling there.

The ban meant journalists could not talk to the leadership of the Liberation Tigers for Tamil Eelam (LTTE), even though the government was involved in negotiations with them through a Norwegian envoy to begin peace talks. The only news of the problems with those negotiations came from the government.

More important, the ban prevented any reporting on the plight of the 500,000 Tamil civilians, 340,000 of them refugees, bottled up in the Vanni suffering under an economic embargo that the government denied existed.

I had travelled through villages in the Vanni and found an unreported humanitarian crisis—people starving, international aid agencies banned from distributing food, no mains electricity, no telephone service, few medicines, no fuel for cars, water pumps or lighting.

I had filed the story and had been trying to leave the Vanni to return to the government-controlled south for three days. This involved walking 30 miles a night through jungle and the knee-deep water and mud of marsh and rice paddies—only to end up sleeping on the same straw mat, on the same dirt floor, in the same mud hut. Even the bugs were starting to look familiar.

Each night I tried to leave, guided by local Tamils. But each time they decided it would be too risky to cross army lines. On Sunday night we came within 50 yards of the border between the two sides. The leader studied the army post we were supposed to slip past. Suddenly, he made a somersaulting motion with his hands and started walking back.

There was no argument; we used hand signals and observed silence until several miles from the army line.

'My mistake,' he said. 'Military alert. Too dangerous.' I watched my guides' tireless brown feet, clad only in black rubber flip-flops, pad unceasingly ahead of me until we reached the base house near the Catholic church at Madhu, home to 10,000 refugees living in tents and huts. At dawn I collapsed into sleep.

Monday night was meant to be third time lucky. As the sun slipped below the horizon, I sat with my guides under a banyan tree, looking out over a silvery lake, waiting for dark in a rare moment of peace and beauty.

We were a motley group. The civilians with me were dressed in a collection of shorts and sarongs. An emaciated old man carried a string shopping bag with two bottles of Pepsi, our only drink. A teenager kept trying the little English that he had learnt in St Patrick's college in Jaffna before the army overran the town and killed his father.

'Tonight, you will be in my father's house sipping milky tea,' the leader of the group said. He was the only one who was armed; he carried an old rifle to protect us from wild boar or elephants. 'We are going a way that is safe and secure.'

The plan was to reach his family's house in the government-controlled area that night. The Tamils would return before dawn and I would get the morning bus to Vavuniya. I had a last cigarette as the sun went down; there would be no smoking, talking or even coughing as we walked the next seven miles to sanctuary.

We trekked single file along narrow jungle trails, sometimes pushing our way through thickets of thorn trees; we waded waist-deep round the edges of a lake, eyeing the lights of an army base on its far edge. They are dotted along the Mannar–Vavuniya road that marks the border we would have to cross. At about 8pm, we crept through dark scrub about half a mile from the road, then waited crouching in a marsh—letting the mosquitoes bite because slapping could alert a soldier—while the group's leader scouted ahead.

I took off my shoes to walk more quietly. At a signal from the leader we followed him to the road. Half-crouched, we negotiated our way through barbed wire on both sides of the road and seemed safely across.

We were running through the last dark field for the line of jungle ahead when the silence was broken by the thunder of automatic weapons fire about 100 yards to the right.

I dived down and began crawling, belly on the ground, for some cover. For a few minutes, someone was crawling on top of me—protection or panic, I don't know. Then I was alone, behind weeds.

A tree was 10 yards away, but it seemed too far. The shooting went on and on. Flashes and light came from an army post nobody had seen.

The shooting stopped and dark and quiet descended. There had not been a sound from my side. I could not tell where anyone was. The only sound was the occasional bellow of a cow which had been hit.

I had a few mad moments of thinking it was over, I had survived. But I knew this was not true. We had been spotted. The army would think this was a Tamil Tiger patrol and would come after us. They would be scared and trigger-happy.

The reputation of the two forces is that the army has superior manpower and weapons, the rebels superior manoeuvrability and commitment. The advantage was to the army this night. I was lying in a field with a decision to make: run for it, lie still or shout.

I lay there for half an hour under the penetrating glare of the flares. I turned my face to the earth when one came drifting down directly above me, worried that my white skin would reveal my hiding place.

Bursts of gunfire began across the road about half a mile away. The search and destroy patrols had come out. I heard soldiers on the road, talking and laughing. One fired a burst from an automatic weapon that scythed down the weeds in front of me and left me covered in green shoots.

If I didn't yell now, they would stumble on me and shoot. I began to shout. 'Journalist! Journalist! American! USA!'

A soldier sighted on the sound and fired. This army was not taking prisoners.

The shot hit me with an impact that stunned me with pain, noise and a sense of defeat. I thought I had been shot in the eye. Blood was pouring from my eye and mouth onto the dirt. I felt a profound sadness that I was going to die.

Then I thought it was taking an awful long time to die if I was really shot in the head (it was actually shrapnel), so I started yelling again. 'English! Anyone speak English?'

There were more shots, but they seemed half-hearted, and lots of hysterical shouting from the soldiers. This was bad. They were as scared as I was. I did not really care because it seemed that I would die anyway, so I just kept shouting.

Searching for a word that non-English speakers might recognise, I fixed on doctor and shouted over and over that I needed one. Finally a voice screamed in English: 'Stand up, stand up!' He fired a few more shots for emphasis.

I stood up slowly, hands in the air, saying, 'Don't shoot, American,' and whatever else I could think of just so that they would keep hearing a foreign voice.

'Take off your jacket,' came the voice. I dropped my blue jacket and stood straight up, hands in the air. Blood poured down my face so I could not see much. Someone yelled, 'Walk to the road.' I stumbled forward.

Every time I fell, feeling faint, they would shout hysterically, afraid that I was pulling some trick, and I would struggle up again. I made it up the incline to the road and was shoved to the ground, flat on my back and kicked by shouting soldiers. A bright light shone in my face. I could not see any of my captors.

I am not sure how long I lay there on my back. I was searched for weapons, then told to walk at gunpoint, prodded by the weapons. The soldiers live in fear of women suicide bombers carrying explosives underneath their clothing.

The LTTE has a ruthless reputation as a result of the activities of the Black Tigers, an elite unit for suicide missions, who have bombed government buildings, assassinated a president and killed Rajiv Gandhi, the Indian leader.

I thought the soldiers were taking me somewhere to shoot me. I remember thinking that they were all scared and that I should act scared and vulnerable. I reached the limit. I could not walk any more and fell, telling them to get a doctor. They relented and put my arms round the shoulders of two men. But they pummelled me again when my hand fell and a soldier shouted that I was going for his grenade.

The nightmare seemed endless. We reached some lighted space outdoors and I was thrown on the ground on my back. A bright light again was in my face and questions shouted in Sinhalese and broken English. Someone ripped open my shirt and pulled it off. They shouted for my weapons. I kept saying, 'Journalist, I need a doctor.'

An officer, or someone in authority, came on the scene and the questions changed into an interrogation: 'Where did you get your training? How many people were with you? Where is your vehicle? Ah, you say you are American but you have no vehicle?'

Things were calming down and my sense of the ridiculous returned. If I had a vehicle, why would I be lying in a field on a dark night?

'Admit that you came to kill us,' he said. 'At least admit that your side fired grenades first. This is true, is it not?'

I said, 'No sir, there was no fire until your soldiers shot at us.'

Then began an endless series of journeys. I was put in the back of a truck and driven, bouncing over potholes, hyperventilating because I could not seem to breathe. I thought it was shock; later, I found that my lungs had been bruised by the shock of the grenade and were filling with fluid.

Someone kind was in the truck. He kept telling me in English: 'We are taking you for medical treatment, you are going to be okay.' I fixed on his voice, and he held my head up so that I could breathe.

At the first hospital I was taken to, the military hospital in Vavuniya, shrapnel was taken out of my head, shoulders and chest. I realised I could not see out of my left eye and I think the doctors panicked the soldiers into some sanity.

I was put in the back of another truck and driven for an hour to Army Victory hospital in Anuradhapura, where an x-ray revealed shrapnel in my eye. A truck took me to a third hospital, the Anuradhapura general hospital. I was never out of army custody.

The doctors seemed scared for me and I asked one to call the American embassy. But an army surgeon kept insisting that they should operate immediately.

'You are going to lose your eye anyway. I can operate now,' he said. I fended him off, but he would appear again, sharpening his imaginary knives, asking to operate.

Telephone calls were being made. It seemed that my request to be taken to Colombo was going to more senior people. At one point I heard a conversation in English. A soldier was saying, 'No, she cannot come to the phone. What is your message?'

I heard him trying to pronounce the name of Steve Holgate, the personable public affairs officer of the American embassy. I shouted: 'Give me the phone.' I had a huge sense of relief that someone knew where I was.

At dawn, someone in the Sri Lankan army hierarchy relented. I was put aboard a military helicopter and flown to Colombo. At the eye hospital, I was shoved on a stretcher against a wall in the crowded emergency room surrounded by hostile soldiers.

Miraculously, Holgate showed up moments later, clipboard in hand, and simply told the soldiers he was taking me into the custody of the American embassy. It was like the moment in a classic Wild West movie when the quiet guy faces down the armed and dangerous gang. I was safe.

Why do I cover wars? I have been asked this often in the past week. It is a difficult question to answer. I did not set out to be a war correspondent. It has always seemed to me that what I write about is humanity in extremis, pushed to the unendurable, and that it is important to tell people what really happens in wars—declared and undeclared.

War has changed remarkably little over the centuries. Do not believe the nice clean videos where Gameboy jets hit Nintendo tanks framed in a satisfying and sanitary 'X'. War is not clean. War is about those who are killed, limbs severed, dirt and rock and flesh torn alike by hot metal. It is terror. It is mothers, fathers, sons and daughters bereft and inconsolable. It is about traumatised children.

My job is to bear witness. I have never been interested in knowing what make of plane had just bombed a village or whether the artillery that fired at it was 120mm or 155mm.

War is also about propaganda. Both sides try to obscure the truth. Foreign journalists arriving in Sri Lanka are told in a government handout that parents in the rebel area keep their children home from school because the Tamil Tigers are recruiting them for service. But the parents told me they keep their children at home because they are hungry and faint in the classroom and do not have money for school supplies.

The Sri Lankan government reacted with anger to my presence in the Tamil-held area of the Vanni. It made no apologies for what had happened to me. I had no permission to go there, the government said, therefore I must have had a 'secret agenda'.

I had no secret agenda. I had a journalist's agenda. I went to the rebel-held areas because talking to the Tamil Tigers and writing about a previously unreported humanitarian crisis are important issues.

I am not going to hang up my flak jacket as a result of this incident. I have been flown to New York, where doctors are going to operate on my injured eye in about a week's time. They have told me it is unlikely I will regain much use of it as a piece of shrapnel went straight through the middle. All I can hope for is a bit of peripheral vision.

Friends have been telephoning to point out how many famous people are blind in one eye. They seem to do fine with only one eye, so I'm not worried. But what I want most, as soon as I get out of hospital, is a vodka martini and a cigarette.

GRATITUDE FROM TAMILS WORLDWIDE

Sri Lanka, April 2001

Colvin was the first foreign reporter in six years to enter Sri Lanka's dangerous northern Vanni region where the Tamil Tigers are waging a civil war against government forces. She went to interview the leaders and was ambushed while trying to walk out of the area last Monday. After reports of Colvin's ordeal last week *The Sunday Times* received many letters from Tamils all over the world offering their support.

This one was typical:

'We Tamils are so proud about your brave foreign correspondent Marie Colvin ... We [are] all aware of the risk she undertook and we appreciate her visit to [the] Vanni area of north Sri Lanka for bringing the news to the outside world. We are deeply concerned about her health and wish her to get well soon. Thank you.'

Signed Elan Ramalingham, Carleton University, Ottawa, Canada.

WAR DOCTOR

David Nott

David Nott (born 1956) is a Welsh vascular surgeon known for volunteering his services in war and crisis zones. Described by *The Times* as the 'Indiana Jones of surgery', Nott has volunteered in places as remote and hostile as Darfur, Iraq, Afghanistan, Syria, Bosnia and Sierra Leone. During the Battle of Aleppo in 2013, he was the only westerner – and one of the few doctors – in the city. His memoir *War Doctor: Surgery on the Front Line* (2019) chronicles his experiences carrying out life-saving operations and field surgery in the most challenging conditions. In 2015, he established the David Nott Foundation dedicated to training medical personnel who operate in disaster zones.

The woman's husband had apparently been making a bomb in his kitchen when it had detonated prematurely. The whole house was destroyed, the bomb-maker killed and his wife rushed to us with a fragment injury to her lower left leg. She was haemorrhaging significantly from the wound, which required a tourniquet to be placed immediately on the thigh.

The anaesthetist took a quick blood sample and put it through our very basic haemoglobinometer, a device which measures the red cell count in blood. It confirmed that she had a haemoglobin of 4 grams per litre – the normal amount of haemoglobin in our bodies, the stuff that carries oxygen in the blood, is between 12 and 15g/L. It was clear she had lost a great deal of blood. He quickly established her blood group and then went to get a pint of fresh blood of the right type from our dwindling supplies. Then, on the other arm, he set up a saline drip to replace some of the fluid that she had lost.

All this happened on the operating table in the dining room. The sister in charge set up the trolley with sterile drapes and instruments as the patient was given general anaesthetic. It was impossible to assess the wound properly as there was arterial bleeding, most likely from the superficial femoral artery in the leg. There was a large dressing on the top, which was acting as a local compression. I scrubbed up and prepared to operate.

One of the Syrian assistants, who didn't speak much English, was helping to lift the leg. As I prepped the limb with iodine I asked the helper to take off the pressure dressing. The bleeding by this time had stopped, and there was a large clot overlying the wound. With the patient now draped and prepped, I started the procedure by making an incision below the tourniquet, high on the leg, so that I could get a clamp on the artery before exploring the wound. After gaining proximal control of the blood vessel I then went down to have a proper look. I tentatively put my finger into a large hole just above her knee joint, and felt an object in there which I assumed was a piece of metal – a fragment from the bomb, or maybe a bit of her house.

In this kind of scenario it is always important to go very carefully, putting your finger into the wound slowly and cautiously because there may be fractured bone which can be as sharp as shards of glass – the last thing you want is a needlestick injury without knowing the blood status of the patient. In this environment there was perhaps less concern about HIV or hepatitis, but it is a common mistake not to assume the worst.

Probing gently with my finger, it didn't appear to be the usual jagged piece of metal or fragment but a smooth, cylindrical object. Very carefully I grabbed it with my fingers and pulled it out. I held it up to examine it, and the Syrian helper who was with me took one look and went pale. He obviously knew what I was holding, and blurted out, '*Mufajir!*' before turning tail and leaving the room.

The anaesthetist and I looked at each other. Was I holding some sort of bomb? In that instant, I froze as I wondered what on earth I should do next. It went extremely quiet – all I could hear was the soft hiss of the ventilator pumping oxygen into the patient's lungs. The anaesthetist shuffled away, moving across to the comer of the room behind one of the cupboards. By now my hands were shaking, I was in danger of dropping whatever it was, and I realized I had to do something. I decided to take a deep breath and walk out of the operating theatre as carefully and slowly as I could. I needed the anaesthetist to open the door for me and jerked my head in its direction to show him what I wanted, hardly daring to speak. He said to wait, as he was sure somebody was going to come very shortly – thankfully he was right, and as I deliberated for a few more seconds the door opened and in came the Syrian helper with a bucket of water. He put the bucket on the floor next to me and he and the anaesthetist ran to the safety of the next room. With my heart pounding, I carefully put the object into the bottom of the bucket, feeling the cold water seeping into the sleeve of my green scrubs, and very gingerly took it outside.

Mufajir means 'detonator'. It was hard to tell if it was live or not. I was told later that it probably would not have killed me, but it would most likely have blown off my hand – not the end of my life, maybe, but certainly the end of my career, and at the time the two were much the same thing.

It wasn't the last time I had a run-in with homespun explosives. Most of the fragmentation wounds from bombs that we were receiving were from the effects of amateur bomb-makers. Several times throughout the mission, we would receive young girls and boys at the hospital who had lost one or both of their hands. Some had severe facial injuries as well, and, even more pitifully, some had dreadful eye damage which rendered them blind. Many times I would go to the ward and hear the sobbing of parents holding their five- or six-year-old, who would never see them again or touch them with their fingers. It was utterly heartbreaking.

Every so often the mission team would change as one of the doctors or nurses would leave and another would arrive. Sometimes it was a relief for the team that people did move on, because there is no doubt that in a very high-stress environment some people do crack with the pressure. You can tell they are not as happy as they were when they arrived; some become a lot quieter, others more vocal. Some even become slightly irrational.

One of the senior nurses who was running the ward was, in my view, beginning to lose sight of the reason we were all there. Pete had done a really difficult operation on a fragmentation injury and was now anxious that there might be a leak from one of his intestinal anastomoses (joins) within the young man's abdomen. In this situation the abdomen becomes extremely painful and its lining, the peritoneum, becomes inflamed. This results in diffuse peritonitis, causing an involuntary spasm of the muscles, which become board-like in nature, like a tabletop. This patient had obvious signs that this had occurred, and Pete and I discussed taking the patient back to theatre. I told the nurse that this was what I was going to do. But she disagreed – she insisted that we needed to take the patient in an ambulance across the border into Turkey for him to have a proper operation there. She started to shout at us in front of other patients and became quite hysterical, demanding we did not operate.

I have no qualms at all about working in a team. From the most junior person up to the most senior, we all have opinions on how best to treat the patient. No one is infallible; it is quite possible to miss a clinical sign or mistake an observation, which will sometimes be brought to light by the most junior person in the room.

When someone actively makes a judgement that is out of their remit, however, then it becomes a problem. Of course, it would have been possible to have called an ambulance to transport the patient to the border, but the journey from Atmeh to the border was slow and protracted, and the pain and distress would have been too great. We discussed the case with Alpha's project manager and in the end took the man back to the operating theatre and fixed him up.

I wasn't sorry to see the back of that nurse, but sometimes it is sad to see people go, as with our German emergency physician, who had been excellent. However, I was delighted that his replacement was to be my previous house officer at Charing Cross Hospital, Natalie Roberts, who has subsequently become a shining light in humanitarian work and with MSF. As our new emergency physician, she would look after our makeshift emergency department in the living room. It looked out onto a patio, on which there was room for about six beds. Here she would examine patients before surgery, and this was also our overspill area if we had a mass-casualty event and needed more space.

Only a few days after her arrival there was another explosion in a house near the hospital. Again, it was the home of an amateur who was making bombs in cylindrical metal containers about the same size as an old-fashioned bottle of washing-up liquid. At 10.30 in the morning we heard the horns start blaring, getting louder and louder as they approached.

An entire family of eight arrived: mother, father and six children. They had been praying together in the back yard of their house. As the father prostrated himself, a device had fallen out of his pocket and detonated. All eight family members were brought through the outside door of the hospital onto the patio, where we began to try to make sense of a tangled mess of body parts.

All six of the children were dead, as well as their mother. The father, the bomb-maker, had severe fragmentation injuries to his arms and legs. Some were fairly superficial but others had gone much deeper. We had no X-ray machine available and all the diagnostics had to be made through clinical findings alone – basically, having a good look and making a decision yourself.

Natalie was examining the patient. A small, highly charged fragment with a lot of energy had burst into his chest and caused internal bleeding. This can come from a fractured rib, from an intercostal vessel which lies between the ribs, from the lung, or – worst-case scenario – from the heart. She correctly diagnosed that the man had a significant amount of blood within the chest cavity, which needed removing using a chest drain. This is a tube that is placed through the armpit area via the ribs into the chest cavity – it allows blood and air that is not within the lung to be evacuated, thereby letting the lungs expand and so improve breathing. Natalie seemed to be getting on with this very well, and was also cleverly taking blood out of the chest using a special filter and placing it in another bag to infuse back into the patient.

When anybody comes in to an emergency department the first thing we do is called the primary survey – looking at the basic parameters that can save lives, which we code as CABCDE.

The first 'C' stands for catastrophic haemorrhage. If the patient is bleeding significantly then the first thing to do is to try to stop it, either by direct pressure or using a tourniquet, a tight binding with a belt or strap between the heart and the open wound. Sometimes the source of the bleeding isn't obvious, though – it might be coming from somewhere hidden like the chest, the abdomen or pelvis. If your clinical examination has ruled out the arms and legs as the source, both front and back, then if the patient is shocked and pale the bleeding must be coming from somewhere we cannot apply pressure. We then make a decision as to whether or not the patient needs to be taken directly to the operating theatre.

The next check is 'A' for airway, to confirm there is no obstruction to the flow of oxygen to the lungs. Then come the lungs themselves, 'B' for breathing, to make sure the lungs can expand properly and provide oxygen to the body. The lungs could be bruised, or be compressed with blood or air, requiring a chest drain.

The second 'C' stands for general circulation and a quick assessment of the blood pressure by feeling the pulses, while 'D' is disability, the most likely being neurological disability caused by head injuries. 'E' stands for exposure – checking the rest of the patient and also understanding the temperature of the environment. To do this, it is important to remove as many of the patient's clothes as possible, so they can be examined front and back.

As I watched Natalie putting in the chest drain I noticed that, although this man's trousers had been partially cut off with a pair of scissors, there was something sticking out of his pocket. It was a cylindrical object, about the size of an aerosol can. Suddenly, it fell out of his pocket onto the floor. It was another bomb.

We all watched as if in slow motion as the object bounced on the hard tiled floor of the emergency department and spun in the air on the rebound. Astonishingly, one of the male Syrian interpreters in attendance displayed lightning reactions and did the best David Beckham kick I have ever seen, booting the bomb straight through the open patio door. It didn't explode, presumably because there was no detonator inside, but it was a lesson for us all. We should have checked the patient more thoroughly first – but we had all been so shocked at the carnage wrought on his family that we just got down to work.

War zones are completely different from routine life at home, and it is very easy to become blinkered and not to take care of oneself, such is the focus on the patients. But extra precautions must be taken at all times to try to limit the risk of getting caught out. You have to have a different bit of your head switched on – you can't take your normal NHS mindset to a war. In more well-established hospitals everyone would be checked for weapons with a hand-held metal detector before being allowed in – in one of the hospitals where I had worked in northern Pakistan a few months before going to Syria, even the volunteer doctors and nurses were checked when they arrived. But we didn't have one in Atmeh, and had to deal with whoever came in, whoever they were.

I was reminded of a bomb-maker I'd had to work on while I was in Pakistan: a Taliban fighter who'd been injured making improvised explosive devices (IEDs) for use against coalition forces across the border in Afghanistan. He came in and I operated on him, and saved his life. I am often asked how I can square my humanitarian work with saving the life of someone who might go on to make something that kills British soldiers or innocent civilians. It's a valid point, of course, and every war surgeon has to wrestle with the conundrum at some stage. But actually it's quite simple: I don't get to choose who I work on. I can only try to intervene to save the life of the person in front of me who is in desperate need of help. Usually I have no idea who they are or what they have done until afterwards anyway – but even if I did know, nothing would change. I rationalize it by thinking, *Well, maybe that Taliban guy or this ISIS fighter will find out his life was saved by a Western, Christian doctor, and that might make him change his outlook.* Some people may consider this is naive, but that's how it is.

AN ACCOUNT FROM MEMORY OF THE RETREAT FROM CABOOL

Dr William Brydon

Dr William Brydon (1811–1873) was an assistant surgeon in the British East India Company Army who, in 1842, fled the military disaster that claimed the lives of nearly 16,000 British troops near Gandamak, Afghanistan. He was one of only a handful of survivors of that deadly encounter and recounted the travails of his escape in a now infamous field report of his epic journey of tragedy and survival. This account of the retreat from Kabul was published as an appendix in **Sir George St Patrick Lawrence**'s memoirs of his time serving in the region with the British Indian Army.

IT was given out to the troops in Cabool on January 5th that arrangements had been completed for a retreat to Hindoostan. Such of the sick and wounded as were unable to march were left under medical charge of Drs. Berwick and Campbell and Lieutenant Evans, her Majesty's 44th, in command. Captains Drummond and Walsh, and Lieutenants J. Conolly, Webb, Warburton, and Airey were placed as hostages in the hands of Mahomed Zeman Khan. The sick were lodged in Timour Shah's fort, the hostages with the new king, Shah Shoojah.

January 6.— The retreat commenced this day about 9 A.M., a temporary bridge having been thrown across the Cabool river for the passage of the infantry; the guns, cavalry, baggage, &c. fording the river, which was about two feet deep. The 5th Native Infantry formed the advance guard, with a hundred sappers, and the guns of the mountain train under Brigadier Anquetil; next came the main body under Brigadier Shelton, followed by the baggage, in rear of which came the 6th regiment Shah Shoojah's force, to which I belonged. We did not leave cantonments till nearly dusk, immediately followed by the rearguard, composed of the 5th Light Company 54th Native Infantry, two Horse Artillery guns, and part of her Majesty's 44th. All the guns, excepting those of the Horse Artillery and Mountain Train, were left in the cantonments, together with a large quantity of magazine stores. I saw Lieutenant Hardyman, 5th Light Company, killed by a shot from the enemy, who had entered the cantonments and fired upon us from the walls immediately the troops left the gates, and in a short time set fire to all the buildings. Each officer carried away what little baggage he could on his own animals. I having six ponies, reserved a favourite chesnut for myself, and mounted all my servants; but before I left cantonments I saw my best horse, which was carrying my boxes of clothing in charge of my groom, seized by the enemy, who dashed in among the baggage and carried off a great quantity of public and private property without resistance between cantonments and the Seeah Sung hill, at which place the two guns with the rearguard were abandoned. We moved so slowly that it was near midnight before we reached our encamping ground across the Loghur river, a distance of only about five miles; but even this short march, with the darkness and deep snow, was too much for the poor native women and children; many lay down and perished, and the cries of others who had lost their way were truly heartrending. On arriving at our ground the scene was sad indeed: the snow several inches deep; only one small tent, saved from the general pillage, was pitched and occupied by the General, and as many more as could find room in it, and the troops lying in the snow, or sitting round fires mostly fed by portions of their own clothing. I rolled myself in my sheepskin cloak, and taking my pony's bridle in my hand, lay down among the men of my regiment and slept.

January 7.— When I awoke in the morning I found the troops preparing to march, so I called to the natives who had been lying near me to get up, which only a very few were able to do; some of them actually laughed at me for urging them, and pointed to their feet, which looked like charred logs of wood. Poor fellows, they were frost-bitten, and had to be left behind. This day, advanced guard 54th Native Infantry, rearguard her Majesty's 44th, and mountain train, our march was to Bootkhak, a distance again of about five miles, and the whole road from Cabool was at this time a dense mass of people. In this march, as in the former, the loss of property was immense, and towards the end of it there was some sharp fighting, in which Lieutenant Shaw, 54th Native Infantry, had his thigh fractured by a shot. The guns of the mountain train were carried off by the enemy, and either two or three of the Horse Artillery were spiked and abandoned. I saw a gallant but fruitless charge made by Lieutenant Macartney to try and recover a horse of his own that was being carried off; he with difficulty rejoined the troops. Few had anything to eat except those who like myself followed the Affghan custom of carrying a bag of parched grain and raisins at their saddle-bow. There was rather less snow than on our former encamping ground, and the night was passed like the former, the pony only getting a few bites of grass from the ground and having the saddle-girths slackened. Up to this time I had seen nothing more of my servants or their ponies. We were tricked into encamping here instead of at once pushing on through the pass by Akhbar Khan, who sent to say he must make arrangements with the chiefs to let us through, but in truth that he might have time to get the hills well manned before we entered the pass, and some of his horsemen who accompanied us are said to have called to the enemy in Persian 'to spare,' and in Pushtoo, which the hillmen speak and few Europeans understood, they exhorted them 'to slay the Kaffirs.'

January 8.— This morning we moved through the Khoord Cabool Pass with great loss of life and property; the heights were in possession of the enemy, who poured down an incessant fire on our column. Great numbers were killed, among them Captain Paton, and Lieutenant Sturt of the Engineers, by a shot in the groin; many more were wounded—of them were Lady Sale and Captain Troup, and when we arrived at our ground at Khoord Cabool Captain Anderson's eldest child was missing. All the stragglers in the rear were cut up by the enemy, who descended as soon as the main body had passed. The pass is about three-and-a-half miles long, with a small stream running through it, which had to be crossed about thirteen times in transit; it was covered with ice, but not strong enough to bear a man, and I had an awkward accident about the middle. I suppose I had forgotten to tighten the girths, and my saddle turned, tumbling me into the water at a place where the enemy's fire was particularly sharp. I managed to get close under a rock to right the saddle, and both I and my horse escaped untouched.

January 9.— We halted at Khoord Cabool this day, scarcely annoyed by the enemy at all; and I was glad to find three of my servants out of the five that had been with me in Cabool, the bheestie (water carrier), khidmutgar (table attendant), and sweeper; those missing being the syce (groom), carried off with the horse the first march, and a tailor. To the bheestie I had given a bag of barley to carry on his pony, and he had filled his mussuck or water-skin with pistachio nuts, so the animals got a feed of corn, and myself and the servants made a hearty meal on the nuts. At this place by treacherous promises Akhhar Khan induced the General to make over to his care all the married officers and their families and some wounded officers.

January 10.— Resumed our march about 10 A.M. and were immediately attacked, and numbers fell in a small rocky gorge, Tareekee Tungee, just outside the camp, before we ascended the Huft-Kotul (Seven Hills). At the moment of starting the Sergeant-major of our regiment brought some eggs and a bottle of wine, which he got from a box left by some of the ladies, and gave them to Captain Hawkins, who divided them with Captain Marshall, Lieutenant Bird, and myself: the eggs were not boiled, but frozen quite hard, and the wine also to the consistency of honey, a little only in the centre being fluid. This was a terrible march, the fire of the enemy incessant, and numbers of officers and men, not knowing where they were going from snow-blindness, were cut up. I led Mr. Baness, a Greek merchant, a great part of the way over the high ground, and often felt so blind myself that from time to time I applied a handful of snow to my eyes, and recommended others to do so, as it gave great relief. Descending towards Tazeen the whiteness was not so intense, and as the sun got low the blindness went off; but the fire of the enemy increased; and as they were able to get very close to us in the passes, which we now again entered, it was very destructive, many of the enemy indeed running across our line and cutting down any they could. So terrible had been the effects of the cold and exposure upon the native troops that they were unable to resist the attacks of the enemy, who all the way pressed hard on our flanks and rear, and on arriving at the valley of Tazeen towards evening a mere handful remained of the native regiments which had left Cabool. Amongst those wounded I saw Drs. Duff and Cardew placed on a gun-carriage, but they did not long survive. Dr. Bryce, just on entering the pass, was shot through the chest, and when dying handed over his will to Captain Marshall. At Tazeen when we halted we found there were killed or missing, besides those named, five officers of the 5th Native Infantry—Swayne, Miles, Dees, Alexander, and Warne; one of the 37th, Ewart of the 54th, and Dr. Magrath. After a short rest, when it was quite dark, our diminished party moved on, setting fire to the carriage and leaving the last of the Horse Artillery guns on the ground, and a great number of the remaining camp followers who would not, or many could not, move further. We passed pretty quietly through the Tazeen valley, but were again fired upon when we entered the hills, especially near a small river where the enemy bad a large sungah on a hill, and where Brigadier Shelton unfortunately halted his men to return their fire, thereby losing severely.

January 11.— We marched all night, the cavalry advanced guard, and arrived at Kutta Sung this morning, having sustained more loss from the enemy firing on us from the heights all the way; we halted a very short time, being only a valley surrounded by hills. Here Captain Dudgin, of her Majesty's 44th, gave me a biscuit and a sardine: very acceptable indeed they were, as I had eaten nothing since the frozen eggs the morning before. As we pushed on to Jugdulluck we found wild liquorice, chewing the roots of which refreshed me much. We reached it about noon, still hard pressed by the enemy from the hills, and close to the camp ground Lieutenant Fortye, of her Majesty's 44th, was killed. Shortly after our arrival the General, Brigadier Shelton, and Captain Johnson went to Akhbar Khan, and were detained as hostages for the march of the troops from Jelialabad and safe-conduct for us.

We were encamped in an old enclosure, which however gave us very little shelter from the enemy, who had possession of the surrounding hills, from which they never ceased firing. Captain Skinner of the commissariat was killed, and many men and officers wounded: among the latter, whose wounds I dressed, were Captains W. Grant and Marshall; the former had his jaw shattered, and Marshall a shot through the chest. Poor Marshall, Bird, and I had dined on a portion of a fat Arab charger that had been shot, slices of which we grilled over a fire of brushwood, during which operation Marshall lost the only bit of rock salt we had, and looked on it as a very bad omen. I suppose it fell in the snow and melted. Shortly afterwards he volunteered to lead a party to try and drive the enemy from a hill from which they specially annoyed us, and he was wounded before he had gone many yards. We did not move from Jugdulluck till about an hour after dark on January 12th, when an order was given to march, owing I believe to a note being received from General Elphinstone, telling us to push on at all hazards, as treachery was suspected. Owing to this unexpected move on our part, the abattis and other impediments which had been thrown across the Jugdulluck pass were undefended by the enemy, who nevertheless pressed on our rear and cut up great numbers. The confusion now was terrible; all discipline was at an end. I started, leading poor Marshall's horse, who was too weak to guide it; and Blair, Bott, and another wounded officer, all of the 5th Light Cavalry, were on a camel close to me. We had not gone far in the dark before I found myself surrounded, and at this moment my khidmutgar rushed up to me saying he was wounded, had lost his pony, and begged me to take him up. I had not time to do so before I was pulled off my horse and knocked down by a blow on the head from an Affghan knife, which must have killed me had I not had a Blackwood's Magazine in my forage cap. As it was, a piece of bone about the size of a wafer was cut from my skull and I was nearly stunned, but managed to rise on my knee; and seeing that a second blow was coming, I met it with the edge of my sword, and I suppose cut off some of my assailant's fingers, as his knife fell to the ground. He bolted one way and I the other, minus my horse, cap, and one of my shoes; the khidmutgar was dead; those who had been with me I never saw again. I regained our troops, scrambled over a barricade of trees made across the pass, where the confusion was awful, and I got a severe blow on the shoulder from a fellow who rushed down from the hill and across the road. Here I picked up a 44th cap. Soon after I came up with Captain Hopkins, and alongside his horse I walked for some distance, holding on to his stirrup. In this way I overtook a Hindoostanee mochee (saddler) of the Shah's cavalry, who told me he was dying, and begged me to take his pony, or some one else would. I tried to encourage him, but he fell off, carrying away one of the stirrups, and I found the poor fellow was shot through the chest and dead. I then mounted his pony, and riding towards the front I met Brigadier Anquetil, who asked me how they were getting on in the rear. When I told him he rode back, and was never seen again. The men were running up the hills on the sides of the road and on in front, throwing away their arms, &c. to lighten themselves, and could not be kept together or controlled, so Captain Bellew, the Quartermaster-General of the force, assembled all he could find mounted, and formed us into an advanced guard, about forty; only one trooper of the 5th Cavalry could be found. We moved steadily on, fired at from the hills, which were blazing with watchfires, but we in the dark, so few hit as we passed along, only once losing our way in the darkness, but regaining it again after a short détour. During this night we all suffered intense thirst, and my shoeless foot, which was unfortunately in my only stirrup, felt as if it were being burnt, so I was glad to find in a bag with other things at my saddle-bow a piece of list which I wound round the iron. At daybreak we found ourselves at Gundumuck, and had lost all trace of those in our rear. Here a dispute arose as to the road, there being two, one over the hills and the other through the Neemlah valley, in which was a large village. I having been encamped for about three months in quiet times in this neighbourhood, and knowing both roads, recommended the former, while a Mr. Bailis, a clerk in one of the public offices, said the other was safest and best. So our small party split, half going each way; those by the valley were attacked and killed by the villagers, except Mr. Bailis, who was taken prisoner (and afterwards I heard taken to Peshawur, where he died of fever). We proceeded over the hills without seeing a single individual. Shortly before entering on the plain we rested a little while in a small grassy glen, and let our horses have a bite of such grass as it was. My saddle was a wooden one, of a kind then common in the Punjaub, with a high peak in front, and from it I now removed the bags, which were very heavy, containing saddler's tools, bullets, a pistol (which none of the bullets would fit), a chain and spike for picketing a horse, &c. All these I threw away except the pistol, which I put in my pocket. On starting after half-an-hour's rest, our party consisted of Captains Bellew, Hopkins, and Collyer, Lieutenants Bird, Steer, and Grey, Dr. Harper, Sergeant Freil, and five or six other European soldiers. We shortly came in sight of the village of Futtehabad, on the plain, and about fifteen miles from Jellalabad; all here seemed quiet, and Captain Bellew said he would go and enquire into the state of the country. In a short time he came back and told us that all was quiet, and that if we would wait he would bring us some bread promised by the headman of the village. In about a quarter of an hour he returned again, and said he was afraid he had ruined us, as from the village, which was on a mound, he could see cavalry coming up on all sides, and he had no doubt that some signal had been given to gather them while we were kept waiting; probably a red flag we saw ourselves. He begged us to keep together and move slowly on, the armed villagers following, and calling to him to come back, as they were friends. Captain Bellew did so, and was immediately killed. At the same time the villagers fired on us and the cavalry charged among us. One fellow cut at me, which I guarded, and he then cut down poor Bird, and it became a case of utter rout, out of which all that got clear were Captains Hopkins and Collyer, Dr. Harper, Lieutenant Steer, and myself. The three former being well mounted left Lieutenant Steer and myself behind, telling us they would soon send us help. After riding on a short distance Lieutenant Steer said he could go no further, as both he and his horse were done (the latter was bleeding from the mouth and nostrils); he would hide till night in one of the many caves we knew were in the hills about half a mile to the right of the road. I tried hard to persuade him to push on, as the plain was sprinkled with people tending sheep and cattle, who must see him, but he would not. So I proceeded alone for a short distance unmolested; then I saw a party of about twenty men drawn up in my road, who when I came near began picking up large stones, with which the plain abounded, so I with difficulty put my pony into a gallop, and taking the bridle in my teeth, cut right and left with my sword as I went through them. They could not reach me with their knives, and I was only hit by one or two stones. A little further on I was met by another similar party, who I tried to pass as I did the former, but was obliged to prick the poor pony with the point of my sword before I could get him into a gallop. Of this party one man on a mound over the road had a gun, which he fired close down upon me, and a large stone broke my sword, leaving about six inches in the handle; but I got clear of them, and then found that the shot had hit the poor pony, wounding him in the loins, and he could now hardly carry me, but I moved on very slowly and saw some fine horsemen, dressed in red. Supposing they were some of our Irregular Cavalry I made towards them, but found they were Affghans, and that they were leading off Captain Collyer's horse. So I tried to get away, but my pony could hardly move, and they sent one of their party after me, who made a cut at me which I guarding with the bit of my sword, it fell from the hilt. He passed me, but turned and rode at me again; this time, just as he was striking, I threw the handle of the sword at his head, in swerving to avoid which he only cut me over the back of the left hand. Feeling it disabled I stretched down the right to pick up the bridle. I suppose my foe thought it was for a pistol, for he turned at once and made off as quick as he could. I then felt for the pistol I had put in my pocket, but it was gone, and I was unarmed, wounded, and on a poor animal I feared could not carry me to Jellalabad, though it was now in sight. Suddenly all energy seemed to forsake me; I became nervous and frightened at shadows, and I really think I should have fallen from my saddle but for the peak of it, and some of our people from the fort coming to my assistance, among the first of whom was Captain Sinclair, of her Majesty's 13th, whose servant gave me one of his own shoes to cover my foot. I was taken to the sappers' mess, my wounds dressed by Dr. Forsyth, and after a good dinner, with great thankfulness enjoyed the luxury of a sound sleep, most hospitably lodged by Captain Francis Cunningham, whose quarters I shared during the whole siege. On examination I found that besides my head and left hand I had a slight sword wound on the left knee, and a ball had gone through my trousers a little higher up, slightly grazing the skin, but how and when these happened I know not. The poor pony directly it was put into a stable lay down and never rose again. Immediately on my telling how things were, General Sale despatched a party to scour the plain in the hopes of picking up any stragglers, but they only found the bodies of Captains Hopkins and Collyer and Dr. Harper. The second night after my arrival poor Mr. Baness was brought in by a fakeer from near Futtrabad, to whom he had done a kindness on a former occasion; in marching up to Cabool he saved the fakeer's mulberry grove from being destroyed by some Sikh soldiers. Mr. Baness only lived one day, being perfectly exhausted by his sufferings from cold and hunger.

TAKEN ON TRUST

Terry Waite

Terry Waite (born 1939) is a humanitarian who was kidnapped and held captive in Lebanon for nearly five years. In 1987, Waite travelled to Beirut as a special envoy for the then Archbishop of Canterbury to try and negotiate the release of Western hostages. There, he was kidnapped and subject to solitary confinement, torture and terrifying mock execution. He now works for many charities including Hostage UK which he founded to support the families of hostages.

I am sinking low. My body aches from repeated coughing which I am quite unable to control. My lungs are congested, and for hours on end I fight for breath. To inhale and expel air requires tremendous energy, and much of the time I am exhausted. There are long periods when it is impossible for me to lie down, and so I remain sitting, cross-legged, attempting to control the spasms which attack me with great frequency. The guards express their concern at times but do little to help. A few days ago they brought me a bottle of cough mixture, which made me vomit.

In my mind I continue writing, and at a deeper level I struggle with my inner contradictions. Somehow I have to do two things. I must, if I am going to make any psychological progress whatsoever, continue my interior dialogue. At the same time I have to bolster myself to an almost ridiculous degree. If I let my inner confidence collapse, I will die. If I don't face my doubts and conflicts I will never progress. I must somehow cultivate a defiant arrogance and yet not believe it. How I long for the love and company of my family. How I yearn with a childish, selfish longing to be understood and cared for. I am frightened. Frightened that, in growing up, my identity may slip away. These are hard days, but I must hold fast to my resolutions: no regrets, no sentimentality, no self-pity. Ridiculous, pompous statements, yet they have some meaning for me.

'You have problem?' the voice of the old man addresses me. I remember him from early in my captivity. He beat the soles of my feet until they were so swollen I couldn't walk unaided.

Yes, my friend, yes. I have a problem, and you have a problem, a massive problem. You are a blind stupid coward. My head swims, I have no breath, I can hardly speak, but I force a response.

'I cannot breathe properly.'

'Why?'

I do my best to explain about the fumes and the necessity for fresh air. He speaks to the guard in Arabic. The door closes. I have a problem, you have a problem, we have a problem, they have a problem. Problem, problem, problem, problem. My head sinks with exhaustion. Death would surely be preferable to this non-life. There is no escape, even sleep is denied me now. Yes, I have a problem, old man, but I'm not finished yet, not yet.

MY LAND AND MY PEOPLE

Dalai Lama

Tenzin Gyatso, 14th Dalai Lama (born 1935) is the spiritual leader, and the exiled political leader of Tibet. Within the tradition of Tibetan Buddhism, he is considered to be a *Bodhisattva*, an enlightened being on the path to Buddhahood. In 1959, following a failed uprising against the occupying forces of the Chinese People's Liberation Army, the Dalai Lama escaped to the northern regions of India where he set up a Tibetan government-in-exile in Dharamsala. In 2011, he retired as the political leader of Tibet in order to establish in his stead a more democratic government. A recipient of the Nobel Peace Prize, the Dalai Lama continues to advocate on behalf of the people of Tibet and their unique cultural identity. He remains actively involved in discussions at the intersection of science and religion, diplomacy, environmentalism, nonviolent protest and economics. His efforts have been actively opposed and his legacy targeted by the Chinese Communist Party throughout its continued occupation of the Tibetan region.

So we went on, and our journey was even sadder than before. I was young and strong, but some of my older companions were beginning to feel the effects of the long journey we had already made so quickly, and the most formidable part of it was still ahead.

But before we left Chongay, I had a most welcome chance to meet some more of the leaders of the Khampas and talk to them frankly. In spite of my beliefs, I very much admired their courage and their determination to carry on the grim battle they had started for our freedom, culture, and religion. I thanked them for their strength and bravery, and also, more personally, for the protection they had given me. I asked them not to be annoyed at the government proclamations which had described them as reactionaries and bandits, and told them exactly how the Chinese had dictated these and why we had felt compelled to issue them. By then, I could not in honesty advise them to avoid violence. In order to fight, they had sacrificed their homes and all the comforts and benefits of a peaceful life. Now they could see no alternative but to go on fighting, and I had none to offer. I only asked them not to use violence except in defending their position in the mountains. And I was able to warn them that our reports from Lhasa showed that the Chinese were planning to attack the part of the mountains where they were camped, so that as soon as they felt they could leave me they should go back to their defenses.

Many monk and lay officials were also waiting to see me there, but we had to cut our time short, because there was still every chance that the Chinese would come round by another route and cut us off before we were near enough to the frontier to have any way of retreat.

For another week we pressed on through the heart of the high mountains, and every day of that week we had to cross a pass. The snow had thawed in the valleys and on the lower passes, and the track was often slippery and muddy. But it sometimes led us to heights of over 19,000 feet, where the snow and ice were still lying. The route had been made by the tough mountain traders of old, and its stages were hard and long for people more used to the sheltered life of Lhasa.

The first night after Chongay we stayed in the monastery of which my Senior Tutor was high lama. The next day, we had to cross the Yarto Tag-la, which is particularly high and steep and difficult. Some of the ponies could not climb the track, and I and most of the others of the party had to dismount and lead them. But on the top, to our surprise, we found a fertile plateau where yaks were grazing, and a lake, thinly covered with ice, with a very high snow mountain to the north of it.

That night, after eleven hours of hard riding and climbing, very tired and saddle-sore, we reached a little place called E-Chhudhogyang. This place is so well known in Tibet that there is a proverb about it: "It's better to be born an animal in a place where there is grass and water than to be born in E-Chhudhogyang." It is a desolate spot, with a population of only four or five hundred. It is always in the grip of storms and gales, and the soil round it is ash-colored sand. So there was no cultivation at all, nor any grass or firewood. The people were almost destitute but happy, for they knew how to look poverty in the face. They welcomed us with open arms, and we were very grateful to share their humble homes. Some of my companions, who could not possibly be packed into the houses, were thankful for shelter in the cattle sheds.

By then we had been traveling for a week. Of course, I knew my friends abroad would be very concerned about the disorder in Lhasa and anxious to know what had happened to me; but none of us, as we struggled on, had any idea that our escape was in headlines in newspapers all over the world, and that people as far away as Europe and America were waiting with interest, and I hope I may say with sympathy, to hear whether I was safe. But even if we had known, there would have been nothing we could do about it, because of course we had no means of communicating with anyone.

But at that stage in our journey we heard that the Chinese had announced that they had dissolved our government, and that was something on which we could take action. Of course, they had no authority, legal or otherwise, to dissolve the government. In fact, by making that announcement they were breaking the only one of their promises in the Seventeen-Point Agreement which had so far nominally remained unbroken: the promise not to alter my status. But now that the announcement had been made, we thought there was some danger that Tibetans in isolated districts might think it had been made with my acquiescence. It seemed to us that the best thing to do was not simply to deny it, but to create a new temporary government; and we decided to do that as soon as we came to Lhuntse Dzong.

That was another two stages on. We left E-Chhudhogyang at five in the morning to face another high pass, the Tag-la, which took us up again, leading our ponies, above the snowline. That was another hard day, ten hours of slippery, stony track before we reached a place called Shopanup; but there we all happily found accommodation, more comfortable than the night before, in a monastery.

The next day, we reached Lhuntse Dzong. A *dzong* is a fort, and Lhuntse Dzong is a vast building on a rock, rather like a smaller Potala. The officials and leaders of the place came out to receive us on the track as we approached, and as we came nearer we were welcomed by an orchestra of monks playing religious music from the terrace of the *dzong*. More than a thousand people, burning incense, stood at the sides of the road. We went into the *dzong* for a ceremony of thanksgiving for our safety.

After that we held the religious ceremony to consecrate the founding of the new temporary government. Monks, lay officials, village headmen, and many other people joined us on the second floor of the *dzong*, bearing the scriptures and appropriate emblems. I received from the monks the traditional emblems of authority, and the lamas who were present, including my tutors, chanted the enthronement prayers. When the religious ceremony was finished, we went down to the floor below, where my ministers and the local leaders were assembled. A proclamation of the establishment of the temporary government was read out to this assembly, and I formally signed copies of it to be sent to various places all over Tibet. The ceremonies ended with the staging of the Droshey, the Dance of Propitious Fortune.

We spent three hours in these pleasant ceremonies, and all of us quite forgot our immediate troubles and tragedies. We felt we were doing something positive for the future of Tibet.

SEVEN YEARS IN TIBET

Heinrich Harrer

Heinrich Harrer (1912–2006) was an Austrian mountaineer, sportsman and author who was arrested by officials of the British Raj in Karachi at the outbreak of the Second World War. He escaped detention alongside his friend and fellow mountaineer, Peter Aufschnaiter, and embarked on a perilous journey towards the Tibetan capital of Lhasa. After running a harrowing gamut of natural catastrophes and run-ins with Tibetan robbers, border patrols and the British authorities, the pair finally arrived in Lhasa. Over the next seven years, Harrer served as a guest and later an official of the Tibetan government and formed a close personal friendship with His Holiness the 14th Dalai Lama. After the Chinese invasion of Tibet in 1950, Harrer fled Lhasa to avoid imprisonment by the People's Liberation Army and made his way back towards India, later meeting up with the Dalai Lama's convoy. Harrer returned to Austria in 1952 and published his bestselling memoir *Seven Years in Tibet*, which recalls his experiences and subsequent years as one of the last Western travellers to see the city before the Chinese invasion and occupation of the country.

Escape

"You made a daring escape. I am sorry, I have to give you twenty-eight days," said the English colonel on our return to the camp. I had enjoyed thirty-eight days of freedom and now had to pass twenty-eight in solitary confinement. It was the regular penalty for breaking out. However, as the English took a sporting view of our bold attempt, I was treated with less than the usual rigour.

When I had finished my spell of punishment I heard that Marchese had endured the same fate in another part of the camp. Later on, we found opportunities to talk over our experiences. Marchese promised to help me in my next attempt to get loose, but would not think of joining me. Without losing any time I at once began to make new maps and to draw conclusions from the experience of my previous flight. I felt convinced that my next attempt would succeed and was determined this time to go alone.

Busy with my preparations I found the winter passing swiftly and by the time the next "escape season" came round I was well equipped. This time I wanted to start earlier, so as to get through the village of Nelang while it was still uninhabited. I had not counted on getting back the kit I had left with the Indian so I supplied myself afresh with the things I most needed. A touching proof of comradeship was the generosity of my companions who, hard up as many of them were, spent their money freely in contributing to my outfit.

I was not the only P.O.W. who wanted to get away. My two best friends, Rolf Magener and Heins von Have, were also engaged in preparing to escape. Both spoke fluent English, and they aimed to work their way through India to the Burma front. Von Have had already escaped two years before with a companion and had almost reached Burma, but was caught just before the frontier. During a second attempt his friend had a fatal accident. Three or four other internees, it was said, planned to escape. Finally the whole seven of us got together and decided to make a simultaneous break-out on the grounds that successive individual attempts increased the vigilance of the guards, and made it more and more difficult to get away as time went on. If the mass escape succeeded each of us, once out of the camp, could follow his own route. Peter Aufschnaiter, who this time had as his partner Bruno Treipel from Salzburg, and two fellows from Berlin, Hans Kopp and Sattler, wished, like me, to escape to Tibet.

Our zero hour was fixed as 2 p.m. on April 29th, 1944. Our plan was to disguise ourselves as a barbed-wire repairing squad. Such working parties were a normal sight. The reason for them was that white ants were always busy eating away the numerous posts which supported the wire and these had to be continually renewed. Working parties consisted of Indians with an English overseer.

At the appointed time we met in a little hut in the neighbourhood of one of the least closely watched wire corridors. Here make-up experts from the camp transformed us in a trice into Indians. Have and Magener got English officers' uniforms. We "Indians" had our heads shaved and put on turbans. Serious as the situation was, we could not help laughing when we looked at one another. We looked like masqueraders bound for a carnival. Two of us carried a ladder, which had been conveyed the night before to an unguarded spot in the wire fencing. We had also wangled a long roll of barbed wire and hung it on a post. Our belongings were stowed away under our white robes and in bundles, which did not look odd as Indians always carry things around with them. Our two "British officers" behaved very realistically. They carried rolls with blue-prints under their arms and swung their swagger-canes.

We had already made a breach in the fence through which we now slipped one after another into the unguarded passage which separated the different sections of the camp. From here it was about 300 yards to the main gate. We attracted no attention and only stopped once, when the sergeant-major rode by the main gate on his bicycle. Our "officers" chose that moment to inspect the wire closely. After that we passed out through the gate without causing the guards to bat an eyelid. It was comforting to see them saluting smartly and obviously suspicious of nobody. Our seventh man, Sattler, who had left his hut rather late, arrived after us. His face was black and he was swinging a tarpot energetically. The sentries let him through and he only caught up with us outside the gate.

As soon as we were out of sight of the guards we vanished into the bush and got rid of our disguises. Under our Indian robes we wore khaki, our normal dress when on outings. In a few words we bade each other goodbye. Have, Magener and I ran for a few miles together and then our ways parted. I chose the same route as last time, and travelled as fast as I could in order to put as long a distance as possible between me and the camp by the next morning. This time I was determined not to depart from my resolve to travel only by night and lie up by day. No! this time I was not going to take any risks. My four comrades, for whom Tibet was also the objective, moved in a party and had the nerve to use the main road which led via Mussoorie into the valley of the Ganges. I found this too risky and followed my former route through the Jumna and Aglar valleys. During the first night I must have waded through the Aglar forty times. All the same, when morning came I lay up in exactly the same place which it had taken me four days to reach in the previous year. Happy to be free, I felt satisfied with my performance, though I was covered with scratches and bruises and owing to my heavy load had walked through the soles of a pair of new tennis shoes in a single night.

I chose my first day-camp between two boulders in the riverbed, but I had hardly unpacked my things when a company of apes appeared. They caught sight of me and began to pelt me with clods. Distracted by their noise I failed to observe a body of thirty Indians who came running up the river-bed. I only noticed them when they had approached dangerously near to my hiding-place. I still do not know if they were fishermen or persons in search of us fugitives. In any case I could hardly believe that they had not spotted me for they were within a few yards of me as they ran by. I breathed again, but took this for a warning and remained in my shelter till evening, not moving till darkness had fallen. I followed the Aglar the whole night long and made good progress. My next camp provided no excitements, and I was able to refresh myself with a good sleep. Towards evening I grew impatient and broke camp rather too early. I had only been walking for a few hundred yards, when I ran into an Indian woman at a water-hole. She screamed with fright, let her waterjar fall and ran towards the nearby houses. I was no less frightened than she was and dashed from the track into a gulley. Here I had to climb steeply and though I knew I was going in the right direction my diversion represented a painful detour that put me back by several hours. I had to climb Nag Tibba, a mountain over 10,000 feet high, which in its upper regions is completely deserted and thickly covered with forest.

As I was loping along in the grey of dawn I found myself facing my first leopard. My heart nearly stopped beating as I was completely defenceless. My only weapon was a long knife which the camp blacksmith had made expressly for me. I carried it sheathed in a stick. The leopard sat on a thick branch fifteen feet or more above the ground, ready to spring. I thought like lightning what was the best thing to do, then, masking my fear, I walked steadily on my way. Nothing happened, but for a long time I had a peculiar feeling in my back.

Up to now I had been following the ridge of Nag Tibba and now at last I tumbled on to the road again. I had not gone far when I got another surprise. In the middle of the track lay some men—snoring! They were Peter Aufschnaiter and his three companions. I shook them awake and we all betook ourselves to a sheltered spot where we recounted what had befallen us on the trek. We were all in excellent shape and were convinced that we should get through to Tibet. After passing the day in the company of my friends I found it hard to go on alone in the evening, but I remained true to my resolve. The same night I reached the Ganges. I had been five days on the run.

At Uttar Kashi, the temple town which I have mentioned in connection with my first escape, I had to run for my life. I had just passed a house when two men came out and started running after me. I fled headlong through fields and scrub down to the Ganges and there hid myself between two great blocks of boulders. All was quiet and it was clear that I had escaped from my pursuers; but only after a longish time did I dare to come out into the bright moonlight. It was a pleasure for me at this stage to travel along a familiar route, and my happiness at such speedy progress made me forget the heavy load I was carrying. It is true that my feet were very sore, but they seemed to recover during my daytime rest. I often slept for ten hours at a stretch.

At length I came to the farmhouse of my Indian friend to whom I had in the previous year entrusted my money and effects. It was now May and we had agreed that he was to expect me at midnight any day during the month. I purposely did not walk straight into the house, and before doing anything else I hid my rucksack, as betrayal was not beyond the bounds of possibility.

The moon shone full upon the farmhouse, so I hid myself in the darkness of the stable and twice softly called my friend's name; The door was flung open and out rushed my friend, threw himself on the ground, and kissed my feet. Tears of joy flowed down his cheeks. He led me to a room lying apart from the house, on the door of which an enormous lock was hanging. Here he lit a pine-wood torch and opened a wooden chest. Inside were all my things carefully sewn up in cotton bags. Deeply touched by his loyalty, I unpacked everything and gave him a reward. You can imagine that I enjoyed the food which he then set before me. I asked him to get me provisions and a woollen blanket before the following night. He promised to do this and in addition made me a present of a pair of hand-woven woollen drawers and a shawl.

The next day I slept in a neighbouring wood and came in the evening to fetch my things. My friend gave me a hearty meal and accompanied me for a part of my way. He insisted on carrying some of my baggage, undernourished as he was and hardly able to keep pace with me. I soon sent him back and after the friendliest parting found myself alone again.

It may have been a little after midnight when I ran into a bear standing on his hind legs in the middle of my path, growling at me. At this point the sound of the swiftly running waters of the Ganges was so loud that we had neither of us heard the other's approach. Pointing my primitive spear at his heart, I backed step by step so as to keep my eyes fixed on him. Round the first bend of the track I hurriedly lit a fire, and pulling out a burning stick, I brandished it in front of me and moved forward to meet my enemy. But coming round the corner I found the road clear and the bear gone. Tibetan peasants told me later that bears are only aggressive by day. At night they are afraid to attack.

I had already been on the march for ten days when I reached the village of Nelang, where last year destiny had wrecked my hopes. This time I was a month earlier and the village was still uninhabited. But what was my delight to find there my four comrades from the camp! They had overtaken me when I was staying with my Indian friend. We took up our quarters in an open house and slept the whole night through. Sattler unfortunately had an attack of mountain sickness; he felt wretched and declared himself unequal to further efforts. He decided to return, but promised not to surrender till two days were up, so as not to endanger our escape. Kopp, who in the previous year had penetrated into Tibet by this route in company with the wrestler Kramer, joined me as a partner.

It took us seven long days' marching, however, before we finally reached the pass which forms the frontier between India and Tibet. Our delay was due to a bad miscalculation. After leaving Tirpani, a well-known caravan camp, we followed the most easterly of three valleys, but eventually had to admit. that we had lost our way. In order to find our bearings Aufschnaiter and I climbed to the top of a mountain from which we expected a good view of the country on the other side. From here we saw Tibet for the first time, but were far too tired to enjoy the prospect and at an altitude of nearly 18,000 feet we suffered from lack of oxygen. To our great disappointment we decided that we must return to Tirpani. There we found that the pass we were bound for lay almost within a stone's throw of us. Our error had cost us three days and caused us the greatest discouragement. We had to cut our rations and felt the utmost anxiety about our capacity to hold out until we had reached the next inhabited place.

From Tirpani our way sloped gently upward by green pastures, through which one of the baby Ganges streams flowed. This brook, which we had known a week back as a raging, deafening torrent racing down the valley, now wound gently through the grasslands. In a few weeks the whole country would be green and the numerous camping-places, recognisable from their fire-blackened stones, made us picture to ourselves the caravans which cross the passes from India into Tibet in the summer season. A troop of mountain sheep passed in front of us. Lightfooted as chamois, they soon vanished from our sight without having noticed us. Alas! our stomachs regretted them. It would have been grand to see one of them stewing in our cooking-pot, thereby giving us a chance, for once, to eat our fill.

At the foot of the pass we camped in India for the last time. Instead of the hearty meat dinner we had been dreaming of, we baked skimpy cakes with the last of our flour mixed with water and laid on hot stones. It was bitterly cold and our only protection against the icy mountain wind that stormed through the valley was a stone wall.

At last on May 17th, 1944, we stood at the top of the Tsangchokla pass. We knew from our maps that our altitude was 17,200 feet.

So here we were on the frontier between India and Tibet, so long the object of our wishful dreams.

Here we enjoyed for the first time a sense of security, for we knew that no Englishman could arrest us here. We did not know how the Tibetans would treat us but as our country was not at war with Tibet we hoped confidently for a hospitable welcome.

On the top of the pass were heaps of stones and prayer-flags dedicated to their gods by pious Buddhists. It was very cold, but we took a long rest and considered our situation. We had almost no knowledge of the language and very little money. Above all we were near starvation and must find human habitation as soon as possible. But as far as we could see there were only empty mountain heights and deserted valleys. Our maps showed only vaguely the presence of villages in this region. Our final objective, as I have already mentioned, was the Japanese lines—thousands of miles away. The route we planned to follow led first to the holy mountain of Kailas and thence along the course of the Brahmaputra till at last it would bring us to Eastern Tibet. Kopp, who had been in Tibet the year before and had been expelled from that country, thought that the indications on our maps were reasonably accurate.

After a steep descent we reached the course of the Optchu and rested there at noon. Overhanging rock walls flanked the valley like a canyon. The valley was absolutely uninhabited and only a wooden pole showed that men sometimes came there. The other side of the valley consisted of slopes of shale up which we had to climb. It was evening before we reached the plateau and we bivouacked in icy cold. Our fuel during the last few days had been the branches of thorn bushes, which we found on the slopes. Here there was nothing growing, so we had to use dry cow-dung, laboriously collected.

Before noon next day we reached our first Tibetan village, Kasapuling, which consisted of six houses. The place appeared to be completely deserted and when we knocked at the doors, nothing stirred. We then discovered that all the villagers were busy sowing barley in the surrounding fields. Sitting on their hunkers they put each individual grain of seed into the ground with the regularity and speed of machines. We looked at them with feelings that might compare with those of Columbus when he met his first Indians. Would they receive us as friends or foes? For the moment they took no notice of us. The cries of an old woman, looking like a witch, were the only sound we heard. They were not aimed at us, but at the swarms of wild pigeons which swooped down to get at the newly planted grain. Until evening the villagers hardly deigned to bestow a glance on us; so we four established our camp near one of the houses, and when at nightfall the people came in from the fields we tried to trade with them. We offered them money for one of their sheep or goats, but they showed themselves disinclined to trade. As Tibet has no frontier posts the whole population is brought up to be hostile to foreigners, and there are severe penalties for any Tibetan who sells anything to a foreigner. We were starving and had no choice but to intimidate them. We threatened to take one of their animals by force if they would not freely sell us one—and as none of the four of us looked a weakling, this method of argument eventually succeeded. It was pitch dark before they handed to us for a shamelessly high price the oldest billy-goat they could put their hands on. We knew we were being blackmailed, but we put up with it, as we wished to win the hospitality of this country.

We slaughtered the goat in a stable and it was not till midnight that we fell to on the half-cooked meat.

We spent the next day resting and looking more closely at the houses. These were stone-built with flat roofs on which the fuel was laid out to dry. The Tibetans who live here cannot be compared with those who inhabit the interior, whom we got to know later. The brisk summer caravan traffic with India has spoilt them. We found them dirty, dark-skinned and shifty-eyed, with no trace of that gaiety for which their race is famous. They went sulkily to their daily work and one felt that they had only settled in this sterile country in order to earn good money from the caravans for the produce of their land. These six houses on the frontier formed, as I later was able to confirm, almost the only village without a monastery.

Next morning we left this inhospitable place without hindrance. We were by now fairly well rested and Kopp's Berlin motherwit, which during the last few days had suffered an eclipse, had us laughing again. We crossed over fields to go downhill into a little valley. On the way up the opposite slope to the next plateau we felt the weight of our packs more than ever. This physical fatigue was mainly caused by a reaction to the disappointment which this long-dreamed-of country had up to now caused us. We had to spend the night in an inhospitable sort of depression in the ground, which barely shielded us from the wind.

At the very beginning of our journey we had detailed each member of the party for special duties. Fetching water, lighting fires and making tea meant hard work. Every evening we emptied our rucksacks in order to use them· as footbags against the cold. When that evening I shook mine out there was a small explosion. My matches had caught fire from friction—a proof of the dryness of the air in the high Tibetan plateau.

By the first light of day we examined the place in which we had camped. We observed that the depression in which we had bivouacked must have been made by the hand of man as it was quite circular and had perpendicular walls. It had perhaps been originally designed as a trap for wild beasts. Behind us lay the Himalayas with Kamet's perfect snow-pyramid; in front forbidding, mountainous country. We went downhill through a sort of loess formation and arrived towards noon in the village of Dusharig. Again we found very few houses and a reception as inhospitable as at Kasapuling. Peter Aufschnaiter vainly showed off all his knowledge of the language acquired in years of study, and our gesticulations were equally unsuccessful.

However we saw here for the first time a proper Tibetan monastery. Black holes gaped in the earthen walls and on a ridge we saw the ruins of gigantic buildings. Hundreds of monks must have lived here once. Now there were only a few living in a more modem house, but they never showed themselves to us. On a terrace in front of the monastery were ordered lines of redpainted tombs.

Somewhat depressed we returned to our tent, which was for us a little home in the midst of an interesting but oddly hostile world.

In Dushang, too, there were—when we came—no officials to whom we might have applied for leave to reside or travel. But this omission was soon to be rectified, for the officials were already on their way to find us. On the next day we resumed our march with Kopp and myself in front, and Aufschnaiter and Treipel a little way behind us. Suddenly we heard the tinkling of bells and two men on ponies rode up and summoned us in the local dialect to return to India by the same way as we had come. We knew that we should not do any good by talking and so to their surprise we just pushed them aside. Luckily they made no use of their weapons, thinking no doubt that we too were armed. After a few feeble attempts to delay us, they rode away and we reached without hindrance the next settlement, which we knew was the seat of a local governor.

The country through which we passed on this day's march was waterless and empty with no sign of life anywhere. Its central point, the little town of Tsaparang, was inhabited only during the winter months and when we went. in search of the governor we learned that he was packing his things for the move to Shangtse, his summer residence. We were not a little astonished to find that he was one of the two armed men who had met us on the way and ordered us to go back. His attitude, accordingly, was not welcoming and we could hardly persuade him to give us a little flour in exchange for medicine. The little medicine chest which I carried in my pack proved our salvation then, and was often to be of good service to us in the future.

At length the governor showed us a cave where we could pass the night, telling us once more that we must leave Tibet, using the road by which we had come. We refused to accept his ruling and tried to explain to him that Tibet was a neutral state and ought to offer us asylum. But his mind could not grasp this idea and he was not competent to take a decision, even if he had understood it. So we proposed to him that we should leave the decision to a high-ranking official, a monk whose official residence was in Thuling, only five miles away.

AS FAR AS MY FEET WILL CARRY ME

Josef M. Bauer

Cornelius Rost (1919–1983) was an Austrian soldier drafted into the Wehrmacht during the Second World War who claimed to have escaped a Soviet Gulag and trekked across Central Asia to the Iranian border. Although the authenticity of Rost's story has been questioned by journalists and historians alike, his tale has served as inspiration for a number of film and television adaptations and Rost is portrayed as a man determined to escape and survive against all odds.

Josef M. Bauer (1901–1970) was a prize-winning novelist and short story writer from Germany. His long series of interviews with Cornelius Rost – who has the alias Clemens Forell in the book – formed the basis of *As Far as Far My Feet Will Carry Me.*

Sunday, 30th October, 1949. At 9 p.m. precisely, Forell opened the back door of the hospital. Immediately, the wind slammed it heavily against his body and a pile of snow collected above the lintel fell, partly into the corridor, making it impossible to shut the door again. An icy draught began to tunnel through the building.

Half-way up the stairs at the far end, Doctor Stauffer was engaging the Russian guard in conversation. For two minutes only he had promised to hold him and in that time, Forell would have to disappear. He had a minute left now in which to free the door. He tore off his fur gloves and started to scoop out the snow. Every trace would have to be removed or its presence inside the corridor would show that someone had left the building. But even then the door still would not shut, and he had to scratch away the snow that had meanwhile frozen in the jamb before at last he could cross the threshold and close the door softly behind him.

Once outside the lee of the hospital hut, Forell caught the full blast of the wind. It was still snowing heavily, and bitterly cold. From time to time, a lump of half-frozen snow slapped against his ears, stinging like a wet hand. He started towards the rock where Stauffer had promised to leave his rucksack.

Mingled with the storm he could hear the steady hiss of snow and then, from the direction of the guard-house, a burst of drunken laughter came past him on the wind. The next moment, one of his gloves blew off. He snatched for it, missed, and found himself flat on his face. Then the wind caught it and sent it scurrying across the snow. Scrambling to his feet, Forell gave chase. By the time he had caught it, the glove was wet inside, but he put it on nevertheless, knowing that his bare hands would quickly freeze.

He had carefully memorized the position of the rock and as he approached and his eyes became accustomed to the night, he could see the dark mass rearing above the snow. Feeling round the base of the rock, he found the rucksack on the far side from the hospital. Ready packed, with the *Kandra* fixed under the flap, shoulder straps – and then his fingers touched something else, made of wood. Taking off his gloves, he felt the shape more carefully: a pair of Siberian skis – Doctor Stauffer's parting gift.

The skis consisted of thin strips of birchwood, each about two feet long and pointed at either end. The points were bent upwards and held in position by lengths of thick catgut stretched between them. In the centre were two straps. The toe of the boot was slipped into one and the other was led up round the heel and fastened over the instep. Forell remembered seeing the Convoy Soldiers wearing skis like this when they had acted as pace-makers for the marching prisoners.

Forell glanced quickly round. Here and there, a lighted window gleamed faintly through the curtain of snow. All seemed deserted. Setting his route as best he could, he began to experiment with the skis, at first without the rucksack, going a short distance each time and then returning to where he had left it on the ground.

The skis, he discovered, were too short to bend at the beginning of a step, but lifted quite easily with the whole foot. It was thus possible to walk with them, or else slide forward keeping them on the ground. In either case, their tracks were shallow and would soon be obliterated by further snow. They were difficult to get used to, because of their length, which was four feet short of the normal. Twice, Forell tipped forward on to his nose, and once the rucksack landed on his head as well, and the experience could be highly unpleasant, with sharp stones lying just below the surface of the snow.

Getting used to the Siberian skis took time and gained him little ground. After thirty minutes of strenuous exertion he could still see lights glowing dimly behind him. By then, at any rate he had learned how to place his weight and adjust his stride, finding that best use could be made out of the skis by treating then simply as outsize soles, enlarging the area of his tread.

Forell then took a compass bearing. He could see the lie of the needle, but its halves were of equal length and without a light he was unable to distinguish north. Again, off came the gloves and he delved for the tinder-box. By the aid of intermittent sparks he saw, this time, that he was heading in the right direction, due west. But there were no landmarks to keep him on his route and after half a mile he checked again with the compass. Still due west.

Then Forell remembered what Doctor Stauffer had said. 'Don't become a slave to your compass. Granted your general direction is west, your route, at any particular moment, lies where the going is easiest and there's least danger, even if it means going due south for a while. And to start with, until you're outside the area of search from the camp, all that matters is to keep on the move. Then, when you're no longer being driven on by fear of pursuit, let yourself be lured on by the horizon and the prospect of the new country that lies beyond. After six weeks, when you've really got into your stride, you can start thinking about a definite objective.'

New country – that was what drew him on now. Though he had already traversed Siberia in the opposite direction and knew that, as yet, the new country was far distant, Forell could not altogether banish the seductive dream that next day, if he stood on tip-toe and craned his neck, he might glimpse another scene.

Meanwhile, progress was hard work and in the dim, milky expanse around him he could see that the ground was steadily rising. He had noticed that from the camp, and also, as he thought, a sprinkling of massive boulders on stony, undulating soil, but of these there was now no sign. Only an occasional shadow – a spur, perhaps – loomed ahead, to hover uncertainly before his eyes and then vanish again.

By now, the sweat was pouring down his cheeks and his earflaps were beginning to trouble him, simulating strange noises as they beat against his jaws in the wind. He pushed them up under his cap. Then frozen snow began to build up round his ears. He pulled the flaps down again and fastened them under his chin. The wind seemed to have dropped a little, though blowing now from his right and then whipping flurries of snow at him from straight ahead, it kept Forell wondering anxiously whether he was maintaining direction.

Sometimes, when the ground fell away from him, the skis threatened to run ahead of his legs. Then he would lean forward and the run would tum into a laborious walk. When the snow began to pile up against him at each step and the going became heavier, he would know he was climbing again. Once, a white wall reared suddenly before his eyes. He flung back his head, put out his hands to shield himself, clutched – nothing but air.

Part of his brain was beginning to long for dawn, and sleep. Twenty, twenty-five miles he should have done by then, a safe enough distance from the camp to justify rest, a good enough record for the first night of his journey to freedom. Twenty-five miles – before dawn. But the night was not over yet, whatever his exhausted body might try to make him believe. And in another part of his brain, he was pleased. That meant, he could still put a mile or two more between himself and the sleighs that sooner or later would set after him from the camp. He felt the urge to have the pistol closer to his hand, then remembered he had promised the doctor to keep it where it was, jolting and scraping between his legs, at any rate for the first two nights. That promise he would fulfil – at least he could do that much for Doctor Stauffer.

Yes, that much – and something more: another few miles before he stopped and made his bivouac. Gathering his ebbing strength, Forell began to count his steps, eyes bent on the few feet of snow ahead. As each foot came down, a number clicked in his brain. One – a hundred – five hundred – a thousand. Again, one to a thousand. And again. And another hundred or so while he was trying to remember whether it was three thousand he had done, or two. Every step, like that, brought its meed of achievement and gradually as he went on counting, peace came to his mind.

At the eight thousandth step – and with his stride that meant he had covered about four and a half miles – Forell stopped to take a bearing. His right hand began to feel for the tinder-box. Then with a shock he realized there was no need. He could see to read the compass without it. With dawn approaching, it was high time to find cover. The storm had slackened and visibility seemed to be improving. Stauffer had warned him that in open country by day, his dark form outlined against the snow would be visible for miles. Moreover, he could take it for granted that his pursuers would always see him first because their eyes would be used to the brilliance of the Siberian snows.

Seeing some rocks ahead of him outlined against the brightening sky, Forell made as fast as he could towards them. Among them was an enormous boulder about fifteen feet high. It looked ideal for his purpose. Blown over the top by the wind the snow was descending on the sheltered side almost vertically and had formed a drift high enough to conceal him from view. Having excavated a trough between the rock and the drift and looked out over the way he had come to make sure his tracks were now buried beneath further snow, he sat down with his back to the rock and rested his aching legs against the side of the mound.

For an hour he waited, not daring to open his rucksack in case he had to move again quickly. The straps had bitten into his shoulders and his leg muscles were twitching continually from the strain of the march. At last, hunger insisting, he took out a crust of bread and began gnawing at it slowly; 'Be hard on yourself,' he could hear Doctor Stauffer saying. 'Remember your food has got to last you a very long time. Ration it out, so much for each day, and then cut the ration by half . . .' He would have bread only, then, for this first meal – no fat. He took out another crust and began chewing again.

Every now and then he stopped to listen, convinced his crunching could be heard for miles.

By now he was thirsty. The doctor had wrapped the tea in chamois leather to keep it dry, and just as well: if the snow continued much longer, the rucksack would be soaked right through. After listening to make sure the coast was clear, Forell stamped down a patch of snow and set up the cooking stand. With a base no bigger than a cigarette packet and the mess-tin on top, it had to be carefully balanced. To melt the snow which he had stuffed into the mess-tin took more meta cubes than expected; another time, he told himself, he would have to wait till he could find water. But it was worth it. The tea warmed him in a way which marching could never do and as he drank it, Forell realized that this was his first pleasurable sensation since he had closed the hospital door behind him.

It would be eight hours before he could be on the move again and meanwhile, he would have to stay where he was, like an animal gone to ground. With the tea inside him and sheltered in his trench, he began to feel pleasantly warm. It was still snowing and before going to sleep, he stuck the skis vertically into the drift on either side and laid the rucksack across the top to form a roof. Then, to the sound of the wind sighing and moaning over the rock above, the prisoner slept.

Some time later, Forell awoke in pitch darkness with the sound of a crash in his ears. He had been dreaming he was skiing down a grassy meadow when suddenly a chasm had opened at his feet and he had dropped down – down – to land with a jolt in the lead mine. He seemed to have hurt his face, for now, fully roused, he could feel it throbbing painfully. He put up a hand – and found the rucksack. The skis had toppled over, depositing it on his head while he slept.

And now it was night again. Forell stood up, aching with cold. His whole body was stiff, as though his muscles had lost their suppleness, like leather that has got wet and been dried out again. Groping in his rucksack for a crust of bread, he came upon a small oblong of bacon. For a moment he held it in his hand, undecided; then he put it back, closed the rucksack and started gnawing the crust of bread. Each time he carried it to his mouth, he caught the savour of bacon, till he was forced to put on his gloves, in case temptation became too strong. Tomorrow morning, he told himself, when he made his bivouac, he would eat just one small shaving of bacon...

Forell shouldered his rucksack and, striking sparks from the tinder-box, took a compass bearing, setting his march due west. Then, a lonely figure, he set off into the freezing void. Only once during the whole of that night did any landmark appear to guide him on his way, and then as no more than a shadow looming ahead, slightly to the north of west. Keeping it on his right, Forell made towards it, only to see it vanish on his approach. Some time later, the same, or another shadow. appeared further northwards.

He was finding it easier to manage the skis now, though for some hours going was slow. After a while, he began counting his steps again. That, at least, convinced him he was making some progress and there was no other means of telling, for hour after hour, the world around him remained without contour or colour, while the only sound was the moaning of the wind. The hundreds he counted on his fingers, stretching them out one at a time. Thousands were represented by eight meta tablets, transferred from left pocket to right and then back again.

Some steps were very short and it seemed to Forell desperately important not to cheat and count half-steps as full digits. If he did, there would be no point in counting at all and nothing certain would be left in the world. One step equals one yard equals one digit: that was the key to the Universe. To restore its validity, he decided to count the whole of a long downhill run on the skis as only one step.

There remained the problem of the actual distance in a straight line that now lay between him and the camp. How many deviations had he made from due west, at what angle and what length? It was impossible to guess, so to allow what seemed a generous margin of error, Forell deducted one-fifth from the distance that his steps had covered. That left him within the radius of search still. Stauffer had put the limit at fifty miles, but it might be twice that distance. At any rate, whenever a man had been missing from the lead mine, the Russians had only abandoned the search after seven days, and in that time they could cover over two hundred miles.

For what remained of the night, Forell tramped on, stepping out the tens, the hundreds and the thousands, with some astronomical target at the back of his mind. With the dawn, the total was fifty-seven thousand. He then decided to make it sixty thousand before he bivouacked for the day, partly as a challenge to himself and partly because he was in open country where, if it stopped snowing and visibility cleared, his form would be seen for miles. And so, once again: ten – twenty – thirty – in whatever rhythm his exhausted body could maintain.

And then he came to a down-slope and he stopped counting as the skis started to run over the snow. After a while, as the slope flattened out into a valley, he began to lose speed. Then the skis were bumping over softer snow – coming to a stop – stopping. And suddenly, all resistance to their movement vanished and they were gliding over ice. Forell looked quickly round – he could hear the ice creaking under foot – saw a watercourse, noted steep banks, poor cover for a bivouac, and then, forgetting that if it stopped snowing, his tracks would remain visible for days, started brushing the ice clear with his skis in an attempt to see which way the water underneath it was flowing. Upstream was the right direction for him; downstream would lead ultimately to the sea. But upstream, according to the compass, was due east; not west. Exhausted and baffled, Forell began to slide further upstream to discover the general lie of the river.

Somewhere close at hand, he could hear water moving freely. He stopped to listen. As he did so, he felt the ice shift under his feet. Suddenly there was a loud crack and the next moment, Forell was knee-deep in freezing water.

Instinct prompted him to lay himself flat, but as he leaned forward on to the moving ice, the surface parted, and a moment later, a second time. By then, he was nearing the bank and, scrambling over some boulders, he managed to wade ashore without sinking into the water beyond his knees. Once on the bank, he could see some distance upstream. The river-bed came down in a series of steps and was liberally strewn with rocks, topped, where they emerged above water, with a crust of frozen spray.

Shivering on the bank, Forell realized now that, all the time he had been struggling in the river, his brain had been busy with a ridiculous childhood rhyme:

'Papa said, Cyril! I forbid it!
And yet the headstrong Cyril did it, Cyril went out skating on the pond.'

And very foolish, too, thought Forell – very foolish of both of us...

He was beginning to feel mortally cold. He tried to undo the ski-straps, so as to get his boots off, but the water in them had already frozen and they wouldn't move. For a moment, Forell thought of carrying straight on with the march, hoping that the movement would restore his circulation. Then he realized it was impossible; the cotton wool padding in his trousers was sodden with water and however long he walked, it would never dry out by the warmth of his body alone. Not only that, but his energy was almost spent. At the outside, he could only keep going for a couple of hours; then he would collapse from exhaustion and in thirty degrees of frost he would be dead almost before he touched the ground.

For some seconds, while the last vestige of warmth seemed to ebb from him, Forell was incapable of thought or action. A picture came before his mind, a picture of a dog-team, a flurry of snow – a sleigh – two soldiers. How he wished they would come...

The wish recoiled into action. He slipped off the rucksack and laid it on a nearby stone. Then he was tearing the laces from his boots. Boots and skis came off together. Then stockings, trousers, underpants. He put the cap on the ground to stand on, stumbled trying to step in and found his bare feet in the snow. But they felt warmer for it, if anything, so he rubbed them over with more snow, then planted them firmly in the cap. Next, off came the jacket to wrap round his legs. And then, behind a conveniently large boulder, a long and desperate attempt to make a fire of sodden meta tablets with sparks from a sodden flint. It was half an hour before he could produce the first low, flickering flame.

When the meta tablets, piled up like a house of cards, finally caught light, he started to dry his clothes, holding them as close to the flame as he could without singeing them, and so warming his hands as well. He was determined to use what heat there was to the utmost; the tablets were too precious to waste on the Siberian sky.

Meanwhile, wearing only the four shirts he had set out with, his bare feet stuck in his cap and the jacket wrapped round his legs, he was agonized with cold. He began to wonder whether he should break up the skis and add them to the fire. But the trouser legs were actually beginning to show some dry patches by now, and that gave him just enough encouragement to consider what would happen if he did burn the skis. It was quite simple. If he burnt the skis, he would probably be caught; and that would mean a much more unpleasant end than quietly freezing to death, here, by himself in the snow.

One heap of methylated tablets had been consumed, a second, and now the third was beginning to burn low. There were enough left for one more heap, but if they went, so would all possibility of brewing a hot drink. And even if he did come across wood, or twigs in this denuded landscape, the fire would be certain to betray him to his pursuers. Forell felt the stockings - they were still wet, frozen stiff by now, so they would have to go. The underpants, despite being held to the flame till they singed, also refused to dry; they too, would be sacrificed. It was impossible to dry the boots, so at the last moment, before the flame expired, he would warm them a little inside. The trousers, he would put on straight away. That done, he could put the jacket on again.

Having got so far, Forell ate a crust of bread and a small piece of bacon. Then he remembered the doctor's vodka, 'for emergency use only'. This was undoubtedly an emergency and it justified one small sip. The sip turned out to be smaller than intended, so Forell took another small sip. Then he put on his boots, lacing them up with difficulty because of his frozen fingers. Then he felt an urge to talk.

'What a fool!' he said, looking at the pistol where he had left it on a stone. 'Forgotten to hoist in the pistol!'

He noticed his voice sounded strangely harsh; still, it was a comfort to hear even himself talking, as Stauffer had foretold. 'Stauffer!' he said loudly. 'Stauffer-r-!' That was better; that had cleared his throat. Now he could talk to the man. 'Like all doctors, you, Herr Stauffer, are just a little bit too fussy.

Surely to God, if I'm going to use the pistol, I'm going to need it quickly? Stands to reason! So why hide it away? No, no, doc-doctor! It's silly! Besides, it's a very fine pistol! From now on. it's going to be in my pocket.'

He picked up the pistol and twirled it on his finger, till the freezing steel began to burn his skin, then, frowning peevishly he thrust it away. The boots felt like blocks of ice and very uncomfortable. The leather seemed to have buckled in places and was chafing the skin. He decided to cut up a shirt and bind the cloth round his feet. For that, he needed the Siberian knife. He grasped for his rucksack, bawling, '*Kandra!* Come here!'

The knife was very sharp and the experience of cutting a shirt off his own back seemed to strike him as funny. He laughed loudly, then once more unlaced his boots, bound up his feet and laced them up again, talking to himself continually. 'Hm! Excellent!' he muttered, as the cloth began to take up the warmth from his feet, then, remembering how near he had been to disaster, said in a tone of sorrowful reproach, 'Oh, you bloody fool...'

Still chattering carelessly away, Forell started stamping round to warm his feet, picking up his belongings as he found them and restoring them to the rucksack. He came on eight meta tablets that must have fallen out of his pocket. They were lying on a stone and quite dry. With an indifferent gesture, he cast them on the fire...

It was time to be going. The snow all round was churned up with his tracks, the site was exposed and before long the daylight would begin to fade. Forell consulted the compass. Due west, if the compass could be trusted, lay straight uphill. All he wanted was sleep, somewhere under cover from wind, weather and Russians where he could lie up in safety until he had regained his strength.

Forell slowly hoisted the rucksack on to his shoulders and stepped out, groaning, on his way. No sleep for almost twenty-four hours, half frozen to death, and now drunk on a sip of vodka, he would never get far. It was all he could do to stand upright, let alone move, let alone count his steps. The bright light was hurting his eyes. Dimly he remembered that Stauffer had included a pair of makeshift sun-glasses with the gear. They would be somewhere in the rucksack. But the very idea of unshouldering it again and then routing about inside made him feel weak. In any case, his eyes would not be troubled for long, soon they would shut of their own accord.

Forell laboured on, trudging slowly uphill. The way was becoming steeper and there seemed to be rocks all around. Suddenly his foot slipped on a stone and he nearly fell over. He stopped – and realized he could do no more. He was spent, and sinking to his knees.

Slowly Forell moved his head, left, and then right, looking for cover, saw some rocks close together on the same level as he, a few yards away. Now he was groping towards them, climbing in between two of the largest, finding room to lie down, slipping out of the pack-straps, feet out of the skis. A hand dropped to the ground, gripped, and lifted some snow to his gums. Hot tea would be better, better than snow – but better than either was sleep.

Forell awoke in full daylight to the sound he had most dreaded to hear: the shrill yapping of a sleigh team. With lurching heart he scrambled to a gap between the rocks and peered out over the snow. Yes, there they were, a full team – fifteen dogs – curving across in a foam of white – two men on the sleigh – soldiers – yes, from the camp! The iron sleigh told him that, the kind they had always used. Forell gripped his pistol.

The sleigh was passing out of sight behind a rock as it rounded the base of the hill. If it kept on turning, it would cross Forell' s tracks of the previous day, and the soldiers would never miss those. They were shouting to each other in a series of yells – on sleigh journeys, everyone yells – and the sounds seemed to be getting louder. Forell jumped to another gap, keeping them in view. They were below him now, coming towards him. He could see their uniforms, and the tommy-gun slung over the shoulder. In a few moments, they would be almost within range. He was tempted to shoot first, then thought it would be better to wait. If he missed either shot, the next would have to be for himself. But if the men spotted his tracks, or saw something suspicious and came up to investigate, he could wait till they were almost on him and make sure of them both.

The dogs seemed to know their route exactly. Heading towards Forell in a wide, right-handed curve, they came to within about seventy yards and then, of their own accord, suddenly changed direction and made off down the left bank of the river below the hill, missing his tracks by at least twice that distance. Two minutes from its first appearance, the sleigh was only a distant speck trailing a pennon of powdery snow. The shouts of the soldiers were still faintly audible, but soon, as their figures dissolved into the distant horizon, the sounds died with them and silence returned once more.

The soldiers had searched in a wide circle and now they were on their way back to camp, that much was clear. The direction they had taken was therefore approximately east. Feeling that there was no need to wait until darkness before resuming the march, Forell made ready and then set off in the opposite direction. To start with, he followed the sleigh tracks westwards, knowing that the dogs would have chosen firm ground.

Having survived the last attempt that the Russians at the lead mine would probably make to recapture him, Forell felt pleasantly relaxed, almost like a tourist. Instead of keeping them all in the rucksack, he put some crusts in his pocket, so that he could eat when he felt like it on the march. When hunger told him it was lunch-time, he seated himself comfortably on a stone and treated himself to the best picnic meal he could devise, none other than the fat little fish that had tasted so nauseating when last he had tried them in hospital. But they were ideal food to give stamina and protection against the cold, for they formed the staple diet of the sleigh dogs, whose work required a full measure of both.

The sleigh had approached in the wide circle and then disappeared behind him, heading east. To keep to a westerly direction, he had, therefore, to cross the tracks at the proper moment and then diverge from them steadily. When later he checked with the compass, it showed that he had guessed the moment correctly and that he had been marching, as he intended, due west.

That evening, for the first time, he was unable to find good cover for the night. Aware of the danger of having to make it in the open, he prepared his bivouac with the utmost care, not forgetting to eat some more of the little fish, so that their fat should help to prevent him freezing to death in his sleep.

THE GULAG ARCHIPELAGO

Aleksandr Solzhenitsyn

Aleksandr Solzhenitsyn (1918–2008) was a Soviet dissident and author whose 1973 publication of *The Gulag Archipelago* exposed the atrocities of the Soviet system of forced labour camps to the eyes of a watchful world. An early convert to Marxism-Leninism, Solzhenitsyn was imprisoned for criticizing Joseph Stalin in a private letter while he was a captain in the Red Army during the Second World War. He later became a fierce critic both of communism – at least in its Soviet manifestation – and Western secular liberalism, which he viewed as equally destructive to the spiritual faculties of man. In addition to *Archipelago*, his books *A Day in the Life of Ivan Denisovich* and *Cancer Ward* – both informed by his experience in the Gulag – contributed to wide-scale disillusionment with Soviet Communism in Western Europe and elsewhere and helped highlight the human rights abuses of the Soviet state. He was hailed as a hero in the West and, following his exile from the Soviet Union, spent sixteen years in the United States where he continued to write and compose poetry. He was awarded the Nobel Prize in Literature for *Archipelago*, which sold tens of millions of copies and served as a crushing indictment of the Soviet State.

And as soon as you have renounced that aim of "surviving at any price," and gone where the calm and simple people go – then imprisonment begins to transform your former character in an astonishing way. To transform it in a direction most unexpected to you.

And it would seem that in this situation feelings of malice, the disturbance of being oppressed, aimless hate, irritability, and nervousness ought to multiply. But you yourself do not notice how, with the impalpable flow of time, slavery nurtures in you the shoots of contradictory feelings.

Once upon a time you were sharply intolerant. You were constantly in a rush. And constantly short of time. And now you have time with interest. You are surfeited with it, with its months and its years behind you and ahead of you

– and a beneficial calming fluid pours thought your blood vessels – patience.

You are ascending...

Formerly you never forgave anyone. You judged people without mercy. And you praised people with equal lack of moderation. And now an understanding mildness has become the basis of your uncategorical judgements. You have come to realize your own weakness – and you can therefore understand the weakness of others. And be astonished at another's strength. And wish to possess it yourself.

The stones rustle beneath our feet. We are ascending.... With the years, armor-plated restraint covers your heart and all your skin. You do not hasten to question and you do not hasten to answer. Your tongue has lost its flexible capacity for easy oscillation. Your eyes do not flash with gladness over good tidings nor do they darken with grief.

For you still have to verify whether that's how it is going to be.

And you also have to work out – what is gladness and what is grief.

And now the rule of your life is this: Do not rejoice when you have found, do not weep when you have lost.

Your soul, which formerly was dry, now ripens from suffering.

And even if you haven't come to love your neighbors in the Christian sense, you are at least learning to love those close to you.

Those close to you in spirit who surround you in slavery. And how many of us come to realize: It is particularly in slavery that for the first rime we have learned to recognize genuine friendship!

And also those close to you in blood, who surrounded you in your former life, who loved you – while you played the tyrant over them ...

Here is a rewarding and inexhaustible direction for your thoughts: Reconsider all your previous life. Remember everything you did that was bad and shameful and take thought – can't you possibly correct it now? Yes, you have been imprisoned for nothing. You have nothing to repent of before the state and its laws.

But... before your own conscience? But... in relation to other individuals?

...Following an operation, I am lying in the surgical ward of a camp hospital. I cannot move. I am hot and feverish, but nonetheless my thoughts do not dissolve into delirium – and I am grateful to Dr. Boris Nikolayevich Kornfeld, who is sitting beside my cot and talking to me all evening. The light has been turned out – so it will not hurt my eyes. He and I – and there is no one else in the ward.

Fervently he tells me the long story of his conversion from Judaism to Christianity. This conversion was accomplished by an educated, cultivated person, one of his cellmates, some good-natured old fellow like Platon Karatayev. I am astonished at the conviction of the new convert, at the ardor of his words.

We know each other very slightly, and he was not the one responsible for my treatment, but there was simply no one here with whom he could share his feelings. He was a gentle and well-mannered person.

It is already late. All the hospital is asleep. Kornfeld is ending up his story thus.

"And on the whole, do you know, I have become convinced that there no punishment that comes to us in this life on earth which is undeserved. Superficially it can have nothing to do with what we are guilty of in actual fact, but if you go over your life with a fine-tooth comb and ponder it deeply, you will always be able to hunt down that transgression of yours for which you have now received this blow."

I cannot see his face. Through the window come only the scattered reflections of the lights of the perimeter outside. And the door from the corridor gleams in a yellow electrical glow. But there is such mystical knowledge in his voice that I shudder.

These were the last words of Boris Kornfeld. Noiselessly he went out into the nighttime corridor and into one of the nearby wards and there lay down to sleep. Everyone slept. And there was no one with whom he could speak even one word. And I went off to sleep myself.

And I was wakened in the morning by running about and tramping in the corridor; the orderlies were carrying Kornfeld's body to the operating room. He had been dealt eight blows on the skull with a plasterer's mallet while he still slept. (In our camp it was the custom to kill immediately after rising time, when the barracks were all unlocked and open and when no one yet had got up, when no one was stirring.) And he died on the operating table, without regaining consciousness.

And so it happened that Kornfeld's prophetic words were his last words on earth. And, directed to me, they lay upon me as an inheritance. You cannot brush off that kind of inheritance by shrugging your shoulders.

But by that time I myself had matured to similar thoughts.

I would have been inclined to endow his words with the significance of a universal law of life. However, one can get all tangled up that way. One would have to admit that on that basis those who had been punished even more cruelly than with prison – those shot, burned at the stake – were some sort of super-evildoers. (And yet ... the innocent are those who get punished most zealously of all.) And what would one then have to say about our so evident torturers: Why does not fate punish them? Why do they prosper?

(And the only solution to this would be that the meaning of earthly existence lies not, as we have grown used to thinking, in prospering, but … in the development of the soul. From *that* point are of view our torturers have been punished most horribly of all: they are turning into swine, they are departing downwards from humanity. From that point of view punishment is inflicted on those whose development… *holds out hope*.)

But there was something in Kornfeld's last words that touched a sensitive chord, and that I accept quite completely *for myself*. And many will accept the same for themselves.

In the seventh year of my imprisonment I had gone over and re-examined my life quite enough and had come to understand why everything has happend to me: both prison and, as an additional piece of ballast, my malignant tumor. And I would not have murmured even if all that punishment had been considered inadequate.

Punishment? But…whose?

Well, just think about that – *whose*?

I lay there a long time in that recovery room from which Kornfeld had gone forth to his death, and all alone during sleepless nights I pondered with astonishment my own life and the turns it had taken. In accordance with my established camp custom I set down my thoughts in rhymed verses – so as to remember them. And the most accurate thing is to cite them here – just as they came from the pillow of a hospital patient, when the hard-labor camp was still shuddering outside the windows in the wake of a revolt.

When was it that I completely
Scattered the good seeds, one and all?
For after all I spent my boyhood
In the bright singing of Thy temples.

Bookish subtleties sparkled brightly,
Piercing my arrogant brain,
The secrets of the world were . . . in my grasp,
Life's destiny ... as pliable as wax.

Blood seethed – and every swirl
Gleamed iridescently before me,
Without a rumble the building of my faith
Quietly crumbled within my heart.

But passing here between being and nothingness,

Stumbling and clutching at the edge,
I look behind me with a grateful tremor
Upon the life that I have lived.

Not with good judgment nor with desire
Are its twists and turns illuminated.
But with the even glow of the Higher Meaning
Which became apparent to me only later on.

And now with measuring cup returned to me,
God of the Universe! I believe again!
Though I renounced You, You were with me!

Looking back I saw that for my whole conscious life I had not understood either myself or my strivings. What had seemed for so long to be beneficial now turned out in actuality to be fatal, and I had been striving to go in the opposite direction to that which was truly necessary to me. But just as the waves of the sea knock the inexperienced swimmer off his feet and keep tossing him back onto the shore, so also was I painfully tossed back on dry land by the blows of misfortune. And it was only because of this that I was able to travel the path which I had always really wanted to travel.

It was granted me to carry away from my prison years on my bent back, which nearly broke beneath its load, this essential experience: *how* a human being becomes evil and *how* good. In the intoxication of youthful successes I had felt myself to be infallible, and I was therefore cruel. In the surfeit of power I was a murderer, and an oppressor. In my most evil moments I was convinced that I was doing good, and I was well supplied with systematic arguments. And it was only when I lay there on rotting prison straw that I sensed within myself the first stirrings of good. Gradually it was disclosed to me that the line separating good and evil passes not through states, nor between classes, nor between political parties either – but right through every human heart – and through all human hearts. This line shifts. Inside us, it oscillates with the years. And even within hearts overwhelmed by evil, one small bridgehead of good is retained. And even in the best of all hearts, there remains… an unuprooted small corner of evil.

Since then I have come to understand the truth of all the religions of the world: They struggle with the *evil inside a human being* (inside every human being). It is impossible to expel evil from the world in its entirety, but it is possible to constrict it within each person.

And since that time I have come to understand the falsehood of all the revolutions in history: They destroy only *those carriers* of evil contemporary with them (and also fail, out of haste, to discriminate the carriers of good as well). And they then take to themselves as their heritage the actual evil itself, magnified still more.

The Nuremberg Trials have to be regarded as one of the special achievements of the twentieth century: they killed the very idea of evil, though they killed very few of the people who had been infected with it. (Of course, Stalin deserves no credit here. He would have preferred to explain less and shoot more.) And if by the twenty-first century humanity has not yet blown itself up and has not suffocated itself – perhaps it is this direction that will triumph?

Yes, and if it does not triumph – then all humanity's history will have turned out to be an empty exercise in marking time, without the tiniest mite of meaning! Whither and to what end will we otherwise be moving? To beat the enemy over the head with a club – even cavemen knew that.

"Know thyself!" There is nothing that so aids and assists the awakening of omniscience within us as insistent thoughts about one's own transgressions, errors, mistakes. After the difficult cycles of such ponderings over many years, whenever I mentioned the heartlessness of our highest-ranking bureaucrats, the cruelty of our executioners, I remember myself in my captain's shoulder boards and the forward march of my battery through East Prussia, enshrouded in fire, and I say: "So were *we* any better?"

When people express vexation, in my presence, over the West's tendency to crumble, its political shortsightedness, its divisiveness, its confusion – I recall too: "Were we before passing though the Archipelago, more steadfast? Firmer in our thoughts?"

And that is why I turn back to the years of my imprisonment and say, sometimes to the astonishment of those about me: "*Bless you, prison!*"

Lev Tolstoi was right when he *dreamed* of being put in prison. At a certain moment that giant began to dry up. He actually needed prison as a drought needs a shower of rain!

All the writers who wrote about prison but who did not themselves serve time there considered it their duty to express sympathy for prisoners and to curse prison. I... have served enough time there. I nourished my soul there, and I say without hesitation:

"*Bless you, prison*, for having been in my life!"

(And from beyond the grave come replies: It is very well for you to say that – when you came out of it alive!)

NO SURRENDER

My Thirty-Year War

Hiroo Onoda

Hiroo Onoda (1922–2014) was an intelligence officer of the Japanese Imperial Army during the Second World War. For three decades Onoda refused to surrender to Allied forces and hid out in the Philippine jungle, surviving on bananas and coconuts, whilst he awaited reinforcements and waged a one-man guerrilla war. Twenty-nine years following the cessation of hostilities between the Allies and Japan, Onoda surrendered to his former commanding officer, who had since become a bookseller. Whilst to the rest of the world he appeared a fanatic, Onoda remained loyal to his oath never to surrender and to die for his emperor, and was received as a national hero back in Japan for his duty and perseverance.

If army uniforms were made out of silk serge, life would have been easier for Kozuka and me. As it was, our clothes were always rotting. During the rainy season on Lubang, it would often pour for several days in a row. Our uniforms, which we wore all the time, gave way faster to rotting than they did to wear and tear.

The trousers would rot first in the knees and the seat, then at the bottoms and in the crotch, until in the last stages, nothing was left but the backs of the legs. The jackets started at the elbows and then the back. The front part usually held together better than the rest.

To patch the holes, we had to make a needle. I found some wire netting somewhere, and we managed to straighten out a piece of the wire, sharpen it at one end, and make an eye in the other. For thread, we used the fibers of a hemplike plant that grew naturally in the forests. We would sew this vertically, horizontally and slantwise over the hole, occasionally making two layers for a quilted effect.

For the first three or four years, when we needed patches, we cut pieces of canvas off the edges of our tents, but this could be carried only so far. After that we "requisitioned" what we needed from the islanders as the opportunity presented itself.

This did not trouble our consciences. It is normal in guerrilla warfare to try to acquire guns, ammunition, food, clothing and other supplies from the enemy. Since the islanders were aiding the enemy task forces that came to look for us, we considered them enemies too.

In the early years, the outfit worn by the islanders consisted of a hand-woven hemp shirt and cotton knee-length shorts, neither of which was of much use to us. The islanders had thick skin, and living as they were on the plain, they required only very light clothing—too light to survive very long in the jungle thickets through which we were always moving.

The "war booty" that we valued most was the equipment that the departing American troops left behind. The islanders also prized this and kept it in a cabin that they guarded fairly closely, but occasionally we would scare them away with gunfire and make off with some of the goods, which included canteens, tents, shoes, blankets and the like. I think it was around 1951 or 1952 when we first acquired manufactured cotton cloth.

The Japanese cloth caps with flaps hanging in the back wore out in about a year. I had an officer's cap made of wool and silk, but even that gave out after about three years. From then on, I had to make my own headgear. There was a war song that started, "Even if my battle cap freezes ..." We changed this to "Even if my battle cap rots ... "

The clothes I had on when I came out of the jungle were some that I had remade after Kozuka's death. The front and back of my jacket were made from the lining of an islander's jumper, and the sleeves from trousers. The islander's trouser legs were not big enough around for my shoulders, but since they were too long, there was enough extra material to space out the shoulders. When I made new trousers for myself, I always reinforced the knees and the seat with leftover parts of the old trousers.

We often had to wade across streams, and to keep from wetting our clothing, we made our trousers so that they came down only a little below the knee, something like riding pants. We fastened our trousers with zippers, which had come to us as part of our war booty. When we were on the move, we left the zippers open to let air in, closing them only when we slept.

We slept in our clothing, of course, and if we put the carryall breast pockets of our jackets too high, they weighed on our chests and tended to keep us from sleeping. We therefore placed this pocket lower than the ordinary shirt pocket. It also had a zipper. Since we were always ducking under tree branches that brushed against our shoulders, we reinforced the shoulders of our jackets.

The shoes I had on when I came out were put together from real shoe leather from the tops of old shoes and rubber soles from an islander's sneakers. I had sewn these together with a thick nylon fishing line. During the early years, I had often worn straw sandals.

Around 1965 synthetic fabrics appeared on Lubang, and we gratefully "accepted" a number of articles of clothing made from them. We were also pleased by the appearance of vinyl plastic, which was useful for rain clothing and for wrapping our guns.

"They must have invented this stuff just for us," laughed Kozuka.

Our principal staple food was bananas. We cut off only the stem, sliced the bananas, skin and all, into rings about a quarter of an inch thick, and then washed them thoroughly in water. That way the green bananas lost much of their bitterness. Then we boiled them with dried meat in coconut milk. The result tasted like overcooked sweet potatoes. It was not good. But we ate this most of the time.

The rats on Lubang, which grow to a length of about eight inches, not counting the tail, eat only the pulp of the bananas, but Kozuka and I could not afford to waste the skins. At mealtime, we always said, "Let's have our feed."

Next to bananas our most important food came from cows that had been turned loose to graze. In 1945 there were about two thousand cows on the island, but their number gradually decreased to the point where it was difficult to find a fat one. Even so, three cows a year were enough to provide meat for one man.

When we could not find cows, we hunted for water buffaloes and horses. Although the water buffaloes are large and furnish a good deal of meat, it does not taste very good. Horsemeat, although tender, has a strong odor and does not taste as good as beef.

It was easiest to find cows in the rainy season. When the Lubang islanders harvest their rice, they leave about nine to twelve inches of stalk for the cows to eat. When the rice stalks are gone, the cows are turned loose at the foot of the mountains to eat grass, which grows best in the rainy season. The cows gradually work their way up the hills toward the forest, as much as to say, "Here we are. Come shoot us."

They usually grazed in herds of about fifteen. We would pick out one and fire at it from a distance of about eighty yards, aiming so that the bullet would enter beneath the backbone and go through the heart. The time to kill a cow was in the evening, after the islanders had gone home from the fields. It was nearly dark then, and if there was rain, it muffled the sound of the shot so that the farmers could not hear it.

If we hit a cow, the others would run away, frightened by the shot. Usually when we approached, the fallen cow still had life enough to move its legs. We would find a stone and smash it into the cow's forehead as hard as possible. Then we would finish it off by stabbing it in the heart with a bayonet. Having pulled it by the legs and tail to an inconspicuous spot under the trees, we cut the aorta to drain the blood.

The cow normally fell on its side, and the first step in dressing an animal was to cut off the front and hind legs on the upper side. Then we would slash down the middle of the belly and strip off the skin to the backbone. After cutting the meat off in hunks, we turned the animal over and repeated the operation on the other side. Finally, we would remove the heart, the liver, the sweetbread and other innards and put them in a sack. It took the two of us about an hour to dismember one cow.

If we left the carcass as it was, the rain and the crows would reduce it to a skeleton, but the remains would tell the enemy where we were. After we cut the cow up, therefore, we moved the carcass along a mountain road to as distant a point as possible. This was done at night, of course. It was really heavy work, because we had to carry all the meat on our backs at the same time.

For the first three days, we would have fresh meat, broiled or stewed, two times a day. Presumably because of the meat's high calory content, as I ate, my body temperature climbed until I felt hot to the soles of my feet. It was hard to breathe when walking and impossible to climb a tree. My head would always feel a little giddy.

I found that if I drank the milk of green coconuts as a vegetable substitute when I ate meat, my temperature would soon return to normal.

On the fourth day we piled as much meat as possible in a pot and boiled it. By heating this up once every day and a half or two days after that, we kept it from spoiling, and the flavor held up for a week or ten days. While we were eating the boiled meat, we dried what was left for future consumption. We called this dried meat "smoked beef."

To prepare the smoked beef, we first built a framework like a table frame. We then skewered the meat on long sticks, placed the skewers across the framework, and built a fire underneath. This had to be done at night in the inner reaches of the jungle; otherwise the islanders might see the smoke or the flame. On the first night, we would keep the fire going all night, so as to harden the outside of the meat a little without causing it to shrink. Afterward, we gradually increased the heat of the fire and cooked the meat about two hours a night for ten nights. By that time it was thoroughly dried. The liver and other innards, we first boiled, then dried.

From one cow, we could make about 250 slices of smoked beef. By eating only one slice apiece each day, we could make the meat last for about four months. It did not always work out this way, however, because when we were moving around a lot to keep out of the way of search parties, we allowed ourselves two slices a day.

We did not eat rice much, because it was so much trouble to hull it. In October and November, however, when the islanders harvested their rice, we usually requisitioned some of it. After pounding it, we separated it with a sieve into chaff, unpolished rice and half-polished rice. Both glutinous rice and nonglutinous rice grow on Lubang. The nonglutioous rice varies greatly in quality. We classified it into four grades, which we called "rice, 'barley' rice, 'millet' rice and fodder rice." The fodder rice was so black and the grains so small that we had trouble thinking of it as rice. When we ate rice, we made soup to go with it. This was concocted from dried meat and the leaves of papaya, eggplant or sweet potatoes, with a dash of salt and powdered pepper. Sometimes we made a gruel with rice and dried meat.

We called salt the "magic medicine." In the days when there were four of us, we had to get by on only about a quart a year. Every once in a while, whoever was cook would say, "It's cold today, so I'll put in a little of the magic medicine." And then he would put in a very tiny pinch. But even that helped the flavor a lot.

At first we had only the briny natural salt that we found on the south shore. Later, when there were only Kozuka and I, we grew more aggressive and invaded the islanders' salt flats in Looc and Tilik, but we never took more than we needed for the foreseeable future. After about 1959 we managed to obtain coffee and a few canned goods from the houses of the islanders. We called the sneak raids to obtain these valuables "stepping out for the evening."

During the thirty years on Lubang, the only thing I always had plenty of was water. The streams on the island were nearly all so clear that you could see the bottom. The only trouble was that the cows and horses that had been turned out to graze would drink water upstream and then relieve themselves in the water. For that reason, we always boiled the water before drinking it, even if it looked perfectly all right.

Having no doctor and no medicine, we were very careful to keep an eye on the condition of our health. We watched for variations in our weight by measuring the girth of our wrists. We also examined our own stool for signs of internal disorders.

I was thinnest just before Akatsu defected. I think this was partially because I was mentally upset at that time, but it was also due to a lack of sufficient nutrition. During this stage, the whites of my fingernails disappeared except for a tiny strip on my thumbs.

I examined my stool every day to see how much there was, how hard it was, and how big the pieces were. If the pieces were too big, it meant that my stomach was not functioning properly. If the stool was soft, my intestines were not absorbing enough.

If something was wrong, I had to decide for myself whether it was because of the weather, because the food I had eaten was not good, or because my body was not in good condition. Every time something went wrong, I thought back over what I had eaten the day before, how the weather had been, and how much I had exerted myself. After I had determined the cause, I adjusted my diet and my activities accordingly.

We ate pretty much the same quantity every day, but there was some variation, simply because some bananas are juicy and others are dry. Also, since good ripe bananas were not always available, we had to make do with green ones a great deal of the time. We tried to adjust the method of cooking to the quality of the food, then judge the effect on our insides by examining our waste. I remember deciding once not to move to a certain spot until the weather was cooler, because the last time I had spent some time there in hot weather, I had come down with diarrhea. There were other places where we could not stay long at one time because the wind chilled us too much at night. When that happened, we invariably suffered from indigestion.

When we were in a place that was too hot, our urine turned yellow, and if we overexerted ourselves, it became more red than yellow. This was a warning to take it easy for a while.

Whenever we settled down in a place, we dug a latrine, leaving the soil beside it to cover it up with when we moved on. The depth of the hole depended on how long we planned to stay. While we encamped, we covered the latrine with a stone; when we left, we filled it with dirt and strewed leaves over it. During our eighteen years together, Kozuka and I spent a great amount of time digging and covering up latrines.

Having no toilet paper, we had to use palm leaves instead. One time Shimada found some paper somewhere, but when he started to use it, Kozuka said, 'You've only got enough for two or three times. Then you'll have to go back to leaves. Why bother?'

When we found leaflets that the enemy had dropped, we would save one of them and leave the rest where they were, so long as they all said the same thing. They made too much smoke to use for starting fires, and we were afraid to use them even for blowing our noses, because one piece of soiled paper might lead the enemy to us.

We often found cartoons or nude photos of women in the mountains. They were not left by the search parties but deliberately distributed by the islanders. I guess they thought we would be tempted to take them, but we did not dare touch them for fear that we would reveal our location.

Fortunately, there was no malaria on Lubang. During my thirty years there, I was sick in bed with a fever only twice. Kozuka impaled his heel on thorns twice. Both times his leg swelled up, but otherwise he had no illnesses.

May is the hottest month in Lubang. In the daytime the thermometer goes up to about 100° Fahrenheit, and even if you sit still in the shade, the sweat pours off. If you have to walk fifty yards to get wood for the fire, you feel as though you were in a hot spring bath.

In June the squalls begin, coming up suddenly almost every day. Then in July, the real rainy season sets in. For two hours at a time it may rain so heavily that you cannot see more than ten yards away. This goes on for about twenty days, and sometimes the rain is accompanied by winds of nearly typhoon force.

In August there are more and more clear days, but the atmosphere is steaming hot. In September there is not much wind, but the rainfall is as heavy as in July. This goes on for about twenty days. Then there are blue skies for a day or two at a time over a period of two or three weeks, and finally in mid October the rainy season ends.

From then until the following April is the dry season. At first it rains a little once or twice a month; then there is no rain for several months. The coolest months are January and February, but even then the thermometer goes up to 85° or so in the daytime. The most comfortable time of the year is about like the hottest part of the summer in Tokyo. During this season only, we wore undershirts underneath our jackets.

In the dry season, we looked around the island carefully and decided where we would spend the next rainy season. There were several conditions that had to be met.

The first, of course, was that the place had to be near a supply of food. There should be banana fields and coconut groves in the neighborhood, and the campsite should not be too far from a place where the cows grazed. At the same time, it had to be a spot to which the islanders did not come.

It also had to be a place where the smoke from our fire was not directly visible to nearby villages, and a little noise not audible. If possible, there should be a breeze. The most desirable location was on the cool eastern side of a mountain.

It was not easy to find a place that had all of the qualifications, and once we had found it, we had to go through a period of anxiety before we built our hut and settled in. The reason was that the rainy season was irregular. Some years it would rain all through May; other years we would be well into June before the first drop fell. If we built our camp before the rain started, there was a danger that the islanders would discover us. We had to wait until we were sure they would not come into the mountains.

When we thought the rainy season was about to start, we would go to the place we had picked to see whether we still thought it was all right. Then we would camp nearby until it began to rain, at which point we would set up our hut as rapidly as possible. We called the hut a *bahai*, the Tagalog word for "house."

The first step in building the *bahai* was to find a large tree to which the whole structure could be anchored. After we selected the tree, we stripped it of branches, which we used to build a frame. Rafters were placed slantwise against the ridgepole and covered with coconut leaves. The latter were folded in two lengthwise and inserted between strips of split bamboo or palm branches. Everything was tied together with vines.

The *bahai* was built on slightly sloping ground, the upper part of the ground serving as the "bedroom." For beds, we first put down a few straight tree branches, then covered them with bamboo matting made by hand, and finally spread over the matting some duck sacks that we had requisitioned from the islanders.

The lower part of the *bahai* was the kitchen. Our "stove" consisted of several flat rocks placed together to form a platform for the fire and a pole above them from which our pot could be hung. Next to the hearth was a sheltered area where we could keep firewood and our rifles. Such walls as the *bahai* had were made of palm leaves, in the same fashion as the roof. Working with bolo knives, Kozuka and I could build the hut in seven or eight hours.

Before we started building huts like this in the early rainy season, we slept in tents, but often the wind blew the rain in until we were soaked and shivering all over. When that happened, we would warm ourselves up by singing army songs at the top of our voices. This was safe, because the noise of wind and the rain drowned out the noise we were making. One of the songs we sang started with the words:

Troops advancing in the snow,
Tramping over the ice ...

We were so cold at times like this that the song seemed appropriate, even on our snowless southern island.

The *bahai* was much more comfortable than the tents, but by the time the dry season approached, the roof had rotted so badly that a good deal of rain leaked through.

When the rainy season ended, we took the *bahai* apart and either burned the pieces or strewed them about in a hollow. Since it would not do merely to leave them there, we covered them with mud and spread branches and fallen trees over the spot. If we piled up a fair number of branches, they were enough camouflage to keep any islanders who might wander past from becoming suspicious.

We washed off the stones near the hut to remove any oil or dirt and covered the ground where the hut had been with branches that we had saved for this purpose. It was particularly important that no one find the place until after we had time to move to a new location, so we pulled vines over the branches to make access as difficult as possible without making the camouflage obvious.

When Shimada, who was an energetic worker with plenty of strength, was still alive, we built our hut deep in the jungle. After his death, we usually settled for a place near the edge of the jungle. We also simplified the hut so that it would be easier to dismantle and hide. Actually, there were not so many satisfactory locations that fulfilled the most important condition, which was that we be near a banana field, and during the whole thirty years, we used the same few places three or four times each.

In the dry season, we slept in tents or in the open. When we stayed in the open, we picked a place with about a ten-degree slope, and Kozuka and I slept side by side. To keep from slipping downhill during the night, we put our baggage or a log below our feet. Our rifles were always within easy reach. We took off our shoes, but during the whole thirty years, I never once took off my trousers at night. I always kept a little pouch with five cartridges in it attached to my belt.

In the early stages, we covered ourselves at night with our tents or clothing. Later we sometimes used the dried hides of cows. At one point we had quilts of a sort that we had made by piecing together bits of rubber that had floated up on the south shore. If it suddenly started to rain while we were sleeping out, we simply got wet. There was nowhere else to go. When this happened, we were cold, and the next day the joints of my legs ached. If my midriff got cold at night, I usually developed a tendency toward diarrhea.

There was a big cave on Snake Mountain, and the islanders had built a number of cabins in the mountains near their fields, but we did not use these as shelters from sudden rains, because there was too much danger of being discovered.

The reason why we slept on sloping ground was simply that this way, if we were suddenly awakened, we could see what was around us without raising up. Actually, during my entire thirty years on Lubang, I never once slept soundly through the night. When we slept on the ground outdoors, I shifted my body often to keep my limbs from getting numb.

I kept a sort of calendar, which after thirty years was only six days off the real calendar. My calendar was based largely on memory and the amount of food left, but I checked it by looking at the moon. For instance, if I picked ten coconuts on the first of the month and we used one a day, the day we used the last one would be the tenth. After making this calculation, I would look at the moon and see whether it was the right size for that day of the month.

When search parties were looking for us and we were moving around every day, I tended to lose track of the date. When that happened, I would start by looking at the moon and then try, in consultation with Kozuka, to figure out how many days had passed since some known date.

The moon was our friend in another respect as well, because we usually moved from one encampment to another at night. Kozuka often said, "The moon isn't on anyone's side, is it? I wish the islanders were the same way!"

Our haircut routine was another help in keeping track of the date. Kozuka once mentioned that back home the people had always celebrated the twenty-eighth day of each month as sacred to a local Buddhist deity. He said they had always eaten a special noodle dish on that day. This reminded me that when I was a boy, the barbers had always put out signs toward the end of the year asking people to have their children's hair cut by December 28 to avoid the year-end scramble. I suggested to Kozuka that we commemorate his hometown's monthly celebration by cutting our hair on the twenty-eighth of every month, and that is what we did. Somehow this seemed all the more appropriate to me because on the days when we did cut each other's hair, we were for a short while children again. Afterward, we often figured the date by counting the number of days since our last haircut. This was easy to remember.

We were particularly careful with the last haircut of the year, because it was a psychological boost to open the year with our heads looking spic and span.

On New Year's Day we made our own version of "red rice," that is, rice cooked with red lentil beans. This is served in Japan on festive occasions. We had no lentil beans, but there was a kind of string bean that grew on Lubang that we used as a substitute. On New Year's Day, we also made a special soup out of meat and papaya leaves, flavored with citron. This was intended as a substitute for the meat and vegetable soup called in Japanese *ozōni,* which is always served in the New Year season.

On the morning of the first day of the year, we bowed in the direction of the emperor's palace, which we considered to be north by northeast. We then formally wished each other a good year, renewed our pledge to do our best as soldiers, and repaired to our feast of "red rice" and *ozōni.*

We also celebrated our birthdays, and I remember one or two birthdays that I commemorated by giving myself a newly made cap.

Every morning I brushed my teeth with fiber from the palm trees. After I washed my face, I usually massaged the skin with kelp. When I wiped off my body, I often washed my underwear and my jacket at the same time. Since we had no soap, this amounted to rinsing the clothing in plain water, but sometimes I removed the grime from the neck and back of my jacket with the lye from ashes. I put the ashes in a pot and poured water over them. When the water cleared, I transferred it to a different pot and soaked the clothes in it. We had to be careful to hang our clothing in an inconspicuous place to dry.

There were several clear rivers, but during the whole thirty years I took a real bath only on those occasions when we cut up a cow and got blood and ooze all over ourselves. There is no such thing in the mountains as a valley with just a stream and nothing else. The valleys are also the roads. Except in the narrowest ravines deep in the mountains, we were afraid to undress to the skin, even at night. In the daytime, we bathed the upper parts of our bodies by pouring water over each other. In the evening we each washed off the lower parts of our bodies just before sunset. I would never have done anything so dangerous as to strip completely.

Translated by Charles S. Taylor

FIRST THEY KILLED MY FATHER

Loung Ung

Loung Ung (born 1970) is a Cambodian–American author and human rights activist. At the age of ten, Ung escaped the Killing Fields of Pol Pot's Khmer Rouge, a regime that murdered an estimated two million Cambodians, nearly a fourth of the country's entire population. At only seven years old, Ung witnessed the forced disappearance of her father and the death of her sister. After a perilous journey disguised as orphans, she and her siblings were assigned to a training camp for child soldiers, where they were indoctrinated into the Khmer Rouge ideology and forced to set landmines in the path of the invading Vietnamese Liberation Army. Ung wrote two books about her experiences and her subsequent journey to the United States. Since 1993, she has worked as an activist and international spokesperson for numerous campaigns for the global eradication of landmines, the prevention of abuse against women and the use of child soldiers.

One month has gone by since Kim was caught stealing corn. The Angkar has increased our food ration and as a result, fewer and fewer people are dying from starvation. Those who have survived the famine are slowly getting stronger. It seems as if every three months the Khmer Rouge has either increased or decreased our food ration without warning or explanation. For two or three months we have food to eat, just enough to keep us alive, then nothing to eat for another few months, then we have a little bit of food again. Kim speculates that it has to do with the rumours of the Youns – the Vietnamese – attacking the borders. Every time the Angkar thinks the Youns will invade Cambodia, the soldiers stock up on food and supplies and ship more rice to China in exchange for guns. When it turns out the Youns are not attacking us, the Angkar stops buying arms and our rations increase.

Even without the pressure to find food for us, Kim is different now and not like the brother I remember from Phnom Penh. He is quieter and rarely says more than a few words. We are all different now: Chou and I have stopped fighting, and Geak, who also has become more and more withdrawn, has stopped asking for Pa. Ma, though, still sits many nights at the door waiting for Pa to return.

Though I am sad and many days wish I am dead, my heart continues to beat with life. My eyes well up at the thought of Pa. 'I miss you so much, Pa,' I whisper to him. 'It is so hard to live without you. I am so sick of missing you.' It is hopeless because no amount of tears will bring him back. I know Pa does not want me to give up, and as hard as it is to endure life here day to day, there is nothing for me to do but go on.

Strange things are going on in the village as entire families disappear overnight. Kim says the Khmer Rouge terror has taken a new toll. The soldiers are executing the entire families of those whom they've taken away, including young children. The Angkar fears the survivors and children of the men they have killed will rise up one day and take their revenge. To eliminate this threat, they kill the entire family. We believe this to be the fate of another one of our neighbours, the Sarrin family.

The Sarrin family lived a few huts down from ours. Like our family, the soldiers also took the father, leaving behind the mother and their three young kids. The kids are our age, ranging from five to ten years old. A few nights back we heard loud cries coming from their direction. Their cries continued for many minutes, then all was quiet again. In the morning I walked to their hut and saw that they were no longer there. Everything they owned was still in the hut: the small pile of black clothes in the corner of the room, the red checked scarves, and their wooden food bowls. It has been maybe three days now and still the hut stands empty. It is as if the family magically disappeared and no one dares to question their whereabouts. We all pretend not to notice their disappearance.

When she returns from work one evening Ma hurriedly gathers Kim, Chou, Geak, and me together; saying she has something to tell us. With all of us sitting in a circle waiting for her, Ma nervously walks around the hut outside to make sure no one can hear us. When she joins us, her eyes are filled with tears.

'If we stay together, we will die together,' she says quietly, 'but if they cannot find us, they cannot kill us.' Her voice shakes when she speaks. 'You three have to leave and go far away. Geak is four and too young to go. She will stay with me.' Her words stab my heart like a thousand daggers. 'You three will each go in different directions. Kim, you go to the south; Chou will head to the north; and Loung to the east. Walk until you come to a work camp. Tell them you are orphans and they will take you in. Change your name; don't even tell each other your new names. Don't let people know who you are.' Ma's voice grows stronger with determination as the words pour out. 'This way if they catch one of you, they cannot get to the rest because you will have no information to give them. You will have to leave tomorrow morning before anyone else is up.' Her mouth says many more words to us, but I cannot hear them. Fear creeps its way into my body, making it tremble. I want to be strong and fearless, to show Ma she does not have to worry about me. 'I don't want to go!' I blurt the words out. Ma looks at me firmly. 'You have no choice,' she says.

The next morning Ma comes to wake me, but I am already up. Chou and Kim are dressed and ready to go. Ma packs my one pair of clothes, wraps my food bowl in a scarf, and ties it diagonally around my back. Slowly I climb down the steps to where Chou and Kim are waiting for me.

'Remember,' Ma whispers, 'don't go together and don't come back.' My heart sinks as I realise Ma is really sending us away.

'Ma, I'm not going!' I plant my feet to the ground, refusing to move.

'Yes, you are!' Ma says sternly. 'Your Pa is gone now, and I just cannot take care of you kids. I don't want you here! You are too much work for me! I want you to leave!' Ma's eyes stare at us blankly.

'Ma,' my arms reach out to her, pleading with her to take me into her arms and tell me I can stay. But she swats them back with a quick slap.

'Now go!' She turns me around by the shoulders and bends down to give me a hard swat on the butt, pushing me away.

Kim is already walking away from us with his eyes looking ahead and his back rigid. Chou follows slowly behind him, her sleeves continuously wiping her eyes. Reluctantly, I drag myself away from Ma and catch up with them. After a few steps, I turn around and see that Ma has already gone back into the hut. Geak sits at the door, watching us leave. She lifts her hand and waves to me silently. We have all learned to be silent with our emotions.

The farther I am away from the village, the more my anger overtakes my sadness. Instead of missing Ma, my blood boils with resentment towards her. Ma doesn't want me around any more. Pa took care of us and kept us together. Ma cannot do this because she is weak, like the Angkar says. The Angkar says women are weak and dispensable. I was Pa's favourite. Pa would have kept me at home. Ma has Geak. She always had Geak. She loves Geak. It is true that Geak is too young to leave, but I am not yet eight. I have nobody. I am completely alone.

The sun climbs to the back of our heads, scorching them. The gravel path burns and digs into the soles of my feet and breaks through the hard calluses. I move off the gravel to walk on the grass. June is only the beginning of the rainy season so the grass is still plump and green. In November, the grass will shrivel up and become sharp like pins. The soles of my feet are so thick and callused that not even the pin grass can cut through them. However, when the grass is tall like it is now, the blades cut my skin like paper. It has been a long time since I have worn shoes. I don't remember when I stopped. I think it was when we arrived in Ro Leap that they burned my red dress. In Phnom Penh, I had black buckle shoes that went with my school uniform; the soldiers burned those too.

Soon it is time for Kim to go off on his own path. He stops us and again repeats Ma's instructions without emotion. Although he is only twelve his eyes have the look of an old man. Without words of goodbye or good luck, he turns and walks away from us. I want to run to him and put my arms around him, hold him the way I held Pa and Keav in my mind. I don't know if or when I will ever see him again. I don't want to bear the sadness of missing him. With my hands clenched into fists by my sides, I stand there and my eyes follow his body until I can no longer see him.

Though it goes against Ma's warnings, Chou and I cannot separate ourselves so we head off in the same direction. With no food or water, we walk in silence all through the morning as the sun beats down on us. Our eyes look everywhere for signs of human life but find none. All around us, the trees are brown, their green leaves, wilted in the heat of the white sky, hang quietly on the branches. The only sound comes from our feet and the pebbles that roll away from our toes. As the sun climbs above our head, our stomachs grumble in unison, asking for food which of course we don't have. In silence, Chou and I follow the red dirt trail winding and stretching before us. As our bodies grow tired and weak, we long to sit and rest in the shade, but we force ourselves on; we do not know where or when our trail will end. It is afternoon when we finally see a camp.

The camp consists of six straw-roofed huts, very much like ours, except they are longer and wider. Opposite them are two open huts that are used as the communal kitchen and three smaller huts where the supervisors live. The camp is surrounded by huge vegetable gardens on all sides. In one, about fifty young children squat in a row, pulling weeds and planting vegetables. Another fifty children lined up at the wells are in the process of watering the gardens. Buckets of water are passed from one person to another, the last person with the bucket pours the water onto the garden and runs the bucket back to the well.

Standing at the gate, we are greeted by the camp supervisor. She is as tall as Ma but much bigger and more intimidating. Her black hair is cut chin-length and square, the same style as the rest of us. From her large, round face, her black eyes peer at us. 'What are you doing here?'

'Met Bong, my sister and I are looking for a place to live.' In Khmer I address the supervisor as 'comrade elder sister' with as much strength in my voice as I can muster.

'This is a children's work camp. Why are you not living with your parents?'

'Met Bong, our parents died a long time ago. We are orphans and have been living with different families, but they no longer want us.' My heart races with guilt as the lies spill out of my mouth. In the Chinese culture it is believed that if you speak of someone's death out loud, it will come true. By telling the comrade sister my parents are dead, I have put a marker on Ma's grave.

'Did they die at the re-education camp?' Met Bong asks. I hear Chou's gasp for breath and warn her not to say anything with my eyes.

'No, Met Bong. We were farmers living in the countryside. I was too young to remember, but I know they died fighting for the Civil War.' I am amazed how easily the lies come out of my mouth. Met Bong seems to believe the lies, or maybe she simply does not care. She is in charge of a hundred kids and does not care if her workforce is increased by two more.

'How old are you and your sister.'

'I am seven, and she is ten.'

'All right, come in.'

This is a girls' camp for those who are considered too weak to work in the rice fields. We are considered useless because we cannot help out the war effort directly. Yet from morning till night we work in the scorching sun, growing food for the army. From sunrise to sunset, we plant crops and vegetables in the garden, stopping only for dinner and lunch. Each night we fall into an exhausted sleep, wedged closely together on a wooden bamboo plank with fifty other girls, the other fifty in another hut.

Nothing at the camp is wasted, especially water. The well water is strictly for the gardens and cooking; to wash ourselves and our clothes we must walk a mile to the pond. After a long day of roasting in the sun, no one is thrilled about the walk for a wash, so we rarely bathe. Everything is collected and reused: old clothes become scarves, old food is dried and saved, and human waste is remixed as topsoil.

After our first evening meal, Chou and I are told to gather around the bonfire for nightly lessons. When we get there we see that all the other children are already there. We squat on the ground waiting for the Met Bong to read the latest news or propaganda from the Angkar. In a voice full of fury and adulation, Met Bong yells out, 'Angkar is all-powerful! Angkar is the saviour and liberator of the Khmer people!' Then one hundred children erupt into four fast claps, their fisted arms raised to the sky, and scream 'Angkar! Angkar! Angkar!' Chou and I follow suit, though we do not understand the propaganda of what Met Bong is saying. 'Today the Angkar's soldiers drove away our enemy, the hated Youn, out of our country!'

'Angkar! Angkar! Angkar!'

'Though there are many more Youns than Khmer soldiers, our soldiers are stronger fighters and will defeat the Youns! Thanks to the Angkar!'

'Angkar! Angkar! Angkar!'

'You are the children of the Angkar! Though you are weak, the Angkar still loves you. Many people have hurt you, but from now on the Angkar will protect you!'

Every night we gather to hear such news and propaganda, and are told of how the Angkar loves us and will protect us. Every night I sit there and imitate their movements while hatred incubates inside me, growing larger and larger. Their Angkar may have protected them, but it never protected me – it killed Keav and Pa. Their Angkar does not protect me when the other children bully Chou and me.

The children despise me and consider me inferior because of my light skin. When I walk by them, my ears ring from their cruel words and their spit eats through my skin like acid. They throw mud at me, claiming it will darken my ugly white skin. Other times, they stick their legs out and trip me, causing me to fall and scrape my knees. Met Bong always turns the other way. At first, I do nothing and take their abuse silently, not wanting to attract any attention to myself. Each time I fall, I dream of breaking their bones. I have not survived this much to be defeated by them.

THE WAY OF HONOUR

Adrian Carton de Wiart

Adrian Carton de Wiart (1880–1963) was a British Army officer who received the Victoria Cross – the highest military honour in the British Empire – for his heroism. He fought in the Boer War, the First World War and the Second World War and over the course of these conflicts received gunshot wounds in his leg, ankle, hip, head and stomach. He was also blinded in his left eye, tunnelled out of a prisoner-of-war camp and survived two plane crashes. He later recalled his time during the First World War as an enjoyable experience.

The Endurance of the Belgian People

A speech delivered on the 17th of May 1915
in the Town Hall of Saint–Etienne

LADIES AND GENTLEMEN, –

Allow me to make a confession to you. When I received the invitation of the authorities of the Department of the Loire and of this great city to address you, honourable and pressing though the invitation was, I hesitated at first to accept it.

We are living to–day through times both serious and pathetic, to which the motto of Hoche, *Res, non verba*, is singularly appropriate. This very morning I read, in a Paris newspaper, this reflection of a humorous writer: "In time of war the first useless mouths are those of the politicians."

In spite of this scruple, I stand, however, before you to-day, and it appears to me that I have more than one duty to accomplish here.

Was it not becoming that, after having paid our homage to Paris, we should also come to salute that French province, and particularly these districts of Forez and Lyonnais, wherein so many of our working people are being so generously entertained?

It seems to us that it is precisely in your part of the country that the fortunate diversity of the French temperament declares itself under certain aspects which we Belgians could desire ourselves to exhibit: the respect, namely, for family traditions, the domestic virtues and hospitality; the maintenance of a sane balance between intelligence and goodwill, initiative and an unwearying diligence in affairs of industry and commerce, a serious outlook upon life and a passion for progress.

It is for these reasons that I have consented, ladies and gentlemen, to be the bearer to you of the greeting of the King's Government and its deep gratitude for what you have done and are doing every day for our refugees. Your public authorities and your private enterprises, the heads of the department and of the town and our Consul and his assistants, all have interested themselves in their cause, adding to that interest, as the President of the Republic promised us when we landed at Havre, "all the warmth of a brotherly affection."

It is this same brotherly affection which makes your hospitality at once so thorough and so tactful. This it is which has brought into being, for the benefit of an unfortunate people who have been driven from their homes by a veritable invasion of Barbarians, and that, one of the most atrocious which History can show, so many ingenious organizations destined to temper for them the bitterness of their poverty, their grief and their exile.

It is this which expresses itself to–day in the courtesy with which you have received me and by the eagerness with which you have come here to listen to yet another voice that is to speak about Belgium.

To this duty which I desire to perform must be added a certain hope in which I know you will pardon any presumption that there may be.

It seems to me, coming as I do directly from the battle front and having been enabled on many occasions, during these latter months of the war, to pass through not only the Belgian lines, but also those of the French and the British, having on the other hand had occasion, to obtain very trustworthy information about the spirit of those of our provinces and of your departments which have been invaded – it seems to me that perhaps, chatting together about those places, we may be able to come to some understanding of that marvellous atmosphere of courage and confidence which surrounds not only the men who are fighting, but also those of our fellow-countrymen who remain prisoners upon their own soil.

Ah! believe me – civilians though we be – we do not mean to allow ourselves to be outdone in tenacity and endurance by our heroes in the trenches. This fair land of France – which superficial critics accused at one time of being degenerate – what a marvellous example of moral resilience and physical health, of simple heroism, of harmonious beauty, she has given during ten months to an admiring world! And how, in every rank of society, the French soul has appeared truly as the mistress of the body which it animates. But this calm energy, this yielding determination, if they are to be found everywhere in France, nowhere are they exhibited so perfectly as in the midst of the Army, among the valiant soldiers of France, England and Belgium, who, at this very moment, only six or seven hundred kilometres from this spot, are at grips with all the difficulties and vicissitudes of their gigantic struggle.

I am mistaken. There is a place where, perhaps, confidence in the final triumph is still more extraordinary and where it is certainly more touching: I mean on the other side of the firing-line, in those occupied regions where our fellow-countrymen, yours as well as ours, in spite of nameless sufferings, have not for one single instant permitted their hope and their endurance to falter.

MISSION
Jimmy Stewart and the Fight for Europe

Robert Matzen

James Maitland 'Jimmy' Stewart (1908–1997) was an American actor and Air Corps pilot who flew twenty brutal combat missions over German-occupied Europe during the Second World War. The Oscar-winning actor was known for his star turns in The *Philadelphia Story, Vertigo* and *It's a Wonderful Life*, but Stewart was deeply affected by his military service and his combat experience often added dramatic emotional weight to his film roles. An avid traveller, Stewart helped smuggle the Pangboche Hand – an artefact from a Buddhist monastery in Nepal believed to belong to the legendary Yeti – out of India. Some years later, the quest to find the Yeti would be taken up by the mountaineer Sir Edmund Hillary, one of the first men to conquer Mount Everest.

Robert Matzen is the author or co-author of nine books. *Mission* forms part of a trilogy of works about Hollywood in World War II. Matzen flew in B–17 and B–24 four-engine bombers, like those piloted by Jimmy Stewart in the war, as part of his research.

"Dear Mrs. Skjeie," began Jim. "I am writing to you to tell you about your husband .... "

Another of his pilots had been lost. Dave Skjeie's father, Endre, was of Scandinavian descent and had served in a U.S. aero squadron in the Great War, inspiring both Dave and his brother to join the Air Corps. Dave had been with Jim since the beginning in Sergeant Bluff and was a fine pilot, but there's just so much you can do when your plane's hit with incendiary bullets, flips over, and explodes, which is what happened to Skjeie's ship yesterday five minutes after bombs away over Gotha. Somehow, two chutes had been seen at the time of the explosion, so Jim's letters to the ten families of the crew would be vague because the odds were two in ten that their loved one had survived. But Dave himself? It was unlikely he would have had time to untether himself from seat belt, oxygen, heating system, and intercom and climb out of the pilot's seat and off the flight deck to the bomb bay. Jim sat there thinking; Dave had celebrated his birthday at Tibenham on December 6; he had turned twenty-two. And now Jim was writing to Dave's widow, Billie, same age.

The 703rd had also lost Pete Abell and the crew of *Star Baby* above Gotha. Reports said they'd been hit with 20mm incendiary rounds from the cannons of an Me 210 that set the ship ablaze. Chutes were seen before *Star Baby* exploded in the sky. Jim had hopes that Pete had bailed out because once it was clear the plane couldn't limp home, they all would have had time to buckle on chutes and jump, even the pilot, and the plane had indeed remained airworthy for a minute or two, said the reports.

It had been only two months since Jim had checked out Abell and crew after their switch from B-17s and late arrival at Tibenham. Pete was twenty-five and had been born in Oklahoma but lately lived in Saratoga, California, which is where Livvie de Havilland had grown up – outside San Francisco.

The Gotha mission had been a bloodbath for the 445th, and Stewart wondered if Station 124 could ever recover. That morning Jim had watched from the control tower as twenty-eight Liberators from the group set out on the day's mission. Three had turned back and landed after mechanical trouble. Late in the afternoon twelve more limped home. He stood at the railing on the tower and watched them come sputtering in, firing flares to signify wounded aboard or mechanical difficulty. One by one the ships taxied back to their hardstands and shut down engines, and the sky grew still and quiet, so still and so quiet that sea gulls began soaring over the runways in the gloom of afternoon. What a feeling that was, standing there on the upper deck of the control tower staring at empty sky, and at so many empty hardstands.

In all, thirteen planes out of the twenty-five that had flown the mission had been lost. So many brave kids had bought the farm or parachuted into enemy hands, and S-2's interrogations, all of which Jim had sat in on, had revealed the horrors of running air battles over Germany that had taken out some of the finest pilots and airmen in the group. Never had he seen such a day; he had heard about them, days like Black Thursday, but he hadn't been this close. Even the lead ship had gone down, Maj. Jim Evans, squadron commander of the 702nd in the pilot's seat and serving as air commander for the day, with Capt. Henry Bussing his copilot, sitting in the seat so often occupied by Stewart of the 703rd.

Wright Lee of 1st Lt. Norm Menaker's crew in the 702nd Squadron reported of this lead ship and Major Evans: "They were attacked head-on by fighters, the bullets pouring into the nose, where they hit bombardier Cassini a dozen times and ripped the thumb from Lieutenant Massey. The blinding flash, which I saw, was a 20mm shell exploding in the cockpit, injuring Evans, the pilot and our squadron commander, but missing Captain Bussing, the copilot." In desperation Evans had put wheels down to signal surrender, then had fallen out of formation. The Luftwaffe had recognized the signal and backed off, allowing surviving crew members to bail out.

Jim heard that Lt. Robert Blomberg of the 700th also had been lost and was most certainly dead – Blomberg had just flown on Jim's right wing on the Brunswick mission five days earlier, and Stewart believed him to be the most promising pilot in the 700. Damn good flier and quality kid. Now he had likely met his fate, as had that go-getting young radio operator, Leone, who had helped Stewart out on Abell's shakedown flight. Jim's eye had snagged on Leone's name as he ran down Blomberg's crew.

From S-2 reports compiled during the questioning of survivors, Stewart knew that sending the Forts and Libs as one air armada had become problematic at the point when they split and the B-17s headed toward Schweinfurt while the B-24s flew east to Gotha. The Eighth Air Force fighters found themselves with two formations to protect instead of one, leaving the Liberator force vulnerable to attack. The result, as he read from S-2 summaries, was devastating:

"For well over 2.5 hours Nazi fighters, rocket ships and dive bombers poured death and destruction into the formation from every angle. All the latest tactics were used by Hitler's supermen to break up and turn back the attack. They used trailing cables with bombs attached that exploded upon contact. Rockets were going off like the Fourth of July, and parachute bombs were used to some extent. At the coast of Holland flak starts. Waves of Focke-Wulf 190s come in, then queue up and come in one at a time. Other fighters stood off at a distance and shot rockets, then came in close and spewed death-dealing 20mm shells.

"One Bf 109 came firing head-on into a Lib and blew the nose off. As the ship went spinning down, the 109 pulled up into a steep climb. The top turret gunner in the Lib behind cut loose and stopped him in midair. The ship fell off on its back and headed down. Halfway through the formation, it crashed into another Lib in the low element and took her whole tail off. The ship, with its tail off and nose pointing toward the sky, headed into the bomb bay of the Liberator above. The two planes crashed together. Both began to expel men from all sections of the ship. Ten men jumped from the almost stationary planes. The planes broke apart, and the one that crashed into the higher plane started sliding down sideways through the formation. One gunner yelled to his pilot, 'Look out from the right!' and the pilot put the ship into a dive as the pilot just in front pulled up to a steep climb, and the doomed Lib sailed between them with inches to spare.

"When one Lib was hit, the Germans pulled up and waited until the crew bailed out before attacking again. One followed the Lib down, while the others rejoined the attack on the formation.

"One lead ship had a fighter, about the seventh in a long line, come in to about 100 yards, roll and climb and release a parachute bomb that the Lib ran into. The explosion tore the top section off the ship back to the wing edge. The plane burst into flames and went tail first toward the ground.

"Another crew member of a doomed Lib bailed out and was floating to earth when another Lib that was shot down picked up his chute on the jagged edge of one of its torn wings and carried him down out of sight of the formation.

"Frantic attempts by remaining ships to dodge parachutes and parts of aircraft that filled the air. Many men who had bailed out held the pulling of rip cords until they thought they had missed the formation only to find they had opened up right in front of another group of ships."

The ink hadn't yet dried as Stewart read more and more sheets of paper fresh off S-2 typewriters recounting the horrifying Gotha mission, now just hours old.

Navigator Wright Lee said of being in the middle of the day's battle, "There was a temptation to 'chuck it all' and bail out...and live ... but you couldn't do that."

Stewart hit the sack with the worst thought of all in the front of his mind: He would be the air commander on tomorrow's Big Week mission if the weather held. He lay there in the blackness with dark visions in his mind, of parachutes caught on jagged metal, of burning Liberators colliding in midair. Of body parts smacking off his windshield.

He got up out of his bunk in the freezing cold darkness. "Walking to the window," he remembered later, "I pulled the blackout curtains and stared into the misty English night. My thoughts raced ahead to morning, all the things I had to do, all the plans I must remember for any emergency. How could I have a clear mind if it were saturated with fear?"

His thoughts drifted back to advice from his father, often-repeated, often-remembered advice about being in a war. Jim had asked his dad if he had been afraid. "Every man is, son," Alex had told him, "but just remember you can't handle fear all by yourself. Give it to God. He'll carry it for you." Jim would have the psalm with him, and his lucky handkerchief. And he would have his wits.

An eternity later, in the pre-dawn morning of Friday, February 25, he heard the door of his quarters in Site 7 click open. With an accompanying blast of cold air came words that shot through him like electricity: "Mission today, Major Stewart."

Big Week had reached day five. Jim's brain scraped itself to some semblance of coherence after little or no sleep – he couldn't be sure he had drifted off even once. He needed to be sharp, but how could he be after the horrors of day four and the visions they had produced – all those fine kids that had met their fate.

He followed the routine: shower, breakfast, briefing, suit up, grab gear, hop a ride to the plane. When the truck pulled out, loaded with crewmen, it turned the wrong way and Stewart screamed, "Where the hell are you going, driver?" One of the crew said he had forgotten his charts. "The hell with the charts!" bellowed Stewart and ordered the truck to make for the hardstands. He looked around him and saw faces that had flown yesterday. Just yesterday. And they would fly today because so many men were dead and there weren't enough crews to go around. He felt bad for yelling, but he was the commander, and they would get over it.

The mission today would be one for the books: A total of 754 B-17s and B-24s, escorted by twenty groups of Eighth Air Force fighters and twelve squadrons of RAF Spitfires and Mustangs, would conduct a mass penetration of southern Germany to attack three Messerschmitt aircraft production centers and a ball-bearing plant. A large group would attack Augsburg and Stuttgart, another would hit Regensburg. The Second Combat Division, including the 445th, would go for Bachmann, von Blumenthal & Co., a final assembly plant and airfield for Me 110s and 410s located at Fürth, just northwest of Nuremburg. The briefing had stressed conservation of fuel because they carried ten hours' worth and would be flying nine and a half at least.

Stewart sat as a second copilot and group commander in ship 447, *Dixie Flyer*, piloted by Neil Johnson with Lieutenant Yandagriff as copilot. They carried forty Ml fragmentation bombs. Wheels up was 0931 into heavy overcast. After assembly they flew south through England, passing London to the west and over the Channel, where they donned helmets and flak vests and tested the guns, then continued south. They reached Dieppe on the French coast at 1142, made landfall, and ran the gauntlet of flak.

As the planes headed inland over France, the skies cleared to CAVU (ceiling and visibility unlimited), offering views of rolling French countryside blanketed in snow. The weather was moderate – only twenty below zero at 18,000. On the flight deck, if they were dwelling on the battle of Gotha just yesterday, Stewart saw no indication. But how could they not think about those poor guys yesterday, flying just like this when out of the literal blue all hell broke loose? Would that happen again today, out of the blue?

Flak reached out to them as they flew by, never a surprise yet always a surprise. Their minds were already so focused, and the flak slapped them even more awake.

The earth below them sparkled in fields of brilliant diamonds. Could this really be wartime with the land down there so beautiful, when there was no flak, when there were no fighters?

As they approached the I.P., Jim put on his flak vest and steel pot and spelled Vandagriff in the right seat. By the time they hit the initial point, everyone was alert. The briefing came sharply into focus: assembly plant with an airfield. Surely, there would be finished fighters sitting there, and however many there were, all must be destroyed. Johnson ordered window to be tossed out to confuse the radar, and goddammit they had better toss out more because the flak hit suddenly! They flew straight into a storm of black flak that knocked the ship around. Ahead they could see smoke from the previous attacks rising around Furth, their target. They flew toward a clot of buildings a little to the north and east of Furth, buildings already afire. Below, even at this height, Stewart could see an airfield with rows of brand-new fighters on the ground, factory fresh and untouched.

Johnson had turned over control of the ship to bombardier Robinson. They flew a straight, true course in Big Week kind of weather with no obstructions of any kind. The black smoke of the Messerschmitt plant reached almost to their altitude of 18,500 as they passed over it. The energy from the smoke kissed the belly of their plane, and they felt the turbulence.

"Bombs away," said Robinson. The ship lightened of its load of fragmentation bombs, and they got the "bombs clear" from aft.

Johnson switched control back from the nose and banked away. Stewart saw out the right window other bombs falling from their altitude, the bombs of many ships, and then the bombs started to hit, the effect shattering; he could see the concussions of the blasts slamming buildings and the airfield, a whirlwind of destruction that ripped into those new airplanes and blew them to flaming dust.

Flak came up again and boomed to the right and left. All at once a loud bang sounded in the flight deck and rocked them so hard that only their safety harnesses kept them in their seats. Then came an explosion right under them. It lifted the ship, Lifted the pilots. Johnson and Stewart took a moment to realize – an .88mm shell had punched up into the bottom of the ship and detonated. They felt frozen air blast straight up into the cockpit from the hole beneath them.

The flight deck cleared of smoke. As it did, Jim looked down to his left and inches from his boot sat a jagged, gaping hole nearly two feet across. He could look down through the fuselage straight to Germany.

"Hydraulics are damaged," said Johnson.

The three punch-drunk men on the flight deck, Johnson, Vandagriff, and Stewart, glanced out the left and right windows to see if propellers continued to spin. All four did, and the ship somehow managed to keep up with those around them.

"Wright to ship 447," they heard in their ears. "Skipper, are you OK?" Wright was just below them in the formation and must have seen what had happened, but radio silence must be maintained and Stewart didn't answer. Just then he realized his map case and parachute were missing; they'd been blown out of the ship and were on their way to Germany. Jim ordered a visual signal to be shot to Wright.

"Fighters, six o'clock high!" Their ship rattled to life with machine guns blazing. Oh no, was this yesterday all over again?

"Where are our damn fighters?" said Johnson into his now-dead oxygen mask. Fighters swarmed about them, Focke-Wulf 190s according to the chatter from the gunners.

The ship convulsed, struck by something, no telling what. "Damn!" said the pilot. "Where are our fighters!"

"Oh, my God!" gasped one of the gunners into the interphone. "Oh, my God." He was watching something out his gun port. "The wing's off that ship! Bail out! Bail out!" Stewart looked; saw the awful sight of a Liberator coming apart. One man had emerged from the crumpling fuselage and his chute opened and caught the air. But one only. How fast they plunged when mortally wounded, these Liberators that had no right to fly in the first place, so ungainly they were.

"No! No!" Another voice. Another 24 hit and careering to earth. Jim didn't see this one, but he knew from the sound of the observer's voice how terrible it was.

Johnson banked hard right again, and the ship responded. They were heading west toward the coast in the frozen air of an open cockpit. Jim glanced out his window and saw a ship right beside him hit – *Nine Yanks and a Jerk*. Jim could have sworn an .88 shell had just gone right through her at the cockpit. But that just couldn't be. *Nine Yanks* pitched and yawed but kept flying; at first Jim thought they were done for. In an instant he no longer worried. That was Mack Williams and he must still be alive.

Below him, Stewart saw the entire target area, buildings aflame with black smoke belching into the sky. To the north, a large building bellowed an angry volcano of smoke thousands of feet in the air. Beyond the buildings sat crumpled, mutilated German fighters that wouldn't fly again. All of a sudden, the airfield's oil tanks caught and erupted in a magnificent explosion so far below Stewart that it might as well have been on a movie screen. A mushroom cloud shot skyward as the target passed out of his view.

The flak kept at them, jostling *Dixie Flyer*. "Crap!" spat Neil Johnson in the pilot's seat. They hung on, hoping for better moments to come.

The fighters never did arrive, but the formation kept together. They flew over what looked to be wintry forest, and then they were shooting west and there wasn't any more flak, and they flew free and clear in the afternoon sky. After yesterday, after taking a flak hit on the flight deck, nobody assumed they would make it home, but minute by minute it seemed they might, their bodies frozen. They reached Pas-de-Calais at 17,000 and more flak, but the Channel beckoned ahead and they shot out over water. The formation continued to hold together as they made land at Beachy Head and followed their flight plan along the coast and then hard left to Tibenham.

The pilots stared at fuel gauges reading near empty. They talked about landing as they descended through 5,000. How were the hydraulics? Nearing Tibenham, the pilots ordered the crew to hand crank the gear down; *Dixie Flyer* remained airborne as the ships with wounded fired flares and landed first. They were running on fumes now. Johnson and Stewart wondered about the structural integrity of a ship hit so squarely by flak? Would she hold together for a landing? They discussed who should sit in the copilot's seat and Stewart said it should be him. As Vandagriff unbuckled and stepped carefully around the hole in the deck, Stewart moved his lanky legs over the seat and settled in beside Johnson. Stewart was the instructor on four-engine bombers after all – the man who had trained many of the pilots in the group.

After the wounded had been landed, Johnson was one of several pilots who ordered a flare fired to say his ship was crippled. When it was 447's turn, Neil aimed for Runway 0-3 with that beautiful base sprawled out beyond. If only they could get down safely, a shot of whiskey, a meal, and a bed awaited. They came in low over the treetops as Johnson told the crew to hold on, and the wheels smacked to earth and she had made it, enjoying the luxury of Tibenham's longest runway. Johnson and Stewart applied the brakes and on she rolled. More brakes, all the muscles they had.

With one final convulsion and physics beyond the capacity of any structure, *Dixie Flyer* came to a scraping stop. Johnson cut the engines and the pilots looked around them. The smell was ungodly. Hot metal and hot rubber. They sat in the open air near the far northern end of the runway with smoke around them and the ship cracked open at the bulkhead behind the cockpit. Muscles ached. The seatbelts had cut into their midsections. But all was still and, after so many hours, quiet. In their ears was the phantom roar of engines and the banshee screech of metal on runway.

As they unbuckled, they knew they didn't have to worry about fire with those empty tanks. So they looked at one another, at faces creased from the rubber edges of oxygen masks and tinged with frostbite, and they slowly made their way off the flight deck and down through the bomb bay as fire crews arrived.

Stewart stood looking at the ship that had gotten them all the way to Nuremburg and all the way back. The flak hadn't just hit under the cockpit. The entire fuselage had been pockmarked by shrapnel, but the crew hadn't said a word into the interphone about it.

The gunners were emerging from the bomb bay now, slowly, full of aches and pains, a group of old men in their twenties.

George Wright landed next and eased his ship around the wounded bird and turned onto the perimeter track before cutting engines. Wright's boys climbed out of their plane and trotted over to see that Stewart and crew were all right, as Liberators still in the air diverted to land on Runway 0-9. As ships continued to touch down, Jim stared at the *Dixie Flyer* lying wounded before him.

"Robbie" Robinson of Wright's crew said, "The tail of the ship was sticking up in the air and the nose was sticking up in the front," with the middle sagging in at the crack in front of the wing.

Said squadron bombardier Jim Myers of Stewart, "He was blue from the cold whistling through the holes in the plane, but he hadn't received a scratch."

Just then Mack Williams managed to land *Nine Yanks and a Jerk* and Stewart rushed over, his legs like soft putty. "Where's Mack?" he asked of the crew emerging from the plane. They pointed him to the far side where Williams stood, face ashen and dragging deeply on a cigarette.

"I thought you were a goner," said Stewart to Williams. They were two men who had survived direct flak hits and somehow still lived and breathed.

A little later, standing in the midst of so many shot-up planes, Jim was heard to murmur, "Somebody sure could get hurt in one of these damned things." But he was down and still in one piece. The Big Week bloodbath that had taken out so many yesterday and again today hadn't claimed James M. Stewart, for whatever reason. This was his tenth completed mission, and meant an Oak Leaf Cluster for his Air Medal. That meant nothing to him; getting his boys back home meant everything. But there were so many gone from his command now, and the strain of today's mission on top of all the others added up. Every time, he felt the weight of leading his squadron and sometimes, like today, his group or even his wing. He felt vulnerable for what the flak and the fighters might do to his body or his plane, sure, but mostly for what they did to his boys. "All I wanted to do was keep them alive and do our job," he said. But in the air he couldn't control who was shot at and hit. He couldn't protect them up there; he could only make the right decisions. But what if he had done better? Would it have saved them? Jim was Jim and that's the way he thought. It was getting to him. All of it. In just eleven weeks, he was ready to unravel.

LOVE AND WAR IN THE APENNINE MOUNTAINS

Eric Newby

Eric Newby (1919–2006) was a British travel writer and infantry officer who was captured during a raid on the coast of Sicily during the Second World War. He conducted a daring escape from a prisoner-of-war camp in Italy and spent two years hiding out in the Apennine Mountains, awaiting the end of the war. While on the run from Italian fascist authorities, Newby was sheltered by a Slovenian woman, Wanda Skof, whom he later married. She became his travelling companion and the subject of his memoir *Love and War in the Apennine Mountains.*

Down the Riverside

[...]

'In the name of God, get in,' he said, 'This place is swarming with Germans. It's like Potsdam.' I got in. There was nothing else to do.

I was not the only passenger. In the back seat of the little car there was a very small, toothless old man, wrapped in a moth-eaten cloak, a garment which is called *tabar* in this part of the world, and with an equally moth-eaten hat to match. Both had once been black, now they were green with age. What he was, guide, someone whom the doctor had recruited to lend verisimilitude to the outing, or simply an old man of the mountains on the way back to them from a black-market expedition I was unable to discover because, during the entire journey, he never uttered a word. He simply sat there in the back with a heavily laden rucksack on his knees, either completely ignoring the doctor when he addressed some remark to him in his dialect, or else uttering what was either a mindless chuckle or something provoked by the workings of a powerful and possibly diabolical intelligence. Whatever he was, the old man was certainly less conspicuous than I was, bolt-upright in the front seat in which I had been put in case, as the doctor said, I had to get out and run for it, absurdly English-looking in spite of my civilian clothes.

Eventually we arrived at the junction of the minor road, on which we had been travelling, with the Via Emilia. Blocking the entrance to it there was a German soldier, probably a military policeman, on a motor cycle. He had his back to us and was watching the main road on which a convoy was moving south towards the front. Knowing what sort of man the doctor was, I was afraid that he might hoot imperiously at him to get out of the way but fortunately he simply switched off the engine and waited for the man to go, which presently he did when a gap occurred in the interminable procession of vehicles, roaring away on his machine in pursuit of them, and we followed him.

Soon we overtook the front part of the convoy and the doctor drove boldly past it on the wrong side of the road, the only part of it available to him, the Via Emilia being narrow and some of the armoured vehicles, the tanks particularly, being enormous. It was a good thing that there were no vehicles coming in the other direction. They must have been halted because of the convoy, because all the time we were on the road we did not meet anything coming in the other direction.

'Sixteenth Panzer Division,' the doctor said. 'Reinforcements.' How he could know this I could not imagine. There were no insignia on the vehicles to proclaim that they were part of Sixteenth Panzer Division, but he said it with such authority that I wouldn't have dreamed of questioning what he said. He was not a man given to making idle remarks.

I was paralysed by the thought of what would happen if he was stopped, if only for having the impertinence to pass part of a German Panzer Division on the move when all other civilian vehicles had been halted in order to allow it to monopolise the road; and I was temporarily hypnotised by the sheer proximity of the enemy in such strength and numbers. I had only to stretch out my right hand through the open window as we went sedately past them to touch their tanks, which were of a sort and size that I had not even seen in diagrams; their self-propelled guns which I recognized; their half-tracked vehicles with anti-aircraft guns at the ready which looked a little like great chariots; and the lorries full of tough-looking Panzer Grenadiers, all ready to peel out of them if the convoy was attacked from the air, who looked down at our tiny vehicle with the red crosses painted on the sides and on the roof, which were making this journey possible, with a complete lack of curiosity, just as our own soldiers would have done in similar circumstances, which I found extremely comforting. Less disinterested were a lot of grumpy-looking officers up in front in a large, open Mercedes, one of them wearing the red tabs of a general on the lapels of his leather coat, who all glared at us as they probably would have done if we had passed them in a Fiat 500, a Mercedes-load of important businessmen, on the autobahn between Ulm and Stuttgart in the years before the war, and, just as I would have then, I had an insane temptation to thumb my nose at them.

Soon, mercifully, we outdistanced the convoy, passed a castle among trees, crossed a wide river with more shingle than water in it by a long bridge, the approaches to which were flanked by allegorical statuary, and after crossing another, shorter bridge, entered the city of Parma, in the centre of which, or what looked like the centre to me, the doctor's motor car broke down under the eyes of Garibaldi, a large statue of whom stood in the *piazza*.

Fortunately it was half past one in the afternoon, according to a large, elaborate clock on the face of a building in the square and the city was in the grip of the siesta. Apart from a couple of German *feldgendarmen* with metal plaques on their chests who were obviously there to direct the convoy on its way when it arrived, and whose presence in the *piazza* was probably enough to cause its depopulation, the place was deserted.

Quite soon the reinforcements for Sixteenth Panzer Division, or whatever they were, began to rumble through it and once again I felt myself the cynosure; while the doctor fiddled with the engine which was fuelled with methane gas, I pretended to help him with my head buried in the engine – as far as anyone can bury his head in the engine of a Fiat 500 – and the old man sat in the back cackling with laughter as if he was enjoying some private and incommunicable joke. 'He! Heh! Heh!'

Eventually the engine started and the rest of the journey to the mountains was without incident.

Haven in a storm

'The Baruffas will look after you,' the doctor said, 'until we can find a place deeper in the mountains. This is a safe house, although it's on the road, but stay away from the windows and don't go outside. I'll be back in a few days.' He said, 'And if you want me to I'll bring Wanda.' And he smiled one of this rare smiles.

'I do want you to,' I said.

We shook hands. I heard him reverse the Fiat out of the farmyard on to the road and drive off. Two days later the Fascists came for him while he was asleep in his bed.

Apart from the slow ticking of a long case-clock, it was very quiet in the kitchen now and sad after all the joking and drinking. The fire which had been stirred up when we arrived had died down.

'You must go,' Signor Baruffa said as soon as the doctor's car was out of earshot. He was not smiling anymore. His wife was not smiling either. She began washing the glasses from which we had been drinking strong, dark wine.

I couldn't believe him.

'Why? You said...'

'I'm afraid. *Ho Paura.*' Literally what he said was, 'I have fear.'

In the three weeks since Italy had collapsed it was an expression I had already heard many times.

'Of what are you afraid?'

In the *pianura*, which was alive with enemies of all sorts, everyone had *paura* and with good reason. Here, in the heart of the Apennines on the road to nowhere, it seemed absurd.

'I am afraid of the Germans. Am afraid of the Fascists. I am afraid of the spies. I am afraid of my neighbours, and I am afraid of having my house burned over my head and of being shot if you are found here. My wife is also afraid. Now go!'

'But where shall I go?'

'I cannot tell you where to go. Only go!'

'You *must* tell me!'

'Then go to Zanoni!' It sounded like an imprecation. 'Zanoni is poor. He has nothing to lose. His house is not like this, by the road. It is up the valley, above the mill. It will only take you an hour. Now go, and do not tell Zanoni that I sent you!'

I went. There was nothing else to do. Neither of them came to the door. As I crossed the threshold there was a long rumble of thunder immediately overhead, a blinding flash of lightning, an apocalyptic wind bent the trees in the yard, and it began to rain heavily. In all my life I had never felt so utterly abandoned and alone.

There was no difficulty in finding the way. The valley was narrow and a stream ran down through it and under a bridge on the road where the Baruffas' farm was, and from the house a path climbed high along the right side of the valley past abandoned terrace fields with mounds of pale stones standing in them, paler still under the lightning, which the people who had cultivated the land had weeded from the earth by hand.

Soon, both the stream and the far side of the valley were invisible, blotted out by the rain which was clouting down, while the thunder boomed and rolled overhead and long barbs of lightning plunged earthwards. There was nowhere to shelter from them but mercifully after a while they ceased and were replaced by sheet lightning which I hoped was less dangerous.

The path was surfaced with long cobbles and the stones were spattered with the dung of sheep and cows and what was more likely to be the dung of mules than horses in such a mountainous place. Some of it was fresh, all of it was now being washed away by the torrents of water which had turned the steeper parts of the path, which were in steps, into a series of waterfalls in which I slipped and fell on all fours, swearing monotonously. Although I did not know it then, this path was the main road to two villages higher up the mountainside, and at any other time I would almost certainly have met other people on it. I was, in fact, lucky without appreciating my luck. This was the last occasion while I was in Italy that I ever used such a public path. From now on, whenever I travelled anywhere, unguided, it was always by more unfrequented tracks or through the woods.

After I had been climbing for about three-quarters of an hour, the path descended the side of the valley to the place where the mill was. The stream was in full spate. It came boiling down over the rocks and under a little hump-backed bridge and surged against the draw-gate which shut off the water from the leat. The mill-house was a tall, narrow building with a steep-pitched roof and it had rusty iron shutters clamped tight over the windows as if for ever, and a rusty iron door. Looking at this sinister building it was difficult to know on such an evening whether it was inhabited or not; but no smoke came from the chimney and no dog barked. The only sounds that could be heard above the thunder, the howling of the wind and the roar of the water, was the furious rattling of a loose paddle on the mill-wheel and the clanking and groaning of the wheel itself as it moved a little, backwards and forwards on its bearings.

By now, although it was only five o'clock it was almost dark.

To the right of the bridge a steep track led away uphill towards a clump of trees beyond which I could just make out some low buildings. This I thought must be casa Zanoni and I squelched up the track towards it.

The house itself was more like an Irish dun than a house, a stone fort built against a rock on the hillside. It was so small that the cowshed, which had a hay loft over it, seemed bigger than the house itself and the cowshed was not large. Every few seconds the house and its outbuildings were illuminated by the lightning so that they looked as if they were coated with silver.

They were roofed with stone slabs and down towards the eaves these great tiles had rocks on top of them, rocks to stop the wind ripping them off. Smoke and sparks were streaming from the chimney of the house which had a cowl on it made from four little piles of stones with a flat piece laid on top of them, so that it looked like a shrine on a mountain with an offering burning in it. Apart from the smoke and sparks the house was as shut up and uninhabited-looking as the mill, but when I got closer I could hear, deep inside it, the sound of a dog barking.

The door of the cowshed was closed but through cracks and holes in it, faint pinpricks of light shone out into the yard in which the mire was boiling under the weight of the rain. I stood on the threshold and said, *'Permesso, non c'é nessuno?'* – 'Excuse me, is there not no one?' – using one of the useful, colloquial, ungrammatical phrases which Wanda had taught me, the sort that everyone used in this part of Italy, and which were so important to me now, and a voice said *'Avanti!'*

I pushed open the door and went in. There was a sweet warm smell of fodder and cows and the light of a lantern was casting huge, distorted shadows of the animals, which I could not yet see, on the whitewashed wall in front of me. It was like Plato's Myth of the Cave. There was the sound of milk spurting into a pail; and now that the door had closed behind me, much more faintly, there was the noise of the storm.

Then the milking stopped and I heard a stool being pulled back over the stone floor and a small man appeared from behind one of the great, looming beasts which were to the left of the doorway. He had a small, dark moustache, wispy hair all over the place, a week's bristle on his face and although, as he told me later, he was only thirty-two, to me, ten years younger, he looked almost old enough to be my father.

He was wearing a suit of what had originally been thick brown corduroy, but it had been repaired so many times with so many different sorts of stuff, old pieces of woollen and cotton cloth in faded reds and blues and greens, and bits of ancient printed material, the kind of thing you see in museums, that it was more like a patchwork quilt with a little bit of corduroy sewn on to it here and there. But how I envied him his suit at this moment. It might be decayed but at least it was warm and dry. In the Sunday best black and white striped trousers, a cotton shirt and the thin black jacket that Wanda's father had given me, all soaked through, I felt as if I had just been fished out of an icy river.

'Signor Zanoni?' I said.

'Yes, I am Zanoni. Who are you?'

'My name is Enrico.' This was the nearest anyone in the country had so far been able to get to my Christian name. My surname was beyond them. Nevertheless, I gave it and then spelt it out phonetically in Italian. 'Newby – *ENNE A DOPPIO V BER IPSILON.*'

'*NEVBU,*' he said, '*Che name strano!*' He raised the lantern above his head and shone the light into my face. 'And what are you, Signor Nevbu, a *Tedesco?*'

He took me for a German deserter from the Wehrmacht of whom there were now said to be a considerable number who had prematurely left their units at the Armistice under the mistaken impression that the war was practically over. What an appropriate word for a German *Tedesco* was. It made me think of some great creature with an armoured shell, a sort of semi-human tank of which the carapace was a living part, the sort of machine that Hieronymous Bosch might have produced if he had been asked to design a fighting vehicle. What I was probably thinking of was a *testudo*.

I told him what I was and where I had come from.

'Now tell me who sent you here,' he said, as soon as I had finished.

I told him this, too. I didn't feel that I owed the Baruffas anything; but I didn't tell him why they had sent me to him. There was no need.

'I know why,' he said. 'It's because old Baruffa has *paura*; but that's all very well, I have *paura*, too.'

I wanted to make an end of it one way or the other.

'Signor Zanoni,' I said, using one of my small store of stock phrases, '*Posso dormire nel vostro fienile?*' 'Can I sleep in your hayloft?'

'Did anyone see you on the road coming here?' he said.

I told him that I had seen no one and that I was as sure as I could be that no one had seen me.

There was a long pause before he answered, which seemed an age.

'No,' he said, finally, 'you can't.'

I knew now that I was done for. I had no food and very little money to buy any, about 100 lire which, at that time, was something like thirty shillings, and I was in no position to go shopping. The only clothes I had, apart from a pullover in the sack which I was carrying, were the ones I stood up in and everything in the sack, including my sleeping-bag was sopping wet, too. Even if I could find another house in the darkness it would be dangerous to knock on the door without knowing who the occupants were. Yet a night on the mountainside in this sort of weather would probably finish me off.

'No, you can't sleep in my hay,' he said after another equally long pause. 'You might set it on fire and where would I be then? But you can sleep in my house in a bed, and you will, too, but before we go in I have to finish with Bella.' And he went back to milking her.

THE CRUEL SEA

Nicholas Monsarrat

Nicholas Monsarrat (1910–1979) was a Lieutenant Commander in the Royal Navy during the Second World War and a prolific author known for his war stories. One of his most famous novels, *The Cruel Sea*, is a semi-autobiographical retelling of his service on Royal Navy corvettes in the North Atlantic during the war. His books are considered an accurate and harrowing description of the trials of naval escorts facing the constant threat of inclement weather and hidden German U-boats.

The torpedo struck *Compass Rose* as she was moving at almost her full speed: she was therefore mortally torn by the sea as well as by the violence of the enemy. She was hit squarely about twelve feet from her bows: there was one slamming explosion, and the noise of ripping and tearing metal, and the fatal sound of sea-water flooding in under great pressure: a blast of heat from the stricken fo'c'sle rose to the bridge like a hideous waft of incense. *Compass Rose* veered wildly from her course, and came to a shaking stop, like a dog with a bloody muzzle: her bows were very nearly blown off, and her stern was already starting to cant in the air, almost before the way was off the ship.

At the moment of disaster, Ericson was on the bridge, and Lockhart, and Wells: the same incredulous shock hit them all like a sickening body-blow. They were masked and confused by the pitch-dark night, and they could not believe that *Compass Rose* had been struck. But the ugly angle of the deck must only have one meaning, and the noise of things sliding about below their feet confirmed it. There was another noise, too, a noise which momentarily paralysed Ericson's brain and prevented him thinking at all; it came from a voice-pipe connecting the fo'c'sle with the bridge – an agonized animal howling, like a hundred dogs going mad in a pit. It was the men caught by the explosion, which must have jammed their only escape: up the voice-pipe came their shouts, their crazy hammering, their screams for help. But there was no help for them: with an executioner's hand, Ericson snapped the voice-pipe cover shut, cutting off the noise.

To Wells he said: 'Call *Viperous* on R/T. Plain Language. Say –' he did an almost violent sum in his brain; 'Say: "Torpedoed in position oh-five-oh degrees, thirty miles astern of you".'

To Lockhart he said: 'Clear away boats and rafts. But wait for the word.'

The deck started to tilt more acutely still. There was a crash from below as something heavy broke adrift and slid down the slope. Steam began to roar out of the safety-valve alongside the funnel.

Ericson thought: God, she's going down already, like *Sorrel*. Wells said: 'The R/T's smashed, sir.'

Down in the wardroom, the noise and shock had been appalling; the explosion was in the very next compartment, and the bulkhead had buckled and sagged towards them, just above the table they were eating at. They all leapt to their feet, and jumped for the doorway: for a moment there were five men at the foot of the ladder leading to the upper deck – Morell, Ferraby, Baker, Carslake, and Tomlinson, the second steward. They seemed to be mobbing each other: Baker was shouting 'My lifebelt – I've left my lifebelt!' Ferraby was being lifted off his feet by the rush, Tomlinson was waving a dish-cloth, Carslake had reached out above their heads and grabbed the hand-rail. As the group struggled, it had an ugly illusion of panic, though it was in fact no more than the swift reaction to danger. Someone had to lead the way up the ladder: by the compulsion of their peril, they had all got there at the same time.

Morell suddenly turned back against the fierce rush, buffeted his way through, and darted into his cabin. Above his bunk was a photograph of his wife: he seized it, and thrust it inside his jacket. He looked round swiftly, but there seemed nothing else he wanted.

He ran out again, and found himself already alone: the others had all got clear away, even during the few seconds of his absence. He wondered which one of them had given way.... Just as he reached the foot of the ladder there was an enormous cracking noise behind him: foolishly he turned, and through the wardroom door he saw the bulkhead split asunder and the water burst in. It flooded towards him like a cataract: quickly though he moved up the ladder, he was waist-deep before he reached the top step, and the water seemed to suck greedily at his thighs as he threw himself clear. He looked down at the swirling chaos which now covered everything – the wardroom, the cabins, all their clothes and small possessions. There was one light still burning under water, illuminating the dark-green, treacherous torrent that had so nearly trapped him. He shook himself, in fear and relief, and ran out into the open, where in the freezing night air the shouting was already wild, the deck already steep under his feet.

The open space between the boats was a dark shambles. Men blundered to and fro, cursing wildly, cannoning into each other, slipping on the unaccustomed slope of the deck: above their heads the steam from the safety-valve was reaching a crescendo of noise, as if the ship, pouring out her vitals, was screaming her rage and defiance at the same time. One of the boats was useless – it could not be launched at the angle *Compass Rose* had now reached: the other had jammed in its chocks, and no effort, however violent, could move it. Tonbridge, who was in charge, hammered and punched at it: the dozen men with him strove desperately to lift it clear: it stuck there as if pegged to the deck, it was immovable. Tonbridge said, for the fourth or fifth time: 'Come on, lads – heave!' He had to roar to make himself heard; but roaring was no use, and heaving was no use either. Gregg, who was by his shoulder, straining at the gunwale, gasped: 'It's no bloody good, Ted... she's fast.... It's the list ...' and Tonbridge called out: 'The rafts, then – clear the rafts!'

The men left the boat, which in their mortal need had failed them and wasted precious minutes, and made for the Carley floats: they blundered into each other once more, and ran full tilt into the funnel-guys, and shouted fresh curses at the confusion. Tonbridge started them lifting the raft that was on the high side of the ship, and bringing it across to the other rail; in the dark, with half a dozen fear-driven men heaving and wrenching at it, it was as if they were already fighting each other for the safety it promised. Then he stood back, looking up at the bridge where the next order – the last order of all – must come from. The bridge was crooked against the sky. He fingered his life-jacket, and tightened the straps. He said, not bothering to make his voice audible:

'It's going to be cold, lads.'

Down in the engine-room, three minutes after the explosion, Watts and E.R.A. Broughton were alone, waiting for the order of release from the bridge. They knew it ought to come, they trusted that it would.... Watts had been 'on the plate' when the torpedo struck home: on his own initiative, he had stopped the engine, and then, as the angle of their list increased, he had opened the safety-valve and let the pressure off the boilers. He had followed what was happening from the noise outside, and it was easy enough to follow. The series of crashes from forward were the bulkheads going, the trampling overhead was the boats being cleared away: the wicked down-hill angle of the ship was their doom. Now they waited, side by side in the deserted engine-room: the old E.R.A. and the young apprentice. Watts noticed that Broughton was crossing himself, and remembered he was a Roman Catholic. Good luck to him tonight.... The bell from the bridge rang sharply, and he put his mouth to the voicepipe:

'Engine-room!' he called.

'Chief,' said the Captain's far-away voice.

'Sir?'

'Leave it, and come up.'

That was all – and it was enough. 'Up you go, lad!' he said to Broughton. 'We're finished here.'

'Is she sinking?' asked Broughton uncertainly. 'Not with me on board.... Jump to it!'

D plus four minutes.... Peace had already come to the fo'c'sle; the hammering had ceased, the wild voices were choked and stilled. The torpedo had struck at a bad moment – for many people, the worst and last moment of their lives. Thirty-seven men of the port watch, seamen and stokers, had been in the mess decks at the time of the explosion: sitting about, or eating, or sleeping, or reading, or playing cards or dominoes; and doing all these things in snug warmth, behind the single closed water-tight door. None of them had got out alive: most had been killed instantly, but a few, lucky or unlucky, had raced or crawled for the door, to find it warped and buckled by the explosion, and hopelessly jammed. There was no other way out, except the gaping hole through which the water was now bursting in a broad and furious jet.

The shambles that followed was mercifully brief; but until the water quenched the last screams and uncurled the last clawing hands, it was as Ericson had heard it through the voice-pipe – a paroxysm of despair, terror, and convulsive violence, all in full and dreadful flood, an extreme corner of the human zoo for which there should be no witnesses.

At the other end of the ship, one peaceful and determined man had gone to his post and set about the job assigned to him under 'Abandon Ship Stations'. This was Wainwright, the leading torpedo-man, who, perched high in the stern which had now begun to tower over the rest of the ship, was withdrawing the primers from the depth-charges, so that they could not explode when the ship went down.

He went about the task methodically. Unscrew, pull, throw away – unscrew, pull, throw away. He whistled as he worked, a tuneless version of 'Roll out the Barrel'. Each primer took him between ten and fifteen seconds to dispose of: he had thirty depth-charges to see to: he reckoned that there would just about be time to finish.... Under his feet, the stern was steadily lifting, like one end of a gigantic see-saw: there was enough light in the gloom for him to follow the line of the ship, down the steep slope that now led straight into the sea. He could hear the steam blowing off, and the voices of the men shouting further along the upper deck. Noisy bastards, he thought, dispassionately. Pity they hadn't got anything better to do.

Alone and purposeful, he worked on. There was an obscure enjoyment in throwing over the side the equipment that had plagued him for nearly three years. The bloody things all had numbers, and special boxes, and check-lists, and history-sheets; now they were just splashes in the dark, and even these need not be counted.

Someone loomed up nearby, climbing the slope with painful effort, and bumped into him. He recognized an officer's uniform, and then Ferraby.

Ferraby said: 'Who's that?' in a strangled voice.

'The L.T., sir, I'm just chucking away the primers.'

He went on with the job, without waiting for a comment.

Ferraby was staring about him as if he were lost in some terrible dream, but presently he crossed to the other depth-charge rail and began, awkwardly, to deal with the depth-charges on that side. They worked steadily, back to back, braced against the slope of the deck. At first they were silent: then Wainwright started to whistle again, and Ferraby, as he dropped one of the primers, to sob. The ship gave a violent lurch under their feet. and the stern rose higher still, enthroning them above the sea.

D plus seven .... Ericson realized that she was going, and that nothing could stop her. The bridge now hung over the sea at an acute forward angle, the stern was lifting, the bows deep in the water, the stem itself just awash. The ship they had spent so much time and care on, their own *Compass Rose*, was pointed for her dive, and she would not be poised much longer.

He was tormented by what he had not been able to do: the signal to *Viperous*, the clearing of the boats, the shoring-up of the wardroom bulkhead, which might conceivably have been caught in time. He thought: the Admiral at Ardnacraish was right – we ought to have practised this more.... But it had all happened too quickly for them: perhaps *nothing* could have saved her, perhaps she was too vulnerable, perhaps the odds were too great, and he could clear his conscience.

Wells, alert at his elbow, said: 'Shall I ditch the books, sir?'

Ericson jerked his head up. Throwing overboard the confidential signal books and ciphers, in their weighted bag, was the last thing of all for them to do, before they went down: it was the final signal for their dissolution. He remembered having watched the man in the U-boat do it – losing his life doing it, in fact. For a moment he held back from the order, in fear and foreboding.

He looked once more down the length of his ship. She was quieter already, fatally past the turmoil and the furious endeavour of the first few minutes:

they had all done their best, and it didn't seem have been any use: now they were simply sweating out the last brief pause, before they started swimming. He thought momentarily of their position, thirty miles astern of the convoy and wondered whether any of the stern escorts would have seen *Compass Rose* catching up on their radar, and then noticed that she had faded out, and guessed what had happened. That was their only chance, on this deadly cold night.

He other said: 'Yes, Wells, throw them over.' Then he turned to another figure waiting at the back of the bridge, and called out: 'Coxswain.'

'Sir,' said Tallow.

'Pipe "Abandon ship".'

He followed Tallow down the ladder and along the steep iron dock hearing his voice bawling 'Abandon ship! Abandon ship!' ahead of him. There was a crowd of men collected, milling around in silence, edging towards the high stern: below them, on the black water, the two Carley floats had been launched and lay in wretched attendance on their peril. A handful of Tonbridge's party, having disposed of the Carleys had turned back to wrestle afresh with the boat, but it had become locked more securely still as their list increased. When Ericson was among his men, he was recognized; the words 'The skipper – the skipper' exploded in a small hissing murmur all around him, and one of the men asked: 'What's the chances, sir?'

Compass Rose trembled under their feet, and slid further forward.

A man by the rails shouted: 'I'm off lads,' and jumped headlong into the sea.

Ericson said: 'It's time to go. Good luck to you all.'

Now fear took hold. Some men jumped straight away, and struck out from the ship, panting with the cold and calling to their comrades to follow them: others held back, and crowded farther towards the stern, on the high side away from the water; when at last they jumped, many of them slid and scraped their way down the barnacled hull, and their clothes and then the softer projections of their bodies – sometimes their faces, sometimes their genitals – were torn to ribbons by the rough plating. The sea began to sprout bobbing red lights as the safetylamps were switched on: the men struck out and away, and then crowded together, shouting and calling encouragement to each other, and turned to watch *Compass Rose*. High out of the water, she seemed to be considering the plunge before she took it: the propeller, bared against the night sky, looked foolish and indecent, the canted mast was like an admonishing finger, bidding them all behave in her absence.

She did not long delay thus: she could not. As they watched, the stern rose higher still: the last man left on board, standing on the tip of the after-rail, now plunged down with a yell of fear. The noise seemed to unloose another: there was a rending crash as the whole load of depth-charges broke loose from their lashings and ploughed wildly down the length of the upper deck, and splashed into the water.

From a dozen constricted throats came the same words: 'She's going.'

There was a muffled explosion, which they could each feel like a giant hand squeezing their stomachs, and *Compass Rose* began to slide down. Now she went quickly, as if glad to be quit of her misery: the mast snapped in a ruin of rigging as she fell. When the stern dipped beneath the surface, a tumult of water leapt upwards: then the smell of oil came thick and strong towards them. It was a smell they had got used to, on many convoys: they had never thought that *Compass Rose* would ever exude the same disgusting stench.

The sea flattened, the oil spread, their ship was plainly gone: a matter of minutes had wiped out a matter of years. Now the biting cold, forgotten before the huge disaster of their loss, began to return. They were bereaved and left alone in the darkness; fifty men, two rafts, misery, fear, and the sea.

There was not room for them all on the two Carleys: there never had been room. Some sat or lay on them, some gripped the ratlines that hung down from their sides, some swam round in hopeful circles, or clung to other luckier men who had found a place. The bobbing red lights converged on the rafts: as the men swam, they gasped with fear and cold, and icy wave hit them in the face, and oil went up their nostrils and down their throats. Their hands were quickly numbed, and then their legs, and then the cold probed deep within them, searching for the main blood of their body. They thrashed about wildly, they tried to shoulder a place at the rafts, and were pushed away again: they swam round and round in the darkness, calling out, cursing their comrades, crying for help, slobbering their prayers.

Some of those grippling with the ratlines found that they could do so no longer, and drifted away. Some of those who had swallowed fuel-oil developed a paralysing cramp, and began to retch up what was poisoning them. Some of those who had torn their bodies against the ship's side were attacked by a deadly and congealing chill.

Some of those on the rafts grew sleepy as the bitter night progressed; and others lost heart as they peered round them at the black and hopeless darkness, and listened to the sea and the wind, and smelt the oil, and heard their comrades giving way before this extremity of fear and cold.

Presently, men began to die.

LETTER FROM A BIRMINGHAM JAIL

Dr Martin Luther King Jr

Dr Martin Luther King Jr (1929–1968) was an African-American civil rights activist and preacher. He served as the de facto leader of the American civil rights movement during the 1950s and 1960s and led a coordinated campaign against racial segregation and white supremacy in the United States, in particular the American South. Inspired by his Christian beliefs and his admiration for the nonviolence movement, King propelled himself into the national spotlight by leading the 1955 Montgomery Bus Boycott, and became the target of the reactionary right, the Ku Klux Klan, and the Federal Bureau of Investigation under Director J. Edgar Hoover. King explicitly tied his fight for racial justice to his opposition to poverty, capitalism and the American war in Vietnam. His shocking assassination in 1968 led to widespread suspicion that his death had been orchestrated in whole or in part by the FBI. King remains a civil rights icon and his legacy is repeatedly invoked to advocate for social justice causes across the political spectrum.

[...]

We have waited for more than 340 years for our constitutional and God given rights. The nations of Asia and Africa are moving with jetlike speed toward gaining political independence, but we still creep at horse and buggy pace toward gaining a cup of coffee at a lunch counter. Perhaps it is easy for those who have never felt the stinging darts of segregation to say, "Wait." But when you have seen vicious mobs lynch your mothers and fathers at will and drown your sisters and brothers at whim; when you have seen hate filled policemen curse, kick and even kill your black brothers and sisters; when you see the vast majority of your twenty million Negro brothers smothering in an airtight cage of poverty in the midst of an affluent society; when you suddenly find your tongue twisted and your speech stammering as you seek to explain to your six year old daughter why she can't go to the public amusement park that has just been advertised on television, and see tears welling up in her eyes when she is told that Funtown is closed to colored children, and see ominous clouds of inferiority beginning to form in her little mental sky, and see her beginning to distort her personality by developing an unconscious bitterness toward white people; when you have to concoct an answer for a five year old son who is asking: "Daddy, why do white people treat colored people so mean?"; when you take a cross county drive and find it necessary to sleep night after night in the uncomfortable corners of your automobile because no motel will accept you; when you are humiliated day in and day out by nagging signs reading "white" and "colored"; when your first name becomes "nigger," your middle name becomes "boy" (however old you are) and your last name becomes "John," and your wife and mother are never given the respected title "Mrs."; when you are harried by day and haunted by night by the fact that you are a Negro, living constantly at tiptoe stance, never quite knowing what to expect next, and are plagued with inner fears and outer resentments; when you are forever fighting a degenerating sense of "nobodiness"—then you will understand why we find it difficult to wait. There comes a time when the cup of endurance runs over, and men are no longer willing to be plunged into the abyss of despair. I hope, sirs, you can understand our legitimate and unavoidable impatience. You express a great deal of anxiety over our willingness to break laws. This is certainly a legitimate concern. Since we so diligently urge people to obey the Supreme Court's decision of 1954 outlawing segregation in the public schools, at first glance it may seem rather paradoxical for us consciously to break laws. One may well ask: "How can you advocate breaking some laws and obeying others?" The answer lies in the fact that there are two types of laws: just and unjust. I would be the first to advocate obeying just laws. One has not only a legal but a moral responsibility to obey just laws. Conversely, one has a moral responsibility to disobey unjust laws. I would agree with St. Augustine that "an unjust law is no law at all."

FRONTIERS OF DISCOVERY

THE KON-TIKI EXPEDITION

By Raft Across the South Seas

Thor Heyerdahl

Thor Heyerdahl (1914–2002) was a Norwegian ethnographer who, in 1947, embarked on an ambitious sailing expedition from Peru to the Tuamotu Islands of French Polynesia on a crude, handmade raft. The perilous journey – known as the Kon-Tiki Expedition – was undertaken to prove an anthropological thesis that ancient peoples could have successfully completed similar long sea voyages. While Heyerdahl's findings were long dismissed by the scientific community, recent DNA evidence has suggested the possibility of extensive contact between South Americans and ancient inhabitants of French Polynesia. This discovery has been taken by some as a vindication of the thesis that inspired Heyerdahl's extraordinary voyage.

The weather became a little more unsettled, with scattered rain squalls, after we had entered the area nearer the South Sea islands, and the trade wind had changed its direction. It had blown steadily and surely from the south-east until we were a good way over in the Equatorial Current; then it had veered round more and more towards due east. We reached our most northerly position on June 10 with latitude 6° 19' south. We were then so close up to the equator that it looked as if we should sail above even the most northerly islands of the Marquesas group, and disappear completely in the sea without finding land. But then the trade wind swung round farther, from east to north-east, and drove us in a curve down towards the latitude of the world of islands.

It often happened that wind and sea remained unchanged for days on end, and then we clean forgot whose steering watch it was, except at night, when the watch was alone on deck. For if sea and wind were steady, the steering oar was lashed fast and the Kon-Tiki sail remained filled without our attending to it. Then the night watch could sit quietly in the cabin door and look at the stars. For if the constellations changed their position in the sky, it was time for him to go out and see whether it was the steering oar or the wind that had shifted.

It was incredible how easy it was to steer by the stars when we had seen them marching across the vault of the sky for weeks on end. Indeed, there was not much else to look at at night. We knew where we could expect to see the different constellations night after night, and when we came up towards the equator the Great Bear rose so clear of the horizon in the north that we were anxious lest we should catch a glimpse of the Pole Star, which appears when one comes from southward and crosses the equator. But as the north-easterly trade winds set in, the Great Bear sank again.

The old Polynesians were great navigators. They took bearings by the sun by day and the stars by night. Their knowledge of the heavenly bodies was astonishing. They knew that the earth was round, and had names for such abstruse conceptions as the equator and the northern and southern tropics. In Hawaii they cut charts of the ocean on the shells of round bottle gourds, and on certain other islands they made detailed maps of plaited boughs to which shells were attached to mark the islands, while the twigs marked particular currents. The Polynesians knew five planets, which they called wandering stars, and distinguished them from the fixed stars, for which they had nearly two hundred different names. A good navigator in old Polynesia knew well in what part of the sky the different stars would rise, and where they would be at different times of the night, and at different times of the year. They knew which stars culminated over the different islands, and there were cases in which an island was named after the star which culminated over it night after night and year after year.

Apart from the fact that the starry sky lay like a glittering giant compass revolving from east to west, they understood that the different stars right over their heads always showed them how far north or south they were. When the Polynesians had explored and brought under their sway their present domain which is the whole of the sea nearest to America, they maintained traffic between some of the islands for many generations to come. Historical traditions relate that when the chiefs from Tahiti visited Hawaii, which lay more than 2,000 sea miles farther north and several degrees farther west, the helmsman steered first due north by sun and stars, till the stars right above their heads told them that they were on the latitude of Hawaii. Then they turned at a right angle and steered due west till they came so near that birds and clouds told them where the group of islands lay.

Whence had the Polynesians obtained their vast astronomical knowledge, and their calendar, which was calculated with astonishing thoroughness? Certainly not from Melanesian or Malayan peoples to the westward. But the same old vanished civilised race, the "white and bearded men", who had taught Aztecs, Mayas and Incas their amazing culture in America, had evolved a curiously similar calendar and a similar astronomical knowledge which Europe in those times could not match.

In Polynesia, as in Peru, the calendar years had been so arranged as to begin on the particular day of the year when the constellation of the Pleiades first appeared above the horizon, and in both areas this constellation was considered patron of agriculture.

In Peru, where the continent slopes down towards the Pacific, there stand to this day in the desert sand the ruins of an astronomical observatory of great antiquity, a relic of the same mysterious civilised people which carved stone colossi, erected pyramids, cultivated sweet potatoes and bottle gourds, and began their year with the rising of the Pleiades. Kon-Tiki knew the stars when he set sail upon the Pacific ocean.

On July 2 the night watch could no longer sit in peace studying the night sky. We had a strong wind and nasty sea after several days of light north-easterly breeze. Late in the night we had brilliant moonlight and a quite fresh sailing wind. We measured our speed by counting the seconds we took to pass a chip flung out ahead on one side of us, and found that we were establishing a speed record. While our average speed was from twelve to eighteen "chips", in the jargon current on board, we were now for a time down to "six chips", and the phosphorescence swirled in a regular wake astern of the raft.

Four men lay snoring in the bamboo cabin while Torstein sat clicking with the Morse key and I was on steering watch. Just before midnight I caught sight of a quite unusual sea which came breaking astern of us right across the whole of my disturbed field of vision, and behind it I could see here and there the foaming crests of two more huge seas like the first following hard on its heels. If we ourselves had not just passed the place, I should have been convinced that what I saw was high surf flung up over a dangerous shoal. I gave a warning shout as the first sea came like a long wall sweeping after us in the moonlight, and wrenched the raft into position to take what was coming.

When the first sea reached us, the raft flung her stern up sideways and rose up over the wave-back which had just broken, so that it hissed and boiled all along the crest. We rode through the welter of boiling foam which poured along both sides of the raft, whilst the heavy sea itself rolled by under us. The bow flung itself up last as the wave passed and we slid stern first down into a broad trough of the waves. Immediately after the next wall of water came on and rose up, while we were again lifted hurriedly into the air and the clear water-masses broke over us aft as we shot over the edge. As a result the raft was flung right broadside on to the seas, and it was impossible to wrench her round quickly enough. The next sea came on and rose out of the stripes of foam like a glittering wall which began to fall along its upper edge just as it reached us. When it plunged down, I saw nothing else for it than to hang on as tight as I could to a projecting bamboo pole of the cabin roof, and there I held my breath while I felt the raft being flung sky-high and everything round me carried away in roaring whirlpools of foam. In a second we and the *Kon-Tiki* were above water again and gliding quietly down a gentle wave-back on the other side. Then the seas were normal again. The three great wave-walls raced on before us, and astern in the moonlight a string of cocoanuts lay bobbing in the water.

The last wave had given the cabin a violent blow, so that Torstein was flung head over heels in the wireless corner and the others woke scared by the noise, while the water gushed up between the logs and in through the wall. On the port side of the foredeck the bamboo wickerwork was blown open like a small crater, and the diving basket had been knocked flat up in the bows, but everything else was as it had been. Where the three big seas came from we have never been able to explain with certainty, unless they were due to disturbances on the sea bottom which are not so uncommon in those regions.

Two days later we had our first storm. It started by the trade wind dying away completely, and the feathery white trade wind clouds which were drifting over our heads up in the topmost blue being suddenly invaded by a thick black cloud-bank which rolled up over the horizon from southward. Then there came gusts of wind from the most unexpected directions, so that it was impossible for the steering watch to keep control. As quickly as we got our stern turned to the new direction of the wind, so that the sail bellied out stiff and safe, just as quickly the gusts came at us from another quarter, squeezed the proud bulge out of the sail, and made it swing round and thrash about to the peril of both crew and cargo. But then the wind suddenly set in to blow straight from the quarter whence the bad weather came, and as the black clouds rolled over us the breeze increased to a fresh wind which worked itself up into a real storm.

In the course of an incredibly short time the seas round about us were flung up to a height of fifteen feet, while single crests were hissing twenty and twenty-five feet above the trough of the sea, so that we had them on a level with our masthead when we ourselves were down in the trough. All hands had to scramble about on deck bent double, while the wind shook the bamboo wall and whistled and howled in all the rigging.

To protect the wireless corner we stretched canvas over the aftermost wall and port side of the cabin. All loose cargo was lashed securely, and the sail was hauled down and made fast round the bamboo yard. When the sky clouded over the sea grew dark and threatening, and in every direction it was white-crested with breaking waves. Long tracks of dead foam lay like stripes to windward down the backs of the long seas, and everywhere where the wave-ridges had broken and plunged down, green patches, like wounds, lay frothing for a long time in the blueblack sea. The crests blew away as they broke, and the spray stood like salt rain over the sea. When the tropical rain poured over us in horizontal squalls and whipped the surface of the sea, invisible all round us, the water that ran from our hair and beards tasted brackish, while we stumbled about the deck bent double, naked and frozen, seeing that all the gear was in order to weather the storm. When the storm rushed up over the horizon and gathered about us for the first time strained anticipation and anxiety were discernible in our looks. But when it was upon us in earnest, and the *Kon-Tiki* took everything that came her way with ease and buoyancy, the storm became an exciting form of sport, and we all delighted in the fury round about us which the balsa raft mastered so stylishly, always seeing that she herself lay on the wave-tops like a cork, while all the main weight of the raging water was always a few inches beneath. The sea had much in common with the mountains in such weather. It was like being out in the wilds in a storm, up in the highest mountain plateaux, naked and grey. Even though we were right in the heart of the tropics, when the raft glided up and down over the smoking waste of sea we always thought of racing downhill among snowdrifts and rock-faces.

The steering watch had to keep its eyes open in such weather. When the steepest seas passed under the forward half of the raft the logs aft rose right out of the water, but the next second they plunged down again to climb up over the next crest. Each time the seas came so close upon one another that the hindmost reached us while the first was still holding the bow in the air; then the solid sheets of water thundered in over the steering watch in a terrifying welter, but the next second the stern went up and the flood disappeared as through the prongs of a fork.

We calculated that in an ordinary calm sea, where there were usually seven seconds between the highest waves, we took in about two hundred tons of water astern in twenty-four hours, which we hardly noticed, because it just flowed quietly round the bare legs of the steering watch, and as quietly disappeared again between the logs. But in a heavy storm more than ten thousand tons of water poured on board astern in the course of twenty-four hours, seeing that loads varying from a few gallons to two or three cubic yards, and occasionally much more, flowed on board every five seconds. It sometimes broke on board with a deafening thunder-clap, so that the helmsman stood in water up to his waist and felt as if he was forcing his way against the current in a swift river. The raft seemed to stand trembling for a moment, but then the cruel load that weighed her down astern disappeared overboard again in great cascades.

Herman was out all the time with his anemometer measuring the squalls of gale force which lasted for twenty-four hours; then they gradually dropped to a stiff breeze with scattered rain squalls, which continued to keep the seas boiling round us as we tumbled on westward, with a good sailing wind. To obtain accurate wind measurements down among the towering seas Herman had, whenever possible, to make his way up to the swaying masthead, where it was all he could to hold on.

When the weather moderated, it was as though the big fish around us had become completely infuriated. The water round the raft was full of sharks, tunnies, dolphins and a few dazed bonitos, all wriggling about close under the timber of the raft and in the waves nearest to it. It was a ceaseless life and death struggle; the backs of big fishes arched themselves over the water and shot off like rockets, one chasing another in pairs while the water round the raft was repeatedly tinged with thick blood. The combatants were mainly tunnies and dolphins, and the dolphins came in big shoals which moved much more quickly and alertly than usual. The tunnies were the assailants; often a fish of 150–200 lbs. would leap high into the air holding a dolphin's bloody head in its mouth. But even if the individual dolphins dashed off with tunnies hard on their heels, the actual shoal of dolphins did not give ground, although there were always several wriggling round with big gaping wounds in their necks. Now and again the sharks, too, seemed to become blind with rage, and we saw them catch and fight with big tunnies, which met in the shark a superior enemy.

Not one single peaceful little pilot fish was to be seen. Either they had been devoured by the furious tunnies, or they had hidden in the chinks under the raft or fled far away from the battlefield. We dared not put our heads down into the water to see.

I had a nasty shock – and could not help laughing afterwards at my own complete bewilderment – when I was aft obeying a call of nature. We were accustomed to a bit of a swell in the water-closet, but it seemed contrary to all reasonable probabilities when I quite unexpectedly received a violent punch astern from something large and cold and very heavy which came butting up against me like a shark's head in the sea. I was actually on my way up the mast-stay, with a feeling that I had a shark hanging on to my hind-quarters before I collected myself. Herman, who was hanging over the steering oar doubled up with laughter, was able to tell me that a huge tunny had delivered a sideways smack at my nakedness with his 160 lbs. or so of cold fish. Afterwards, when Herman and then Torstein were on watch, the same fish tried to jump on board with the seas from astern, and twice the big fellow was right up on the end of the logs, but each time it flung itself overboard again before we could get a grip of the slippery body.

After that a stout bewildered bonito came right on board with a sea, and with that, and a tunny caught the day before, we decided to fish, to bring order into the sanguinary chaos that surrounded us.

Our diary says:

"A six-foot shark was hooked first and hauled on board. As soon as the hook was out again it was swallowed by an eight-foot shark, and we hauled that on board. When the hook came out again we got a fresh six-foot shark. and had got it over the edge of the raft when it broke loose and dived. The hook went out again at once, and an eight-foot shark came on to it and gave us a hard tussle. We had its head over the logs when all four steel lines were cut through and the shark dived into the depths. New hook out, and a seven-foot shark was hauled on board. It was now dangerous to stand on the slippery logs after fishing, because the three sharks kept on throwing up their heads and snapping, long after one would have thought they were dead. We dragged the sharks forward by the tail into a heap on the foredeck, and soon afterwards a big tunny was hooked and gave us more of a fight than any shark before we got it on board. It was so fat and heavy tht none of us could lift it by the tail.

"The sea was just as full of furious fish-backs. Another shark was hooked, but broke away when it was just being got onboard. But then we got a six-foot shark safely on board. After that a five-foot shark, which also came on board. Then we caught yet another six-foot shark and hauled it up. When the hook came our again, we hauled in a seven-foot shark."

Wherever we walked on deck, there were big sharks lying in the way, beating their tails convulsively on the deck or thrashing against the bamboo cabin as they snapped round them. Tired and worn out already when we began to fish after the nights of the storm, we became completely at sea as to which sharks were quite dead, which were still snapping convulsively if we went near them, and which were quite alive and were lying in ambush for us with their green cat's eyes. When we had nine big sharks lying round us in every direction, we were so weary of hauling on heavy lines and fighting with recalcitrant sharks, that we gave up after five hours' toil.

Next day there were fewer dolphins and tunnies but just as many sharks. We began to fish and haul them in again, but soon stopped when we perceived that all the fresh shark's blood that ran off the raft only attracted still more sharks. We threw all the dead sharks overboard and washed the whole deck clean of blood. The bamboo mats were torn by shark's teeth and rough sharkskin, and we threw the bloodiest and most torn of them overboard and replaced them with new golden-yellow bamboo mats, several layers of which were lashed fast on the foredeck.

When we turned in on these evenings, in our mind's eye we saw greedy open shark's jaws and blood. And the smell of shark meat stuck in our nostrils.

We could eat shark; it tasted like haddock if we just got the ammoniac out of the pieces of fish by putting them in the sea water for twenty-four hours. But bonito and tunny were infinitely better.

That evening, for the first time, I heard one of the fellows say that it would soon be pleasant to be able to stretch oneself out comfortably on the green grass on a palm island; he would be glad to see something else than cold fish and rough sea.

The weather had become quite quiet again, but it was never as constant and dependable as before. Incalculable, violent gusts of wind from time to time brought with them heavy showers, which we were glad to see because a large part of our water supply had begun to go bad and tasted like evil-smelling marsh water. When it was pouring its hardest, we collected water from the cabin roof and stood on deck naked, thoroughly to enjoy the luxury of having the salt washed off with fresh water.

The pilot fish were wriggling along again in their usual places, but whether they were the same old ones which had returned after the blood-bath, or whether they were new followers taken over in the heat of battle, we could not say.

On July 21 the wind suddenly died away again. It was oppressive and absolutely still, and we knew from previous experience what this might mean. And right enough, after a few violent gusts from east and west and south, the wind freshened up to a breeze from southward, where black, threatening clouds had again rushed up over the horizon. Herman was out with his anemometer all the time, measuring already fifty feet and more per second, when suddenly Torstein's sleeping bag went overboard. And what happened in the next few seconds took a much shorter time than it takes to tell it.

Herman tried to catch the bag as it went, took a rash step and fell overboard. We heard a faint cry for help amid the noise of the waves, and saw Herman's head and a waving arm, as well as some vague green object twirling about in the water near him. He was struggling for life to get back to the raft through the high seas which had lifted him out from the port side. Torstein, who was at the steering oar aft, and I myself, up in the bows, were the first to perceive him, and we went cold with fear. We bellowed "man overboard!" at the pitch of our lungs as we rushed to the nearest life-saving gear. The others had not heard Herman's cry at all because of the noise of the sea, but in a trice there was life and bustle on deck. Herman was an excellent swimmer, and though we realised at once that his life was at stake, we had a fair hope that he would manage to crawl back to the edge of the raft before it was too late.

Torstein, who was nearest, seized the bamboo drum round which was the line we used for the lifeboat, for this was within his reach. It was the only time on the whole voyage that this line got caught up. The whole thing happened in a few seconds Herman was now on a level with the stem of the raft, but a few yards away, and his last hope was to crawl to the blade of the steering oar and hang on to it. As he missed the end of the logs, he reached out for the oar-blade, but it slipped away from him. And there he lay, just where experience had shown we could get nothing back. While Bengt and I launched the dinghy, Knut and Erik threw out the lifebelt. Carrying a long line, it hung ready for use on the corner of the cabin roof, but to-day the wind was so strong that when they threw the lifebelt it was simply blown back to the raft. After a few unsuccessful throws Herman was already far astern of the steering oar, swimming desperately to keep up with the raft, while the distance increased with each gust of wind. He realised that henceforth the gap would simply go on increasing, but he set a faint hope on the dinghy, which we had now got into the water. Without the line which acted as a brake, it would perhaps have been practicable to drive the rubber raft to meet the swimming man, but whether the rubber raft would ever get back to the *Kon-Tiki* was another matter. Nevertheless, three men in a rubber dinghy had some chance, one man in the sea had none.

Then we suddenly saw Knut take off and plunge head first into the sea. He had the lifebelt in one hand and was heaving himself along. Every time Herman's head appeared on a waveback Knut was gone, and every time Knut came up Herman was not there. But then we saw both heads at once; they had swum to meet each other and both were hanging on to the lifebelt. Knut waved his arm, and as the rubber raft had meanwhile been hauled on board, all four of us took hold of the line of the lifebelt and hauled for dear life, with our eyes fixed on the great dark object which was visible just behind the two men. This same mysterious beast in the water was pushing a big greenish-black triangle up above the wave-crests; it almost gave Knut a shock when he was on his way over to Herman. Only Herman knew then that the triangle did not belong to a shark or any other sea monster. It was an inflated comer of Torstein's watertight sleeping bag. But the sleeping bag did not remain floating for long after we had hauled the two men safe and sound on board. Whatever dragged the sleeping bag down into the depths had just missed a better prey.

"Glad I wasn't in it," said Torstein, and took hold of the steering oar where he had let it go.

But otherwise there were not many cheery cracks that evening. We all felt a chill running through nerve and bone for a long time afterwards. But the cold shivers were mingled with a warm thankfulness that there were still six of us on board.

We had a lot of nice things to say to Knut that day, Herman and the rest of us too.

But there was not much time to think about what had happened already, for as the sky grew black over our heads the gusts of wind increased in strength, and before night came a new storm was upon us. At last we got the lifebelt to hang astern of the raft on a long line, so that we had something behind the steering oar towards which to swim if one of us should fall overboard again in a squall. Then it grew pitch dark around us as night fell and hid the raft and the sea, and bouncing wildly up and down in the darkness, we only heard and felt the gale howling in masts and guy-ropes, while the gusts pressed with smashing force against the springy bamboo cabin, till we thought it would fly overboard. But it was covered with canvas and well guyed. And we felt the *Kon-Tiki* tossing with the foaming seas, while the logs moved up and down with the waves' movement like the keys of an instrument. We were just as astonished every time that cascades of water did not gush up through the wide chinks in the floor, but they only acted as regular bellows through which damp air rushed up and down.

For five whole days the weather varied between full storm and light gale; the sea was dug up into wide valleys filled with the smoke from foaming grey-blue seas which seemed to have their backs pressed out long and flat under the onset of the wind. Then on the fifth day the heavens split to show a glimpse of blue, and the malignant black cloud-cover gave place to the ever victorious blue sky as the storm passed on. We had come through the gale with the steering oar smashed and the sail rent, and the centre-boards hung loose and banged about like crowbars among the logs, because all the ropes which had tightened them up under water were worn through. But we ourselves and the cargo were completely undamaged.

After the two storms the *Kon-Tiki* had become a good deal weaker in the joints. The strain of working over the steep wave-backs had stretched all the ropes, and the continuously working logs had made the ropes eat into the balsa wood. We thanked Providence that we had followed the Inca's prescription and had not use wire ropes, which would simply have sawn the whole raft into match-wood in the gale. And if we had used bone-dry, high-floating balsa at the start, the raft would long ago have sunk into the sea under us, saturated with sea water. It was the sap in the fresh logs which served as an impregnation and hindered the water from filtering in through the porous balsa wood. But now the ropes had become so loose that it was dangerous to let one's foot slip down between two logs, for it could be crushed when they came together violently. Forward and aft, where there was no bamboo deck, we had to give at the knees when we stood with our feet wide apart on two logs at the same time. The logs aft were as slippery as banana leaves with wet seaweed, and even though we had made a regular path through the greenery where we usually walked, and had laid down a broad plank for the steering watch to stand on, it was not easy to keep one's foothold when a sea struck the raft. And on the port side one of the nine giants bumped and banged against the cross-beams with dull, wet thuds both by night and by day. There came also new and fearful creakings from the ropes which held the two sloping masts together at the masthead, for the steps of the masts worked about independently of each other, because they rested on two different logs.

We got the steering oar spliced and lashed with long billets of mangrove wood, as hard as iron, and with Erik and Bengt as sailmakers Kon-Tiki soon raised his head again and swelled his breast in a stiff bulge towards Polynesia, while the steering oar danced behind in seas which the fine weather had made soft and gentle. But the centreboards never again became quite what they had been; they did not meet the pressure of the water with their full strength, because they gave way and hung dangling loose and unguyed under the raft. It was useless to try to inspect the ropes on the underside, for they were completely overgrown with seaweed. On taking up the whole bamboo deck we found only three of the main ropes broken; they had been lying crooked and pressed against the cargo, which had worn them away. It was evident that the logs had absorbed a great weight of water, but the cargo had been lightened, so this was roughly cancelled out. Most of our provisions and drinking water was already used up, likewise the wireless operators' dry batteries.

Nevertheless, after the last storm it was clear enough that we should both float and hold together for the short distance that separated us from the islands ahead. Now quite another problem came into the foreground – how would the voyage end?

VOYAGES ROUND THE WORLD

Captain James Cook

Captain James Cook (1728–1779) was a British Royal Navy Officer and explorer. His three voyages through largely uncharted Pacific territory yielded some of the earliest known geographical and cultural information related to Australia, New Zealand and various Pacific islands. On his third and final voyage, he was attacked and killed while trying to detain a ruling chief in the Hawaiian islands. While his legacy of colonization and exploitation remains controversial, Cook was long admired for his courage, leadership and exemplary cartography and navigation skills.

On the 31st in the evening, our latitude being nearly that of 50° south, we passed by a large island of ice, which at that instant crumbled to pieces with a tremendous explosion. The next morning a bundle of sea weeds was seen floating past the sloop; and in the afternoon, captain Furneaux in the Adventure having hailed us, acquainted captain Cook that he had seen a number of divers, resembling those in the English seas, and had past a great bed of floating rock-weeds. In consequence of these observations we stood off and on during the night, and continued an easterly course the next morning. We saw many petrels and black shear-waters, some rock-weed, and a single tern (*flerna*) or as the seamen call it an egg-bird, which had a forked tail. At noon we observed in 48° 36' south latitude, which was nearly the same in which the French discoveries are said to be situated. After noon we flood fourth-westward, but the next day the gale encreased to such a degree, as obliged us to hand our topsails, and stand on under the courses all night: however, at eight o'clock on the 4th, we found a smooth sea again, and set more sail, changing our course to the north-westward at noon. On the 6th our latitude at noon was nearly 48 degree south, about 60 degrees east from Greenwich, when not seeing any land, we gave over the attempt to stand in search of it, and directed our course of our voyage. The smoothness of the sea, whilst we had strong earsterly gales, however persuaded us, that there was probably some land near us to the eastward, and the situation given to the French discoveries, in M.Vaughondy's late chart, has confirmed our supposition; for, according to it, we must have been at least 2 degrees of longitude to the west of it, on the second of February, when we were farthest to the east in the given latitude. Though we did not fall in with the land itself, yet we have done so much service to geography by our track, as to put it beyond a doubt, that the French discovery is a small island, and not, what it was supposed at first to be, the north cape of a great sourthern continent.

On the 8th in the morning, we had an exceeding thick fog, during which we lost sight of the Adventure, our consort. We fired guns all that day and the next, at first every half hour, and afterwards every hour, without receiving any answer; and at night we burnt false fires, which like-wise proved ineffectual.

On the 10th in the morning; notwithstanding all our endevours to recover our consort, we were obliged to proceed alone on a dismal course to the southward, and to expose ourselves once more to the dangers of that frozen climate, without the hope of being saved by our fellow-voyagers, in case of losing our own vessel. Our parting with the Adventure, was almost universally regretted amoung our crew, and none of them ever looked around the ocean without expressing some concern on seeing our ship alone on this vast and unexplored expanse, where the appearance of a companion seemed to alleviate our toils, and inspired cheerfulness and comfort. We were likewise not entirely without apprehensions, that the Adventure might have fallen in with land, as the fight of penguins, of little diving petrels, and especially of a kind of grebe, seemed to vindicate its vicinity. Indeed, according to the chart of M. Vaugondy we must have been but very little to the south of it at that time.

On the 17th we were near 58 degrees south, and took up a great quantity of small ice, with which we filled our water-casks. A variety of petrels and albatrosses, had attended us continually; and from time to time the skua, or great northen gull *(larus catarractes)*, which our people called a Port Egmont hen, many pinguins, some seals, and some whales had made their appearance near us. A beautiful phaenomenon was observed during the preceding night, which appeared again this and several following nights. It consisted of long columns of a clear white light, shooting up from the horizon to the eastward, almost to the zenith, and gradually spreading on the whole southern part of the sky. These columns sometimes were bent sideways at their upper extremity, and though in most respects similar to the northen lights (*aurora borealis*) of our hemisphere, yet differed from them, in being always of a whitish colour, whereas ours affume various tints, especially those of a fiery, and purple hue. The stars were sometimes hid by, and sometimes faintly to be seen through the substance of these southern lights, (*aurora auftralis*), which have hitherto, as far as I can find, escaped the notice of voyagers. The sky was generally clear when they appeared, and the air sharp and cold, the thermometer standing at the freezing point.

On the 24th, being in about 62 degrees south latitude, we fell in once more with a solid field of ice, which confined our progress to the south, very much to the satisfaction of every body on board. We had now been long at sea, without receiving any refreshment; the favorable season for making discoveries towards the frozen zone, drew to an end; the weather daily became more sharp, and uncomfortable, and presaged a dreadful winter in these seas; and, lastly, the nights lengthened apace, and made our navigation more dangerous than it had hithero been. It was therefore very natural, that our people, exhausted by fatigues and the want of wholesome food, should wish for a place of refreshment, and rejoice to leave a part of the world, where they could not expect to meet with it. We continued however from this day till the 17th March to run the eastward, between 61° and 58° of south latitude, during which time we had a great share of easterly winds, which commonly brought fogs, and rains with them, and repeatedly exposed us to the most imminent danger of being wrecked against huge islands of ice. The shapes of these large frozen masses, were frequently singularly ruinous, and so far picturesque enough; among them we passed one a great size, with a hollow in the middle, resembling a grotto or cavern, which was pierced through, and admitted the light from the other side. Some had the appearance of a spire or steeple; and many others gave full scope to our imagination, which compared them to several known objects, by that means attempting to overcome the tediousness of our cruise, which the sight of birds, porpesses, seals, and whales, now too familiar to our eyes, could not prevent from falling heavily upon us. Notwithstanding our excellent preservatives, especially the four-krout, several of our people had now strong symptoms of sea-scurvy, such as bad gums, difficulty breathing, livid blotches, eruptions, contracted limbs, and greenish greasy filaments in the urine. Wort was therefore prescribed to them, and those who were the moft affected drank five pints of it a day; the contracted limbs were bathed in it, and the warm grains applied to them. By this means we succeeded to mitigate, and in some individuals entirely to remove the symptoms of this horrid disease. The rigours of the climate likewife yiolently afftected the live sheep, which we had embarked at the Cape of Good Hope. They were covered with eruptions, dwindled to mere skeletons, and would hardly take any nourishment. Our goats and sows too, miscarried in the tempestuous weather, or their offspring were killed by the cold. In short, we felt, from the numerous concurrent circumstances, that it was time to abandon the high southern latitudes, and retire to some port, where our crew might obtain refreshments, and where we might have the few sheep, which were intended as presents to the natives of the South-sea islands.

On the 16th, being in about 58 degrees of south latitude, we saw the sea luminous at night, though not to such a degree as we had observed it near the Cape, but only by means of some scattered sparks. This phaenomenon was however remarkable, on account of the high latitude we were in, and the cold weather, our thermometer being at 33¼° at noon. We saw the southern lights again during the nights of the 16th and 19th; and this last time, the coloumns formed an arch across the sky, rather brighter than any we had hitherto seen. We now stood to the north-eastward, in order to reach the south end of New-Zeeland; and on this course we had strong gales, and frequently saw weeds, especially rock-weeds, together with numbers of petrels, and other birds. We were much amused by a singular chace of several skuas or great grey gulls, after a large white albatross. The skuas seemed to get the better of this bird, notwithstanding its length of wings, and whenever they overtook it, they endeavoured to attack it under the belly; probably knowing that to be the most defenceless part; the albatross on these occasions had no other method of escaping than by settling on the water, where its formidable beak seemed to keep them at bay. The skuas are in general very strong and rapacious birds, and in the Ferro Islands frequently tear lambs to pieces, and carry them away to their nests. The albatrosses do not seem to be so rapacious, but live upon small marine animals, especially of the *mollusca,* or blubber class. They appeared in great numbers around us, as we came to the northward of 50 degrees south, only few solitary birds having gone so far to the south as we had penetrated; from whence it may be inferred, that they are properly inhabitants of the temperate zone.

As we stood to the northward, we also observed more seals every day, which came from the coast of New Zeeland. A large trunk of tree, a several bunches of weeds were seen on the 25th, and greatly exhilarated the spirits of our sailors. Soon after, the land was decried, bearing N.E. by E. at a vast distance. About five o'clock in the afternoon we were within a few miles of it, and saw some high mountains inland, and a broken rocky coast before us, where several inlets seemed to indicate an extensive bay or found. We tried soundings in 30 fathoms, but found none; however, at the mast-head they observed sunken rocks close to us, on which we immediately tacked, and stood off shore, as the weather was growing dark and misty. The next morning we found this part of New Zeeland lay to the southward of Cape West, and had not been explored by captain Cook, in the Endeavour.

This ended our first cruise in the high southern latitudes, after a space of four months and two days, out of sight of land, during which we had experienced no untoward accident, and had been safely led through numerous dangers by the by the guiding hand of Providence, which preserved our crew in good health during the whole time, a few individuals excepted. Our whole course from the Cape of Good Hope to New Zeeland, was a series of hardships, which had never been experienced before: all the disagreeable circumstances of the sails and rigging shattered to pieces, the vessel rolling gunwale to, and her upper works torn by the violence of the strain; the concomitant effects of storms, which have been painted with such strong expression, and blackness of *Colorit*, by the able writer of Anson's Voyage, were perhaps the least distressing occurences of ours. We had the perpetual feverities of a rigourous climate to cope with; our seamen and officers were exposed to rain, fleet, hail, and snow; our rigging was constantly encrusted with ice, which cut the hands of those who were obliged to touch it; our provision of fresh water was to be collected in lumps of ice floating on the sea, where the cold, and the sharp saline element alternately numbed, and scarifeced the sailors' limbs; we were perpetually exposed to the danger of running against huge masses of ice, which filled the immense Southern ocean: the frequency and sudden appearance of these perils required an almost continual exertion of the whole crew, to manage the ship with the greatest degree of precision and dispatch. The length of time which we remained out of sight of land, and the long abstinence from any sort of refreshment were equally distressful; for our hooks and lines distributed in November had hitherto been of no service, on account of our navigation in high southern latitudes, and across an unfathomable ocean, where we saw no fish except whales, and where it is well known no others can be expected; the torrid zone being the only one where they may be caught out of foundings.

——Atrum
Defendens pisces hiemat mare. HORAT.

We may add to these the dismal gloominess which always prevailed in the southern latitudes, where we had impenetrable fogs lasting for weeks together, and where we rarely saw the cheering face of the sun; a circumstance which alone is sufficient to deject the most undaunted, and to sour the spirits of the most cheerful. It is therefore justly to be wondered at, and ought to be considered as a distinguishing mark of divine protection, that we had not felt those ill effects which might have been expected, and justly dreaded as the result of such accumulated distresses.

BOOK OF THE MARVELS OF THE WORLD

Rustichello da Pisa

Marco Polo (1254–1324) was an Italian merchant who, in the late thirteenth and early fourteenth centuries, travelled East along the Silk Road, becoming one of the first Westerners to document the cultural riches of the Chinese Yuan Dynasty and the Mongol Empire under Kublai Khan for European readers. During his travels, he passed through much of modern-day India, Japan, Sri Lanka, Indonesia, Myanmar and Vietnam. His autobiography, dictated to the Italian romance-writer **Rustichello da Pisa**, contained the first ever written record of gunpowder, paper currency and porcelain in the West. It inspired generations of later travellers, including Christopher Columbus, and remains a valuable source of insight into medieval Western encounters with the Far East. The extracts that follow, and which are not sequential, give a sense of the breadth of Marco Polo's experiences, whether that be crossing deserts, witnessing great battles or encountering foreign cities for the first time, such as Cambaluc, on the site of what is now Beijing.

Prologue

GREAT PRINCES, Emperors, and Kings, Dukes and Marquises, Counts, Knights, and Burgesses! and People of all degrees who desire to get knowledge of the various races of mankind and of the diversities of the sundry regions of the World, take this Book and cause it to be read to you. For ye shall find therein all kinds of wonderful things, and the divers histories of the Great Hermenia, and of Persia, and of the Land of the Tartars, and of India, and of many another country of which our Book doth speak, particularly and in regular succession, according to the description of Messer Marco Polo, a wise and noble citizen of Venice, as he saw them with his own eyes. Some things indeed there be therein which he beheld not; but these he heard from men of credit and veracity. And we shall set down things seen as seen, and things heard as heard only, so that no jot of falsehood may mar the truth of our Book, and that all who shall read it or hear it read may put full faith in the truth of all its contents.

For let me tell you that since our Lord God did mould with his hands our first Father Adam, even until this day, never hath there been Christian, or Pagan, or Tartar, or Indian, or any man of any nation, who in his own person hath had so much knowledge and experience of the divers parts of the World and its Wonders as hath had this Messer Marco! And for that reason he bethought himself that it would be a very great pity did he not cause to be put in writing all the great marvels that he had seen, or on sure information heard of, so that other people who had not these advantages might, by his Book, get such knowledge. And I may tell you that in acquiring this knowledge he spent in those various parts of the World good six-and-twenty years. Now, being thereafter an inmate of the Prison at Genoa, he caused Messer Rusticiano of Pisa, who was in the said Prison likewise, to reduce the whole to writing; and this befell in the year 1298 from the birth of Jesus.

Of the City of Lop and the Great Desert

Lop is a large town at the edge of the Desert, which is called the Desert of Lop, and is situated between east and north-east. It belongs to the Great Kaan, and the people worship Mahommet. Now, such persons as propose to cross the Desert take a week's rest in this town to refresh themselves and their cattle; and then they make ready for the journey, taking with them a month's supply for man and beast. On quitting this city they enter the Desert.

The length of this Desert is so great that 'tis said it would take a year and more to ride from one end of it to the other. And here, where its breadth is least, it takes a month to cross it. 'Tis all composed of hills and valleys of sand, and not a thing to eat is to be found on it. But after riding for a day and a night you find fresh water, enough mayhap for some 50 or 100 persons with their beasts, but not for more. And all across the Desert you will find water in like manner, that is to say, in some 28 places altogether you will find good water, but in no great quantity; and in four places also you find brackish water.

Beasts there are none; for there is nought for them to eat. But there is a marvellous thing related of this Desert, which is that when travellers are on the move by night, and one of them chances to lag behind or to fall asleep or the like, when he tries to gain his company again he will hear spirits talking, and will suppose them to be his comrades. Sometimes the spirits will call him by name; and thus shall a traveller ofttimes be led astray so that he never finds his party. And in this way many have perished. Even in the day-time one hears those spirits talking. And sometimes you shall hear the sound of a variety of musical instruments, and still more commonly the sound of drums.

So thus it is that the Desert is crossed.

Of the Battle that the Great Kaan fought with Nayan

What shall I say about it? When day had well broken, there was the Kaan with all his host upon a hill overlooking the plain where Nayan lay in his tent, in all security, without the slightest thought of any one coming thither to do him hurt. In fact, this confidence of his was such that he kept no vedettes whether in front or in rear; for he knew nothing of the coming of the Great Kaan, owing to all the approaches having been completely occupied as I told you. Moreover, the place was in a remote wilderness, more than thirty marches from the Court, though the Kaan had made the distance in twenty, so eager was he to come to battle with Nayan.

And what shall I tell you next? The Kaan was there on the hill, mounted on a great wooden bartizan, which was borne by four well-trained elephants, and over him was hoisted his standard, so high aloft that it could be seen from all sides. His troops were ordered in battles of 30,000 men apiece; and a great part of the horsemen had each a foot-soldier armed with a lance set on the crupper behind him (for it was thus that the footmen were disposed of); and the whole plain seemed to be covered with his forces. So it was thus that the Great Kaan's army was arrayed for battle.

When Nayan and his people saw what had happened, they were sorely confounded, and rushed in haste to arms. Nevertheless they made them ready in good style and formed their troops in an orderly manner. And when all were in battle array on both sides as I have told you, and nothing remained but to fall to blows, then might you have heard a sound arise of many instruments of various music, and of the voices of the whole of the two hosts loudly singing. For this is a custom of the Tartars, that before they join battle they all unite in singing and playing on a certain two-stringed instrument of theirs, a thing right pleasant to hear. And so they continue in their array of battle, singing and playing in this pleasing manner, until the great Naccara of the Prince is heard to sound. As soon as that begins to sound the fight also begins on both sides; and in no case before the Prince's Naccara sounds dare any commence fighting.

So then, as they were thus singing and playing, though ordered and ready for battle, the great Naccara of the Great Khan began to sound. And that of Nayan also began to sound. And thenceforward the din of battle began to be heard loudly from this side and from that. And they rushed to work so doughtily with their bows and their maces, with their lances and swords, and with the arblasts of the footmen, that it was a wondrous sight to see. Now might you behold such flights of arrows from this side and from that, that the whole heaven was canopied with them and they fell like rain. Now might you see on this side and on that full many a cavalier and man-at-arms fall slain, insomuch that the whole field seemed covered with them. From this side and from that such cries arose from the crowds of the wounded and dying that had God thundered, you would not have heard Him! For fierce and furious was the battle, and quarter there was none given.

But why should I make a long story of it? You must know that it was the most parlous and fierce and fearful battle that ever has been fought in our day. Nor have there ever been such forces in the field in actual fight, especially of horsemen, as were then engaged—for, taking both sides, there were not fewer than 760,000 horsemen, a mighty force! and that without reckoning the footmen, who were also very numerous. The battle endured with various fortune on this side and on that from morning till noon. But at the last, by God's pleasure and the right that was on his side, the Great Khan had the victory, and Nayan lost the battle and was utterly routed. For the army of the Great Kaan performed such feats of arms that Nayan and his host could stand against them no longer, so they turned and fled. But this availed nothing for Nayan; for he and all the barons with him were taken prisoners, and had to surrender to the Kaan with all their arms.

Now you must know that Nayan was a baptized Christian, and bore the cross on his banner; but this nought availed him, seeing how grievously he had done amiss in rebelling against his Lord. For he was the Great Kaan's liegeman, and was bound to hold his lands of him like all his ancestors before him.

Concerning the City of Cambaluc

Now there was on that spot in old times a great and noble city called Cambaluc, which is as much as to say in our tongue "The city of the Emperor". But the Great Kaan was informed by his Astrologers that this city would prove rebellious, and raise great disorders against his imperial authority. So he caused the present city to be built close beside the old one, with only a river between them. And he caused the people of the old city to be removed to the new town that he had founded; andthis is called Taidu.

As regards the size of this (new) city you must know that it has a compass of 24 miles, for each side of it hath a length of 6 miles, and it is four-square.

And it is all walled round with walls of earth which have a thickness of full ten paces at bottom, and a height of more than 10 paces; but they are not so thick at top, for they diminish in thickness as they rise, so that at top they are only about 3 paces thick. And they are provided throughout with loop-holed battlements, which are all whitewashed

There are 12 gates, and over each gate there is a great and handsome palace, so that there are on each side of the square three gates and five palaces; for (I ought to mention) there is at each angle also a great and handsome palace. In those palaces are vast halls in which are kept the arms of the city garrison.

The streets are so straight and wide that you can see right along them from end to end and from one gate to the other. And up and down the city there are beautiful palaces, and many great and fine hostelries, and fine houses in great numbers. Moreover, in the middle of the city there is a great clock – that is to say, a bell –which is struck at night. And after it has struck three times no one must go out inthe city, unless it be for the needs of a woman in labour, or of the sick. And those who go about on such errands are bound to carry lanterns with them. Moreover, the established guard at each gate of the city is 1,000 armed men; not that you are to imagine this guard is kept up for fear of any attack, but only as a guard of honour for the Sovereign, who resides there, and to prevent thieves from doing mischief in the town.

THE VALLEY OF THE ASSASSINS

Freya Stark

Freya Stark (1893–1993) was an English author and explorer who wrote extensively about her travels throughout the Middle East. From an early age, Stark was fascinated by 'the East', and her early interests served as the impetus for her later travels. In 1927, she traveled to Beirut and began her eastern wanderings with a secret journey through French-controlled Lebanon and Syria, later writing in British magazines about the brutality of colonial rule in the French Mandate. She became the first Western traveler to explore parts of the remote Iranian wilderness, which she documented in her 1934 memoir *The Valley of the Assassins*. During the Second World War, Stark assisted the British propaganda effort in the Middle East by attempting to persuade Arab tribes to oppose the Axis, or at least maintain neutrality in the conflict. Dogged by accusations of antisemitism throughout her career, Stark's fierce opposition to Zionism and lack of sympathy for the plight of European Jews has tarnished her reputation as a critical examiner of late colonial interactions between East and West. Nevertheless, her writing remains a fascinating glimpse into the mind of one of the first Westerners to explore previously uncharted regions of the Middle East.

Sitt Zeinabar's Tomb

[...]

We saw ahead of us the first red pinnacles of the Alamut gorge, naked rock piled in chaos and rounded by weather, without a blade of grass upon it. Most of the bridges were washed away, but we found one, sagging in the middle but still fairly solid, and crossed to the south bank of the Shah Rud below a village called Kandichal.

Here there was no salt in the ground, and a kinder nature appeared; we rode along an overhanging cliff, high above the brown snow-water. But here

I felt too ill to continue. We came to a small solitary corrie where a white-washed shrine or *Imamzadeh* slept peacefully in front of a sloping field or two of corn. A brook and a few tangled fruit trees were on one side of it in a hollow. A grey-bearded priest, dressed in blue peasant garb and black skull cap, gave permission to stay; and Ismail put up my bed in the open, under a pear and *sanjid* tree overgrown with vines near the brook.

For nearly a week I lay there, not expecting to recover, and gazed through empty days at the barren Rudbar hills across the river, where shadows of the clouds threw patterns, the only moving things in that silent land. To look on its nakedness was in itself a preparation for the greater nakedness of death, so that gradually the mind was calmed of fear and filled with austerity and peace.

I lived on white of egg and sour milk, and had barley cooked in my water so that the taste might tell me if it were boiled, since the little stream running from the village on the hill was probably not as pure as it looked. It was an incredible effort to organize oneself for illness with only Ismail to rely on and the women of Kandichal, whose dialect was incomprehensible. One of them, called Zora, used to look after me for fourpence a day. With her rags, which hung in strips about her, she had the most beautiful and saddest face I have ever seen. She would sit on the grass by my bedside with her knees drawn up, silent by the hour, looking out with her heavy-lidded eyes to the valley below and the far slopes where the shadows travelled, like some saint whose Eternity is darkened by the remote voice of sorrow in the world. I used to wonder what she thought, but was too weak to ask, and slipped from coma to coma, waking to see rows of women squatting round my bed with their children in their arms, hopeful of quinine.

The whole Shah Rud valley is riddled with malaria and desperately poor, with no doctor. Even soap was an unknown luxury. A man of Kandichal once brought a wife from Qazvin, who remained a year before she fled back to civilization, and left a memory of soap as one among the marvels of her trousseau. But the women brought me eggs and curds in blue bowls from Hamadan to pay for my doctoring, and looked at me pityingly as they sat round in their long eastern silences. Behind us rose the mountains which cut us off from Qazvin and motor roads and posts: they were ten hours' ride away, as inaccessible as if they had been in another world, as indeed they were.

A little way off, under another patch of trees, the two mules browsed, and Ismail sat through the day smoking discontented pipes and anxious to be off. There the old Seyid used to join him, with his sickle under his arm, for it was harvest time. He would pause as he passed my bed, and with his back carefully turned out of a sense of propriety, would ask how I did and tell me that Sitt Zeinabar, the patroness of the shrine, was good for cures. He was a fine old man, descended from a venerable Seyid Tahir, and evidently much looked up to round about. Sitt Zeinabar, he said, was a daughter of the Imam Musa of Kadhimain in Iraq. I was pleased to have happened upon a female saint–so rare in these lands–and I promised her Seyid the sacrifice of a black kid if I recovered under her auspices. Her little well of water, which they called the Spring of Healing, sounded clean and pure: I made a vow to use none other for my food or drink or washing, and Zora, favourably impressed, would toil every evening across the fields with a two-handled jar, from which she poured handfuls over my face and arms, murmuring blessings in her unknown speech.

As the dusk fell, the old priest would come in from his harvests, lay down his sickle, and sit and smoke a pipe beside Ismail, while he told of his difficulties with his flock–how they had tried to take this land away from him, and Sitt Zeinabar had punished them, sending the Shah Rud down in flood for two successive years, so that their low-lying rice-fields were carried away–until they repented and gave him back his land. As it grew dark, he would get up to light his little oil wick in the shrine, which always burned the whole night through, and would borrow my matches for the purpose in the place of his flint.

By the third day I was no better, and my heart began to give trouble. I decided to send Ismail and one of the mules across the mountain range to get a prescription from some doctor in Qazvin. This he did, and came back on the afternoon of the next day with a bottle of digitalis and a letter in good English from some unknown well-wisher who "hoped that I now realized the gravity of my situation and would abandon this foolish idea of wandering unprotected over Persia." I had, as a matter of fact, very nearly abandoned any idea of wandering altogether, and was envisaging eternity under the shadow of Sitt Zeinabar' s tomb. But on the fifth day my temperature dropped, the pain ceased: I had long ago abandoned the thought of King Solomon's Throne; but I thought I could now make shift to be carried over the mountain range and find a car next evening to take me to a Teheran hospital.

In spite of myriads of mosquitoes I slept peacefully that night, soothed by the fact of having been able to decide on something. I woke now and then, and looked at Cassiopeia between the pear leaves and the vine, and finally roused myself in the gentle light of the dawn because Ismail was already packing the saddle-bags. He made a smooth platform on the mule's back, and spread my quilt on top of the luggage so that I could ride half reclining. A few early reapers and Zora and the old Seyid came to wish us good-bye. And then in the morning light I looked up at the mountains. I had not been able to see them all the days of my illness: and now they appeared beyond Alamut in the east as a vision ethereal and clean. If only I could get up among them, I thought, in the good hill air away from these mosquitoes, I would get well. Suddenly I decided not to make for hospital, but to trust myself to the hills and try to reach Solomon's Throne after all. I was already mounted by this time; all Ismail had to do was to tum the mules round and start in the opposite direction.

A Doctor in Alamut

When I reached Alamut the year before the stream was in flood, and we penetrated into the valley by a mule track above the cliff and defile of Shirkuh. This was now beyond my strength and was luckily not necessary. It was August, and the water low enough for fording, so that we could follow zig-zagging from bank to bank the defile through which the Alamut River pours itself into the Shah Rud.

Cliffs pile themselves on either hand and make a cool winding passage hardly touched by the sun. On the left, red precipices such as I had looked on through my illness; on the right, black and grey granite where the mass of Shirkuh, or Bidalan, as this part of the promontory is called, tumbles in stony ribs hundreds of feet to the water. Somewhere at the top is the Assassin castle of Durovon.

But I had enough to do without thinking of Assassins. Even on the level ground it took us three hours to reach the far side of the defile from Kandichal, and when we had done so, I lay on my quilt, injected camphor to steady my troublesome heart, and fed myself on white of egg and brandy, the only food I dared risk. We were in the last of the shadow cast by the defile, where it was filled with the pleasantness of running water that travelled there like light. The boulders by the river were covered with mauve flowers belonging to some creeping plant, and in the damper crevices a scented, milky-leaved shrub about five feet high, with bell clusters of pink flowers veined with red, swayed in the breeze of the river, and filled the place with a secret loveliness.

Having rested here, we rode for another two hours along the first hot stretch of the Alamut valley until the open lands of Shahrak appeared, green with walnuts and poplars and meadows. Under the shadow of trees, people were harvesting. The black oxen trod in a slow circle round heaps of corn, pressing out the grain with heavy wooden rollers.

"The years, like great black oxen, tread the world."

At a little distance, where the young men worked with forks, hillocks of chaff were rising, tossed and carried to one side by the wind as the heavier grain dropped down.

We dismounted and I lay under the walnut trees in the grass. Here, too, as at Sitt Zeinabar, I found myself under female jurisdiction, for the squire of Shahrak is a woman, though possibly not a saint. One of the villagers soon came to ask me to call on her; but this I was unable to do, and lay with closed eyes while the hill-women gathered round, their bright clothes and air of prosperity noticeable in comparison to the poverty outside the valley, a thing I remembered observing the year before. They were full of pity, and sat fanning the flies from my face, while a young girl, seizing my head in her two palms, pressed the temples gently and firmly, with a slowly increasing pressure, amazingly restful, that seemed to transfer her youthful strength to me.

We left again at three-thirty, hoping to reach the head of the valley and 'Aziz my guide before nightfall, in a district free of mosquitoes. It was not to be, however. The hot sandstone reaches were almost unbearable in the afternoon. I was tortured with thirst. Water seemed to draw me as if I were bewitched: I thought of Ulysses and the Sirens: it was all I could do not to slide down and lie in the streams as we waded through them. Towards five o'clock we saw the trees of the village of Shutur Khan appearing round a bend, and I decided to stop there with my friends of last year, and go no farther that night.

The first man to greet us was the owner of a little melon patch outside the village. From his small platform, a thing perched on four poles to be out of reach of mosquitoes (a fond idea), he came running to welcome us.

They were all expecting me, said he. We turned the comer, and saw the Assassin fortress, the Rock of Alamut, in the sunset, shining from its northern valley, and the Squire of Shutur Khan, owner of the Rock, standing with all his family to greet me on his doorstep.

All were very kind, and nothing had changed from the year before, except that the baby had died and a new one was coming, and the pretty daughter whose husband had deserted her was wasting away with a strange disease due, they said, to her having swallowed the shell of a nut by mistake. The two boys were as jolly as ever, and the wife had a new blue bow to her hair. As I crept to my bed on the terrace, she, with that Persian insight into beauty which redeems so many faults, told one of the servants to tum the brook into the little garden below, so that its murmur might soothe me through the night.

I was, as a matter of fact, too tired to sleep, and lay enjoying the quiet noise of the poplar leaves that moved one against the other in the moonlight. In spite of the long day, no pain or fever had returned. I felt wonderfully happy to be out of the deathly Shah Rud, and up again in the hills. The stillness of the mountain valley lay like an empty theatre round our village and its waters; the night was full of peace; when suddenly a new and strange crisis seized me: every ounce of life seemed to be sucked away; I was shrivelled up with a withering dryness, soon succeeded by floods of perspiration; and I knew by a slight unpleasant shiver that this must be malaria.

This added complication was the last straw, and made it impossible to leave again next morning. I lay gloomily in bed, while various village acquaintances came to greet me, among them 'Aziz, my good guide, with pleasure and concern all over his face and with the surprising news that a Persian doctor from the Caspian shore was spending a summer holiday in a village only five hours' ride away. He had not brought him because it would be so expensive, said 'Aziz. The doctor refused to ride for ten hours during his holiday for less than five tomans (10*s*.). But "Health is more than gold," said I, or words to that effect, and sent Ismail off at once to fetch him.

He returned in the dusk with a young man neatly dressed in European fashion, all but a collar and the shoes, which were white cotton Persian *givas*. He had a pleasant, big-nosed face, with one wall eye, over which a shock of hair continually came drooping, and a mouth which seemed always on the edge of a smile in some secret amusement of its own. He questioned me capably, and diagnosed malaria and dysentery; "diseases we are used to," he remarked.

"To-morrow, I will take you to my village, and get you well in a week," said he, while injecting camphor, emetine, and quinine in rapid succession, and in the most surprising quantities. "Now would you like a morphia injection to make you sleep!"

His ideas on quinine ran to three times the maximum marked in my medical guide, and I thought that a similar experiment with morphia might have too permanent an effect altogether. I refused, and turned my attention to a bowl of a soup called *harira*, made of rice, almonds and milk.

"Almonds," said the doctor, who had specially ordered this delicacy for me, "are most excellent for dysentery. They scrub one out like soap. Pepper is good also."

He caught a dubious look, and begged me to have confidence. "We know more than your doctors do about these diseases," he said again.

Supper was now brought and laid on a round mat on the floor by the head of my bed, where my host and the doctor sat down to it in the light of a small oil lamp. Having dispatched it, they settled to the business of opium, handing the pipe to and fro over a small charcoal brazier, a scene of dissipation in the flickering light that made me think of the "Rake's Progress," which I used to wonder over at Madame Tussaud's in my childhood. Here it was all in action, so to say, and I myself, rather surprisingly, in the picture, with the opium smokers squatting at my bedside in the Assassins' valley.

"I can see that you disapprove," said the doctor, looking up suddenly with one of his whimsical smiles. "I disapprove myself, but I do it all the same."

"It will make you die young," said I.

He shrugged his shoulders with the melancholy fatalism which is all that the East promises to retain in the absence of religion.

I was so weak that next morning I could scarcely walk across the terrace to my room, and did not think myself fit to go away. To dress and pack my few things was difficult. I fainted twice on to the saddle-bags, and finally emerged for the five hours' ride feeling anything but confident. But the doctor was cheerful if I was not. He hoisted me on to my mule, my sunshade was put in my hand, the kind people of Shutur Khan waved good-bye, and I was led, drooping and passive, up the valley, which is barren and hot for some way above the village.

We crossed and kept to the southern bank of the Alamut stream, and looked at last year's path on the other side, wondering at its extreme narrowness as it clung to red, slanting cliffs. But I was unable to notice much, and lay half reclined on my jogging platform, seeing little except the doctor in the immediate foreground who, with feet dangling below his saddle and the *Pahlevi* hat at a rakish angle over a handkerchief draped against the sun, was humming Persian love songs, and swinging a stick, while the long ears of his mule bobbed up and down before him.

After about three hours we came again to green parts of the valley, and to Zavạrak, its loveliest village, in the shade of trees. It is the largest village of all Alamut, and the brother of Nasir-ud-Din Shah seized on it as a royal gift, built a castle, and held it for twenty-five years, in spite of protests from the peasants who had never had an overlord since the days of the Assassins. When the late Shah was dethroned, the men of Zavarak took and razed the castle and returned to their independence. They are, as may be imagined, staunch supporters of the new *régime*.

They were all out now in the meadows, threshing and winnowing–a scene of prosperity in Arcadia.

Here they lifted me down and laid me on felt *Mazanderani* mats in a small room. They gave me glasses of tea, injected more camphor, and threw a cloth over me to keep away the flies, while the doctor chatted to the family and heard the village news. After three hours or so, we started again.

We climbed now southward, up the face of the Elburz range, which here hangs out an immense terrace, running parallel with the valley but about 1,000 feet above, and intersected at more or less regular intervals by wide, deep, and nearly perpendicular gullies. On this terrace are three villages. Painrud, Balarud, Verkh, each cut off from its neighbours by these gullies, each with the shoulders of Elburz behind it, and with Alamut and all the eastern hills, even to the Throne of Solomon himself supreme on the skyline, spread in a semicircle before it.

We climbed for one and a half hours, first zig-zagging up the wall from the Alamut valley and then making at a gentler but still very steep gradient over the stubble-fields of the shelf, till we reached Balarud, tilted towards the north among fenced gardens, with a brook running through its scattered houses, and every sort of fruit tree, walnut, cherry, apple, pear, medlar, and poplars and willows, throwing shade upon it.

'Aziz, who had abandoned his affairs in his village of Garmrud to attend to me, now spurred up his mule.

"Which house would you like?" said he.

I selected a high cottage with two rooms on the roof and open spaces on three sides of it. 'Aziz went to turn out the inhabitants.

With the unquestioning hospitality of the East, they cleared away most of their belongings in fifteen minutes, swept the reed matting on the floor with an inadequate brush of leaves, and allowed me to install myself while they settled in what looked like a hen-house down below. And while 'Aziz and Ismail busied themselves with the furnishing, I stood at the window and looked at King Solomon's Throne, its black arms high and sharp in the distant sky, but nearer than I had ever thought again to see it.

TRAVELS IN ASIA AND AFRICA

Ibn Battuta

Ibn Battuta (1304–1368/9) was a Moroccan scholar and adventurer whose extensive travels throughout Europe, Asia and Iberia distinguished him as the most well-travelled pre-modern explorer, surpassing even renowned travellers like Marco Polo and Zheng He. His autobiography *The Rihla* provided a wealth of information about distant lands to generations of explorers, cartographers, merchants and travellers.

I left Tangier, my birthplace, on Thursday, 2nd Rajab 725 [June 14, 1325], being at that time twenty-two years of age [22 lunar years; 21 and 4 months by solar reckoning], with the intention of making the Pilgrimage to the Holy House [at Mecca] and the Tomb of the Prophet [at Medina].

I set out alone, finding no companion to cheer the way with friendly intercourse, and no party of travellers with whom to associate myself. Swayed by an overmastering impulse within me, and a long-cherished desire to visit those glorious sanctuaries, I resolved to quit all my friends and tear myself away from my home. As my parents were still alive, it weighed grievously upon me to part from them, and both they and I were afflicted with sorrow.

On reaching the city of Tilimsan [Tlemsen], whose sultan at that time was Abu Tashifin, I found there two ambassadors of the Sultan of Tunis, who left the city on the same day that I arrived. One of the brethren having advised me to accompany them, I consulted the will of God in this matter, and after a stay of three days in the city to procure all that I needed, I rode after them with all speed. I overtook them at the town of Miliana, where we stayed ten days, as both ambassadors fell sick on account of the summer heats. When we set out again, one of them grew worse, and died after we had stopped for three nights by a stream four miles from Miliana. I left their party there and pursued my journey, with a company of merchants from Tunis.

Ibn Battuta travels overland from Algiers to Tunis

On reaching al-Jaza'ir [Algiers] we halted outside the town for a few days, until the former party rejoined us, when we went on together through the Mitija [the fertile plain behind Algiers] to the mountain of Oaks [Jurjura] and so reached Bijaya [Bougiel].

The commander of Bijaya at this time was the chamberlain Ibn Sayyid an-Nas. Now one of the Tunisian merchants of our party had died leaving three thousand dinars of gold, which he had entrusted to a certain man of Algiers to deliver to his heirs at Tunis. Ibn Sayyid an-Nas came to hear of this and forcibly seized the money. This was the first instance I witnessed of the tyranny of the agents of the Tunisian government.

At Bijaya I fell ill of a fever, and one of my friends advised me to stay there till I recovered. But I refused, saying, "If God decrees my death, it shall be on the road with my face set toward Mecca." "If that is your resolve," he replied, "sell your ass and your heavy baggage, and I shall lend you what you require. In this way you will travel light, for we must make haste on our journey, for fear of meeting roving Arabs on the way." I followed his advice and he did as he had promised – may God reward him!

On reaching Qusantinah [Constantine] we camped outside the town, but a heavy rain forced us to leave our tents during the night and take refuge in some houses there. Next day the governor of the city came to meet us. Seeing my clothes all soiled by the rain he gave orders that they should be washed at his house, and in place of my old worn headcloth sent me a headcloth of fine Syrian cloth, in one of the ends of which he had tied two gold dinars. This was the first alms I received on my journey.

From Qusantinah we reached Bona [Bone] where, after staying in the town for several days, we left the merchants of our party on account of the dangers of the road, while we pursued our journey with the utmost speed. I was again attacked by fever, so I tied myself in the saddle with a turban-cloth in case I should fall by reason of my weakness. So great was my fear that I could not dismount until we arrived at Tunis.

Ibn Battuta and his party arrive at Tunis

The population of the city came out to meet the members of our party, and on all sides greetings and question were exchanged, but not a soul greeted me as no one there was known to me. I was so affected by my loneliness that I could not restrain my tears and wept bitterly, until one of the pilgrims realized the cause of my distress and coming up to me greeted me kindly and continued to entertain me with friendly talk until I entered the city.

The Sultan of Tunis at that time was Abu Yahya, the son of Abu' Zakariya IL, and there were a number of notable scholars in the town. During my stay the festival of the Breaking of the Fast fell due, and I joined the company at the Praying-ground. The inhabitants assembled in large numbers to celebrate the festival, making a brave show and wearing their richest apparel. The Sultan Abu Yahya arrived on horseback, accompanied by all his relatives, courtiers, and officers of state walking on foot in a stately procession. After the recital of the prayer and the conclusion of the Allocution the people returned to their homes.

Ibn Battuta leaves Tunis with the annual pilgrim caravan

Some time later the pilgrim caravan for the Hijaz was formed, and they nominated me as their qadi [judge]. We left Tunis early in November [1325], following the coast road through Susa Sfax, and Qabis, where we stayed for ten days on account of incessant rains. Thence we set out for Tripoli, accompanied for several stages by a hundred or more horsemen as well as a detachment of archers, out of respect for whom the Arabs [brigands] kept their distance.

I had made a contract of marriage at Sfax with the daughter of one of the syndics at Tunis, and at Tripoli she was conducted to me, but after leaving Tripoli I became involved in a dispute with her father, which necessitated my separation from her. I then married the daughter of a student from Fez, and when she was conducted to me I detained the caravan for a day by entertaining them all at a wedding party.

Arrival at Alexandria

At length on April 5th (1326) we reached Alexandria. It is a beautiful city, well-built and fortified with four gates and a magnificent port. Among all the ports in the world I have seen none to equal it except Kawlam [Quilon] and Calicut in India, the port of the infidels [Genoese] at Sudaq [Sudak, in the Crimea] in the land of the Turks, and the port of Zaytun [Canton?] in China, all of which will be described later.

The famous lighthouse, one of the "wonders of the ancient world"

I went to see the lighthouse on this occasion and found one of its faces in ruins. It is a very high square building, and its door is above the level of the earth. Opposite the door, and of the same height, is a building from which there is a plank bridge to the door; if this is removed there is no means of entrance. Inside the door is a place for the lighthouse-keeper, and within the lighthouse there are many chambers. The breadth of the passage inside is nine spans and that of the wall ten spans; each of the four sides of the lighthouse is 140 spans in breadth. It is situated on a high mound and lies three miles from the city on a long tongue of land which juts out into the sea from close by the city wall, so that the lighthouse cannot be reached by land except from the city. On my return to the West in the year 750 [1349] I visited the lighthouse again, and found that it had fallen into so ruinous a condition that it was not possible to enter it or climb up to the door.

Al-Malik an-Nasir had started to build a similar lighthouse alongside it but was prevented by death from completing the work. Another of the marvellous things in this city is the awe-inspiring marble column [an obelisk] on its outskirts which they call the Pillar of Columns. It is a single block, skilfully carved, erected on a plinth of square stones like enormous platforms, and no one knows how it was erected there nor for certain who erected it.

Two holy men of the city

One of the learned men of Alexandria was the qadi, a master of eloquence, who used to wear a turban of extraordinary size. Never either in the eastern or the western lands have I seen a more voluminous headgear.

Another of them was the pious ascetic Burhan ad-Din, whom I met during my stay and whose hospitality I enjoyed for three days. One day as I entered his room he said to me "I see that you are fond of travelling through foreign lands." I replied "Yes, I am" (though I had as yet no thought of going to such distant lands as India or China). Then he said "You must certainly visit my brother Farid ad-Din in India, and my brother Rukn ad-Din in Sind, and my brother Burhan ad-Din in China, and when you find them give them greeting from me." I was amazed at his prediction and the idea of going to these countries having been cast into my mind, my journeys never ceased until I had met these three that he named and conveyed his greeting to them.

A visit to a holy man in the country

During my stay at Alexandria I had heard of the pious Shaykh al-Murshidi, who bestowed gifts miraculously created at his desire. He lived in solitary retreat in a cell in the country where he was visited by princes and ministers. Parties of men in all ranks of life used to come to him every day and he would supply them all with food. Each one of them would desire to eat some flesh or fruit or sweetmeat at his cell, and to each he would give what he had suggested, though it was frequently out of season. His fame was carried from mouth to mouth far and wide, and the Sultan too had visited him several times in his retreat. I set out from Alexandria to seek this shaykh and passing through Damanhur came to Fawwa [Fua], a beautiful township, close by which, separated from it by a canal, lies the shaykh's cell. I reached this cell about mid-afternoon, and on saluting the shaykh I found that he had with him one of the sultan's aides-de-camp, who had encamped with his troops just outside. The shaykh rose and embraced me, and calling for food invited me to eat. When the hour of the afternoon prayer arrived he set me in front as prayer-leader, and did the same on every occasion when we were together at the times of prayer during my stay. When I wished to sleep he said to me "Go up to the roof of the cell and sleep there" (this was during the summer heats). I said to the officer "In the name of God," but he replied [quoting from the Koran] "There is none of us but has an appointed place." So I mounted to the roof and found there a straw mattress and a leather mat, a water vessel for ritual ablutions, a jar of water and a drinking-cup, and I lay down there to sleep.

A dream of travels to come

That night, while I was sleeping on the roof of the cell, I dreamed that I was on the wing of a great bird which was flying with me towards Mecca, then to Yemen, then eastwards and thereafter going towards the south, then flying far eastwards and finally landing in a dark and green country, where it left me. I was astonished at this dream and said to myself "If the shaykh can interpret my dream for me, he is all that they say he is." Next morning, after all the other visitors had gone, he called me and when I had related my dream interpreted it to me saying: "You will make the pilgrimage [to Mecca] and visit [the Tomb of] the Prophet, and you will travel through Yemen, Iraq, the country of the Turks, and India. You will stay there for a long time and meet there my brother Dilshad the Indian, who will rescue you from a danger into which you will fall." Then he gave me a travelling-provision of small cakes and money, and I bade him farewell and departed. Never since parting from him have I met on my journeys aught but good fortune, and his blessings have stood me in good stead.

THE COURT OF THE KING

Margaret Benson

Margaret Benson (1865–1916) was an English Egyptologist and author who became the first woman to be granted a licence to excavate in Egypt, where she notably contributed to the ongoing excavations at Karnak. With a family history of mental illness, Benson suffered from frail health for most of her life and had to abandon her work. She experienced a severe mental breakdown in 1907 and spent the remainder of her life in a series of mental institutions until her death in 1916.

The great square doorway of the tomb showed inky black on the face of the cliff, golden in the moonlight; the shaft plunged steeply downwards into the rock, with short, high steps roughly cut against one wall. Down these we slowly made our way, the utter darkness pricked here and there by the flame of a candle in some one's hand. A flame shone for a moment on the little shelf cut back into the rock, where the string bed and wooden pillow of the guard still wait his return, just where he went out and left them so many thousand years ago. The steps stopped suddenly on the edge of a pit deep and broad; by the light of a candle held high we could dimly see the red and blue patterns painted on its plastered walls. A hole had been broken through them on the opposite side of the chasm, and crossing by a little plank bridge we crept through, still deeper into the heart of the cliff. On the other side of the wall the tunnel still went downwards, but the faint light showed a deep alcove to the right. On the rocky floor lay a man, bound upon a crumbling wooden boat; the painful bonds still held the brown and shrivelled limbs, his knees drawn up, his head pressed back.

Again down the steep stairway we climbed, feeling along the rough-cut wall, and again at the bottom a chamber opened to the right. A man, a woman, and a girl lie here, side by side in the middle of the floor. They have suffered the indignity of stripping; wounds are in their breasts; the thick black hair upon their heads makes the small faces and limbs seem the more withered and unhuman. It is a pitiful sight.

For the third time the rock-hewn ladder led us down to the square-cut doorway which opened to the presence-chamber of a king of Egypt. The great hall stretched back into the darkness, dimly lighted by hidden candles, heavy with the silence of three thousand years. The faint gleam fell upon the painted walls and pillars of the eternal dwelling-place, the work of such far-off hands clear and fresh with the freshness of yesterday. On the great square pillars Amenhetep still feels the fullness of his earthly life and draws strength from mysterious communing with the life-giving god. On the walls a huge papyrus seems unrolled where the spirit of the King, in the depth of the nether world, may learn to wrestle with and overthrow the serpent-monsters brought by each gloomy Hour. At the back of the hall two steps lead down to the high vaulted space where stands the great rose-granite sarcophagus. In the darkness and the silence the lid of the inner coffin was raised and we were in the presence of the King.

The dim-veiled figure lay before us, wrapt in an inexpressible mystery, the impress of his kingship still upon him, crowned with the greater dignity of death. Far from the loved Egyptian sunshine, from the sweet breath of the north wind, from the fleeting ways of men, the inhabitant of the rock holds his solemn court through the centuries which have no power upon him, with the records of his life and warfare around him and the mimosa wreaths upon his breast.

THE TOMB OF TUT ANKH AMEN, Volume 1

Howard Carter

Howard Carter (1874–1939) was an English Egyptologist who became famous for his 1922 discovery of the tomb of the Egyptian Pharaoh Tutankhamun of the 18th Dynasty. A series of mysterious deaths and illnesses of prominent members of the expedition led to a popular fascination with the idea of a Pharaoh's Curse, which Carter colourfully dismissed as 'tommy-rot'. Carter is credited with reviving popular interest in Egyptology and archaeology, and his discovery of the tomb in the Valley of the Kings yielded the best-preserved mummified remains in the history of Egyptology.

This was to be our final season in The Valley. Six full seasons we had excavated there, and season after season had drawn a blank; we had worked for months at a stretch and found nothing, and only an excavator knows how desperately depressing that can be; we had almost made up our minds that we were beaten, and were preparing to leave The Valley and try our luck elsewhere; and then – hardly had we set hoe to ground in our last despairing effort than we made a discovery that far exceeded our wildest dreams. Surely, never before in the whole history of excavation has a full digging season been compressed within the space of five days.

Let me try and tell the story of it all. It will not be easy, for the dramatic suddenness of the initial discovery left me in a dazed condition, and the months that have followed have been so crowded with incident that I have hardly had time to think. Setting it down on paper will perhaps give me a chance to realize what has happened and all that it means.

I arrived in Luxor on October 28th, and by November 1st I had enrolled my workmen and was ready to begin. Our former excavations had stopped short at the north-east corner of the tomb of Rameses VI, and from this point I started trenching southwards. It will be remembered that in this area there were a number of roughly constructed workmen's huts, used probably by the labourers in the tomb of Rameses. These huts, built about three feet above bed-rock, covered the whole area in front of the Ramesside tomb, and continued in a southerly direction to join up with a similar group of huts on the opposite side of The Valley, discovered by Davis in connexion with his work on the Akh·en·Aten cache. By the evening of November 3rd we had laid bare a sufficient number of these huts for experimental purposes, so, after we had planned and noted them, they were removed, and we were ready to clear away the three feet of soil that lay beneath them.

Hardly had I arrived on the work next morning (November 4th) than the unusual silence, due to the stoppage of the work, made me realize that something out of the ordinary had happened, and I was greeted by the announcement that a step cut in the rock had been discovered underneath the very first hut to be attacked. This seemed too good to be true, but a short amount of extra clearing revealed the fact that we were actually in the entrance of a steep cut in the rock, some thirteen feet below the entrance to the tomb of Rameses VI, and a similar depth from the present bed level of The Valley. The manner of cutting was that of the sunken stairway entrance so common in The Valley, and I almost dared to hope that we had found our tomb at last. Work continued feverishly throughout the whole of that day and the morning of the next, but it was not until the afternoon of November 5th that we succeeded in clearing away the masses of rubbish that overlay the cut, and were able to demarcate the upper edges of the stairway on all its four sides.

It was clear by now beyond any question that we actually had before us the entrance to a tomb, but doubts, born of previous disappointments, persisted in creeping in. There was always the horrible possibility, suggested by our experience in the Thothmes III Valley, that the tomb was an unfinished one, never completed and never used: if it had been finished there was the depressing probability that it had been completely plundered in ancient times. On the other hand, there was just the chance of an untouched or only partially plundered tomb, and it was with ill-suppressed excitement that I watched the descending steps of the staircase, as one by one they came to light. The cutting was excavated in the side of a small hillock, and, as the work progressed, its western edge receded under the slope of the rock until it was, first partially, and then completely, roofed in, and became a passage, 10 feet high by 6 feet wide. Work progressed more rapidly now; step succeeded step, and at the level of the twelfth, towards sunset, there was disclosed the upper part of a doorway, blocked, plastered, and sealed.

A sealed doorway – it was actually true, then! Our years of patient labour were to be rewarded after all, and I think my first feeling was one of congratulation that my faith in The Valley had not been unjustified. With excitement growing to fever heat I searched the seal impressions on the door for evidence of the identity of the owner, but could find no name: the only decipherable ones were those of the well-known royal necropolis seal, the jackal and nine captives. Two facts, however, were clear: first, the employment of this royal seal was certain evidence that the tomb had been constructed for a person of very high standing; and second, that the sealed door was entirely screened from above by workmen's huts of the Twentieth Dynasty was sufficiently clear proof that at least from that date it had never been entered. With that for the moment I had to be content.

While examining the seals I noticed, at the top of the doorway, where some of the plaster had fallen away, a heavy wooden lintel. Under this, to assure myself of the method by which the doorway had been blocked, I made a small peephole, just large enough to insert an electric torch, and discovered that the passage beyond the door was filled completely from floor to ceiling with stones and rubble – additional proof this of the care with which the tomb had been protected.

It was a thrilling moment for an excavator. Alone, save for my native workmen, I found myself, after years of comparatively unproductive labour, on the threshold of what might prove to be a magnificent discovery. Anything, literally anything, might lie beyond that passage, and it needed all my self-control to keep from breaking down the doorway, and investigating then and there.

One thing puzzled me, and that was the smallness of the opening in comparison with the ordinary Valley tombs. The design was certainly of the Eighteenth Dynasty. Could it be the tomb of a noble buried here by royal consent? Was it a royal cache, a hiding-place to which a mummy and its equipment had been removed for safety? Or was it actually the tomb of the king for whom I had spent so many years in search?

Once more I examined the seal impressions for a clue, but on the part of the door so far laid bare only those of the royal necropolis seal already mentioned were clear enough to read. Had I but known that a few inches lower down there was a perfectly clear and distinct impression of the seal of Tut·ankh·Amen, the king I most desired to find, I would have cleared on, had a much better night's rest in consequence, and saved myself nearly three weeks of uncertainty. It was late, however, and darkness was already upon us. With some reluctance I re-closed the small hole that I had made, filled in our excavation for protection during the night, selected the most trustworthy of my workmen – themselves almost as excited as I was – to watch all night above the tomb, and so home by moonlight, riding down The Valley.

Naturally my wish was to go straight ahead with our clearing to find out the full extent of the discovery, but Lord Carnarvon was in England, and in fairness to him I had to delay matters until he could come. Accordingly, on the morning of November 6th I sent him the following cable: – "At last have made wonderful discovery in Valley; a magnificent tomb with seals intact; re-covered same for your arrival; congratulations."

My next task was to secure the doorway against interference until such time as it could finally be re-opened. This we did by filling our excavation up again to surface level, and rolling on top of it the large flint boulders of which the workmen's huts had been composed. By the evening of the same day, exactly forty-eight hours after we had discovered the first step of the staircase, this was accomplished. The tomb had vanished. So far as the appearance of the ground was concerned there never had been any tomb, and I found it hard to persuade myself at times that the whole episode had not been a dream.

I was soon to be reassured on this point. News travels fast in Egypt, and within two days of the discovery congratulations, inquiries, and offers of help descended upon me in a steady stream from all directions. It became clear, even at this early stage, that I was in for a job that could not be tackled single-handed, so I wired to Callender, who had helped me on various previous occasions, asking him if possible to join me without delay, and to my relief he arrived on the very next day. On the 8th I had received two messages from Lord Carnarvon in answer to my cable, the first of which read, "Possibly come soon," and the second, received a little later, "Propose arrive Alexandria 20th."

We had thus nearly a fortnight's grace, and we devoted it to making preparations of various kinds, so that when the time of re-opening came, we should be able, with the least possible delay, to handle any situation that might arise. On the night of the 18th I went to Cairo for three days, to meet Lord Carnarvon and make a number of necessary purchases, returning to Luxor on the 21st. On the 23rd Lord Carnarvon arrived in Luxor with his daughter, Lady Evelyn Herbert, his devoted companion in all his Egyptian work, and everything was in hand for the beginning of the second chapter of the discovery of the tomb. Callender had been busy all day clearing away the upper layer of rubbish, so that by morning we should be able to get into the staircase without any delay.

By the afternoon of the 24th the whole staircase was clear, sixteen steps in all, and we were able to make a proper examination of the sealed doorway. On the lower part the seal impressions were much clearer, and we were able without any difficulty to make out on several of them the name of Tut-ankh-Amen. This added enormously to the interest of the discovery. If we had found, as seemed almost certain, the tomb of that shadowy monarch, whose tenure of the throne coincided with one of the most interesting periods in the whole of Egyptian history, we should indeed have reason to congratulate ourselves.

With heightened interest, if that were possible, we renewed our investigation of the doorway. Here for the first time a disquieting element made its appearance. Now that the whole door was exposed to light it was possible to discern a fact that had hitherto escaped notice – that there had been two successive openings and re-closings of a part of its surface: furthermore, that the sealing originally discovered, the jackal and nine captives, had been applied to the re-closed portions, whereas the sealings of Tut-ankh-Amen covered the untouched part of the doorway, and were therefore those with which the tomb had been originally secured. The tomb then was not absolutely intact, as we had hoped. Plunderers had entered it, and entered it more than once – from the evidence of the huts above, plunderers of a date not later than the reign of Rameses VI – but that they had not rifled it completely was evident from the fact that it had been re-sealed.

Then came another puzzle. In the lower strata of rubbish that filled the staircase we found masses of broken potsherds and boxes, the latter bearing the names of Akh·en·Aten, Smenkh·ka·Re and Tut·ankh·Amen, and, what was much more upsetting, a scarab of Thothmes Ill and a fragment with the name of Amen·hetep III. Why this mixture of names? The balance of evidence so far would seem to indicate a cache rather than a tomb, and at this stage in the proceedings we inclined more and more to the opinion that we were about to find a miscellaneous collection of objects of the Eighteenth Dynasty kings, brought from Tell el Amarna by Tut·ankh·Amen and deposited here for safety.

So matters stood on the evening of the 24th. On the following day the sealed doorway was to be removed, so Callender set carpenters to work making a heavy wooden grille to be set up in its place. Mr. Engelbach, Chief Inspector of the Antiquities Department, paid us a visit during the afternoon, and witnessed part of the final clearing of rubbish from the doorway.

On the morning of the 25th the seal impressions on the doorway were carefully noted and photographed, and then we removed the actual blocking of the door, consisting of rough stones carefully built from floor to lintel, and heavily plastered on their outer faces to take the seal impressions.

This disclosed the beginning of a descending passage (not a staircase), the same width as the entrance stairway, and nearly seven feet high. As I had already discovered from my hole in the doorway, it was filled completely with stone and rubble, probably the chip from its own excavation. This filling, like the doorway, showed distinct signs of more than one opening and re-closing of the tomb, the untouched part consisting of clean white chip, mingled with dust, whereas the disturbed part was composed mainly of dark flint. It was clear that an irregular tunnel had been cut through the original filling at the upper corner on the left side, a tunnel corresponding in position with that of the hole in the doorway.

As we cleared the passage we found, mixed with the rubble of the lower levels, broken potsherds, jar sealings, alabaster jars, whole and broken, vases of painted pottery, numerous fragments of smaller articles, and water skins, these last having obviously been used to bring up the water needed for the plastering of the doorways. These were clear evidence of plundering, and we eyed them askance. By night we had cleared a considerable distance down the passage, but as yet saw no sign of second doorway or of chamber.

The day following (November 26th) was the day of days, the most wonderful that I have ever lived through, and certainly one whose like I can never hope to see again. Throughout the morning the work of clearing continued, slowly perforce, on account of the delicate objects that were mixed with the filling. Then, in the middle of the afternoon, thirty feet down from the outer door, we came upon a second sealed doorway, almost an exact replica of the first. The seal impressions in this case were less distinct, but still recognizable as those of Tut·ankh·Amen and of the royal necropolis. Here again the signs of opening and re-closing were clearly marked upon the plaster. We were firmly convinced by this time that it was a cache that we were about to open, and not a tomb. The arrangement of stairway, entrance passage and doors reminded us very forcibly of the cache of Akh·en·Aten and Tyi material found in the very near vicinity of the present excavation by Davis, and the fact that Tut·ankh·Amen's seals occurred there likewise seemed almost certain proof that we were right in our conjecture. We were soon to know. There lay the sealed doorway, and behind it was the answer to the question.

Slowly, desperately slowly it seemed to us as we watched, the remains of passage debris that encumbered the lower part of the doorway were removed, until at last we had the whole door clear before us. The decisive moment had arrived. With trembling hands I made a tiny breach in the upper left hand corner. Darkness and blank space, as far as an iron testing-rod could reach, showed that whatever lay beyond was empty, and not filled like the passage we had just cleared. Candle tests were applied as a precaution against possible foul gases, and then, widening the hole a little, I inserted the candle and peered in, Lord Carnarvon, Lady Evelyn and Callender standing anxiously beside me to hear the verdict. At first I could see nothing, the hot air escaping from the chamber causing the candle flame to flicker, but presently, as my eyes grew accustomed to the light, details of the room within emerged slowly from the mist, strange animals, statues, and gold – everywhere the glint of gold. For the moment – an eternity it must have seemed to the others standing by – I was struck dumb with amazement, and when Lord Carnarvon, unable to stand the suspense any longer, inquired anxiously, "Can you see anything?" it was all I could do to get out the words, "Yes, wonderful things." Then widening the hole a little further, so that we both could see, we inserted an electric torch.

TRAVELS IN THE INTERIOR DISTRICT OF AFRICA

Mungo Park

Mungo Park (1771–1806) was a Scottish explorer who became the first Westerner to explore the interior regions of the Niger River on an expedition to find the source of the Niger and Congo rivers. He was killed on his second expedition, but his writings introduced many in the West to the first geographical and cultural narratives regarding the vast, unexplored continent that would inflame the colonial desires of Europe for the next two centuries.

Chapter XIV

Journey Continued; Arrival at Wawra

IT is impossible to describe the joy that arose in my mind when I looked around and concluded that I was out of danger. I felt like one recovered from sickness; I breathed freer; I found unusual lightness in my limbs; even the desert looked pleasant; and I dreaded nothing so much as falling in with some wandering parties of Moors, who might convey me back to the land of thieves and murderers from which I had just escaped.

I soon became sensible, however, that my situation was very deplorable, for I had no means of procuring food nor prospect of finding water. About ten o'clock, perceiving a herd of goats feeding close to the road, I took a circuitous route to avoid being seen, and continued travelling through the wilderness, directing my course by compass nearly east-south-east, in order to reach as soon as possible some town or village of the kingdom of Bambarra.

A little after noon, when the burning heat of the sun was reflected with double violence from the hot sand, and the distant ridges of the hills, seen through the ascending vapour, seemed to wave and fluctuate like the unsettled sea, I became faint with thirst, and climbed a tree in hopes of seeing distant smoke, or some other appearance of a human habitation—but in vain: nothing appeared all around but thick underwood and hillocks of white sand.

About four o'clock I came suddenly upon a large herd of goats, and pulling my horse into a bush, I watched to observe if the keepers were Moors or negroes. In a little time I perceived two Moorish boys, and with some difficulty persuaded them to approach me. They informed me that the herd belonged to Ali, and that they were going to Deena, where the water was more plentiful, and where they intended to stay until the rain had filled the pools in the desert. They showed me their empty water-skins, and told me that they had seen no water in the woods. This account afforded me but little consolation; however, it was in vain to repine, and I pushed on as fast as possible, in hopes of reaching some watering-place in the course of the night. My thirst was by this time become insufferable; my mouth was parched and inflamed; a sudden dimness would frequently come over my eyes, with other symptoms of fainting; and my horse being very much fatigued, I began seriously to apprehend that I should perish of thirst. To relieve the burning pain in my mouth and throat I chewed the leaves of different shrubs, but found them all bitter, and of no service to me.

A little before sunset, having reached the top of a gentle rising, I climbed a high tree, from the topmost branches of which I cast a melancholy look over the barren wilderness, but without discovering the most distant trace of a human dwelling. The same dismal uniformity of shrubs and sand everywhere presented itself, and the horizon was as level and uninterrupted as that of the sea.

Descending from the tree, I found my horse devouring the stubble and brushwood with great avidity; and as I was now too faint to attempt walking, and my horse too much fatigued to carry me I thought it but an act of humanity, and perhaps the last I should ever have it in my power to perform, to take off his bridle and let him shift for himself, in doing which I was suddenly affected with sickness and giddiness, and falling upon the sand, felt as if the hour of death was fast approaching. Here, then, thought I, after a short but ineffectual struggle, terminate all my hopes of being useful in my day and generation; here must the short span of my life come to an end. I cast, as I believed, a last look on the surrounding scene, and whilst I reflected on the awful change that was about to take place, this world with its enjoyment seemed to vanish from my recollection. Nature, however, at length resumed its functions, and on recovering my senses, I found myself stretched upon the sand, with the bridle still in my hand, and the sun just sinking behind the trees. I now summoned all my resolution, and determined to make another effort to prolong my existence; and as the evening was somewhat cool, I resolved to travel as far as my limbs would carry me, in hopes of reaching—my only resource—a watering-place. With this view

I put the bridle on my horse, and driving him before me, went slowly along for about an hour, when I perceived some lightning from the north-east—a most delightful sight, for it promised rain. The darkness and lightning increased very rapidly, and in less than an hour I heard the wind roaring among the bushes. I had already opened my mouth to receive the refreshing drops which I expected, but I was instantly covered with a cloud of sand, driven with such force by the wind as to give a very disagreeable sensation to my face and arms, and I was obliged to mount my horse and stop under a bush to prevent being suffocated. The sand continued to fly in amazing quantities for nearly an hour, after which I again set forward, and travelled with difficulty until ten o'clock. About this time I was agreeably surprised by some very vivid flashes of lightning, followed by a few heavy drops of rain. In a little time the sand ceased to fly, and I alighted and spread out all my clean clothes to collect the rain, which at length I saw would certainly fall. For more than an hour it rained plentifully, and I quenched my thirst by wringing and sucking my clothes.

There being no moon, it was remarkably dark, so that I was obliged to lead my horse, and direct my way by the compass, which the lightning enabled me to observe. In this manner I travelled with tolerable expedition until past midnight, when the lightning becoming more distant, I was under the necessity of groping along, to the no small danger of my hands and eyes. About two o'clock my horse started at something, and looking round, I was not a little surprised to see a light at a short distance among the trees; and supposing it to be a town, I groped along the sand in hopes of finding corn-stalks, cotton, or other appearances of cultivation, but found none. As I approached I perceived a number of other lights in different places, and began to suspect that I had fallen upon a party of Moors. However, in my present situation, I was resolved to see who they were, if I could do it with safety. I accordingly led my horse cautiously towards the light, and heard by the lowing of the cattle and the clamorous tongues of the herdsmen, that it was a watering-place, and most likely belonged to the Moors. Delightful as the sound of the human voice was to me, I resolved once more to strike into the woods, and rather run the risk of perishing of hunger than trust myself again in their hands; but being still thirsty, and dreading the approach of the burning day, I thought it prudent to search for the wells, which I expected to find at no great distance.

In this purpose I inadvertently approached so near to one of the tents as to be perceived by a woman, who immediately screamed out. Two people came running to her assistance from some of the neighbouring tents, and passed so very near to me that I thought I was discovered, and hastened again into the woods.

About a mile from this place I heard a loud and confused noise somewhere to the right of my course, and in a short time was happy to find it was the croaking of frogs, which was heavenly music to my ears. I followed the sound, and at daybreak arrived at some shallow muddy pools, so full of frogs, that it was difficult to discern the water. The noise they made frightened my horse, and I was obliged to keep them quiet, by beating the water with a branch, until he had drunk. Having here quenched my thirst, I ascended a tree, and the morning being calm, I soon perceived the smoke of the watering-place which I had passed in the night, and observed another pillar of smoke east-south-east, distant twelve or fourteen miles. Towards this I directed my route, and reached the cultivated ground a little before eleven o'clock, where, seeing a number of negroes at work planting corn, I inquired the name of the town, and was informed that it was a Foulah village belonging to Ali, called Shrilla. I had now some doubts about entering it; but my horse being very much fatigued, and the day growing hot—not to mention the pangs of hunger, which began to assail me—I resolved to venture; and accordingly rode up to the dooty's house, where I was unfortunately denied admittance, and could not obtain oven a handful of corn either for myself or horse. Turning from this inhospitable door, I rode slowly out of the town, and, perceiving some low, scattered huts without the walls, I directed my route towards them, knowing that in Africa, as well as in Europe, hospitality does not always prefer the highest dwellings. At the door of one of these huts an old motherly-looking woman sat, spinning cotton. I made signs to her that I was hungry, and inquired if she had any victuals with her in the hut. She immediately laid down her distaff, and desired me, in Arabic, to come in. When I had seated myself upon the floor, she set before me a dish of kouskous that had been left the preceding night, of which I made a tolerable meal; and in return for this kindness I gave her one of my pocket-handkerchiefs, begging at the same time a little corn for my horse, which she readily brought me.

Whilst my horse was feeding the people began to assemble, and one of them whispered something to my hostess which very much excited her surprise. Though I was not well acquainted with the Foulah language, I soon discovered that some of the men wished to apprehend and carry me back to Ali, in hopes, I suppose, of receiving a reward. I therefore tied up the corn; and lest any one should suspect I had run away from the Moors, I took a northerly direction, and went cheerfully along, driving my horse before me, followed by all the boys and girls of the town. When I had travelled about two miles, and got quit of all my troublesome attendants, I struck again into the woods, and took shelter under a large tree, where I found it necessary to rest myself, a bundle of twigs serving me for a bed, and my saddle for a pillow.

July 4.—At daybreak I pursued my course through the woods as formerly; saw numbers of antelopes, wild hogs, and ostriches, but the soil was more hilly, and not so fertile as I had found it the preceding day. About eleven o'clock I ascended an eminence, where I climbed a tree, and discovered, at about eight miles' distance, an open part of the country, with several red spots, which I concluded were cultivated land, and, directing my course that way, came to the precincts of a watering-place about one o'clock. From the appearance of the place, I judged it to belong to the Foulahs, and was hopeful that I should meet a better reception than I had experienced at Shrilla. In this I was not deceived, for one of the shepherds invited me to come into his tent and partake of some dates. This was one of those low Foulah tents in which there is room just sufficient to sit upright, and in which the family, the furniture, &c., seem huddled together like so many articles in a chest. When I had crept upon my hands and knees into this humble habitation, I found that it contained a woman and three children, who, together with the shepherd and myself, completely occupied the floor. A dish of boiled corn and dates was produced, and the master of the family, as is customary in this part of the country, first tasted it himself, and then desired me to follow his example. Whilst I was eating, the children kept their eyes fixed upon me, and no sooner did the shepherd pronounce the word *Nazarani*, than they began to cry, and their mother crept slowly towards the door, out of which she sprang like a greyhound, and was instantly followed by her children. So frightened were they at the very name of a Christian, that no entreaties could induce them to approach the tent. Here I purchased some corn for my horse, in exchange for some brass buttons, and having thanked the shepherd for his hospitality, struck again into the woods. At sunset I came to a road that took the direction for Bambarra, and resolved to follow it for the night; but about eight o'clock, hearing some people coming from the southward, I thought it prudent to hide myself among some thick bushes near the road. As these thickets are generally full of wild beasts, I found my situation rather unpleasant, sitting in the dark, holding my horse by the nose with both hands, to prevent him from neighing, and equally afraid of the natives without and the wild beasts within. My fears, however, were soon dissipated; for the people, after looking round the thicket, and perceiving nothing, went away, and I hastened to the more open parts of the wood, where I pursued my journey east-south-east, until past midnight, when the joyful cry of frogs induced me once more to deviate a little from my route, in order to quench my thirst. Having accomplished this from a large pool of rain-water, I sought for an open place, with a single tree in the midst, under which I made my bed for the night. I was disturbed by some wolves towards morning, which induced me to set forward a little before day; and having passed a small village called Wassalita, I came about ten o'clock (July 5th), to a negro town called Wawra, which properly belongs to Kaarta, but was at this time tributary to Mansong, King of Bambarra.

ON THE REVOLUTIONS OF THE HEAVENLY SPHERES

Nicolaus Copernicus

Nicolaus Copernicus (1473–1543) was a Renaissance Polish astronomer and mathematician. His book *On the Revolutions of the Heavenly Spheres* (*De revolutionibus orbium coelestium*) triggered the Copernican Revolution in astronomy which marked a shift away from the Ptolemaic model of the universe, which placed Earth at the centre of the universe, towards his heliocentric model that placed the sun at the centre. A devoted Catholic till his death, Copernicus wrote against the backdrop of a Europe divided by the Reformation, and his work was criticized by both the Catholic Church and various Protestant theologians. His theory overturned conventional thought at a time when such departures often resulted in severe and possibly even fatal consequences. His work inspired later astronomers like Tycho Brahe, Johannes Kepler, Galileo Galilei and Isaac Newton and launched a new scientific revolution that would mark the dawn of the secular age in Western Europe.

I can readily imagine, Holy Father, that as soon as some people hear that in this volume, which I have written about the revolutions of the spheres of the universe, I ascribe certain motions to the terrestrial globe, they will shout that I must be immediately repudiated together with this belief. For I am not so enamored of my own opinions that I disregard what others may think of them. I am aware that a philosopher's ideas are not subject to the judgement of ordinary persons, because it is his endeavor to seek the truth in all things, to the extent permitted to human reason by God. Yet I hold that completely erroneous views should be shunned. Those who know that the consensus of many centuries has sanctioned the conception that the earth remains at rest in the middle of the heaven as its center would, I reflected, regard it as an insane pronouncement if I made the opposite assertion that the earth moves. Therefore I debated with myself for a long time whether to publish the volume which I wrote to prove the earth's motion or rather to follow the example of the Pythagoreans and certain others, who used to transmit philosophy's secrets only to kinsmen and friends, not in writing but by word of mouth, as is shown by Lysis' letter to Hipparchus. And they did so, it seems to me, not, as some suppose, because they were in some way jealous about their teachings, which would be spread around; on the contrary, they wanted the very beautiful thoughts attained by great men of deep devotion not to be ridiculed by those who are reluctant to exert themselves vigorously in any literary pursuit unless it is lucrative; or if they are stimulated to the nonacquisitive study of philosophy by the exhortation and example of others, yet because of their dullness of mind they play the same part among philosophers as drones among bees. When I weighed these considerations, the scorn which I had reason to fear on account of the novelty and unconventionality of my opinion almost induced me to abandon completely the work which I had undertaken.

But while I hesitated for a long time and even resisted, my friends drew me back. Foremost among them was the cardinal of Capua, Nicholas Schönberg, renowned in every field of learning. Next to him was a man who loves me dearly, Tiedemann Giese, bishop of Chelmno, a close student of sacred letters as well as of all good literature. For he repeatedly encouraged me and, sometimes adding reproaches, urgently requested me to publish this volume and finally permit it to appear after being buried among my papers and lying concealed not merely until the ninth year but by now the fourth period of nine years. The same conduct was recommended to me by not a few other very eminent scholars. They exhorted me no longer to refuse, on account of the fear which I felt, to make my work available for the general use of students of astronomy. The crazier my doctrine of the earth's motion now appeared to most people, the argument ran, so much the more admiration and thanks would it gain after they saw the publication of my writings dispel the fog of absurdity by most luminous proofs. Influenced therefore by these persuasive men and by this hope, in the end I allowed my friends to bring out an edition of the volume, as they had long besought me to do.

THE LOST WORLD

Sir Arthur Conan Doyle

Professor George Edward Challenger is the fictional protagonist of **Sir Arthur Conan Doyle's** *The Lost World.* The character is based in part on Doyle's friend Percy Fawcett, a British adventurer who disappeared into the Amazon in 1925 in search of a mythical lost city (see page 388). In the novel, Challenger leads an expedition to a remote plateau in South America that is a haven for prehistoric creatures, including dinosaurs. One of his companions is young, plucky newspaper reporter, Edward Malone – the narrator of this extract, who sets out to explore the plateau one night by himself. The character of Professor Challenger and the novel itself served as the inspiration for later works of fiction, including Michael Crichton's *Jurassic Park.*

Chapter XII

'It was dreadful in the forest'

I have said—or perhaps I have not said, for my memory plays me sad tricks these days—that I glowed with pride when three such men as my comrades thanked me for having saved, or at least greatly helped, the situation. As the youngster of the party, not merely in years, but in experience, character, knowledge, and all that goes to make a man, I had been overshadowed from the first. And now I was coming into my own. I warmed at the thought. Alas! for the pride which goes before a fall! That little glow of self-satisfaction, that added measure of self-confidence, were to lead me on that very night to the most dreadful experience of my life, ending with a shock which turns my heart sick when I think of it.

It came about in this way. I had been unduly excited by the adventure of the tree, and sleep seemed to be impossible. Summerlee was on guard, sitting hunched over our small fire, a quaint, angular figure, his rifle across his knees and his pointed, goat-like beard wagging with each weary nod of his head. Lord John lay silent, wrapped in the South American poncho which he wore, while Challenger snored with a roll and rattle which reverberated through the woods. The full moon was shining brightly, and the air was crisply cold. What a night for a walk! And then suddenly came the thought, "Why not?" Suppose I stole softly away, suppose I made my way down to the central lake, suppose I was back at breakfast with some record of the place—would I not in that case be thought an even more worthy associate? Then, if Summerlee carried the day and some means of escape were found, we should return to London with first-hand knowledge of the central mystery of the plateau, to which I alone, of all men, would have penetrated. I thought of Gladys, with her "There are heroisms all round us." I seemed to hear her voice as she said it. I thought also of McArdle. What a three column article for the paper! What a foundation for a career! A correspondentship in the next great war might be within my reach. I clutched at a gun—my pockets were full of cartridges—and, parting the thorn bushes at the gate of our zareba, quickly slipped out. My last glance showed me the unconscious Summerlee, most futile of sentinels, still nodding away like a queer mechanical toy in front of the smouldering fire.

I had not gone a hundred yards before I deeply repented my rashness. I may have said somewhere in this chronicle that I am too imaginative to be a really courageous man, but that I have an overpowering fear of seeming afraid. This was the power which now carried me onwards. I simply could not slink back with nothing done. Even if my comrades should not have missed me, and should never know of my weakness, there would still remain some intolerable self-shame in my own soul. And yet I shuddered at the position in which I found myself, and would have given all I possessed at that moment to have been honorably free of the whole business.

It was dreadful in the forest. The trees grew so thickly and their foliage spread so widely that I could see nothing of the moon-light save that here and there the high branches made a tangled filigree against the starry sky. As the eyes became more used to the obscurity one learned that there were different degrees of darkness among the trees—that some were dimly visible, while between and among them there were coal-black shadowed patches, like the mouths of caves, from which I shrank in horror as I passed. I thought of the despairing yell of the tortured iguanodon—that dreadful cry which had echoed through the woods. I thought, too, of the glimpse I had in the light of Lord John's torch of that bloated, warty, blood-slavering muzzle. Even now I was on its hunting-ground. At any instant it might spring upon me from the shadows—this nameless and horrible monster. I stopped, and, picking a cartridge from my pocket, I opened the breech of my gun. As I touched the lever my heart leaped within me. It was the shot-gun, not the rifle, which I had taken!

Again the impulse to return swept over me. Here, surely, was a most excellent reason for my failure—one for which no one would think the less of me. But again the foolish pride fought against that very word. I could not—must not—fail. After all, my rifle would probably have been as useless as a shot-gun against such dangers as I might meet. If I were to go back to camp to change my weapon I could hardly expect to enter and to leave again without being seen. In that case there would be explanations, and my attempt would no longer be all my own. After a little hesitation, then, I screwed up my courage and continued upon my way, my useless gun under my arm.

The darkness of the forest had been alarming, but even worse was the white, still flood of moonlight in the open glade of the iguanodons. Hid among the bushes, I looked out at it. None of the great brutes were in sight. Perhaps the tragedy which had befallen one of them had driven them from their feeding-ground. In the misty, silvery night I could see no sign of any living thing. Taking courage, therefore, I slipped rapidly across it, and among the jungle on the farther side I picked up once again the brook which was my guide. It was a cheery companion, gurgling and chuckling as it ran, like the dear old trout-stream in the West Country where I have fished at night in my boyhood. So long as I followed it down I must come to the lake, and so long as I followed it back I must come to the camp. Often I had to lose sight of it on account of the tangled brush-wood, but I was always within earshot of its tinkle and splash.

As one descended the slope the woods became thinner, and bushes, with occasional high trees, took the place of the forest. I could make good progress, therefore, and I could see without being seen. I passed close to the pterodactyl swamp, and as I did so, with a dry, crisp, leathery rattle of wings, one of these great creatures—it was twenty feet at least from tip to tip—rose up from somewhere near me and soared into the air. As it passed across the face of the moon the light shone clearly through the membranous wings, and it looked like a flying skeleton against the white, tropical radiance. I crouched low among the bushes, for I knew from past experience that with a single cry the creature could bring a hundred of its loathsome mates about my ears. It was not until it had settled again that I dared to steal onwards upon my journey.

The night had been exceedingly still, but as I advanced I became conscious of a low, rumbling sound, a continuous murmur, somewhere in front of me. This grew louder as I proceeded, until at last it was clearly quite close to me. When I stood still the sound was constant, so that it seemed to come from some stationary cause. It was like a boiling kettle or the bubbling of some great pot. Soon I came upon the source of it, for in the center of a small clearing I found a lake—or a pool, rather, for it was not larger than the basin of the Trafalgar Square fountain—of some black, pitch-like stuff, the surface of which rose and fell in great blisters of bursting gas. The air above it was shimmering with heat, and the ground round was so hot that I could hardly bear to lay my hand on it. It was clear that the great volcanic outburst which had raised this strange plateau so many years ago had not yet entirely spent its forces. Blackened rocks and mounds of lava I had already seen everywhere peeping out from amid the luxuriant vegetation which draped them, but this asphalt pool in the jungle was the first sign that we had of actual existing activity on the slopes of the ancient crater. I had no time to examine it further for I had need to hurry if I were to be back in camp in the morning.

It was a fearsome walk, and one which will be with me so long as memory holds. In the great moonlight clearings I slunk along among the shadows on the margin. In the jungle I crept forward, stopping with a beating heart whenever I heard, as I often did, the crash of breaking branches as some wild beast went past. Now and then great shadows loomed up for an instant and were gone—great, silent shadows which seemed to prowl upon padded feet. How often I stopped with the intention of returning, and yet every time my pride conquered my fear, and sent me on again until my object should be attained.

At last (my watch showed that it was one in the morning) I saw the gleam of water amid the openings of the jungle, and ten minutes later I was among the reeds upon the borders of the central lake. I was exceedingly dry, so I lay down and took a long draught of its waters, which were fresh and cold. There was a broad pathway with many tracks upon it at the spot which I had found, so that it was clearly one of the drinking-places of the animals. Close to the water's edge there was a huge isolated block of lava. Up this I climbed, and, lying on the top, I had an excellent view in every direction.

The first thing which I saw filled me with amazement. When I described the view from the summit of the great tree, I said that on the farther cliff I could see a number of dark spots, which appeared to be the mouths of caves. Now, as I looked up at the same cliffs, I saw discs of light in every direction, ruddy, clearly-defined patches, like the port-holes of a liner in the darkness. For a moment I thought it was the lava-glow from some volcanic action; but this could not be so. Any volcanic action would surely be down in the hollow and not high among the rocks. What, then, was the alternative? It was wonderful, and yet it must surely be. These ruddy spots must be the reflection of fires within the caves—fires which could only be lit by the hand of man. There were human beings, then, upon the plateau. How gloriously my expedition was justified! Here was news indeed for us to bear back with us to London!

For a long time I lay and watched these red, quivering blotches of light. I suppose they were ten miles off from me, yet even at that distance one could observe how, from time to time, they twinkled or were obscured as someone passed before them. What would I not have given to be able to crawl up to them, to peep in, and to take back some word to my comrades as to the appearance and character of the race who lived in so strange a place! It was out of the question for the moment, and yet surely we could not leave the plateau until we had some definite knowledge upon the point.

Lake Gladys—my own lake—lay like a sheet of quicksilver before me, with a reflected moon shining brightly in the center of it. It was shallow, for in many places I saw low sandbanks protruding above the water. Everywhere upon the still surface I could see signs of life, sometimes mere rings and ripples in the water, sometimes the gleam of a great silver-sided fish in the air, sometimes the arched, slate-colored back of some passing monster. Once upon a yellow sandbank I saw a creature like a huge swan, with a clumsy body and a high, flexible neck, shuffling about upon the margin. Presently it plunged in, and for some time I could see the arched neck and darting head undulating over the water. Then it dived, and I saw it no more.

My attention was soon drawn away from these distant sights and brought back to what was going on at my very feet. Two creatures like large armadillos had come down to the drinking-place, and were squatting at the edge of the water, their long, flexible tongues like red ribbons shooting in and out as they lapped. A huge deer, with branching horns, a magnificent creature which carried itself like a king, came down with its doe and two fawns and drank beside the armadillos. No such deer exist anywhere else upon earth, for the moose or elks which I have seen would hardly have reached its shoulders. Presently it gave a warning snort, and was off with its family among the reeds, while the armadillos also scuttled for shelter. A new-comer, a most monstrous animal, was coming down the path.

For a moment I wondered where I could have seen that ungainly shape, that arched back with triangular fringes along it, that strange bird-like head held close to the ground. Then it came back, to me. It was the stegosaurus—the very creature which Maple White had preserved in his sketch-book, and which had been the first object which arrested the attention of Challenger! There he was—perhaps the very specimen which the American artist had encountered. The ground shook beneath his tremendous weight, and his gulpings of water resounded through the still night. For five minutes he was so close to my rock that by stretching out my hand I could have touched the hideous waving hackles upon his back. Then he lumbered away and was lost among the boulders.

Looking at my watch, I saw that it was half-past two o'clock, and high time, therefore, that I started upon my homeward journey. There was no difficulty about the direction in which I should return for all along I had kept the little brook upon my left, and it opened into the central lake within a stone's-throw of the boulder upon which I had been lying. I set off, therefore, in high spirits, for I felt that I had done good work and was bringing back a fine budget of news for my companions. Foremost of all, of course, were the sight of the fiery caves and the certainty that some troglodytic race inhabited them. But besides that I could speak from experience of the central lake. I could testify that it was full of strange creatures, and I had seen several land forms of primeval life which we had not before encountered. I reflected as I walked that few men in the world could have spent a stranger night or added more to human knowledge in the course of it.

I was plodding up the slope, turning these thoughts over in my mind, and had reached a point which may have been half-way to home, when my mind was brought back to my own position by a strange noise behind me. It was something between a snore and a growl, low, deep, and exceedingly menacing. Some strange creature was evidently near me, but nothing could be seen, so I hastened more rapidly upon my way. I had traversed half a mile or so when suddenly the sound was repeated, still behind me, but louder and more menacing than before. My heart stood still within me as it flashed across me that the beast, whatever it was, must surely be after ME. My skin grew cold and my hair rose at the thought. That these monsters should tear each other to pieces was a part of the strange struggle for existence, but that they should turn upon modern man, that they should deliberately track and hunt down the predominant human, was a staggering and fearsome thought. I remembered again the blood-beslobbered face which we had seen in the glare of Lord John's torch, like some horrible vision from the deepest circle of Dante's hell. With my knees shaking beneath me, I stood and glared with starting eyes down the moonlit path which lay behind me. All was quiet as in a dream landscape. Silver clearings and the black patches of the bushes—nothing else could I see. Then from out of the silence, imminent and threatening, there came once more that low, throaty croaking, far louder and closer than before. There could no longer be a doubt. Something was on my trail, and was closing in upon me every minute.

I stood like a man paralyzed, still staring at the ground which I had traversed. Then suddenly I saw it. There was movement among the bushes at the far end of the clearing which I had just traversed. A great dark shadow disengaged itself and hopped out into the clear moonlight. I say "hopped" advisedly, for the beast moved like a kangaroo, springing along in an erect position upon its powerful hind legs, while its front ones were held bent in front of it. It was of enormous size and power, like an erect elephant, but its movements, in spite of its bulk, were exceedingly alert. For a moment, as I saw its shape, I hoped that it was an iguanodon, which I knew to be harmless, but, ignorant as I was, I soon saw that this was a very different creature. Instead of the gentle, deer-shaped head of the great three-toed leaf-eater, this beast had a broad, squat, toad-like face like that which had alarmed us in our camp. His ferocious cry and the horrible energy of his pursuit both assured me that this was surely one of the great flesh-eating dinosaurs, the most terrible beasts which have ever walked this earth. As the huge brute loped along it dropped forward upon its fore-paws and brought its nose to the ground every twenty yards or so. It was smelling out my trail. Sometimes, for an instant, it was at fault. Then it would catch it up again and come bounding swiftly along the path I had taken.

Even now when I think of that nightmare the sweat breaks out upon my brow. What could I do? My useless fowling-piece was in my hand. What help could I get from that? I looked desperately round for some rock or tree, but I was in a bushy jungle with nothing higher than a sapling within sight, while I knew that the creature behind me could tear down an ordinary tree as though it were a reed. My only possible chance lay in flight. I could not move swiftly over the rough, broken ground, but as I looked round me in despair I saw a well-marked, hard-beaten path which ran across in front of me. We had seen several of the sort, the runs of various wild beasts, during our expeditions. Along this I could perhaps hold my own, for I was a fast runner, and in excellent condition. Flinging away my useless gun, I set myself to do such a half-mile as I have never done before or since. My limbs ached, my chest heaved, I felt that my throat would burst for want of air, and yet with that horror behind me I ran and I ran and ran. At last I paused, hardly able to move. For a moment I thought that I had thrown him off. The path lay still behind me. And then suddenly, with a crashing and a rending, a thudding of giant feet and a panting of monster lungs the beast was upon me once more. He was at my very heels. I was lost.

Madman that I was to linger so long before I fled! Up to then he had hunted by scent, and his movement was slow. But he had actually seen me as I started to run. From then onwards he had hunted by sight, for the path showed him where I had gone. Now, as he came round the curve, he was springing in great bounds. The moonlight shone upon his huge projecting eyes, the row of enormous teeth in his open mouth, and the gleaming fringe of claws upon his short, powerful forearms. With a scream of terror I turned and rushed wildly down the path. Behind me the thick, gasping breathing of the creature sounded louder and louder. His heavy footfall was beside me. Every instant I expected to feel his grip upon my back. And then suddenly there came a crash—I was falling through space, and everything beyond was darkness and rest.

As I emerged from my unconsciousness—which could not, I think, have lasted more than a few minutes—I was aware of a most dreadful and penetrating smell. Putting out my hand in the darkness I came upon something which felt like a huge lump of meat, while my other hand closed upon a large bone. Up above me there was a circle of starlit sky, which showed me that I was lying at the bottom of a deep pit. Slowly I staggered to my feet and felt myself all over. I was stiff and sore from head to foot, but there was no limb which would not move, no joint which would not bend. As the circumstances of my fall came back into my confused brain, I looked up in terror, expecting to see that dreadful head silhouetted against the paling sky. There was no sign of the monster, however, nor could I hear any sound from above. I began to walk slowly round, therefore, feeling in every direction to find out what this strange place could be into which I had been so opportunely precipitated.

It was, as I have said, a pit, with sharply-sloping walls and a level bottom about twenty feet across. This bottom was littered with great gobbets of flesh, most of which was in the last state of putridity. The atmosphere was poisonous and horrible. After tripping and stumbling over these lumps of decay, I came suddenly against something hard, and I found that an upright post was firmly fixed in the center of the hollow. It was so high that I could not reach the top of it with my hand, and it appeared to be covered with grease.

TWENTY THOUSAND LEAGUES UNDER THE SEA

Jules Verne

Captain Nemo is the fictional antagonist of **Jules Verne**'s classic science fiction novel *Twenty Thousand Leagues Under the Sea.* He is portrayed as a brilliant scientist and inventor who has constructed an electric-powered submarine which he uses to roam the seas conducting scientific research. The novel recalls Homer's *Odyssey,* with the crew experiencing a variety of tests and trials on their seemingly endless nautical journey. The name 'Nemo' itself is a reference to the pseudonym used by Odysseus to escape the clutches of the cyclops. Like Odysseus, Captain Nemo seems doomed to endlessly traverse the oceans in a state of exile.

Chapter VI

At Full Steam

At this cry the whole ship's crew hurried towards the harpooner—commander, officers, masters, sailors, cabin boys; even the engineers left their engines, and the stokers their furnaces.

The order to stop her had been given, and the frigate now simply went on by her own momentum. The darkness was then profound, and, however good the Canadian's eyes were, I asked myself how he had managed to see, and what he had been able to see. My heart beat as if it would break. But Ned Land was not mistaken, and we all perceived the object he pointed to. At two cables' length from the Abraham Lincoln, on the starboard quarter, the sea seemed to be illuminated all over. It was not a mere phosphoric phenomenon. The monster emerged some fathoms from the water, and then threw out that very intense but mysterious light mentioned in the report of several captains. This magnificent irradiation must have been produced by an agent of great SHINING power. The luminous part traced on the sea an immense oval, much elongated, the centre of which condensed a burning heat, whose overpowering brilliancy died out by successive gradations.

"It is only a massing of phosphoric particles," cried one of the officers.

"No, sir, certainly not," I replied. "That brightness is of an essentially electrical nature. Besides, see, see! it moves; it is moving forwards, backwards; it is darting towards us!"

A general cry arose from the frigate.

"Silence!" said the captain. "Up with the helm, reverse the engines."

The steam was shut off, and the Abraham Lincoln, beating to port, described a semicircle.

"Right the helm, go ahead," cried the captain.

These orders were executed, and the frigate moved rapidly from the burning light.

I was mistaken. She tried to sheer off, but the supernatural animal approached with a velocity double her own.

We gasped for breath. Stupefaction more than fear made us dumb and motionless. The animal gained on us, sporting with the waves. It made the round of the frigate, which was then making fourteen knots, and enveloped it with its electric rings like luminous dust.

Then it moved away two or three miles, leaving a phosphorescent track, like those volumes of steam that the express trains leave behind. All at once from the dark line of the horizon whither it retired to gain its momentum, the monster rushed suddenly towards the Abraham Lincoln with alarming rapidity, stopped suddenly about twenty feet from the hull, and died out—not diving under the water, for its brilliancy did not abate—but suddenly, and as if the source of this brilliant emanation was exhausted. Then it reappeared on the other side of the vessel, as if it had turned and slid under the hull. Any moment a collision might have occurred which would have been fatal to us. However, I was astonished at the manoeuvres of the frigate. She fled and did not attack.

On the captain's face, generally so impassive, was an expression of unaccountable astonishment.

"Mr. Aronnax," he said, "I do not know with what formidable being I have to deal, and I will not imprudently risk my frigate in the midst of this darkness. Besides, how attack this unknown thing, how defend one's self from it? Wait for daylight, and the scene will change."

"You have no further doubt, captain, of the nature of the animal?"

"No, sir; it is evidently a gigantic narwhal, and an electric one."

"Perhaps," added I, "one can only approach it with a torpedo."

"Undoubtedly," replied the captain, "if it possesses such dreadful power, it is the most terrible animal that ever was created. That is why, sir, I must be on my guard."

The crew were on their feet all night. No one thought of sleep. The Abraham Lincoln, not being able to struggle with such velocity, had moderated its pace, and sailed at half speed. For its part, the narwhal, imitating the frigate, let the waves rock it at will, and seemed decided not to leave the scene of the struggle. Towards midnight, however, it disappeared, or, to use a more appropriate term, it "died out" like a large glow-worm. Had it fled? One could only fear, not hope it. But at seven minutes to one o'clock in the morning a deafening whistling was heard, like that produced by a body of water rushing with great violence.

The captain, Ned Land, and I were then on the poop, eagerly peering through the profound darkness.

"Ned Land," asked the commander, "you have often heard the roaring of whales?"

"Often, sir; but never such whales the sight of which brought me in two thousand dollars. If I can only approach within four harpoons' length of it!"

"But to approach it," said the commander, "I ought to put a whaler at your disposal?"

"Certainly, sir."

"That will be trifling with the lives of my men."

"And mine too," simply said the harpooner.

Towards two o'clock in the morning, the burning light reappeared, not less intense, about five miles to windward of the Abraham Lincoln. Notwithstanding the distance, and the noise of the wind and sea, one heard distinctly the loud strokes of the animal's tail, and even its panting breath. It seemed that, at the moment that the enormous narwhal had come to take breath at the surface of the water, the air was engulfed in its lungs, like the steam in the vast cylinders of a machine of two thousand horse-power.

"Hum!" thought I, "a whale with the strength of a cavalry regiment would be a pretty whale!"

We were on the qui vive till daylight, and prepared for the combat. The fishing implements were laid along the hammock nettings. The second lieutenant loaded the blunder busses, which could throw harpoons to the distance of a mile, and long duck-guns, with explosive bullets, which inflicted mortal wounds even to the most terrible animals. Ned Land contented himself with sharpening his harpoon—a terrible weapon in his hands.

At six o'clock day began to break; and, with the first glimmer of light, the electric light of the narwhal disappeared. At seven o'clock the day was sufficiently advanced, but a very thick sea fog obscured our view, and the best spy glasses could not pierce it. That caused disappointment and anger.

I climbed the mizzen-mast. Some officers were already perched on the mast-heads. At eight o'clock the fog lay heavily on the waves, and its thick scrolls rose little by little. The horizon grew wider and clearer at the same time. Suddenly, just as on the day before, Ned Land's voice was heard:

"The thing itself on the port quarter!" cried the harpooner.

Every eye was turned towards the point indicated. There, a mile and a half from the frigate, a long blackish body emerged a yard above the waves. Its tail, violently agitated, produced a considerable eddy. Never did a tail beat the sea with such violence. An immense track, of dazzling whiteness, marked the passage of the animal, and described a long curve.

The frigate approached the cetacean. I examined it thoroughly.

The reports of the Shannon and of the Helvetia had rather exaggerated its size, and I estimated its length at only two hundred and fifty feet. As to its dimensions, I could only conjecture them to be admirably proportioned. While I watched this phenomenon, two jets of steam and water were ejected from its vents, and rose to the height of 120 feet; thus I ascertained its way of breathing. I concluded definitely that it belonged to the vertebrate branch, class mammalia.

The crew waited impatiently for their chief's orders. The latter, after having observed the animal attentively, called the engineer. The engineer ran to him.

"Sir," said the commander, "you have steam up?"

"Yes, sir," answered the engineer.

"Well, make up your fires and put on all steam."

Three hurrahs greeted this order. The time for the struggle had arrived. Some moments after, the two funnels of the frigate vomited torrents of black smoke, and the bridge quaked under the trembling of the boilers.

The Abraham Lincoln, propelled by her wonderful screw, went straight at the animal. The latter allowed it to come within half a cable's length; then, as if disdaining to dive, it took a little turn, and stopped a short distance off.

This pursuit lasted nearly three-quarters of an hour, without the frigate gaining two yards on the cetacean. It was quite evident that at that rate we should never come up with it.

"Well, Mr. Land," asked the captain, "do you advise me to put the boats out to sea?"

"No, sir," replied Ned Land; "because we shall not take that beast easily."

"What shall we do then?"

"Put on more steam if you can, sir. With your leave, I mean to post myself under the bowsprit, and, if we get within harpooning distance, I shall throw my harpoon."

"Go, Ned," said the captain. "Engineer, put on more pressure."

Ned Land went to his post. The fires were increased, the screw revolved forty-three times a minute, and the steam poured out of the valves. We heaved the log, and calculated that the Abraham Lincoln was going at the rate of 18½ miles an hour.

But the accursed animal swam at the same speed.

For a whole hour the frigate kept up this pace, without gaining six feet. It was humiliating for one of the swiftest sailers in the American navy. A stubborn anger seized the crew; the sailors abused the monster, who, as before, disdained to answer them; the captain no longer contented himself with twisting his beard—he gnawed it.

The engineer was called again.

"You have turned full steam on?"

"Yes, sir," replied the engineer.

The speed of the Abraham Lincoln increased. Its masts trembled down to their stepping holes, and the clouds of smoke could hardly find way out of the narrow funnels.

They heaved the log a second time.

"Well?" asked the captain of the man at the wheel.

"Nineteen miles and three-tenths, sir."

"Clap on more steam."

The engineer obeyed. The manometer showed ten degrees. But the cetacean grew warm itself, no doubt; for without straining itself, it made 19 3/10 miles.

What a pursuit! No, I cannot describe the emotion that vibrated through me. Ned Land kept his post, harpoon in hand. Several times the animal let us gain upon it.—"We shall catch it! we shall catch it!" cried the Canadian. But just as he was going to strike, the cetacean stole away with a rapidity that could not be estimated at less than thirty miles an hour, and even during our maximum of speed, it bullied the frigate, going round and round it. A cry of fury broke from everyone!

At noon we were no further advanced than at eight o'clock in the morning.

The captain then decided to take more direct means.

"Ah!" said he, "that animal goes quicker than the Abraham Lincoln. Very well! we will see whether it will escape these conical bullets. Send your men to the forecastle, sir."

The forecastle gun was immediately loaded and slewed round. But the shot passed some feet above the cetacean, which was half a mile off.

"Another, more to the right," cried the commander, "and five dollars to whoever will hit that infernal beast."

An old gunner with a grey beard—that I can see now—with steady eye and grave face, went up to the gun and took a long aim. A loud report was heard, with which were mingled the cheers of the crew.

The bullet did its work; it hit the animal, and, sliding off the rounded surface, was lost in two miles depth of sea.

The chase began again, and the captain, leaning towards me, said:

"I will pursue that beast till my frigate bursts up."

"Yes," answered I; "and you will be quite right to do it."

I wished the beast would exhaust itself, and not be insensible to fatigue like a steam engine. But it was of no use. Hours passed, without its showing any signs of exhaustion.

However, it must be said in praise of the Abraham Lincoln that she struggled on indefatigably. I cannot reckon the distance she made under three hundred miles during this unlucky day, November the 6th. But night came on, and overshadowed the rough ocean.

Now I thought our expedition was at an end, and that we should never again see the extraordinary animal. I was mistaken. At ten minutes to eleven in the evening, the electric light reappeared three miles to windward of the frigate, as pure, as intense as during the preceding night.

The narwhal seemed motionless; perhaps, tired with its day's work, it slept, letting itself float with the undulation of the waves. Now was a chance of which the captain resolved to take advantage.

He gave his orders. The Abraham Lincoln kept up half steam, and advanced cautiously so as not to awake its adversary. It is no rare thing to meet in the middle of the ocean whales so sound asleep that they can be successfully attacked, and Ned Land had harpooned more than one during its sleep. The Canadian went to take his place again under the bowsprit.

The frigate approached noiselessly, stopped at two cables' lengths from the animal, and following its track. No one breathed; a deep silence reigned on the bridge. We were not a hundred feet from the burning focus, the light of which increased and dazzled our eyes.

At this moment, leaning on the forecastle bulwark, I saw below me Ned Land grappling the martingale in one hand, brandishing his terrible harpoon in the other, scarcely twenty feet from the motionless animal. Suddenly his arm straightened, and the harpoon was thrown; I heard the sonorous stroke of the weapon, which seemed to have struck a hard body. The electric light went out suddenly, and two enormous waterspouts broke over the bridge of the frigate, rushing like a torrent from stem to stern, overthrowing men, and breaking the lashings of the spars. A fearful shock followed, and, thrown over the rail without having time to stop myself, I fell into the sea.

NARRATIVE OF A SECOND VOYAGE IN SEARCH OF A NORTH-WEST PASSAGE

John Ross, James Clark Ross

John Ross (1777–1856) was a Royal Navy officer who launched various expeditions in search of a Northwest Passage linking the Atlantic and Pacific oceans in the Arctic. It was during the second voyage that his nephew and fellow navy officer, **James Clark Ross** (1800–1862), and writing in the extract below, located the north magnetic pole. Both men would launch voyages – James in 1848 and John in 1850 – to find Sir John Franklin's lost expedition to discover the Northwest Passage. James Ross participated in six Arctic expeditions led by others and his own Antarctic expedition in 1839. He was the last explorer to complete a polar expedition of this kind completely under sail.

We were now within fourteen miles of the calculated position of the magnetic pole; and my anxiety, therefore, did not permit me to do or endure any thing which might delay my arrival at the long wished-for spot. I resolved, in consequence, to leave behind the greater part of our baggage and provisions, and to take onwards nothing more than was strictly necessary, lest bad weather or other accidents should be added to delay, or lest unforseen circumstances, still more untoward, should deprive me entirely of the high gratification which I could not but look to in accomplishing this most desired object.

June l*st*. We commenced, therefore, a rapid march, comparatively disencumbered as we now were; and, persevering with all our might, we reached the calculated place at eight in the morning of the first of June. I believe I must leave it to others to imagine the elation of mind with which we found ourselves now at length arrived at this great object of our ambition: it almost seemed as if we had accomplished every thing that we had come so far to see and to do; as if our voyage and all its labours were at an end, and that nothing now remained for us but to return home and be happy for the rest of our days. They were after-thoughts which told us that we had much yet to endure and much to perform, and they were thoughts which did not then intrude; could they have done so, we should have cast them aside, under our present excitement: we were happy, and desired to remain so as long as we could.

The land at this place is very low near the coast, but it rises into ridges of fifty or sixty feet high about a mile inland. We could have wished that a place so important had possessed more of mark or note. It was scarcely censurable to regret that there was not a mountain to indicate a spot to which so much of interest must ever be attached; and I could even have pardoned any one among us who had been so romantic or absurd as to expect that the magnetic pole was an object as conspicuous and mysterious as the fabled mountain of Sinbad, that it even was a mountain of iron, or a magnet as large as Mont Blanc. But Nature had here erected no monument to denote the spot which she had chosen as the centre of one of her great and dark powers; and where we could do little ourselves towards this end, it was our business to submit, and to be content in noting by mathematical numbers and signs, as with things of far more importance in the terrestrial system, what we could but ill distinguish in any other manner.

We were, however, fortunate in here finding some huts of Esquimaux, that had not long been abandoned. Unconscious of the value which not only we, but all the civilized world, attached to this place, it would have been a vain attempt on our part to account to them for our delight, had they been present. It was better for us that they were not; since we thus took possession of their works, and were thence enabled to establish our observations with the greater ease; encamping at six in the evening on a point of land about half a mile to the westward of those abandoned snow houses.

The necessary observations were immediately commenced, and they were continued throughout this and the greater part of the following day. Of these, the details for the purposes of science have been since communicated to the Royal Society; as a paper containing all that philosophers require on the subject has now also been printed in their Transactions. I need not therefore repeat them here, even had it not been the plan of the whole of this volume to refer every scientific matter which had occurred to Captain Ross and myself, to a separate work, under the name of an appendix.

But it will gratify general curiosity to state the most conspicuous results in a simple and popular manner. The place of the observatory was as near to the magnetic pole as the limited means which I possessed enabled me to determine. The amount of the dip, as indicated by my dipping-needle, was 89° 59', being thus within one minute of the vertical; while the proximity at least of this pole, if not its actual existence where we stood, was further confirmed by the action, or rather by the total inaction of the several horizontal needles then in my possession. These were suspended in the most delicate manner possible, but there was not one which showed the slightest effort to move from the position in which it was placed: a fact, which even the most moderately informed of readers must now know to be one which proves that the centre of attraction lies at a very small horizontal distance, if at any.

As soon as I had satisfied my own mind on this subject, I made known to the party this gratifying result of all our joint labours; and it was then, that amidst mutual congratulations, we fixed the British flag on the spot, and took possession of the North Magnetic Pole and its adjoining territory, in the name of Great Britain and King William the Fourth. We had abundance of materials for building, in the fragments of limestone that covered the beach; and we therefore erected a cairn of some magnitude, under which we buried a canister, containing a record of the interesting fact: only regretting that we had not the means of constructing a pyramid of more importance, and of strength sufficient to withstand the assaults of time and of the Esquimaux. Had it been a pyramid as large as that of Cheops, I am not quite sure that it would have done more than satisfy our ambition, under the feelings of that exciting day. The latitude of this spot is 70° 5' 17', and its longitude 96° 46' 45" west.

This subject is much too interesting, even to general readers, to permit the omission of a few other remarks relating to the scientific part of this question, desirous as I have been of passing over or curtailing these. During our absence, Professor Barlow had laid down all the curves of equal variation to within a few degrees of the point of their concurrence; leaving that point, of course, to be determined by observation, should such observation ever fall within the power of navigators. It was most gratifying to find, on our return, that the place which I had thus examined was precisely that one where these curves should have coincided in a centre, had they been protracted on his magnetic chart; and if I do not here state these particulars in a more full and scientific manner, it is because of the limits which I have drawn for myself, and because I can refer to his paper, which was read to the Royal Society six months before our arrival in England.

One further remark I must yet be permitted to make: since in relating what has been done, it would leave an important question imperfect did I not also note what remains to be effected.

It has been seen, that as far as our instruments can be trusted, we had placed ourselves within one minute of the magnetic pole, but had not fixed on the precise spot; presuming that this precise point could be determined by such instruments as it is now within the power of mechanics to construct. The scientific reader has been long aware of this: if popular conversation gives to this voyage the credit of having placed its flag on the very point, on the summit or that mysterious pole which it perhaps views as a visible and tangible reality, it can now correct itself as it may please; but in such a case, while a little laxity is of no moment, the very nonsense of the belief gives an interest to the subject which the sober truth could not have done.

To determine that point, with greater, or with absolute precision (if indeed such precision be attainable), it would be necessary to have the co-operation of different observers, at different distances, and in different directions, from the calculated place; while, to obtain all the interesting results which these must be expected to furnish, such labours should also be carried on for a considerable time. What these several expectations are, I need not here say, since the subject is, in this view, somewhat too abstruse for popular readers; though I may barely allude to the diurnal and annual motions of the needle, and to the variations in the place of the pole itself, with the consequent deductions that might be made as to the future in this respect: all of them being of the highest importance in the theory of magnetism.

Having this therefore stated, however briefly, what yet remains for future observation, having pointed out what, I may fearlessly say, is still wanting, and which as such, claims the attention of those who have the the power of promoting a work of this nature I can only express my wishes, if I dare not indulge in hopes, that the same nation which has already carried its discoveries so far, that our own Britain which has already established its supremacy in scientific and geographical researches, will not now abandon them, and leave to others to reap the crop of which it has in this case sown the seeds. That the place for the needful observations is now far more accessible than it was once supposed, has been proved by our own voyage and its results; so that the main difficulty is levelled, and the readiest excuse that could have been offered is no longer of any weight.

The chief object of our present expedition having thus been accomplished in a manner even more satisfactory than we could have expected, and in a shorter time also than we had much right to anticipate, I became desirous to extend our knowledge of the country as much further to the northward as the state of our time, and of our finances, if I may give this name to our provisions, would permit. Unluckily, the latter would not allow me to devote more than one day to this object. I could only wish that we had been better stored with the means of travelling: but, as on all former occasions of a similar nature, it was idle to regret what no contrivance on our part could have remedied. Oh that men could live without food! was a wish that had never failed to obtrude itself on every occasion of this nature.

I therefore left the party in their little snow camp, under the care of Blanky, and proceeded with Abernethy, at eleven in this our day-like night, along that shore which here stretches to the northward. After some very quick walking, we arrived, by three in the morning of *June 2d*, at a point of more than ordinary elevation. We dared not venture further, for the reasons just assigned: but hence we saw the line of the coast stretching out due north to the distance of ten or twelve miles; while I then also concluded that it preserved, in all probability, the same direction as far as Cape Walker in Lat. 74° 15'. Here we erected a cairn of stones, to mark the utmost limits of our investigations in this quarter, and returning homewards, rejoined our companions at eight in the morning.

In our absence, a hole had been cut through the ice for the purpose of examining its thickness. which was found to be six feet and eight inches. The time of high water had been observed to be a quarter of an hour after noon, and the rise and fall of the tide somewhat less than three feet.

We had not been an hour in our hut before the wind shifted to the southward, bringing on thick weather, with snow; on which the thermometer rose to the freezing point. The cold, therefore, no longer annoyed us; but the consequence was as vexatious, or even more tormenting, since the snow of our huts melted under this temperature and that of our bodies, so as to wet us in a very disagreeable manner. It soon also blew a hard gale; but as that became more moderate about eleven o'clock, we commenced our return to the ship.

For this haste in setting out, we had the best of reasons; being without any thing to eat, as we had departed supperless, until we could reach the place where we had left our baggage and provisions; hoping all the while, and not without ample cause, that no bear, or no equally hungry and more gormandizing native, had discovered that store on which we depended for many suppers and many breakfasts. We reached it, and found all intact, on the morning of the third, at seven o'clock.

The gale bad now renewed itself; and it at length blew a storm, with so much drifting snow that it was impossible to think of proceeding for the present. About one in the morning of the fourth, it however moderated so far as to permit us to move; and as we had examined all the shore on this route, in our progress forward, we now met with no cause to interfere with such rapidity as we could exert. Thus we reached the place of our former encampment at ten in the morning of the fifth.

There was now less than ever to delay us, as we had seen all that this line of coast could offer, and had done every thing that was to be effected. Our walk was, therefore, as much without note as without interruption, during two days; nor was I sorry that I had not to record occurrences and remarks which had long ceased to interest myself, as they must often have appeared tiresome to the readers, equally of my journal, and of that of Captain Ross, indispensable as their relation has been.

But I must nevertheless note, that on the sixth, in the morning, we encamped on the spot where we had formerly been detained by the blindness of some of our party, already noticed, and that I here repeated the magnetic observations which I had made in the same place during our progress forward, confirming by them that accuracy of which it was so important to be assured. Here also I had an opportunity of examining my chronometer; and was gratified to find that it had preserved a steady rate, since it was the watch by which I had determined the longitudes on the coast which we had now quitted.

At nine in the evening we crossed over to the south-east point of the inlet; but the ice being very rugged, and some of the party lame, we did not reach it till seven in the morning of the seventh. At two on this morning the thermometer was at only four degrees above zero: that being a severity of temperature which we had never before experienced at the same period of the year.

On the evening of this day, at seven, we set forward once more towards the now well-known Neitchillee, having chosen this road for returning to the ship. During this route, and early on the following morning, we arrived at a place where we found a large party of the natives assembled; the situation in question being about three miles westward of Cape Isabella. They were busily occupied in fishing; and their prey consisted of the two species of cod, described in the Appendix of Natural History, by the names Gadus Mochica, and Callarias. These they took through some holes which they had made in the ice for that purpose; and we discovered from them, that this fishery was a very productive one. Our application for a supply was readily granted, and it proved a welcome one to all of us, limited, both in quantity and quality, as we had been for some days.

9th. From this, after resting two hours, we proceeded onwards to Cape Isabella, and encamped at eight in the morning. But a dense fog now came on, with the effect of rendering our route very uncertain, as it also made the travelling difficult. This we endured as we could, entertaining better hopes for the following morning; when, at six, we again set out, being as soon as was practicable, and encamped near Padliak; having found it utterly impossible to travel any further at this time, in consequence of the increased density of the fog.

But towards noon it cleared away; and this horrible mist, bad enough in a known country, but incredibly worse amid such obstructions as the surface here for ever presents, and where there is no guide but a compass, was succeeded by bright and brilliant weather. The sun shone forth, in consequence, with such power, that we obtained abundance of water from the streams which ran from the rocks and lodged in the pools formed among them: a far more acceptable supply than it is easy for readers to conceive, as it may, perhaps, surprise them to be told that it was the first natural water that we had obtained during this year, though it now wanted but a few days of midsummer. Is there aught that can convey a deeper impression of the state and nature of this most atrocious climate? If there be, I know not well what it is.

If I here also obtained some magnetic observations, as I had before done at Cape Isabella during this returning jouruey, they are matter for the Appendix, not for this place. There at least they can be consulted by the scientific reader, among much more, whether in meteorology or in the other branches of natural history, which it has been judged most convenient to place in such a supplement: but as far as the present observations are concerned, the paper in the Transactions of the Royal Society, to which I have already alluded, will give complete information to all those who may be interested in this subject. I have, however, attached to the end of this narrative, the means of the· observations in question, that they who are inclined may see at least the general results. It is for this simple journal to say, that we proceeded along the valley of Padliak at ten o'clock, and reached the great middle lake, so often described, about midnight. Then coasting along its southern shore till nine in the morning of the tenth, we halted on the northern point of a small inlet, putting up some grouse, and seeing a number of deer under the pursuit of a wolf.

At ten in the evening, according to our usual plan, which advantageously turned day into night, we directed our course to the north-east corner of this lake, in order to ascertain whether there was any river which communicated between it and its neighbour, so as to discharge this collection of waters into the sea. Thus it proved, and we thence ascertained that to be a fact which bad formerly been only a matter of conjecture.

11*th*. At three in the morning of the eleventh, we arrived, in this our homeward progress, at another place, now familiar from its having been a spot of rest during more than one of our former journeys; but it presented at this time a very different appearance from what it had done on the corresponding day in the preceding year. At the same place, during that journey, we had been obliged to wade knee-deep in water for nearly two miles, in crossing to the head of the inlet of Shag-a-voke. At present all was solid ice, there was not a drop of water any where to be seen, nor was there the slightest mark to indicate the commencement of a thaw. Can it be believed that there were but ten days to midsummer, that all was still hard winter, and that winter in the middle, I may almost say, of summer: a season such as the January of our own native land seldom sees.

It was no small satisfaction for hard-worked men and hungry stomachs, to find on the opposite shore of this inlet, some provisions which had been deposited for us by Captain Ross; and, taking possession of them, we crossed the two next lakes and encamped, at six in the evening, near the head of the bay into which their water finds its exit.

12*th*. Here we were detained by a heavy storm from the southwest until noon on the twelfth of June, when it began to moderate, and tempted us to proceed on our now last day of labour; the ship being at length within our reach. But our attempt proved vain. The gale was soon renewed with increased violence, and the snow drifted so densely as to entirely blind us to our way, so that we were compelled, in spite of all our efforts and wishes, to halt and encamp at nine on the following morning. It was an unusual disappointment. If we bad on many former occasions been as wearied, as hungry, and as anxious to reach our companions and our home, we had now more interesting news to relate than had ever occurred to us before; but we were to exert our patience, at least this once more, and exerted it was.

But this trial of our tempers was not destined to be very durable. The gale at length moderated so far, that we could contrive to see and find our way; and having but ten miles remaining, we bestirred ourselves in proportion, even till midnight; when, after as much hard labour as we could well manage, and might not have endured if not under such a stimulus, we neared our home; still labouring with all our power till we found ourselves at length, and once more, on board the Victory, at five in the morning of the thirteenth of June. We had been absent twenty-eight days. If we were fatigued and attenuated, who could be surprised? but excepting petty grievances, we were all in good health.

THE SAGA OF ERIK THE RED

Unknown

Gudrid ThorBjarnardóttir (980–1019) was an Icelandic explorer famous for her voyage to Vinland (contemporary Newfoundland) as recalled in the *Vinland Sagas*. Gudrid's son Snorri was believed to have been the first European born on North American soil. Gudrid's voyage to the New World and her family's conversion to Christianity have become legendary events in the folk imagination of the Icelandic people. She is remembered as an intrepid explorer and a symbol of Iceland's broader transition from a pagan nation to a Christian one.

Chapter 9

When summer was at hand they discussed about their journey, and made an arrangement. Thorhall the Sportsman wished to proceed northwards along Furdustrandir, and off Kjalarnes, and so seek Vinland; but Karlsefni desired to proceed southwards along the land and away from the east, because the land appeared to him the better the further south he went, and he thought it also more advisable to explore in both directions. Then did Thorhall make ready for his journey out by the islands, and there volunteered for the expedition with him not more than nine men; but with Karlsefni there went the remainder of the company. And one day, when Thorhall was carrying water to his ship, he drank, and recited this verse:

> "The clashers of weapons did say when I came here that I should have the best of drink (though it becomes me not to complain before the common people). Eager God of the war-helmet! I am made to raise the bucket; wine has not moistened my beard, rather do I kneel at the fountain."

Afterwards they put to sea, and Karlsefni accompanied them by the island. Before they hoisted sail Thorhall recited a verse:

"Go we back where our countrymen are. Let us make the skilled hawk of the sand-heaven explore the broad ship-courses; while the dauntless rousers of the sword-storm, who praise the land, and cook whale, dwell on Furdustrandir."

Then they left, and sailed northwards along Furdustrandir and Kjalarnes, and attempted there to sail against a wind from the west. A gale came upon them, however, and drove them onwards against Ireland, and there were they severely treated, enthralled, and beaten. Then Thorhall lost his life.

Chapter 10

Karlsefni proceeded southwards along the land, with Snorri and Bjarni and the rest of the company. They journeyed a long while, and until they arrived at a river, which came down from the land and fell into a lake, and so on to the sea. There were large islands off the mouth of the river, and they could not come into the river except at high flood-tide.

Karlsefni and his people sailed to the mouth of the river, and called the land Hop. There they found fields of wild wheat wherever there were low grounds; and the vine in all places were there was rough rising ground. Every rivulet there was full of fish. They made holes where the land and water joined and where the tide went highest; and when it ebbed they found halibut in the holes. There was great plenty of wild animals of every form in the wood. They were there half a month, amusing themselves, and not becoming aware of anything. Their cattle they had with them. And early one morning, as they looked around, they beheld nine canoes made of hides, and snout-like staves were being brandished from the boats, and they made a noise like flails, and twisted round in the direction of the sun's motion.

Then Karlsefni said, "What will this betoken?" Snorri answered him, "It may be that it is a token of peace; let us take a white shield and go to meet them." And so they did. Then did they in the canoes row forwards, and showed surprise at them, and came to land. They were short men, ill-looking, with their hair in disorderly fashion on their heads; they were large-eyed, and had broad cheeks. And they stayed there awhile in astonishment. Afterwards they rowed away to the south, off the headland.

Chapter 11

They had built their settlements up above the lake. And some of the dwellings were well within the land, but some were near the lake. Now they remained there that winter. They had no snow whatever, and all their cattle went out to graze without keepers.

Now when spring began, they beheld one morning early, that a fleet of hide-canoes was rowing from the south off the headland; so many were they as if the sea were strewn with pieces of charcoal, and there was also the brandishing of staves as before from each boat. Then they held shields up, and a market was formed between them; and this people in their purchases preferred red cloth; in exchange they had furs to give, and skins quite grey. They wished also to buy swords and lances, but Karlsefni and Snorri forbad it. They offered for the cloth dark hides, and took in exchange a span long of cloth, and bound it round their heads; and so matters went on for a while. But when the stock of cloth began to grow small, then they split it asunder, so that it was not more than a finger's breadth. The Skrælingar (Esquimaux) gave for it still quite as much, or more than before.

Chapter 12

Now it came to pass that a bull, which belonged to Karlsefni's people, rushed out of the wood and bellowed loudly at the same time. The Skrælingar, frightened thereat, rushed away to their canoes, and rowed south along the coast. There was then nothing seen of them for three weeks together. When that time was gone by, there was seen approaching from the south a great crowd of Skrælingar boats, coming down upon them like a stream, the staves this time being all brandished in the direction opposite to the sun's motion, and the Skrælingar were all howling loudly. Then took they and bare red shields to meet them. They encountered one another and fought, and there was a great shower of missiles. The Skrælingar had also war-slings, or catapults.

Then Karlsefni and Snorri see that the Skrælingar are bringing up poles, with a very large ball attached to each, to be compared in size to a sheep's stomach, dark in colour; and these flew over Karlsefni's company towards the land, and when they came down they struck the ground with a hideous noise. This produced great terror in Karlsefni and his company, so that their only impulse was to retreat up the country along the river, because it seemed as if crowds of Skrælingar were driving at them from all sides. And they stopped not until they came to certain crags. There they offered them stern resistance.

Freydis came out and saw how they were retreating. She called out, "Why run you away from such worthless creatures, stout men that ye are, when, as seems to me likely, you might slaughter them like so many cattle? Let me but have a weapon, I think I could fight better than any of you." They gave no heed to what she said. Freydis endeavoured to accompany them, still she soon lagged behind, because she was not well; she went after them into the wood, and the Skrælingar directed their pursuit after her. She came upon a dead man; Thorbrand, Snorri's son, with a flat stone fixed in his head; his sword lay beside him, so she took it up and prepared to defend herself therewith.

Then came the Skrælingar upon her. She let down her sark and struck her breast with the naked sword. At this they were frightened, rushed off to their boats, and fled away. Karlsefni and the rest came up to her and praised her zeal. Two of Karlsefni's men fell, and four of the Skrælingar, notwithstanding they had overpowered them by superior numbers. After that, they proceeded to their booths, and began to reflect about the crowd of men which attacked them upon the land; it appeared to them now that the one troop will have been that which came in the boats, and the other troop will have been a delusion of sight. The Skrælingar also found a dead man, and his axe lay beside him. One of them struck a stone with it, and broke the axe. It seemed to them good for nothing, as it did not withstand the stone, and they threw it down.

Chapter 13

[Karlsefni and his company] were now of opinion that though the land might be choice and good, there would be always war and terror overhanging them, from those who dwelt there before them. They made ready, therefore, to move away, with intent to go to their own land. They sailed forth northwards, and found five Skrælingar in jackets of skin, sleeping [near the sea], and they had with them a chest, and in it was marrow of animals mixed with blood; and they considered that these must have been outlawed. They slew them. Afterwards they came to a headland and a multitude of wild animals; and this headland appeared as if it might be a cake of cow-dung, because the animals passed the winter there. Now they came to Straumsfjordr, where also they had abundance of all kinds. It is said by some that Bjarni and Freydis remained there, and a hundred men with them, and went not further away. But Karlsefni and Snorri journeyed southwards, and forty men with them, and after staying no longer than scarcely two months at Hop, had come back the same summer. Karlsefni set out with a single ship to seek Thorhall, but the (rest of the) company remained behind. He and his people went northwards off Kjalarnes, and were then borne onwards towards the west, and the land lay on their larboard-side, and was nothing but wilderness. And when they had proceeded for a long time, there was a river which came down from the land, flowing from the east towards the west. They directed their course within the river's mouth, and lay opposite the southern bank.

Chapter 14

One morning Karlsefni's people beheld as it were a glittering peak above the open space in front of them, and they shouted at it. It stirred itself, and it was a being of the race of men that have only one foot, and he came down quickly to where they lay. Thorvald, son of Eirik the Red, sat at the tiller, and the One-footer shot him with an arrow in the lower abdomen. He drew out the arrow. Then said Thorvald, "Good land have we reached, and fat is it about the paunch." Then the One-footer leapt away again northwards. They chased after him, and saw him occasionally, but it seemed as if he would escape them. He disappeared at a certain creek. Then they turned back, and one man spake this ditty:

> "Our men chased (all true it is) a One-footer down to the shore; but the wonderful man strove hard in the race.... Hearken, Karlsefni."

Then they journeyed away back again northwards, and saw, as they thought, the land of the One-footers. They wished, however, no longer to risk their company. They conjectured the mountains to be all one range; those, that is, which were at Hop, and those which they now discovered; almost answering to one another; and it was the same distance to them on both sides from Straumsfjordr. They journeyed back, and were in Straumsfjordr the third winter. Then fell the men greatly into backsliding. They who were wifeless pressed their claims at the hands of those who were married.

Snorri, Karlsefni's son, was born the first autumn, and he was three winters old when they began their journey home. Now, when they sailed from Vinland, they had a southern wind, and reached Markland, and found five Skrælingar; one was a bearded man, two were women, two children. Karlsefni's people caught the children, but the others escaped and sunk down into the earth. And they took the children with them, and taught them their speech, and they were baptized. The children called their mother Vætilldi, and their father Uvægi. They said that kings ruled over the land of the Skrælingar, one of whom was called Avalldamon, and the other Valldidida. They said also that there were no houses, and the people lived in caves or holes. They said, moreover, that there was a land on the other side over against their land, and the people there were dressed in white garments, uttered loud cries, bare long poles, and wore fringes. This was supposed to be Hvitramannaland (whiteman's land). Then came they to Greenland, and remained with Eirik the Red during the winter.

Chapter 15

Bjarni, Grimolf's son, and his men were carried into the Irish Ocean, and came into a part where the sea was infested by ship-worms. They did not find it out before the ship was eaten through under them; then they debated what plan they should follow. They had a ship's boat which was smeared with tar made of seal-fat. It is said that the ship-worm will not bore into the wood which has been smeared with the seal-tar. The counsel and advice of most of the men was to ship into the boat as many men as it would hold. Now, when that was tried, the boat held not more than half the men. Then Bjarni advised that it should be decided by the casting of lots, and not by the rank of the men, which of them should go into the boat; and inasmuch as every man there wished to go into the boat, though it could not hold all of them; therefore, they accepted the plan to cast lots who should leave the ship for the boat. And the lot so fell that Bjarni, and nearly half the men with him, were chosen for the boat. So then those left the ship and went into the boat who had been chosen by lot so to do.

And when the men were come into the boat, a young man, an Icelander, who had been a fellow-traveller of Bjarni, said, "Dost thou intend, Bjarni, to separate thyself here from me." "It must needs be so now," Bjarni answered. He replied, "Because, in such case, thou didst not so promise me when I set out from Iceland with thee from the homestead of my father." Bjarni answered, "I do not, however, see here any other plan; but what plan dost thou suggest?" He replied, "I propose this plan, that we two make a change in our places, and thou come here and I will go there." Bjarni answered, "So shall it be; and this I see, that thou labourest willingly for life, and that it seems to thee a grievous thing to face death." Then they changed places. The man went into the boat, and Bjarni back into the ship; and it is said that Bjarni perished there in the Worm-sea, and they who were with him in the ship; but the boat and those who were in it went on their journey until they reached land, and told this story afterwards.

Chapter 16

The next summer Karlsefni set out for Iceland, and Snorri with him, and went home to his house in Reynines. His mother considered that he had made a shabby match, and she was not at home the first winter. But when she found that Gudrid was a lady without peer, she went home, and their intercourse was happy. The daughter of Snorri, Karlsefni's son, was Hallfrid, mother of Bishop Thorlak, the son of Runolf. (Hallfrid and Runolf) had a son, whose name was Thorbjorn; his daughter was Thorun, mother of Bishop Bjarn. Thorgeir was the name of a son of Snorri, Karlsefni's son; he was father of Yngvild, the mother of the first Bishop Brand. And here ends this story.

THE JOURNALS OF LEWIS AND CLARK

Meriwether Lewis, William Clark

Meriwether Lewis and **William Clark** were American explorers who, in 1803, traversed the newly acquired western portion of the United States following the Louisiana Purchase. The pair encountered numerous Native American tribes on their journey, the majority of whom were helpful and supportive to the expedition, offering their knowledge of wilderness survival and methods for acquiring food. The three-year-long expedition helped establish a legal claim by the fledgling United States to the western North American territories versus rival Spanish and British colonial endeavours, and formed diplomatic ties between the United States and at least two dozen Native American tribes. The presence the mission created in the newly acquired territory opened the gateway for increased settlement, colonization and trade with the native tribes. However, it also accelerated the westward expansion of the American empire and fanned the nascent vision of Manifest Destiny that would ultimately bring disastrous consequences for the Native American population.

Lewis, April 12, 1805

Friday April the 12th 1805. Set out at an early hour. our peroge and the Canoes passed over to the Lard side in order to avoid a bank which was rappidly falling in on the Stard. the red perogue contrary to my expectation or wish passed under this bank by means of her toe line where I expected to have seen her carried under every instant. I did not discover that she was about to make this attempt untill it was too late for the men to reembark, and retreating is more dangerous than proceeding in such cases; they therefore continued their passage up this bank, and much to my satisfaction arrived safe above it. this cost me some moments of uneasiness, her cargo was of much importance to us in our present advanced situation—We proceeded on six miles and came too on the lower side of the entrance of the little Missouri on the Lard shore in a fine plain where we determined to spend the day for the purpose of celestial observation. we sent out 10 hunters to procure some fresh meat. at this place made [some] observations.

The night proved so cloudy that I could make no further observations. George Drewyer shot a Beaver this morning, which we found swiming in the river a small distance below the entrance of the little Missouri. the beaver being seen in the day, is a proof that they have been but little hunted, as they always keep themselves closly concealed during the day where they are so.—found a great quantity of small onions in the plain where we encamped; had some of them collected and cooked, found them agreeable. the bulb grows single, is of an oval form, white, and about the size of a small bullet; the leaf resembles that of the shive, and the hunters returned this eving with one deer only. the country about the mouth of this river had been recently hunted by the Minetares, and the little game which they had not killed and frightened away, was so extreemly shy that the hunters could not get in shoot of them.

The little Missouri disembogues on the S. side of the Missouri 1693 miles from the confluence of the latter with the Mississippi. it is 134 yards wide at it's mouth, and sets in with a bould current but it's greatest debth is not more than 21/2 feet. it's navigation is extreemly difficult, owing to it's rapidity, shoals and sand bars. it may however be navigated with small canoes a considerable distance. this river passes through the Northern extremity of the black hills where it is very narrow and rapid and it's banks high an perpendicular. it takes it's rise in a broken country West of the Black hills with the waters of the yellow stone river, and a considerable distance S. W. of the point at which it passes the black hills. the country through which it passes is generally broken and the highlands possess but little timber. there is some timber in it's bottom lands, which consists of Cottonwood red Elm, with a small proportion of small Ash and box alder. the under brush is willow, red wood, (sometimes called red or swamp willow-) the red burry, and Choke cherry the country is extreamly broken about the mouth of this river, and as far up on both sides, as we could observe it from the tops of some elivated hills, which stand betwen these two rivers, about 3 miles from their junction. the soil appears fertile and deep, it consists generally of a dark rich loam intermixed with a small proportion of fine sand. this river in it's course passed near the N. W. side of the turtle mountain, which is said to be no more than 4 or 5 leagues distant from it's entrance in a straight direction, a little to the S. of West.—this mountain and the knife river have therefore been laid down too far S. W. the colour of the water, the bed of the river, and it's appearance in every respect, resembles the Missouri; I am therefore induced to believe that the texture of the soil of the country in which it takes it's rise, and that through which it passes, is similar to the country through which the Missouri passes after leaving the woody country, or such as we are now in.—on the side of a hill not distant from our camp I found some of the dwarf cedar of which I preserved a specimen (See No. 2). this plant spreads it's limbs alonge the surface of the earth, where they are sometimes covered, and always put forth a number of roots on the under side, while on the upper there are a great number of small shoots which with their leaves seldom rise higher than 6 or eight inches. they grow so close as perfectly to conceal the eath. it is an evergreen; the leaf is much more delicate than the common Cedar, and it's taste and smell the same. I have often thought that this plant would make very handsome edgings to the borders and walks of a garden; it is quite as handsom as box, and would be much more easily propegated.—the appearance of the glauber salts and Carbonated wood still continue.

Clark, April 12, 1805

12th April Friday 1805 a fine morning Set out verry early, the murcery Stood 56° above 0. proceeded on to the mouth of the Little Missouri river and formed a Camp in a butifull elivated plain on the lower Side for the purpose of takeing Some observations to fix the Latitude & Longitude of this river. this river falls in on the L. Side and is 134 yards wide and 2 feet 6 Inches deep at the mouth, it takes its rise in the N W extremity of the black mountains, and through a broken countrey in its whole course washing the N W base of the Turtle Mountain which is Situated about 6 Leagues S W of its mouth, one of our men Baptiest who came down this river in a canoe informs me that it is not navagable, he was 45 days descending.

One of our men Shot a beaver Swimming below the mouth of this river.

I walked out on the lower Side of this river and found the countrey hilley the Soil composed of black mole & a Small perportion of Sand containing great quantity of Small peable Some limestone, black flint, & Sand Stone I killed a Hare Changeing its Colour Some parts retaining its long white fur & other parts assumeing the Short grey, I Saw the Magpie in pars, flocks of Grouse, the old field lark & Crows, & observed the leaf of the wild Chery half grown, many flowers are to be seen in the plains, remains of Minetarra & Ossinneboin hunting Camps are to be Seen on each Side of the two Missouris

The wind blew verry hard from the S. all the after part of the day, at 3 o'Clock P M. it became violent & flowey accompanied with thunder and a little rain. We examined our canoes &c found Several mice which had already commenced cutting our bags of corn & parched meal, the water of the little Missouri is of the Same texture Colour & quallity of that of the Big Missouri the after part of the day so Cloudy that we lost the evening observation.

Lewis, April 13, 1805

Saturday April 13th Being disappointed in my observations of yesterday for Longitude, I was unwilling to remain at the entrance of the river another day for that purpose, and therefore determined to set out early this morning; which we did accordingly; the wind was in our favour after 9 A.M. and continued favourable untill three 3 P.M. we therefore hoisted both the sails in the White Perogue, consisting of a small squar sail, and spritsail, which carried her at a pretty good gate, untill about 2 in the afternoon when a suddon squall of wind struck us and turned the perogue so much on the side as to allarm Sharbono who was steering at the time, in this state of alarm he threw the perogue with her side to the wind, when the spritsail gibing was as near overseting the perogue as it was possible to have missed. the wind however abating for an instant I ordered Drewyer to the helm and the sails to be taken in, which was instant executed and the perogue being steered before the wind was agin placed in a state of security. this accedent was very near costing us dearly. beleiving this vessell to be the most steady and safe, we had embarked on board of it our instruments, Papers, medicine and the most valuable part of the merchandize which we had still in reserve as presents for the Indians. we had also embarked on board ourselves, with three men who could not swim and the squaw with the young child, all of whom, had the perogue overset, would most probably have perished, as the waves were high, and the perogue upwards of 200 yards from the nearest shore; however we fortunately escaped and pursued our journey under the square sail, which shortly after the accident I directed to be again hoisted. our party caught three beaver last evening; and the French hunters 7. as there was much appearance of beaver just above the entrance of the little Missouri these hunters concluded to remain some days; we therefore left them without the expectation of seeing them again.—just above the entrance of the Little Missouri the great Missouri is upwards of a mile in width, tho immediately at the entrance of the former it is not more than 200 yards wide and so shallow that the canoes passed it with seting poles. at the distance of nine miles passed the mouth of a creek on the Stard. side which we called onion creek from the quantity of wild onions which grow in the plains on it's borders. Capt. Clark who was on shore informed me that this creek was 16 yards wide a mile & a half above it's entrance, discharges more water than creeks of it's size usually do in this open country, and that there was not a stick of timber of any discription to be seen on it's borders, or the level plain country through which it passes. at the distance of 10 miles further we passed the mouth of a large creek; discharging itself in the center of a deep bend. of this creek and the neighbouring country, Capt Clark who was on shore gave me the following discription "This creek I took to be a small river from it's size, and the quantity of water which it discharged. I ascended it 11/2 miles, and found it the discharge of a pond or small lake, which had the appearance of having formerly been the bed of the Missouri. several small streems discharge themselves into this lake. the country on both sides consists of beautifull level and elivated plains; asscending as they recede from the Missouri; there were a great number of Swan and gees in this lake and near it's borders I saw the remains of 43 temperary Indian lodges, which I presume were those of the Assinniboins who are now in the neighbourhood of the British establishments on the Assinniboin river-" This lake and it's discharge we call Boos Egg from the circumstance of Capt Clark shooting a goose while on her nest in the top of a lofty cotton wood tree, from which we afterwards took one egg. the wild gees frequently build their nests in this manner, at least we have already found several in trees, nor have we as yet seen any on the ground, or sand bars where I had supposed from previous information that they most commonly deposited their eggs.- saw some Buf-haloe and Elk at a distance today but killed none of them. we found a number of carcases of the Buffaloe lying along shore, which had been drowned by falling through the ice in winter and lodged on shore by the high water when the river broke up about the first of this month. we saw also many tracks of the white bear of enormous size, along the river shore and about the carcases of the Buffaloe, on which I presume they feed. we have not as yet seen one of these anamals, tho their tracks are so abundant and recent. the men as well as ourselves are anxious to meet with some of these bear. the Indians give a very formidable account of the strengh and ferocity of this anamal, which they never dare to attack but in parties of six eight or ten persons; and are even then frequently defeated with the loss of one or more of their party. the savages attack this anamal with their bows and arrows and the indifferent guns with which the traders furnish them, with these they shoot with such uncertainty and at so short a distance, that they frequently mis their aim & fall a sacrefice to the bear. two Minetaries were killed during the last winter in an attack on a white bear. this anamall is said more frequently to attack a man on meeting with him, than to flee from him. When the Indians are about to go in quest of the white bear, previous to their departure, they paint themselves and perform all those superstitious rights commonly observed when they are about to make war uppon a neighbouring nation. Oserved more bald eagles on this part of the Missouri than we have previously seen saw the small hawk, frequently called the sparrow hawk, which is common to most parts of the U States. great quantities of gees are seen feeding in the praries. saw a large flock of white brant or gees with black wings pass up the river; there were a number of gray brant with them; from their flight I presume they proceed much further still to the N. W.—we have never been enabled yet to shoot one of these birds, and cannot therefore determine whether the gray brant found with the white arc their brude of the last year or whether they are the same with the grey brant common to the Mississippi and lower part of the Missouri.—we killed 2 Antelopes today which we found swiming from the S. to the N. side of the river; they were very poor.—We encamped this evening on the Stard. shore in a beautiful plain, elivated about 30 feet above the river.

THE CONQUEST OF THE NEW SPAIN

Bernal Díaz del Castillo

Bernal Díaz del Castillo (*c.*1496–1584) was a Spanish conquistador and soldier of fortune who wrote one of the most comprehensive firsthand accounts of the conquest of Mexico under Hernán Cortés. Del Castillo wrote critically of Cortés as both a leader and an individual, and was cynical of the project of colonization and conquest of the New World, even while actively participating in it. In this extract, having made landfall on the Yucatán peninsula and come under attack by the people living there, the soldiers decide to retreat – heavily wounded and desperate for water – to Cuba. His account is considered one of the more accurate and reliable descriptions of the trials and tribulations overcome by the conquistadors during the Spanish conquest and provides important firsthand insights into Mesoamerican culture before its downfall.

We resolve to return to Cuba. The extreme thirst we suffered, and all the fatigues we underwent until our arrival in the port of Havannah.

After we had got into our vessels, as above related, and returned thanks to God for our preservation, we commenced dressing our wounds. None of us had escaped without two, three, or four wounds. Our captain had as many as twelve, and there was only one single soldier who came off whole. We therefore determined to return to Cuba; but as most of the sailors who had accompanied us on shore were also wounded, we had not sufficient hands to work the sails, we were therefore forced to set fire to our smallest vessel and leave it to the mercy of the waves, after taking out all the ropes, sails, and anchors, and distributing the sailors, who were not wounded equally among the two other vessels. We had, however, to struggle with another far greater evil. This was our great want of fresh water; for although we had filled our barrels and casks near Potonchan, we did not succeed to bring them off, owing to the furious attack of the natives and the hurry we were in to get on board: thus we had been compelled to leave them behind and return without a single drop of water. We suffered most intensely from thirst, and the only way we could in some measure refresh our parched tongues was to hold the edges of our axes between our lips. Oh, what a fearful undertaking it is to venture out on the discovery of new countries, and place one's life in danger, as we were obliged to do! Those alone can form any idea of it who have gone through the hard school of experience.

We now kept as close into the shore as possible, to look out for some stream or creek where we might meet with fresh water. After thus continuing our course for three days we espied an inlet or mouth of some river as we thought, and sent a few hands on shore in the hopes of meeting with water. These were fifteen sailors who had remained on board during the battle at Potonchan, and three soldiers who had been only slightly wounded. They carried along with them pickaxes and three small casks. But the water in the inlet was salt, and wherever they dug wells it was equally bad. They nevertheless filled the casks with it, but it was so bitter and salty as to be unfit for use. Two soldiers who drank of it became ill of the consequences. The water here swarmed with lizards; we therefore gave this place the name of Lizard Bay, under which name it stands on the sea charts.

But, to continue my history, I must not forget to mention that while our boats were on shore in search of water, there suddenly arose such a violent tempest from the north-east, that our ships were nigh being cast on shore. For, as we were forced to lay to, the wind blowing hard from the north and north-east, our position was extremely dangerous, from a scarcity of ropes.

When the men who had gone on shore with our boats perceived the danger we were in, they hastened to our assistance, and cast out additional anchors and cables. In this way we lay for two days and two nights. After the expiration of that time we again heaved our anchors and steered in the direction of Cuba. Our pilot Alaminos here held a consultation with the two others, when they concluded that the best plan would be to get, if possible, into the latitude of Florida, which, according to their charts and furthest measurement, could not be more than 210 miles distant; for they assured us if we could get into the latitude of Florida, we should have a better and speedier sail to the Havannah. It turned out exactly as they had said; for Alaminos had been in these parts before, having accompanied Juan de Leon when he discovered Florida, about ten or twelve years previously. After four days' sail we crossed this gulf and came in sight of Florida.

THE VOYAGE OF THE BEAGLE

Charles Darwin

Charles Darwin (1809–1882) was a British naturalist and explorer who developed the theory of evolution by natural selection. It was during his five-year voyage on the *HMS Beagle* that he observed generational changes in various species that led him to believe that natural selection over an extraordinarily long period of time led to variations in species. Following the publication of *On the Origin of Species*, his theory of evolution took the scientific world by storm, and by the end of his life, it was the dominant theory regarding human origins.

On the 19th of August we finally left the shores of Brazil. I thank God, I shall never again visit a slave-country. To this day, if I hear a distant scream, it recalls with painful vividness my feelings, when passing a house near Pernambuco, I heard the most pitiable moans, and could not but suspect that some poor slave was being tortured, yet knew that I was as powerless as a child even to remonstrate. I suspected that these moans were from a tortured slave, for I was told that this was the case in another instance. Near Rio de Janeiro I lived opposite to an old lady, who kept screws to crush the fingers of her female slaves. I have stayed in a house where a young household mulatto, daily and hourly, was reviled, beaten, and persecuted enough to break the spirit of the lowest animal. I have seen a little boy, six or seven years old, struck thrice with a horse-whip (before I could interfere) on his naked head, for having handed me a glass of water not quite clean; I saw his father tremble at a mere glance from his master's eye. These latter cruelties were witnessed by me in a Spanish colony, in which it has always been said, that slaves are better treated than by the Portuguese, English, or other European nations. I have seen at Rio de Janeiro a powerful negro afraid to ward off a blow directed, as he thought, at his face. I was present when a kind-hearted man was on the point of separating forever the men, women, and little children of a large number of families who had long lived together. I will not even allude to the many heart-sickening atrocities which I authentically heard of;—nor would I have mentioned the above revolting details, had I not met with several people, so blinded by the constitutional gaiety of the negro as to speak of slavery as a tolerable evil. Such people have generally visited at the houses of the upper classes, where the domestic slaves are usually well treated, and they have not, like myself, lived amongst the lower classes. Such inquirers will ask slaves about their condition; they forget that the slave must indeed be dull, who does not calculate on the chance of his answer reaching his master's ears.

It is argued that self-interest will prevent excessive cruelty; as if self-interest protected our domestic animals, which are far less likely than degraded slaves, to stir up the rage of their savage masters. It is an argument long since protested against with noble feeling, and strikingly exemplified, by the ever-illustrious Humboldt. It is often attempted to palliate slavery by comparing the state of slaves with our poorer countrymen: if the misery of our poor be caused not by the laws of nature, but by our institutions, great is our sin; but how this bears on slavery, I cannot see; as well might the use of the thumb-screw be defended in one land, by showing that men in another land suffered from some dreadful disease. Those who look tenderly at the slave owner, and with a cold heart at the slave, never seem to put themselves into the position of the latter; what a cheerless prospect, with not even a hope of change! picture to yourself the chance, ever hanging over you, of your wife and your little children—those objects which nature urges even the slave to call his own—being torn from you and sold like beasts to the first bidder! And these deeds are done and palliated by men, who profess to love their neighbours as themselves, who believe in God, and pray that his Will be done on earth! It makes one's blood boil, yet heart tremble, to think that we Englishmen and our American descendants, with their boastful cry of liberty, have been and are so guilty: but it is a consolation to reflect, that we at least have made a greater sacrifice, than ever made by any nation, to expiate our sin.

On the last day of August we anchored for the second time at Porto Praya in the Cape de Verd archipelago; thence we proceeded to the Azores, where we stayed six days. On the 2nd of October we made the shore, of England; and at Falmouth I left the Beagle, having lived on board the good little vessel nearly five years.

Our Voyage having come to an end, I will take a short retrospect of the advantages and disadvantages, the pains and pleasures, of our circumnavigation of the world. If a person asked my advice, before undertaking a long voyage, my answer would depend upon his possessing a decided taste for some branch of knowledge, which could by this means be advanced. No doubt it is a high satisfaction to behold various countries and the many races of mankind, but the pleasures gained at the time do not counterbalance the evils. It is necessary to look forward to a harvest, however distant that may be, when some fruit will be reaped, some good effected.

Many of the losses which must be experienced are obvious; such as that of the society of every old friend, and of the sight of those places with which every dearest remembrance is so intimately connected. These losses, however, are at the time partly relieved by the exhaustless delight of anticipating the long wished-for day of return. If, as poets say, life is a dream, I am sure in a voyage these are the visions which best serve to pass away the long night. Other losses, although not at first felt, tell heavily after a period: these are the want of room, of seclusion, of rest; the jading feeling of constant hurry; the privation of small luxuries, the loss of domestic society and even of music and the other pleasures of imagination. When such trifles are mentioned, it is evident that the real grievances, excepting from accidents, of a sea-life are at an end. The short space of sixty years has made an astonishing difference in the facility of distant navigation. Even in the time of Cook, a man who left his fireside for such expeditions underwent severe privations. A yacht now, with every luxury of life, can circumnavigate the globe. Besides the vast improvements in ships and naval resources, the whole western shores of America are thrown open, and Australia has become the capital of a rising continent. How different are the circumstances to a man shipwrecked at the present day in the Pacific, to what they were in the time of Cook! Since his voyage a hemisphere has been added to the civilized world.

If a person suffer much from sea-sickness, let him weigh it heavily in the balance. I speak from experience: it is no trifling evil, cured in a week. If, on the other hand, he take pleasure in naval tactics, he will assuredly have full scope for his taste. But it must be borne in mind, how large a proportion of the time, during a long voyage, is spent on the water, as compared with the days in harbour. And what are the boasted glories of the illimitable ocean. A tedious waste, a desert of water, as the Arabian calls it. No doubt there are some delightful scenes. A moonlight night, with the clear heavens and the dark glittering sea, and the white sails filled by the soft air of a gently blowing trade-wind, a dead calm, with the heaving surface polished like a mirror, and all still except the occasional flapping of the canvas. It is well once to behold a squall with its rising arch and coming fury, or the heavy gale of wind and mountainous waves. I confess, however, my imagination had painted something more grand, more terrific in the full-grown storm. It is an incomparably finer spectacle when beheld on shore, where the waving trees, the wild flight of the birds, the dark shadows and bright lights, the rushing of the torrents all proclaim the strife of the unloosed elements. At sea the albatross and little petrel fly as if the storm were their proper sphere, the water rises and sinks as if fulfilling its usual task, the ship alone and its inhabitants seem the objects of wrath. On a forlorn and weather-beaten coast, the scene is indeed different, but the feelings partake more of horror than of wild delight.

Let us now look at the brighter side of the past time. The pleasure derived from beholding the scenery and the general aspect of the various countries we have visited, has decidedly been the most constant and highest source of enjoyment. It is probable that the picturesque beauty of many parts of Europe exceeds anything which we beheld. But there is a growing pleasure in comparing the character of the scenery in different countries, which to a certain degree is distinct from merely admiring its beauty. It depends chiefly on an acquaintance with the individual parts of each view. I am strongly induced to believe that as in music, the person who understands every note will, if he also possesses a proper taste, more thoroughly enjoy the whole, so he who examines each part of a fine view, may also thoroughly comprehend the full and combined effect. Hence, a traveller should be a botanist, for in all views plants form the chief embellishment. Group masses of naked rock, even in the wildest forms, and they may for a time afford a sublime spectacle, but they will soon grow monotonous. Paint them with bright and varied colours, as in Northern Chile, they will become fantastic; clothe them with vegetation, they must form a decent, if not a beautiful picture.

When I say that the scenery of parts of Europe is probably superior to anything which we beheld, I except, as a class by itself, that of the intertropical zones. The two classes cannot be compared together; but I have already often enlarged on the grandeur of those regions. As the force of impressions generally depends on preconceived ideas, I may add, that mine were taken from the vivid descriptions in the Personal Narrative of Humboldt, which far exceed in merit anything else which I have read. Yet with these high-wrought ideas, my feelings were far from partaking of a tinge of disappointment on my first and final landing on the shores of Brazil.

Among the scenes which are deeply impressed on my mind, none exceed in sublimity the primeval forests undefaced by the hand of man; whether those of Brazil, where the powers of Life are predominant, or those of Tierra del Fuego, where Death and decay prevail. Both are temples filled with the varied productions of the God of Nature:—no one can stand in these solitudes unmoved, and not feel that there is more in man than the mere breath of his body. In calling up images of the past, I find that the plains of Patagonia frequently cross before my eyes; yet these plains are pronounced by all wretched and useless. They can be described only by negative characters; without habitations, without water, without trees, without mountains, they support merely a few dwarf plants. Why, then, and the case is not peculiar to myself, have these arid wastes taken so firm a hold on my memory? Why have not the still more level, the greener and more fertile Pampas, which are serviceable to mankind, produced an equal impression? I can scarcely analyze these feelings: but it must be partly owing to the free scope given to the imagination. The plains of Patagonia are boundless, for they are scarcely passable, and hence unknown: they bear the stamp of having lasted, as they are now, for ages, and there appears no limit to their duration through future time. If, as the ancients supposed, the flat earth was surrounded by an impassable breadth of water, or by deserts heated to an intolerable excess, who would not look at these last boundaries to man's knowledge with deep but ill-defined sensations?

Lastly, of natural scenery, the views from lofty mountains, through certainly in one sense not beautiful, are very memorable. When looking down from the highest crest of the Cordillera, the mind, undisturbed by minute details, was filled with the stupendous dimensions of the surrounding masses.

Of individual objects, perhaps nothing is more certain to create astonishment than the first sight in his native haunt of a barbarian—of man in his lowest and most savage state. One's mind hurries back over past centuries, and then asks, could our progenitors have been men like these?—men, whose very signs and expressions are less intelligible to us than those of the domesticated animals; men, who do not possess the instinct of those animals, nor yet appear to boast of human reason, or at least of arts consequent on that reason. I do not believe it is possible to describe or paint the difference between savage and civilized man. It is the difference between a wild and tame animal: and part of the interest in beholding a savage, is the same which would lead every one to desire to see the lion in his desert, the tiger tearing his prey in the jungle, or the rhinoceros wandering over the wild plains of Africa.

Among the other most remarkable spectacles which we have beheld, may be ranked, the Southern Cross, the cloud of Magellan, and the other constellations of the southern hemisphere—the water-spout—the glacier leading its blue stream of ice, overhanging the sea in a bold precipice—a lagoon-island raised by the reef-building corals—an active volcano—and the overwhelming effects of a violent earthquake. These latter phenomena, perhaps, possess for me a peculiar interest, from their intimate connection with the geological structure of the world. The earthquake, however, must be to every one a most impressive event: the earth, considered from our earliest childhood as the type of solidity, has oscillated like a thin crust beneath our feet; and in seeing the laboured works of man in a moment overthrown, we feel the insignificance of his boasted power.

It has been said, that the love of the chase is an inherent delight in man—a relic of an instinctive passion. If so, I am sure the pleasure of living in the open air, with the sky for a roof and the ground for a table, is part of the same feeling, it is the savage returning to his wild and native habits. I always look back to our boat cruises, and my land journeys, when through unfrequented countries, with an extreme delight, which no scenes of civilization could have created. I do not doubt that every traveller must remember the glowing sense of happiness which he experienced, when he first breathed in a foreign clime, where the civilized man had seldom or never trod.

There are several other sources of enjoyment in a long voyage, which are of a more reasonable nature. The map of the world ceases to be a blank; it becomes a picture full of the most varied and animated figures. Each part assumes its proper dimensions: continents are not looked at in the light of islands, or islands considered as mere specks, which are, in truth, larger than many kingdoms of Europe. Africa, or North and South America, are well-sounding names, and easily pronounced; but it is not until having sailed for weeks along small portions of their shores, that one is thoroughly convinced what vast spaces on our immense world these names imply.

From seeing the present state, it is impossible not to look forward with high expectations to the future progress of nearly an entire hemisphere. The march of improvement, consequent on the introduction of Christianity throughout the South Sea, probably stands by itself in the records of history. It is the more striking when we remember that only sixty years since, Cook, whose excellent judgment none will dispute, could foresee no prospect of a change. Yet these changes have now been effected by the philanthropic spirit of the British nation.

In the same quarter of the globe Australia is rising, or indeed may be said to have risen, into a grand centre of civilization, which, at some not very remote period, will rule as empress over the southern hemisphere. It is impossible for an Englishman to behold these distant colonies, without a high pride and satisfaction. To hoist the British flag, seems to draw with it as a certain consequence, wealth, prosperity, and civilization.

In conclusion, it appears to me that nothing can be more improving to a young naturalist, than a journey in distant countries. It both sharpens, and partly allays that want and craving, which, as Sir J. Herschel remarks, a man experiences although every corporeal sense be fully satisfied. The excitement from the novelty of objects, and the chance of success, stimulate him to increased activity. Moreover, as a number of isolated facts soon become uninteresting, the habit of comparison leads to generalization. On the other hand, as the traveller stays but a short time in each place, his descriptions must generally consist of mere sketches, instead of detailed observations. Hence arises, as I have found to my cost, a constant tendency to fill up the wide gaps of knowledge, by inaccurate and superficial hypotheses.

But I have too deeply enjoyed the voyage, not to recommend any naturalist, although he must not expect to be so fortunate in his companions as I have been, to take all chances, and to start, on travels by land if possible, if otherwise, on a long voyage. He may feel assured, he will meet with no difficulties or dangers, excepting in rare cases, nearly so bad as he beforehand anticipates. In a moral point of view, the effect ought to be, to teach him good-humoured patience, freedom from selfishness, the habit of acting for himself, and of making the best of every occurrence. In short, he ought to partake of the characteristic qualities of most sailors. Travelling ought also to teach him distrust; but at the same time he will discover, how many truly kind-hearted people there are, with whom he never before had, or ever again will have any further communication, who yet are ready to offer him the most disinterested assistance.

THE WORLD ENCOMPASSED

Sir Francis Drake

Sir Francis Drake (1540–1596) was an English explorer and naval officer famous for circumnavigating the world in 1580. An early example of the 'pirate' archetype, Drake's privateering and voyages of exploration have given rise to a multitude of legends relating to buried treasure, ghost tales and hoards of gold in both Europe and the New World. A hero to the English, he was condemned by the Spanish as *El Draque* (The Dragon) for his part in the defeat of the Grand Armada in 1588. Drake's legacy has been embroiled in recent controversies that highlight his participation in the English Slave Trade, and he remains a complicated symbol of the British Empire in its earliest Elizabethan expansions.

In the time of this incredible storme, the 15 of *September*, the moone was ecclipsed in Aries, and darkened about three points, for the space of two glasses; which being ended, might seeme to give us some hope of alteration and change of weather to the better. Notwithstanding, as the ecliptical conflict could adde nothing to our miserable estate, no more did the ending thereof ease us anything at all, nor take away any part of our troubles from us: but our ecclipse continued still in its full force, so prevailing against us, that, for the space of full 52 dayes together, we were darkened more then the moone by 20 parts, or more then we by any meanes could ever have preserved or recovered light of ourselves againe, if the Sonne of God, which layed this burden upon our backs, had not mercifully borne it up with his owne shoulders, and upheld us in it by his owne power, beyond any possible strength or skill of man. Neither indeed did we at all escape, but with the feeling of great discomforts through the same.

For these violent and extraordinarie flawes (such as seldome have beene seene) still continuing, or rather increasing, *September* 30, in the night, caused the sorrowfull separation of the *Marigold* from us; in which was captaine *John Thomas*, with many others of our deare friends, who by no means that we could conceive could helpe themselves, but by spooming along before the sea. With whom, albeit wee could never meet againe, yet (our Generall having aforehand given order, that if any of our fleet did loose company, the place of resort to meet againe should be in 30 deg. or thereabouts, upon the coast of Peru, toward the Equinoctiall), wee long time hoped (till experience shewed our hope was vaine) that there we should joyfully meet with them: especially for that they were well provided of victuals, and lackt no skilfull and sufficient men (besides their Captaine) to bring forwards the ship to the place appointed.

From the seventh of *September* (in which the storme began) till the seventh of *October*, we could not by any meanes recover any land (having in the meane time beene driven so farre South as to the 57 deg. and somewhat better) on this day towards night, somewhat to the Northward of that cape of America (whereof mention is made before, in the description of our departure from the straite into this sea), with a sorrie saile wee entred a harbour: where hoping to enjoy some freedome and ease till the storme was ended, we received within few houres after our comming to anchor so deadly a stroake and hard entertainement, that our Admirall left not onely an anchor behind her, through the violence and furie of the flawe, but in departing thence, also lost the company and sight of our Vice-admirall, the *Elizabeth*, partly through the negligence of those that had the charge of her, partly through a kind of desire that some in her had to be out of those troubles, and to be at home againe; which (as since is knowne) they thenceforward by all meanes assayed and performed. For the very next day, *October* 8, recovering the mouth of the straits againe (which wee were now so neere unto) they returned backe the same way by which they came forward, and so coasting Brasill, they arrived in England *June* 2 the yeare following.

So that now our Admirall, if she had retained her old name of *Pellican*, which she bare at our departure from our country, she might have beene now indeed said to be as a pellican alone in the wildernesse. For albeit our Generali sought the rest of his fleet with great care, yet could we not have any sight or certaine newes of them by any meanes.

From this day of parting of friends, we were forcibly driven backe againe into 55 deg. towards the pole Antarticke. In which height we ranne in among the Ilands before mentioned, lying to the Southward of America, through which we passed from one sea to the other, as hath beene declared. Where, comming to anchor, wee found the waters there to have their indraught and free passage, and that through no small guts or narrow channels, but indeed through as large frets or straights as it hath at the supposed straights of Magellane, through which we came.

Among these Ilands making our abode with some quietnesse for a very little while (viz. two dayes), and finding divers good and wholesome herbs, together with fresh water; our men, which before were weake, and much empaired in their health, began to receive good comfort, especially by the drinking of one herbe (not much unlike that herbe which wee commonly call Pennyleafe), which purging with great facilitie, affoorded great helpe and refreshing to our wearied and sickly bodies. But the winds returning to their old wont, and the seas raging after their former manner, yea everything as it were setting itselfe against our peace and desired rest, here was no stay permitted us, neither any safety to be looked for.

For such was the present danger by forcing and continuall flawes, that we were rather to looke for present death then hope for any delivery, if God Almightie should not make the way for us. The winds were such as if the bowels of the earth had set all at libertie, or as if all the clouds under heaven had beene called together to lay their force upon that one place. The seas, which by nature and of themselves are heavie, and of a weightie substance, were rowled up from the depths, even from the roots of the rockes, as if it had beene a scroll of parchment, which by the extremity of heate runneth together; and being aloft were carried in most strange manner and abundance, as feathers or drifts of snow, by the violence of the winds, to water the exceeding tops of high and loftie mountaines. Our anchors, as false friends in such a danger, gave over their holdfast, and as if it had beene with horror of the thing, did shrinke downe to hide themselves in this miserable storme, committing the distressed ship and helpelesse men to the uncertaine and rowling seas, which tossed them, like a ball in a racket. In this case, to let fall more anchors would availe us nothing; for being driven from our first place of anchoring, so unmeasurable was the depth, that 500 fathome would fetch no ground. So that the violent storme without intermission; the impossibility to come to anchor; the want of opportunitie to spread any sayle; the most mad seas; the lee shores; the dangerous rocks; the contrary and most intollerable winds; the impossible passage out; the desperate tarrying there; and inevitable perils on every side, did lay before us so small likelihood to escape present destruction, that if the speciall providence of God himselfe had not supported us, we could never have endured that wofull state: as being invironed with most terrible and most fearefull judgements round about. For truly, it was more likely that the mountaines should have beene rent in sunder from the top to the bottome, and cast headlong into the sea, by these unnaturall winds, then that we, by any helpe or cunning of man, should free the life of any one amongst us.

Notwithstanding, the same God of mercy which deliuered *Ionas* out of the Whales belly, and heareth all those that call upon him faithfully in their distresse, looked downe from heaven, beheld our teares, and heard our humble petitions, joyned with holy vowes. Even God (whom not the winds and seas alone, but even the divels themselues and powers of hell obey) did so wonderfully free us, and make our way open before us, as it were by his holy Angels still guiding and conducting us, that, more than the affright and amaze of this estate, we received no part of damage in all the things that belonged unto us.

But escaping from these straites and miseries, as it were through the needles ey (that God might have the greater glory in our delivery), by the great and effectuall care and travell of our Generall, the Lord's instrument therein; we could now no longer forbeare, but must needes find some place of refuge, as well to provide water, wood, and other necessaries, as to comfort our men, thus worne and tired out by so many and so long intollerable toyles; the like whereof, its to be supposed, no traveller hath felt, neither hath there ever beene such a tempest (that any records make mention of), so violent and of such continuance since *Noahs* flood; for, as hath beene sayd, it lasted from *September* 7 to *October* 28th, full 52 dayes.

Not many leagues therefore to the Southwards of our former anchoring, we ranne in againe among these Ilands, where we had once more better likelihood to rest in peace; and so much the rather, for that wee found the people of the countrie travelling for their living from one Iland to another, in their canowes, both men, women, and young infants wrapt in skins, and hanging at their mothers backs; with whom we had traffique for such things as they had, as chaines of certaine shells and such other trifles. Here the Lord gave us three dayes to breath ourselves and to prouide such things as we wanted, albeit the same was with continuall care and troubles to avoid imminent dangers, which the troubled seas and blustering windes did every houre threaten unto us.

But when we seemed to have stayed there too long, we were more rigorously assaulted by the not formerly ended but now more violently renewed storme, and driven thence also with no small danger, leaving behind us the greater part of our cable with the anchor; being chased along by the winds and buffeted incessantly in each quarter by the seas, (which our Generall interpreted as though God had sent them of purpose to the end which ensued), till at length wee fell with the uttermost part of land towards the South Pole, and had certainely discovered how farre the same doth reach Southward from the coast of America aforenamed.

The uttermost cape or hedland of all these llands, stands neere in 56 deg., without which there is no maine nor lland to be seene to the Southwards, but that the Atlanticke Ocean and the South Sea, meete in a most large and free scope.

It hath beene a dreame through many ages, that these Ilands have beene a maine, and that it hath beene *terra incognita*, wherein many strange monsters lived. Indeed, it might truly before this time be called *incognota*, for howsoever the mappes and generall descriptions of *cosmographers*, either upon the deceiveable reports of other men, or the deceitfull imaginations of themselves (supposing never herein to be corrected), have set it downe, yet it is true, that before this time, it was never discovered or certainely knowne by any traveller that wee have heard of.

PERSONAL NARRATIVE OF TRAVELS TO THE EQUINOCTIAL REGIONS OF AMERICA DURING THE YEARS 1799–1804

Alexander von Humboldt

Alexander von Humboldt (1769–1859) was a German naturalist and explorer who travelled extensively in the New World on various scientific expeditions. The brother of an influential German statesman, von Humboldt's early success in his botanical studies gained him the notoriety and financial support needed to travel extensively through North and South America. It was here that he first developed the theory that South America and Africa had once been a single landmass, an observation that would later inform the Pangea Hypothesis. He also generated the first known descriptions of the phenomenon of man-made climate variability, and is considered the father of the modern conception of climate change.

After our journey to the Orinoco, we left a part of these collections at the island of Cuba, intending to take them on our return from Peru to Mexico. The rest followed us during the space of five years, on the chain of the Andes, across New Spain, from the shores of the Pacific to the coasts of the Caribbean Sea. The conveyance of these objects, and the minute care they required, occasioned embarrassments scarcely conceiveable even by those who have traversed the most uncultivated parts of Europe. Our progress was often retarded by the necessity of dragging after us, during expeditions of five or six months, twelve, fifteen, and sometimes more than twenty loaded mules, exchanging these animals every eight or ten days, and superintending the Indians who were employed in driving the numerous caravan. Often, in order to add to our collections of new mineral substances, we found ourselves obliged to throw away others, which we had collected a considerable time before. These sacrifices were not less vexatious than the losses we accidentally sustained. Sad experience taught us but too late, that from the sultry humidity of the climate, and the frequent falls of the beasts of burden, we could preserve neither the skins of animals hastily prepared, nor the fishes and reptiles placed in phials filled with alcohol. I enter into these details, because, though little interesting in themselves, they serve to show that we had no means of bringing back, in their natural state, many objects of zoology and comparative anatomy, of which we have published descriptions and drawings. Notwithstanding some obstacles, and the expense occasioned by the carriage of these articles, I had reason to applaud the resolution I had taken before my departure, of sending to Europe the duplicates only of the productions we collected. I cannot too often repeat, that when the seas are infested with privateers, a traveller can be sure only of the objects in his own possession. A very few of the duplicates, which we shipped for Europe during our abode in America, were saved; the greater part fell into the hands of persons who feel no interest for science. When a ship is condemned in a foreign port, boxes containing only dried plants or stones, instead of being sent to the scientific men to whom they are addressed, are put aside and forgotten. Some of our geological collections taken in the Pacific were, however, more fortunate. We were indebted for their preservation to the generous activity of Sir Joseph Banks, President of the Royal Society of London, who, amidst the political agitations of Europe, unceasingly laboured to strengthen the bonds of union between scientific men of all nations.

In our investigations we have considered each phenomenon under different aspects, and classed our remarks according to the relations they bear to each other. To afford an idea of the method we have followed, I will here add a succinct enumeration of the materials with which we were furnished for describing the volcanoes of Antisana and Pichincha, as well as that of Jorullo: the latter, during the night of the 20th of September, 1759, rose from the earth one thousand five hundred and seventy-eight French feet above the surrounding plains of Mexico. The position of these singular mountains in longitude and latitude was ascertained by astronomical observations. We took the heights of the different parts by the aid of the barometer, and determined the dip of the needle and the intensity of the magnetic forces. Our collections contain the plants which are spread over the flanks of these volcanoes, and specimens of different rocks which, superposed one upon another, constitute their external coat. We are enabled to indicate, by measures sufficiently exact, the height above the level of the ocean, at which we found each group of plants, and each volcanic rock. Our journals furnish us with a series of observations on the humidity, the temperature, the electricity, and the degree of transparency of the air on the brinks of the craters of Pichincha and Jorullo; they also contain topographical plans and geological profiles of these mountains, founded in part on the measure of vertical bases, and on angles of altitude. Each observation has been calculated according to the tables and the methods which are considered most exact in the present state of our knowledge; and in order to judge of the degree of confidence which the results may claim, we have preserved the whole detail of our partial operations.

It would have been possible to blend these different materials in a work devoted wholly to the description of the volcanoes of Peru and New Spain. Had I given the physical description of a single province, I could have treated separately everything relating to its geography, mineralogy, and botany; but how could I interrupt the narrative of a journey, a disquisition on the manners of a people, or the great phenomena of nature, by an enumeration of the productions of the country, the description of new species of animals and plants, or the detail of astronomical observations. Had I adopted a mode of composition which would have included in one and the same chapter all that has been observed on one particular point of the globe, I should have prepared a work of cumbrous length, and devoid of that clearness which arises in a great measure from the methodical distribution of matter. Notwithstanding the efforts I have made to avoid, in this narrative, the errors I had to dread, I feel conscious that I have not always succeeded in separating the observations of detail from those general results which interest every enlightened mind. These results comprise in one view the climate and its influence on organized beings, the aspect of the country, varied according to the nature of the soil and its vegetable covering, the direction of the mountains and rivers which separate races of men as well as tribes of plants; and finally, the modifications observable in the condition of people living in different latitudes, and in circumstances more or less favourable to the development of their faculties. I do not fear having too much enlarged on objects so worthy of attention: one of the noblest characteristics which distinguish modern civilization from that of remoter times is, that it has enlarged the mass of our conceptions, rendered us more capable of perceiving the connection between the physical and intellectual world, and thrown a more general interest over objects which heretofore occupied only a few scientific men, because those objects were contemplated separately, and from a narrower point of view.

THE MYSTERY OF EASTER ISLAND

Katherine Routledge

Katherine Routledge (1866–1935) was a British archaeologist who completed the first successful, comprehensive survey of Easter Island. A devoted anthropologist, Routledge abandoned family wealth and influence to pursue her anthropological interests. She moved with her husband to British East Africa to live among the Kikuyu, a Bantu people who form the predominant ethnic group in modern Kenya and parts of Tanzania. Her insights, revealed in *With A Prehistoric People* (1910), proved a valuable anthropological and ethnographic resource for the scientific community. Routledge's expedition to Easter Island landed in 1914 and her book records how the inhabitants lived, their oral history, legends and beliefs, as well as her journey back to Southampton, England, via the Pitcairn Islands, Hawaii and San Francisco.

Easter Island at last! It was in the misty dawn of Sunday, March 29th, 1914, that we first saw our destination, just one week in the year earlier than the Easter Day it was sighted by Roggeveen and his company of Dutchmen. We had been twenty days at sea since leaving Juan Fernandez, giving a wide berth to the few dangerous rocks which constitute Salo-y-Gomez and steering directly into the sunset. It was thirteen months since we had left Southampton, out of which time we had been 147 days under way, and here at last was our goal. As we approached the southern coast we gazed in almost awed silence at the long grey mass of land, broken into three great curves, and diversified by giant molehills. The whole looked an alarmingly big land in which to find hidden caves. The hush was broken by the despairing voice of Bailey, the ship's cook. "I don't know how I am to make a fire on that island, there is no wood!" He spoke the truth; not a vestige of timber or even brushwood was to be seen. We swung round the western headland with its group of islets and dropped anchor in Cook's Bay. A few hundred yards from the shore is the village of Hanga Roa, the native name for Cook's Bay. This is the only part of the island which is inhabited, the two hundred and fifty natives, all that remain of the population, having been gathered together here in order to secure the safety of the livestock, to which the rest of the island is devoted. The yacht was soon surrounded by six or seven boat-loads of natives, clad in nondescript European garments, but wearing a head-covering of native straw, somewhat resembling in appearance the high hat of civilisation.

The Manager, Mr. Edmunds, shortly appeared, and to our relief, for we had not been sure how he would view such an invasion, gave us a very kind welcome. He is English, and was, to all intent, at the time of our arrival, the only white man on the island; a French carpenter, who lived at Hanga Roa with a native wife, being always included in the village community. His house is at Mataveri, a spot about two miles to the south of the village, surrounded by modern plantations which are almost the only trees on the island; immediately behind it rises the swelling mass of the volcano Rano Kao. The first meal on Easter Island, taken here with Mr. Edmunds, remains a lasting memory. It was a large plain room with uncarpeted floor, scrupulously orderly; a dinner table, a few chairs, and two small book-cases formed the whole furniture. The door on to the veranda was open, for the night was hot, and the roar of breakers could be heard on the beach; while near at hand conversation was accompanied by a never-ceasing drone of mosquitoes. The light of the unshaded lamp was reflected from the clean rough-dried cloth of the table round which we sat, and lit up our host's features, the keen brown face of a man who had lived for some thirty years or more, most of it in the open air and under a tropical sun. He was telling us of events which one hardly thought existed outside magazines and books of adventure, but doing it so quietly that, with closed eyes, it might have been fancied that the entertainment was at some London restaurant, and we were still at the stage of discussing the latest play.

"This house," said our host, "was built some fifty years ago by Bornier, who was the first to exploit the island. He was murdered by the natives: they seized the moment when he was descending from a ladder; one spoke to him and another struck him down. They buried him on the hillock near the cliff just outside the plantation: you will see his grave, when the grass is not so long; it is marked by a circle of stones. A French warship arriving almost immediately afterwards, they explained that he had been killed by a fall from his horse, and this is the version still given in some of the accounts of the island, but murder will always out. After that another manager had trouble: it was over sheep-stealing. There were three or four white men here at the time, and they all rode down to the village to teach the natives a lesson, but the ponies turned restive at the sound of gun-fire, and the rifles themselves were defective, so the boot was on the other foot, and they had to retreat up here followed by the mob; for months they lived in what was practically a state of siege, with one man always on guard for fear of attack.

"My latest guests were a crew of shipwrecked mariners, Americans, who landed on the island last June. A fortnight earlier the barometer here had been extraordinarily low, but we did not get much wind; further to the south, however, the gale was terrific, and the *El Dorado* was in the midst of it. The captain, who had been a whaler in his day, said that he had never seen anything approaching it, the sea was simply a seething mass of crested waves. The ship was a schooner, trading between Oregon and a Chilean port; she was a long way from land, as sailing vessels make a big semicircle to get the best wind. She had a deck load of timber, 15 feet high, which of course shifted in such a sea; she sprang leaks in every direction, and it was obvious that she must soon break up. The crew took to their boat, not that they had much hope of saving their lives, but simply because there was nothing else to be done. They got some tins of milk and soup on board, and a box of biscuits, and a cask holding perhaps twenty gallons of water. The captain managed to secure his sextant, but when he went back for his chronometers, the chart-room was too deep in water for him to be able to reach them. They saw by the chart that the nearest land was this island: it was seven hundred miles off, and as they had no chronometer, and could take no risks, they would have to go north first in order to get their latitude, which would add on another two hundred. There was nothing for it, however, but to do the best they could; they had more gales too, and only saved the boat from being swamped by making a sea-anchor of their blankets. The spray of course kept washing over them, and as the boat was only 20 feet long and there were eleven of them, there was no room for them to lie down. Each day they had between them a tin of the soup and one of milk, and an allowance of water, but the sea got into the water-cask and made it brackish, and before the end their sufferings from thirst were so great that one or two of them attempted to drink salt water; the mate stopped that by saying that he would shoot the first man who did it.

"After nine days they sighted this island, but then luck was against them, for the wind changed, and it was forty-eight hours, after they saw the coast, before they were able to beach the boat. They got on shore at the other end of the island, which is uninhabited. They were pretty much at the last stage of exhaustion, and their skin was in a terrible condition with salt water; their feet especially were so bad that they could hardly walk. One of them fell down again and again, but struggled on saying, 'I won't give up, I won't give up.' At last my man, who looks after the cattle over there, saw them and brought me word. The officers were put up here, you must really forgive the limitations of my wardrobe, for I had to give away nearly everything that I had in order to clothe them.

"The most curious part of the whole business was that after they had been here three or four months the captain took to the boat again. I believe that he was buying his house at home on the instalment plan, and that if he did not get in the last payment by the end of the year the whole would be forfeited; anyway, as soon as the fine weather came on he had out the boat and patched her up. He got two of his men to go with him. I lent him a watch for navigation purposes, and we did all we could for him in the way of food; there were no matches on the island, so he learnt how to make fire with two pieces of wood native fashion. Anyway, off he started last October for Mangareva, sixteen hundred miles from here; he must have got there safely, for you brought me an answer to a letter that I gave him to post. But," and here for the first time the eyes of our host grew animated, and he raised his voice slightly, "it is maddening to think of that cargo drifting about in the Pacific. I do trust that next time a ship breaks up with a deck-load of timber, she will have at least the commonsense to do so near Easter Island." Then, after a pause, "I wish you no ill, but the yacht would make a splendid wreck."

We kept *Mana* for nearly two months while learning our new surroundings. Not only were we anxious to find if we had the necessary camp gear and stores, but we were engaged in agonised endeavours to foresee the details of excavation and research, in case essential tools or equipment had been forgotten, which the yacht could fetch from Chile. The time, however, arrived when she must go. Mr. Ritchie was now on shore with us for survey work, but as his service with the Expedition was limited, the vessel had to return in time to take him back to civilisation by the correct date. Mr. Gillam had from this time sole charge of the navigation of *Mana*. Instructions for him had to be written, and correspondence grappled with; business letters, epistles for friends, and reports to Societies were hurriedly dealt with; and an article which had been promised to the *Spectator*, "First Impressions of Easter Island," was written in my tent, by the light of a hurricane-lamp, during the small hours of more than one morning.

When the mail-bag was finally sealed, there was great difficulty in getting hold of *Mana*. The position of a skipper of a boat off Easter Island, unless she has strong steam-power, is not a happy one. Mr. Gillam used to lie in his berth at Cook's Bay hearing the waves break on the jagged reaches of lava, and the longer he listened the less he liked it. The instant that the wind shows signs of going to the west, a ship must clear out. It is reported that on one occasion there were some anxious moments on board: a sudden change of wind and tide were setting the yacht steadily on the rocks; the engineer was below in the engine-room, and Mr. Gillam shouted to him down the hatchway, "If you can't make that motor of yours go round in three minutes, you will know whether there is a God or not."

To get in touch with the yacht was like a game of hide-and-seek, for often by the time those on shore arrived at one side of the island, the wind had shifted, and she had run round to the other. She was on the north coast when we managed to catch her, and to get back to Mataveri necessitated retracing our steps, as will be seen from the map, over the high central ground of the island, and down on the other side; the track was rough, and the ride would ordinarily take from two to three hours. It was 4 p.m. before all work was done on board, the good-byes said, and we were put on shore; the sandy cove, the horses and men, with *Mana* in the offing, formed a delightful picture in the evening light, but there the charms of the situation ended. There was only one pack-horse, and a formidable body of last collections sat looking at us in a pile on the grass. In addition we had not, in the general pressure, sufficiently taken into account that we were bringing off the engineer, now to be turned into photographer; there he was, and not he alone but his goods and bedding. The sun set at five o›clock, and it would be dark at half-past five; it seemed hopeless to get back that night.

A neighbouring cave was first investigated as a possible abiding-place, but proved full of undesirable inhabitants, so everyone set to work and the amount stowed on that wretched pack-horse was wonderful. Then each attendant was slung round with some remaining object, S. took the additional member on his pony, and off we set. Before we got to the highest point all daylight had gone, and there was only just enough starlight to keep to the narrow track by each man following a dim vision of the one immediately in front. My own beast had been chosen as "so safe" that it was most difficult to keep him up with the others, let alone on his four legs. The packhorse, too, began pointing out that he was not enjoying the journey; the load was readjusted more than once, but when we were on the down grade again he came to a full stop and we all dismounted. There in the creepy darkness we had a most weird picnic; not far off was a burial-place, with a row of fallen statues, while the only light save that of the stars was the striking of an occasional match. S. produced a tin of meat, which he had brought from the yacht, and which was most acceptable, as he and I had had no substantial food, save a divided tin of sardines, since breakfast at 7 o'clock. He shared it out between the party amid cries from our retainers of "Good food, good Pappa," for we were, as in East Africa, known as "Pap-pa" and "Mam-ma" to a large and promising family. By some inducement the pack-horse was then deluded into proceeding, and we finally reached Mataveri at nine o'clock, relieved to find we had not been given up and that supper awaited us. So did we cut our last link with civilisation, and were left in mid-Pacific with statues and natives.

The next part of this story deals with the island, the conditions of life on it, and our experience during the sixteen months we were to spend there. Such scientific work as the Expedition was able to accomplish will be recounted later.

ROAD TO THE STARS

Yuri Gagarin

Yuri Gagarin (1934–1968) was a Soviet cosmonaut who, in April of 1961, became the first human to enter outer space, marking a major victory for the Soviet Union in the Space Race against the United States. Gagarin later served as a backup crew member for the launch of the Soyuz 1 mission, which ended in disaster and claimed the life of Gagarin's devoted friend Vladimir Komarov. The mission was undertaken despite the protests of Gagarin, who was banned from further space flights due to the political ramifications of potentially losing a national hero. Nevertheless, the Soviet Union did lose him several years later when Gagarin died in a plane crash a mere five weeks after completing his flight certifications.

The bus sped along the highway. We were still far away when I caught sight of the upward-pointed silvery hull of the rocket equipped with six engines developing a total thrust of 20 million horse power. The closer we came to the launching site the larger the rocket became, as though it actually grew in size. It resembled a giant beacon and the first ray of the rising sun played on its pointed head.

The weather favoured the flight.

The sky was clear and only somewhere very far away one could see fleecy clouds.

"A million kilometres of altitude, a million kilometres of visibility," I heard someone say. Only a flyer could have said that.

On the launching site I saw the Theoretician of Cosmonautics and the Chief Designer. For them that was the most difficult day. As always, they stood side by side. Their expressive faces were illumined, to the very last wrinkle, by the morning light. There, too, were members of the State Commission for the First Space Flight, cosmodrome officials, chiefs of the starting crew, scientists, leading designers, my true friend Cosmonaut Two and other fellow cosmonauts. Everything all around was flooded with the light of the breaking day.

"What a cheerful sun!" I exclaimed.

I recalled my first flight in the North, the pink snowcovered knolls floating past me under the plane, the earth besprinkled with the bluish drops of lakes and the dark-blue, cold sea beating against the granite cliffs.

"How beautiful!" I exclaimed involuntarily.

"Keep your eyes on the instruments," snapped the flight commander Vasilyev.

This had been long ago, yet it came to my mind now, before the flight. But business before emotions ....

We were becoming increasingly more impatient. Everybody kept looking at chronometers. Finally, it was reported that the rocket and ship were ready for the space flight. All that had to be done now was to place the cosmonaut in the cabin, check up on all systems for the last time and start the rocket off.

I walked over to the Chairman of the State Commission, one of the industrial leaders well known in our country, and reported:

"Senior Lieutenant Gagarin is ready for the first flight in the spaceship *Vostok*."

"Happy landings! Good luck!" he answered and firmly shook my hand. His voice was not strong, but it was merry and kind, like that of my father.

I looked at the ship in which I had to start on an unusual trip. The ship was beautiful, more beautiful than a locomotive, steamship, plane, palaces and bridges, all put together. It occurred to me that that beauty was eternal and would live for the people of all countries for ever. I beheld not only a remarkable piece of machinery, but also an impressive work of art.

Before being lifted into the cabin of the ship I made a statement for the press and the radio. I felt an extraordinary elation. With every fibre of my being I heard the music of nature: the quiet rustling of the grass, then the noise of the wind which merged into the roar of the waves beating against the shore during a storm. This music was born within me and reflected a whole complex range of emotions: it put unusual words into my mouth the like of which I had never used in my life before.

"Dear friends, intimate and unknown, fellow citizens, people of all countries and all continents!" I said. "In a few minutes the mighty spaceship will carry me off into the distant spaces of the universe. What can I say to you in these last minutes before the start? All my life now appears as a single beautiful moment to me..."

I paused, collecting my thoughts. All my life flashed before my eyes. I saw myself as a barefooted boy helping shepherds tend the collective-farm cattle. A school boy writing the word Lenin for the first time. A pupil of a vocational school who had made his first mould... A student working on his diploma thesis.... A flyer guarding the state frontiers....

"All I have done and lived for has been done and lived for this moment." In those words I said all I had thought about during the last few days after I had been told that I would be the first to fly.

"You understand very well yourself how difficult it is for me to analyse my feelings now that the hour of trial for which we prepared so long and passionately is so near. It is hardly worth talking about the feelings I experienced when I was asked to make this first space flight in history. Did I experience joy? No, it was not only joy. Pride? No it was not only pride. I was immensely happy. To be the first in outer space, to meet nature face to face in this unusual single-handed encounter – could I possibly have dreamed of more?"

It was quiet all around. The recording tape rustled like a light breeze in the grass.

"Then I thought about the tremendous responsibility I had assumed. To be the first to accomplish what generations of people dreamed of, to be the first to pave the way for humanity to outer space.... Can you name a more complex task than the one I am undertaking? This is a responsibility not to one, not to scores of people and not even to a large single group of people. This is a responsibility to all the Soviet people, to all of humanity, to its present and future. And if I have decided to make this flight, it is only because I am a Communist, because I have behind me models of unexampled heroism of my countrymen – Soviet people."

And in my mind's eye I saw Chapayev and Chkalov, Pokryshkin and Kantaria, Kurchatov and Gaganova, Tursunkulov and Mamai.... They, and not only they, but all Soviet people have always drawn their vital forces from the same deep and clear well – Lenin's teachings. We cosmonauts and our entire young generation educated by the Leninist Party of Communists, have greedily drunk from this well.

For a brief moment I was lost in reverie, but quickly collected my thoughts and continued:

"I know that I shall collect all my will power to carry out my assignment to the best of my ability. I understand the importance of my mission and shall do all I can to fulfil the assignment of the Communist Party and the Soviet people.

"Am I happy, starting on my space flight? Of course, I am. Indeed, at all times and epochs people considered it the greatest happiness to take part in new discoveries.... "

Looking above the microphone, while I spoke, I saw the attentive faces of my tutors and friends – the Chief Designer, the Theoretician of Cosmonautics, Nikolai Kamanin, thr kind and friendly Yevgeny Anatolyevich, and Cosmonaut Two ....

"I want to dedicate this first space flight to the people of communism, of the society which our Soviet people are building and which, I am sure, all the people of the world will build."

I noticed that the Chief Designer took a furtive glance at his watch. It was time I finished.

"Only a few minutes are left before the start," I said. "I am saying good-bye to you, dear friends, as people always say to each other when leaving for a long journey. I should love to embrace all of you, friends and strangers, near and far!"

And already on the iron landing before entering the cabin, taking leave of the comrades remaining on the earth, I raised both my arms in salute and said:

"See you soon!"

I entered the cabin which smelled of the wind of the fields and was placed in my seat; the hatch was closed noiselessly. I was alone with the instruments now illumined not by sunlight but by artificial lighting. I heard everything that was being done outside the ship, on the earth which had become still dearer to my heart. The iron girders were removed and silence ensued. I reported:

"Hallo, 'Earth'! I am 'Cosmonaut'. I have tested the communications. The tumblers on the control panel are in the assigned initial position. The globe is at the point of division. Pressure in the cabin – unity, humidity – GS per cent, temperature – 19°C., pressure in compartment – 1.2, pressure in orientation systems – normal. I feel fine and am ready for the start."

The technical flight supervisor gave one-and-a-half-hour's notice for the flight. Then an hour's and lastly a half-hour's notice. Several minutes before the start I was told that my face could well be seen on the screen of the television device and that my cheerfulness made everybody happy. I was also told that I had a pulse of 64 and a respiratory rate of 24.

I answered:

"My heart beats normally. I feel fine; I have put on the gloves, closed the helmet and am ready for the start."

All the starting commands were also transmitted to me.

Finally, the technical flight supervisor ordered:

"Lift off!"

I answered:

"Off we go! Everything is normal."

I glanced at my watch. It was seven minutes past nine Moscow time. I heard a shrill whistle and a mounting roar. The giant ship shuddered, and slowly, very slowly lifted from the launching pad. The roar was not louder than that one hears in the cabin of a jet aircraft, but it contained a multitude of new shades and pitches that no composer has ever put down in music or any musical instrument or human voice can as yet reproduce. The rocket's huge engines were fashioning the music of the future, which, I imagine, is much more exciting and beautiful than the greatest creations of the past.

The acceleration of G-forces began to increase. I felt as though some uncompromising force were riveting me to my seat. Although the seat had been positioned to minimise the tremendous gravity pull that now brought all its weight to bear on me, I could hardly move my hands or feet. I knew that this state would not last long, that it would pass as soon as the ship, after accelerating, would begin to orbit.

The earth called my attention:

"Seventy seconds have passed since the take-off."

I replied:

"I hear you: seventy. I feel well. Am continuing flight. The G-forces are increasing. Everything is all right."

I replied in a cheerful voice, but to myself I thought: Is it really only seventy seconds? The seconds seem like minutes.

The earth again asked:

"How do you feel?"

"All right. How are things going with you?"

The reply from the earth was:

"Everything's normal."

I maintained two-way radio link with the earth through three channels. My short-wave transmitters worked on a frequency of 9.019 and 20.006 megacycles and on an ultra-short wave frequency of 143.025 megacycles. The voices of the comrades working at the radio stations were so distinct that it seemed they were sitting beside me.

As soon as the ship passed the dense layers of the atmosphere the head fairlead was automatically discarded and it flew away. The surface of the earth loomed in the distance through the portholes. Just then the *Vostok* was flying over a broad Siberian river. I could clearly see the sunlit, taiga-overgrown islands and banks.

"How magnificent!" I exclaimed again, unable to restrain myself, and stopped short: my job was to transmit flight data and not admire nature's beauties, the more so that the earth was again calling me.

"I hear you distinctly," I replied. "I feel well. The flight is proceeding normally. The G-forces are increasing. I see the earth, forests, clouds...."

The G's were indeed mounting. But I gradually got used to them and was even struck by the thought that they were greater in the centrifuge. The vibration, too, was considerably smaller than what I had had to experience during training. In short, the devil is not so terrible as he is painted.

A multi-stage space rocket is such an intricate piece of machinery that it is hard to find a comparison for it, and yet people perceive things by comparison. When each stage burns up its fuel it becomes useless, it is automatically separated and discarded, while the remaining part of the rocket goes on building up the speed of the flight. I never had occasion to meet the scientists and engineers who developed the splendid, portable fuel for the engines of Soviet rockets. But as I was climbing ever higher to my predetermined orbit I wanted to thank them and gratefully shake hands with them. The intricate engines worked marvellously, with the precision of the Kremlin Chimes.

Using up their fuel, the stages dropped off one by one and the moment came when I could report:

"The rocket-carrier has separated in conformity with the programme. I feel well. The cabin parameters are: pressure – one; humidity – 65 per cent; temperature – 20 degrees; pressure in the compartment – one; the orientation systems are working normally."

The ship entered its orbit, a broad space route, and I found myself in the state of weightlessness, about which I had read in books by Konstantin Tsiolkovsky when I was a boy. At first I felt uncomfortable, but I soon got used to it, found how to get about and continued with the flight programme. "I wonder what the people on earth will say when they learn of my flight?" I thought.

For all the inhabitants of the earth, weightlessness is a strange phenomenon. But one quickly adapts oneself to it and all that remains is an extraordinary lightness in every limb. What happened to me during this time? I left my seat and hung in the air between the ceiling and floor of the cabin. The change to this state was very smooth.

I felt wonderful when the gravity pull began to disappear. I suddenly found I could do things much more easily than before. And it seemed as though my hands and legs and my whole body did not belong to me. They did not weigh anything. You neither sit nor lie, but just keep floating in the cabin. All the loose objects likewise float in the air and you watch them as in a dream. The map-case, the pencil, the notepad.... Drops of a liquid that dripped out of a tube took on a spherical shape and floated about freely in space until they came into contact with the porthole glass and clung to it like dew on a flower.

Weightlessness does not affect one's capacity for work. I worked all the time. I kept an eye on the instruments in the cabin, maintained observation through the portholes and made records in the log. I made my notes with an ordinary lead pencil without taking off the pressurised gloves. I found no difficulty in writing and the words appeared clearly in the flight log-book. For a moment forgetting where and in what position I was, I put the pencil down beside me and it sailed away. I made no attempt to catch it, only commenting loudly on everything I saw. A tape-recorder took down everything I said on its narrow running tape. At the same time, I was in constant radio contact with the earth through several telephone and telegraph channels.

The earth wanted to know what I saw below me. I reported back that the view I had of our planet was approximately the same as what can be seen from a jet aircraft flying at a high altitude. I could distinctly see mountain ranges, large rivers, big tracts of forests, islands and the coastline of seas.

The *Vostok* sped over the spaces of my country and a feeling of filial love swelled in me. It is only natural that we, her children, should love her. The peoples of the whole world are looking to my country with hope. Poor and backward a short time ago, she has become a mighty industrial and collective-farm power. Organised and reared by the Communist Party, the Soviet people have shaken off the dust of the old world, squared their titanic shoulders and moved forward along the path that had been blazed by Lenin. Led by the Party, our mighty people established the rule of the working people, founding the first Soviet state in the world.

With the example of her heroes, our Motherland inculcated the best and loftiest of human feelings into us from our childhood.

The *Lay of Prince Igor's Host*, the oldest Russian chronicle of patriotic deeds, thrilled me when I was a boy. I remember staying behind in the classroom at breaks, gazing at a map showing the great Russian rivers – the Volga, the Dnieper, the Ob, the Yenisei and the Amur forming a network of blue veins round the mighty body of our country, and dreaming of distant countries and of journeys. And here I was making the greatest journey of my life, a journey round the earth.

I saw clouds and the light shadows they cast on the dear, distant earth. For a fleeting moment the son of a collective farmer awakened in me and the pitch black sky looked like a ploughed field sown with star-like grain.

That grain was bright and clean as though it bad been winnowed again and again. The sun, too, is remarkably bright in outer space and you cannot look at it with your naked eye even if you squint it. It is many tens and perhaps hundreds of times brighter than when you look at it from the earth. It is brighter than the molten metal I had handled when I worked in a foundry. To subdue its glare I had to cover the portholes from time to time with special filters.

I wanted to observe the moon, to see how it looks in space. Unfortunately, it was not in my field of vision during the flight. "Never mind," I said to myself, "I'll see it next time."

My observations covered not only the sky but also the earth. What does the water surface look like? Darkish. with faintly gleaming spots. Do you get the feeling that our planet is round? Indeed, you do. On the horizon I could see the sharp, contrasting change from the light surface of the earth to the inky blackness of the sky. The earth was gay with a lavish palette of colours. It had a pale blue halo around it. Then this band gradually darkened, becoming turquoise, blue, violet and then coal-black. This change is very beautiful and pleasing to the eye.

The notes of my country's music floated in my cabin and I heard *The Amur Waves*, which is one of my favourite songs. I recalled some Americans writing that nobody can foretell what will be the effect of space on man; but one thing is certain, they said, in space man will be bored and lonely. I felt nothing of the kind. In space I worked and lived the life of my people. The radio connected me like a navel cord with the earth. I received commands and transmitted reports on the work of all the ship's systems, and in each word from the earth I felt the support of the people, the Government and the Party.

With my eye constantly on the instruments, I determined that the *Vostok*, which was flying strictly in a predetermined orbit, was about to enter the shaded side of our planet. This happened quickly, it suddenly became pitch dark. Evidently, I was flying over an ocean because I could not see even the golden dust of illumined cities.

As I crossed the Western Hemisphere I thought of Columbus, of his hardships and sufferings before he discovered the New World. But this New World was named America after Amerigo Vespucci, who earned immortality by his *Description of New Lands*, a book of thirty-two pages. I read the story of this historical mistake in a book by Stelan Zweig.

While musing about America, I could not help thinking of the lads who were planning to follow us into space. Somehow I felt it would be done by Alan Shepard. Would the American astronauts serve peace as we were, or would they become the slaves of those who were preparing for war? What a wonderful thing it would be if the peoples of the globe heeded the wise voice of Nikita Sergeyevich Khrushchov, and directed all their energies towards the achievement of lasting, universal peace.

The automatic orientation system was turned on at 09 hours 51 minutes. After the *Vostok* emerged from the shadow, the system oriented the ship at the sun. Its rays were piercing the earth's atmosphere, the horizon blazed in a bright orange which gradually changed into all the colours of the rainbow: to light blue, blue, violet, black. Words fail to describe this colour range. It reminded me of the canvases of the artist Nikolai Rörich.

The time was 09 hours 52 minutes. I was flying over the region of Cape Horn and sent the report:

"The flight is proceeding normally. I feel well. The onboard apparatus is working faultlessly."

I checked with the flight time-table. It was being rigidly kept to. The *Vostok* was flying at a speed of nearly 28,000 kilometres an hour. It is a speed that one can hardly imagine. During the flight I felt neither hungry nor thirsty. But at a definite time, keeping to the flight programme, I had a meal and drank water from a special water-supply system. The food that I ate was prepared according to a recipe prescribed by the Academy of Medical Sciences. I ate just as I did on earth, but the only trouble was that I could not open my mouth wide. Although I knew that the behaviour of my organism was being watched on earth, I sometimes could not help listening to my own heart. In the state of weightlessness the pulse and respiration were normal, I felt comfortable and I had full command of my capacity for thinking and for work.

Leads from light convenient transducers that converted physiological parameters – biocurrents of the heart, the pulse oscillations of the vascular wall, and the respiratory movement of the chest – into electric signals were built into my spacesuit. Special amplifiers and measuring systems radioed to earth the impulses showing my respiration and blood circulation at all stages of the flight. The earth therefore knew more of how I felt than I did myself.

From the moment the rocket lifted from the launching pad, control over all of the ship's intricate mechanisms was taken over by clever automatic systems. They guided the rocket along a predetermined trajectory. They controlled the engines' speed. They discarded the burnt out stages of the rocket. Automatic arrangements maintained the necessary temperature in the ship, oriented it in space, kept the measuring instruments operating and carried out many other difficult tasks. At the same time I had manual ship controls at my disposal. All I had to do to take over the control of the flight and landing of the *Vostok* was to turn on the required tumbler. I would have had to re-check my position by the instruments on board the *Vostok* hurtling above the earth at terrific speed. Then I would have had to calculate the landing, control the ship's orientation and turn on the retrorockets at the required moment. But there was no need for all this – the automatic instruments were working faultlessly. Everything had been considered and weighed by scientists.

The Chief Designer had told me of what was being done to reduce the weight and size of each part of spaceships, and that Soviet automation experts were developing systems consisting of many thousands of components and creating self-setting devices that can adapt themselves to changing conditions. With youthful enthusiasm he had told us of control arrangements that ensure reliability of a system in spite of a large number of elements.

These memories flashed through my mind. And having remembered it all, I began to think of the Chief Designer. The scientists who had put their intelligence, energy and work into the spaceship could be proud of it.

I tried to picture the people who helped build the ship and I mentally saw rank upon rank of working people passing by as at a May Day demonstration in Red Square. I wished I could see them at their work in the laboratories and in factory workshops, shake their hands and thank them. When you come to think of it, the most wonderful thing on earth is man engaged in work.

With a thrill of joy I closely observed the world around me, endeavouring to see and understand everything. Through the portholes I saw a diamond-field of shining, bright, cold stars. They were far, oh, so far away, perhaps tens of years of flight, and yet from my orbit I was much closer to them than I would have been on the earth. It was gratifying and yet a little frightening to be conscious that I had been entrusted with a spaceship, a priceless treasure of the state into which so much of the people's labour and money had been put.

Despite my complex work I could not help thinking. I thought of my mother, recalling how she kissed me between the shoulder-blades, putting me to sleep when I was a baby. Does she know where I am? Did Valya tell her about my flight? And thinking of my mother I could not help thinking of my Motherland. It is not without reason that people call their native land Motherland, for it is eternally alive, it is immortal. Man owes all his achievements in life to his Motherland. "Our socialist Motherland is the finest in the world, and all I have achieved I owe to my Motherland," I thought.

Many different thoughts entered my head, but they were all joyful, holiday-like thoughts. I remembered how we had gone into the collective-farm orchard as boys and shaken apples off the trees, how on the eve of the flight I had wandered through the noisy, merry streets of Moscow, had ended up in Red Square and stood for a long time beside the Mausoleum. The thought then struck me that the spaceship was carrying Lenin's ideas round the globe.

"What is Cosmonaut Two doing now?" I wondered and felt the strength and warmth of the embrace he gave me on parting. Indeed, he will have to experience everything I am experiencing now.

Countries sped by below me in a single whole undivided by state boundaries.

At 10 hours 15 minutes while the ship was approaching Africa the automatic programme arrangement issued commands preparing the ship's apparatus for switching on the braking rockets. I radioed my routine report: "Flight proceeding normally. Am bearing state of weightlessness well."

It occurred to me that somewhere below was the mountain glorified by Ernest Hemingway in his story "The Snows of Kilimanjaro".

Then I thought that the ship was flying over the Congo, where the imperialists had foully murdered Patrice Lumumba, the courageous fighter against colonialism, fighter for the happiness of his people.

But there was no time for reflections. I was now entering the final and perhaps more responsible stage of the flight than when the ship was placed in orbit and began to circle the earth, for this was the homing stage. I began to prepare for it. Ahead of me was the change from the state of weightlessness to new and perhaps bigger deceleration G-forces and the colossal temperature to which the ship's external plating would be heated when it entered the dense layers of the atmosphere. So far everything had progressed approximately as we had anticipated during our training on earth. But what would the last, final stage of the flight be like? Would all the systems operate normally? Was some unforeseen danger lying in wait for me? With all my respect for automation, I determined the ship's position and held myself in readiness to take over the controls and, if need be, to guide the descent to earth in a suitable region selected by me.

My ship had a solar orientation system fitted with special transducers. These transducers "caught" the sun and, orientating themselves by it, kept the ship in a definite flight attitude with the retrorockets facing in the direction of the flight. The braking system was turned on automatically at 10 hours 25 minutes and worked without a flaw, at exactly the time set in advance. Having climbed to a high altitude, the *Vostok* now had to descend the same distance. It gradually lost speed, going over from an orbit into a descending trajectory. The ship began to enter the dense layers of the atmosphere. Its external skin rapidly became red-hot and through the porthole filters I saw the frightening crimson reflection of the flames raging all round the ship. But in the cabin it was only twenty degrees above zero in spite of the fact that the ship was hurtling down towards the earth like a ball of fire.

The state of weightlessness had passed long ago, and the growing G-forces pinned me to my seat. They kept increasing and were much greater than the stresses I had experienced at the take-off. The ship began to rotate and I reported this to earth. But the rotation which worried me quickly stopped and the further descent proceeded normally. All the systems were obviously working faultlessly and the ship was making for the predetermined landing area. Overcome with joy I broke into my favourite song at the top of my voice:

The country hears,
The country knows....

The ship descended lower and lower. Making sure that it would safely reach the earth, I prepared for the landing.

Ten thousand metres.... Nine thousand.... Eight.... Seven....

The silvery Volga flashed past below. I at once recognised the great Russian river and its banks. This was where Dmitry Martyanov had taught me to fly. Everything here was familiar to me: the spring fields, the thickets, the roads and Saratov, whose houses I could make out in the distance.

At 10 hours 55 minutes, the *Vostok*, having completed its orbit round the earth, landed safely in a predetermined area, on a ploughed field of the Leninsky Put (Lenin's Path) Collective Farm, which is situated to the south-west of the town of Engels, near the village of Smelovka. Events unfolded as in an exciting novel. I returned from outer space to exactly the same spot where I first learned to fly an aircraft. How much time had passed since then? Not more than six years. But how the yardstick had changed! On this day I flew two hundred times faster and at an altitude that was two hundred times higher.

FIRST ON THE MOON

Neil Armstrong, Michael Collins, Edwin E. Aldrin Jr

Neil Armstrong (1930–2012) was an American astronaut and pilot who became the first man to walk on the surface of the moon. A talented naval aviator and engineer, Armstrong served in the Korean War and later joined the NASA astronaut corps. On 20 July 1969, he and his lunar module pilot, **Buzz Aldrin** (born 1930), became the first people in history to walk on the surface of the moon, successfully fulfilling President Kennedy's assertion that the United States would have a man on the moon by the end of the decade. Armstrong and Aldrin, along with fellow astronaut **Michael Collins** (1930–2021), provided the United States with a strategic public relations victory in the highly publicized Space Race with the Soviet Union, as most of the previous landmarks in Space Race had been achieved by Soviet cosmonauts.

"Don't forget one in the command module"

[...]

At 9:56 P.M., Houston time, Neil Armstrong stepped out of the dish-shaped landing pad and onto the surface of the moon: "THAT'S ONE SMALL STEP FOR A MAN, ONE GIANT LEAP FOR MANKIND."

The quality of the television transmission delighted everyone in the world – everyone, perhaps, except Jon Eisele, age four, the son of the astronaut Donn Eisele. He had been put to bed and was sound asleep long before Armstrong and Aldrin stepped onto the moon. But his mother Harriet, who was working for a television company during the moon flight, left word with her neighbor and best friend, Faye Stafford, the wife of Tom Stafford, to wake the children up when it came time to watch TV from the moon. Faye Stafford awakened Jon and plunked him on the floor in front of the television set. Jon sat there for about five minutes and said, "I can't see, the picture is all fuzzy. I'm going *outside* to see." He went out, looked up at the moon, came back and said, outside "I still can't see. Where's the flashlight?"

In Houston there was a new class of temporarily unemployed: the long-distance telephone operators. When the moon walk was on, they had little to do. The number of long-distance calls handled by Houston operators through the Southwestern Bell exchange between 9 P.M. and midnight on July 20 was 6,408, compared with 8,968 during the same three hours on the preceding Sunday.

At Honeysuckle Creek in Australia, where it was 12:56 P.M. Monday, July 21, there was a sign on the notice board in the operations room which bore the title of the pop song, "Fly Me to the Moon." Someone had written in two additional words: ". . . and back."

It has a stark beauty all its own. It's like much of the high desert of the United States.

NEIL ARMSTRONG

I don't believe any pair of people had been more removed physically from the rest of the world than we were.

EDWIN E. ALDRIN JR.

"Magnificent desolation"

IT WAS TIME for Neil Armstrong to walk on the moon, and as Armstrong waited for Aldrin to follow him out in nineteen minutes his first reaction to the environment was a favourable one. He was able immediately to discard the theory, once widely held, that the windless surface of the moon was overlaid with a dangerously deep coating of dust in which men and manmade machines would founder. The lunar module's footpads had made only a shallow penetration, and Neil's boots sank only a fraction of an inch: "Maybe an eighth of an inch, but I can see footprints of my boots and the treads in the fine sandy particles." And he could move around: "There seems to be no difficulty.... It's even perhaps easier than the simulations at one-sixth G....It's actually no trouble to walk about. The descent engine did not leave a crater of any size. There's about one foot clearance on the ground. We're essentially on a level place here. I can see some evidence of rays emanating from the descent engine, but very insignificant amounts... Okay, Buzz, we're ready to bring down the camera."

"I'm all ready," Aldrin answered... "Okay, you'll have to pay out all the LEC [Lunar equipment conveyor]. It looks like it's coming out nice and evenly."

"Okay, it's quite dark here in the shadow and a little hard for me to see if I have good footing. I'll work my way over into the sunlight here without looking directly into the sun."

'Okay, it's taut now," Aldrin said. "Don't hold it quite so tight."

Armstrong looked up at the LM and messaged: "I'm standing directly in the shadow now, looking up at Buzz in the window. And I can see everything clearly. The light is sufficiently bright, backlighted into the front of the LM, that everything is very clearly visible."

"Okay, I'm going to be changing this film magazine," Aldrin said.

"Okay," Armstrong said. "Camera installed on the RCU bracket [remote control unit]. I'm storing the LEC on the secondary strut.... I'll step out and take some of my first pictures here."

HOUSTON (McCandless): Roger, Neil, we're reading you loud and clear. We see you getting some pictures and the contingency sample.

"This is very interesting," Armstrong said. "It's a very soft surface, but here and there where I plug with the contingency sample collector, I run into a very hard surface... I'll try to get a rock in here. Here's a couple."

"That looks beautiful from here, Neil," Aldrin said from Eagle's cabin.

"It has a stark beauty all its own. It's like much of the high desert of the United States. It's different, but it's very pretty out here.

In El Lago Jan Armstrong was ticking off the minutes, but not out of any particular safety concern; her concern was still the one she had expressed much earlier: would they be able to do all they had been assigned to do on this first lunar landing mission? But there was tension in the small talk. Jan volunteered, "Buzz will not come down until Neil gets the contingency sample – which will fall out of his pocket." There was discussion about how to tell who was saying what; Armstrong and Aldrin had voices of similar timbre, and even people who knew them well had had difficulty with their voice identifications throughout the flight. Ricky Armstrong suggested, "You can recognize Daddy – he always says 'Uhhh.'" In Nassau Bay Pat Collins commented: "Look at Neil move. He looks like he's dancing – that's the kangaroo hop." Waiting for Buzz to come out, Joan Aldrin reacted to Armstrong's physical descriptions of the lunar surface with amazement: "He *likes* it!" Later Neil Armstrong said...

« The most dramatic recollections I had were the sights themselves. Of all the spectacular views we had, the most impressive to me was on the way to the moon when we flew through its shadow. We were still thousands of miles away, but close enough so that the moon almost filled our circular window. It was eclipsing the sun, from our position, and the corona of the sun was visible around the limb of the moon as a gigantic lens-shaped or saucer-shaped light, stretching out to several lunar diameters. It was magnificent, but the moon was even more so. We were in its shadow, so there was no part of it illuminated by the sun. It was illuminated only by the earthshine. It made the moon appear blue-gray, and the entire scene looked decidedly three-dimensional.

I was really aware, visually aware, that the moon was in fact a sphere, not a disc. It seemed almost as if it were showing us its roundness, its similarity in shape to our earth, in a sort of welcome. I was sure that it would be a hospitable host. It had been awaiting its first visitors for a long time. »

"You can really throw things a long way up here," Armstrong said. "That pocket open, Buzz?"

"Yes, it is, but it's not up against your suit though. Hit it back once more. More toward the inside. Okay, that's good."

"That in the pocket?"

"Yes, push down," Aldrin said. "Got it? No, it's not all the way in. Push it. There you go."

"Contingency sample is in the pocket," Armstrong confirmed. "...Are you getting a TV picture now, Houston?"

HOUSTON (McCandless): Neil, yes we are getting a TV picture... You're not in it at the present time. We can see the bag on the LEC being moved by Buzz, though. Here you come into our field of view.

ALDRIN: "Okay. Are you ready for me to come out?" ARMSTRONG: "Yes. Just stand by a second. I'll move this over the handrail. Okay?" ALDRIN: "All right. That's got it. Are you ready?" ARMSTRONG: "All set. Okay, you saw what difficulties I was having. I'll try to watch your PLSS down. There you go.... About an inch clearance on top of your PLSS...Okay, you're right at the edge of the porch... Looks good." ALDRIN: "Now I want to back up and partially close the hatch, making sure not to lock it on my way out." ARMSTRONG: "A particularly good thought." ALDRIN: "That's our home for the next couple of hours and I want to take good care of it."

Watching television in Nassai Bay, Rusty Schweickart encouraged: "Don't close it all the way, Buzz. What do you mean? You forgot the key?" As Buzz started down the ladder Jan Armstrong wondered aloud, "Wouldn't that be something if they locked themselves out?" In the Aldrin home the astronaut Fred Haise, watching Armstrong move across the screen, laughed and said, "That's the fastest I've ever seen Neil move!" He warned Joan Aldrin, "Buzz is about to come out now, Joan." She said, "It's like making an entrance on stage." But as seconds passed with no sight of Buzz, she said, "You see, he's doing it just like on Gemini 12. He's going to explain eery single thing he does." Then Buzz's legs did appear. Joan screamed and kicked her own legs up in the air. As he came slowly down the ladder she said, "He's going to analyze every step at a time."

ALDRIN: "It's a very simple matter to hop down from one step to the next." ARMSTRONG: "Yes, I found it to be very comfortable, and walking is also very comfortable. You've got three more steps and then a long one." ALDRIN: "Okay, I'm going to leave that one floor up there and both hands down to about the fourth rung up." ARMSTRONG: "There you go." ALDRIN: "Okay. Now I think I'll do the same." ARMSTRONG: "A little more. About another inch. There you got it. That's a good step, about a three-footer." ALDRIN: "Beautiful view." *Ten-fifteen in the evening, Houston time...They were now both on the moon, and for a few seconds they were awed in spite of themselves...*

"Isn't that something?" Armstrong asked. "Magnificent sight out here."

"Magnificent desolation," Aldrin said. And later...

"Neil and I are both fairly reticent people, and we don't go in for free exchanges of sentiment. Even during our long training we didn't have many free exchanges. But there was that moment on the moon, a brief moment, in which we sort of looked at each other and slapped each other on the shoulder – that was about the space available – and said, 'We made it. Good show.' Or something like that..."

Written with Gene Farmer and Dora Jane Hamblin

ENDURANCE

My Year in Space, A Lifetime of Discovery

Scott Kelly

Scott Kelly (born 1964) is an American astronaut who served as commander of the International Space Station on three separate missions. In 2015, he undertook spending a whole year in space – twice the length of a regular stay – at the International Space Station as part of a research project into the effects of long-term space flight on the human body. During the ISS year-long mission he read *Endurance: Shackleton's Incredible Voyage* as a way of framing the challenges of his own journey in the context of Shackleton's travails. The book inspired his own memoir *Endurance: A Year in Space, a Lifetime of Discovery.*

Paul McCartney is singing over the crackly communication system. So far we've heard Coldplay, Bruce Springsteen, Roberta Flack. I happen to like "Killing Me Softly," but I can't help but think it's inappropriate considering the circumstances. I'm crammed in the righthand seat of the Soyuz, acutely aware of the 280 tons of explosive propellant under me. In an hour, we will tear into the sky. For now, soft rock is distracting us from the pain of sitting in the cramped capsule.

When we got off the bus at the launch site, it was fully dark, floodlights illuminating the launch vehicle so it could be seen from miles around. Though I've done it three times before, approaching the rocket I was about to climb into is still an unforgettable experience. I took in the size and power of this machine, the condensation from the hypercooled fuel billowing eerily in a giant cloud, enveloping our feet and legs. As always, the number of people around the launchpad surprised me, considering how dangerous it is to have a fully fueled rocket – basically a bomb – sitting there. At the Kennedy Space Center, the area was always cleared of nonessential personnel for three miles around, and even the closeout crew drove to a safe viewing site after strapping us into our seats. Today, dozens of people were milling around, some of them smoking, and a few of them will watch the launch from dangerously close. Once, I watched a Soyuz launch while serving as backup for one of the crew, I was standing outside the bunker, just a few hundred yards away. As the engines ignited, the manager of the launchpad said in Russian, "Open your airway and brace for shock."

In 1960, an explosion on the launchpad killed hundreds of people, an incident that would have caused a full investigation and an array of new regulations for NASA. The Soviets pretended it hadn't happened and sent Yuri Gagarin to space the following year. The Soviet Union acknowledged the disaster only after the information about it was declassified in 1989.

By tradition, there is one last ritual: Gennady, Misha, and I climb the first few stairs heading toward the elevator, then turn to say goodbye to the assembled crowd, waving to the people of Earth one last time.

Now we wait in the Soyuz, something we've all experienced before, so we know our roles and know what to expect. I anticipate the excruciating pain in my knees that nothing seems to alleviate. I try to distract myself with work: I check our communication systems and introduce oxygen into the capsule with a series of valves – one of my primary responsibilities as the flight engineer 2, a position that I like to describe as the copilot of the copilot of the spacecraft. Gennady and Misha murmur to each other in Russian, and certain words jump out: "ignition," "dinner," "oxygen," "whore" (the all-purpose Russian swear). The capsule heats up as we wait. The music we hear now is "Time to Say Goodbye" by Sarah Brightman, who was going to travel to the International Space Station later this year but has had to cancel her plans. A Russian pop song, "Aviator," follows.

The activation of the launch escape system wakes us up with a loud thunk. The escape system is a separate rocket connected to the top of the spacecraft, much like the one on the old Apollo/Saturn that was designed to pull the capsule free in case of an explosion on the pad or a failure during launch. (The Soyuz escape rocket was used once, saving two cosmonauts from a fireball, in 1983.) The fuel and oxidizer turbo pumps spin up to speed with a screaming whine – they will feed massive amounts of liquid oxygen and kerosene to the engines during ascent.

Russian mission control warns us it's one minute to launch. On an American spacecraft, we would already know because we'd see the countdown clock ticking backward toward zero. Unlike NASA, the Russians don't feel the drama of the countdown is necessary. On the space shuttle, I never knew whether I was really going to space that day until I felt the solid rocket boosters light under me; there were always more scrubs than launches. On Soyuz, there is no question. The Russians haven't scrubbed a launch after the crew was strapped in since 1969.

"*My gotovy,*" Gennady responds into his headset. We are ready.

"*Zazhiganiye,*" mission control says. Ignition.

The rocket engines of the first stage roar to full capacity. We sit rumbling on the launchpad for a few seconds, vibrating with the engines' power – we need to burn off some of the propellant to become light enough to lift off. Then our seats push hard into our backs. Some astronauts use the term "kick in the pants" to describe this moment. The slam of acceleration – going from still to the speed of sound in a minute – is heart pounding and addictive, and there is no question that we are going straight up.

It's night, but we wouldn't be able to see anything out our windows even if it were broad daylight. The capsule is encased in a metal cylinder, called a fairing, which protects it from aerodynamic stress until we are out of the atmosphere. Inside, it's dark and loud and we are sweaty in our Sokol suits. My visor fogs up, and I have trouble reading my checklist.

The four strap-on boosters of four engines each fall away smoothly after two minutes, leaving the four remaining engines of the second stage to push us into space. As we accelerate to three times the Earth's gravity, the crushing force smashes me into my seat and makes it difficult to breathe.

Gennady reports to the control center that we are all feeling fine and reads off data from the monitors. My knees hurt, but the excitement of launch has masked the pain some. The second-stage rockets fire for three minutes, and as we are feeling their thrust, the fairing is jettisoned away from us in two pieces by explosive charges. We can see outside for the first time. I look out the window at my elbow, but I see only the same black we launched into.

Suddenly, we are thrown forward against our straps, then slammed back into our seats. The second stage has finished, and the third stage has taken over. After the violence of staging, we feel some roll oscillations, a mild sensation of rocking back and forth, which isn't alarming. Then the last engine cuts off with a bang and there is a jolt, like a minor car crash. Then nothing.

Our zero-g talisman, a stuffed snowman belonging to Gennady's youngest daughter, floats on a string. We are in weightlessness. This is the moment we call MECO, pronounced "mee-ko," which stands for "main engine cutoff." It's always a shock. The spacecraft is now in orbit around the Earth. After having been subjected to such strong and strange forces, the sudden quiet and stillness feel unnatural.

We smile at one another and reach up for a three-handed high-five, happy to have survived this far. We won't feel the weight of gravity again for a very long time.

Something seems out of the ordinary, and after a bit I realize what it is. "There's no debris," I point out to Gennady and Misha, and they agree it's strange. Usually MECO reveals what junk has been lurking in the spacecraft, held in their hiding places by gravity – random tiny nuts and bolts, staples, metal shavings, plastic flotsam, hairs, dust – what we call foreign object debris, and of course NASA has an acronym for it:

FOD. There were people at the Kennedy Space Center whose entire job was to keep this stuff out of the space shuttles. Having spent time in the hangar where the Soyuz spacecraft are maintained and prepared for flight, and having observed that it's not very clean compared to the space shuttle's Orbiter Processing Facility, I'm impressed that the Russians have somehow maintained a high standard of FOD avoidance.

The Soyuz solar arrays unfurl themselves from the sides of the instrumentation module, and the antennas are deployed. We are now a fully functional spacecraft in orbit. It's a relief, but only briefly.

We open our helmets. The fan noise and pump noise blending together are so loud we have trouble hearing one another. I had remembered this about my previous mission to the ISS, of course, but still I can't believe it's so noisy. I can't believe I'll ever get used to it.

"I realized a few minutes ago, Misha," I say, "that our lives without noise have ceased to exist."

"Guys," Gennady says. "*Tselyi god!*" An entire year!

"*Ne napominai, Gena*," answers Misha. Gena, don't remind me.

"*Vy geroi blya*." You're freaking heroes.

"*Yep*," Misha agrees. "Totally screwed."

Now we are in the rendezvous stage. Joining two objects in two different orbits traveling at different speeds (in this case, the Soyuz and the ISS) is a long process. It's one we understand well and have been through many times, but still it's a delicate maneuver. We pick up a strange broadcast over Europe:

...scattered one thousand four hundred feet. Temperature one nine. Dew point one seven. Altimeter two niner niner five. ATIS information Oscar ...

It's some airport's terminal broadcast, a recording giving pilots information about weather and approaches. We shouldn't be receiving this, but the Soyuz comm system is horrible. Every time Russian mission control talks to us we can hear the characteristic *dit-dit-dit* of cell phone interference. I want to yell at them to turn off their cell phones, but in the name of international cooperation I don't.

A few hours into the flight my vision is still good, with no blurring – a positive sign. I do start to feel congested, though, which is a symptom I've experienced in space before. I feel my legs cramp, from being crammed into this seat for hours, and there is the never-ending knee pain. After MECO, we can unstrap ourselves, but there isn't really anywhere to go.

Gennady opens the hatch to the orbital module, the other habitable part of the Soyuz, where the crew can stay if it takes more than a few hours to get to the station, but this module doesn't have much more space. I disconnect my medical belt, a strap that goes around my chest to monitor my respiration and heartbeat during launch, and float up to the orbital module to use the toilet. It's nearly impossible to pee while still halfway in my pressure suit. I can't imagine how the women do it. After I get back into my seat, mission control yells at me to plug my medical belt back in. We get strapped back in a few hours before we will be docking. Gennady scrolls through the checklist on his tablet and starts inputting commands to the Soyuz systems. The process is largely automated, but he needs to stay on top of it in case something goes wrong and he has to take over.

When it's time for the docking probe to activate, nothing happens. We wait. Gennady says something to Russian mission control in rapidfire Russian. They respond, sounding annoyed, then garble into static. We are not sure if they heard us. We are still a long way from ISS.

"Fucking *blya*," Gennady groans. Fucking bullshit.

Still no indication the docking probe has deployed. This could be a problem.

The process of docking two spacecraft together has remained pretty much unchanged from the Gemini days: one spacecraft sticks out a probe (in this case, us), inserts it into a receptacle called a drogue in the other spacecraft (the ISS), a connection is made, everyone cracks sex jokes, we leak-check the interface before opening the hatch and greeting our new crewmates. The process has been reliable for the past fifty years, but this time the probe doesn't appear to have worked.

The three of us give one another a look, an international I-can't-fucking-believe-this look. Soon, ISS will be looming in the window, its eight solar array wings glinting in the sun like the legs of a giant insect. But without the docking probe, we won't be able to connect to it and climb aboard. We'll have to return to Earth. Depending on when the next Soyuz will be ready, we might have to wait weeks or months. We could miss our chance altogether.

We contemplate the prospect of coming back to Earth, how ridiculous we'll feel climbing out of this capsule, saying hello again to people we've just said the biggest good-bye in the world to. Comm with the ground is intermittent, so they can't help us much in our efforts to figure out what's going on. I turn to see Misha's face. He is shaking his head in disappointment.

Once Gennady and Misha transition the computer software to a new mode, we see that the probe is in fact deployed. It was just a software "funny."

All three of us sigh with relief. This day hasn't been for nothing. We are still going to the space station.

I watch the fuzzy black-and-white image on our display as the docking port on ISS inches closer and closer. I wonder if it's true that the probe is actually okay. The last part of the rendezvous is exciting, much more dynamic than the space shuttle docking ever was. The shuttle had to be docked manually, so it was a slow ballet with little room for error. But the Soyuz normally docks with ISS automatically, and in the last minutes of the approach it whips itself around quickly to do an adjustment burn. Even though we'd known to expect this, it's still attention getting, and I watch out the window as the station comes flying into view, its brilliant metal sparkling in the sunlight as if it's on fire. The engines fire briefly, and we hear and feel the acceleration. Leftover fuel vents outside, glinting in the sun. With the burn complete, we snap back into position to move toward the docking port.

When we finally make contact with the station, we hear and feel the eerie sound of the probe hitting, then scratching its way into the drogue, a grinding metal-on-metal sound that ends with a satisfying clunk. Now both ISS and the Soyuz are commanded to free drift – they are no longer controlling their attitude and are rotating freely in space until a more solid connection can be made. The probe is retracted to draw the two vehicles closer together, then hooks are driven through the docking port to reinforce the connection. We've made it. We slap one another on the arms.

I join Gennady in the orbital module, where we struggle out of the Sokol suits we put on nearly ten hours earlier. We are tired and sweaty but excited to be attached to our new home. I take off the diaper I've been wearing since I left Earth and put it in a Russian wet trash bag for later disposal on the ISS. I get into the blue flight suit I call my Captain America suit because of the huge American flag emblazoned across the front. I hate these flight suits – the Russian who has been making them for years can't be made to understand that we stretch an inch or two in space, so within a few weeks I will no longer be able to wear the Captain America suit without having my balls crushed.

As eager as we are to greet our new crewmates, we need to make sure the seal between the Soyuz and ISS is good. The leak checks take nearly two hours. The space between the two docking compartments has to be filled with air, which we then test to determine whether its pressure is dropping. If it is, we don't have a good seal, and opening the hatch will cause ISS and Soyuz to lose their atmosphere. Occasionally, as we wait, we hear the crew on the other side banging on the hatch in a friendly greeting. We bang back.

The leak check finally complete, Gennady opens the hatch on our side. Anton Shkaplerov, the only cosmonaut on board the ISS, opens the Russian hatch on their side. I smell something strangely familiar and unmistakable, a strong burned metal smell, like the smell of sparklers on the Fourth of July. Objects that have been exposed to the vacuum of space have this unique smell on them, like the smell of welding – the smell of space.

There are three people up here already: the commander and the only other American, Terry Virts (forty-seven); Anton (forty-three); and an Italian astronaut representing the European Space Agency, Samantha Cristoforetti (thirty-seven). I know them all, some much better than others. Soon, we will all know one another much better.

I've known Terry since he was selected as an astronaut in 2000, though we haven't overlapped much in our work. Anton and Samantha I've only gotten to know well since we've been preparing for this mission over the last year. The last time I hung out with Anton was in Houston, before my last flight. We both got pretty drunk at my neighborhood bar, Boondoggles, and later ended up spending the night at a friend's house nearby since neither of us was in shape to drive.

Over the course of this year in space, Misha and I will see a total of thirteen other people come and go. In June, a Soyuz will leave with Terry, Samantha, and Anton, to be replaced by a new crew of three in July. In September, three more will join us, bringing our total to nine – an unusual number – for just ten days. Then, in December, three will leave, to be replaced a few days later. Misha and I hope that the change in crew members will help break up the mission and the monotony to make our year less challenging.

Unlike the early days of spaceflight, when piloting skill was what mattered, twenty-first-century astronauts are chosen for our ability to perform a lot of different jobs and to get along well with others, especially in stressful and cramped circumstances for long periods of time. Each of my crewmates is not only a close coworker in an array of different high-intensity jobs but also a roommate and a surrogate for all humanity.

Gennady floats through the hatch first and hugs Anton. These greetings are always jubilant – we know exactly who we're going to see when we open the hatch, but still it's somehow startling to launch off the Earth, travel to space, and find friends already living up here. The big hugs and big smiles you see if you watch the hatch opening live on NASA TV are completely sincere. As Gennady and Anton say their hellos, Misha and I are waiting our turns. We know that many people on the ground are watching, including our families. There is a live feed playing for everyone at Baikonur, as well as in mission control in Houston and online. The video signal is bounced off a satellite and then down to Earth, as with all of our communications. Suddenly I get an idea and turn to Misha.

"Let's go through together," I suggest. "As a show of solidarity."

"Good idea, my brother. We are in this together."

It's a bit awkward floating through the small hatch together, but the gesture gets a big smile from everyone on the other side. Once we're through, I shake hands with Anton.

Next I give Terry Virts a hug, then Samantha Cristoforetti. She is the first Italian woman to fly in space, and soon she'll be the record holder for the longest single spaceflight by a woman.

Our families in Baikonur are waiting to have a conference call with us, which we'll do from the Russian service module. I float down there and make a wrong turn. It's weird to be back here – floating through the station is so familiar, but it's also disorienting. It's only day one.

As the smell of space dissipates, I'm starting to detect the unique smell of the ISS, as familiar as the smell of my childhood home. The smell is mostly the off-gassing from the equipment and everything else, which on Earth we call the new car smell. Up here the smell is stronger because the plastic particles are weightless, as is the air, so they mingle in every breath. There is also the faint scent of garbage and a whiff of body odor. Even though we seal up the trash as well as we can, we only get rid of it every few months when a resupply craft reaches us and becomes a garbage truck after we empty it of cargo.

The sound of fans and the hum of electronics are both loud and inescapable. I feel like I have to raise my voice to be heard above the noise, though I know from experience I'll get used to it. This part of the Russian segment is especially loud. It's dark and a bit cold as well. I feel a shiver of realization: I'm going to be up here for nearly a year. What exactly have I gotten myself into? It occurs to me for a moment that this might be one of the stupider things I've ever done.

When we reach the service module, I notice right away that it's much brighter than when I was here last. Apparently the Russians have improved their lightbulbs. It's also much better organized than I remember, which I suspect is a result of Anton trying to impress Gennady with his organizational skills. Gennady is a stickler for keeping the Russian segment neat and tidy.

During the conference call, our families can see and hear us, but we can only hear them. There is a loud echo. The comm configuration up here is slightly off. I hear Charlotte telling me what the launch was like, then I talk briefly to my daughter Samantha and then to Amiko. It's great to hear their voices. But I'm conscious that my Russian colleagues are waiting to talk to their families, too.

Once we finish the call, I head down to the U.S. segment with Terry and Samantha Cristoforetti, where I'm going to spend the better part of the year to come. Though ISS is all one facility, for the most part the Russians live and work on their side and everyone else lives and works on the other side – "the U.S. segment." I notice it's much darker than I remember – burned-out lightbulbs haven't been replaced. This isn't Terry and Samantha's fault, but a reflection of the conservative way the control center has come to manage our consumables since I was last here. I decide to make it a project over the coming months to improve how we use our resources, since I'm going to be up here for so long, and good lighting will be critical to my well-being.

Terry and Samantha show me around, reminding me how things work up here now. They start with the most important piece of equipment to master: the toilet, also known as the Waste and Hygiene Compartment, or WHC. We also run through a quick safety brief that we will redo more thoroughly in a couple of days once I'm more settled. An emergency could strike at any time – fire, ammonia leak, depressurization – and I'll have to be ready to deal with whatever comes, even on day one.

We head back to the Russian segment for a traditional welcoming party – special dinners are held there every Friday night and on other special occasions, including holidays, birthdays, and good-bye dinners before each Soyuz leaves. Welcoming parties are one of those occasions, and Terry has warmed up my favorite, barbecued beef, which I stick to a tortilla using the surface tension of the barbecue sauce (we eat tortillas because of their long shelf life and lack of crumbs). We also have the traditional foods we share at Friday night dinners – lump crabmeat and black caviar. Everyone is in a festive mood. It's been a long, tough day for the three of us who just arrived. Technically, two days. Eventually we say our good nights, and Terry, Samantha, and I head back to the U.S. segment.

I find my crew quarters, or CQ, the one part of the space station that will belong just to me. It's about the size of an old-fashioned phone booth. Four CQs are arranged in Node 2: floor, ceiling, port side, starboard side. I'm on the port wall this time; last time I was on the ceiling. The CQ is clean and empty, and I know that over the course of the next year it will fill with clutter, like any other home. I zip myself into my sleeping bag, making a special point to appreciate that it's brand new. Though I will replace the liner a couple of times, the bag itself won't be cleaned or replaced over the next year. I turn off the light and close my eyes. Sleeping while floating isn't easy, especially when you're out of practice. Even though my eyes are closed, cosmic flashes occasionally light up my field of vision, the result of radiation striking my retinas, creating the illusion of light. This phenomenon was first noticed by astronauts during the Apollo era, and its cause still isn't thoroughly understood. I'll get used to this, too, but for now the flashes are an alarming reminder of the radiation zipping through my brain. After trying unsuccessfully to sleep for a while, I bite off a piece of a sleeping pill. As I drift off into a restless haze, it occurs to me that this is the first of 340 times I will have to fall asleep here.

PIONEERS & MAVERICKS

ENDURANCE
Shackleton's Incredible Voyage
Alfred Lansing

Ernest Shackleton (1874–1922) was a British explorer who led three expeditions to Antarctica, the most famous of which resulted in his ship *Endurance* being trapped in the ice and slowly crushed. His crew escaped on the ship's lifeboats, in which they made an arduous sea journey of 720 nautical miles to South Georgia Island. Shackleton is remembered as one of the greats of the Heroic Age of Exploration for his coolness under pressure and superb leadership qualities. All of his crew survived and were rescued after living for just over a year on the ice. He died deeply in debt and without experiencing much success outside of his voyages, but his legacy as one of the greatest British explorers lives on.

Alfred Lansing (1921–1975) was an American journalist and writer. When researching *Endurance* he interviewed the expedition's surviving members and had access to numerous diaries and journals detailing their experiences.

In all the world there is no desolation more complete than the polar night. It is a return to the Ice Age — no warmth, no life, no movement. Only those who have experienced it can fully appreciate what it means to be without the sun day after day and week after week. Few men unaccustomed to it can fight off its effects altogether, and it has driven some men mad.

By coincidence, the man who had once been a partner in the *Endurance*, M. le Baron de Gerlache, had himself been beset in the Weddell Sea aboard a vessel called the *Belgica* in 1899. With the coming of the night, the *Belgica's* crew became infected with a strange melancholy. As the weeks went by this slowly deepened into depression and then despair. In time they found it almost impossible to concentrate or even to eat. In order to offset the terrifying symptoms of insanity they saw in themselves, they took to walking in a circle around the ship. The route came to be known as 'madhouse promenade.'

One man died of a heart ailment brought on partly by his unreasoning terror of the darkness. Another was seized with the idea that the rest of the crew intended to kill him, and whenever he slept he squeezed himself into a tiny recess of the ship. Still another gave way to hysteria which left him temporarily deaf and dumb.

But there was very little depression on board the *Endurance*. The coming of the polar night somehow drew the men closer together.

When the *Endurance* sailed from England, there could hardly have been a more heterogeneous collection of individuals. They varied from Cambridge University dons to Yorkshire fishermen. But after nine months of being together almost constantly and living and working in the same close quarters, the men had built up a backlog of shared experiences that offset the vast differences between them. During these nine months, the men on board the *Endurance* had come to know one another very well indeed. And with few exceptions, they had come to like one another, too.

Nobody much thought of Blackboro as a stowaway any more. The stocky dark-haired young Welshman was a regular member of the crew now. Blackboro was an extremely quiet individual but nonetheless quick-witted and well liked, a cheerful, willing shipmate, who helped Green in the galley.

They all knew Bobbie Clark, the biologist, to be a dour, hardworking, almost humorless Scot. But they knew also that he could be counted on to do his share and more whenever all hands were called to duty. He got excited only when the dredge he lowered through the ice each day fetched up a new species of creature for his collection of bottled specimens. The crew once tricked him into great excitement by placing some pieces of cooked spaghetti in one of his jars of formaldehyde. Clark kept his own counsel, and never mentioned to a single man anything about his personal life.

Tom Crean – tall, almost gaunt – was exactly what he appeared to be – a heavy-handed sailorman, forthright and tactless, who spoke with a sailorman's rough vocabulary. He was certainly not a very warm personality, but he knew the sea and he knew his job, and the others respected him for it. Shackleton was personally quite fond of Crean. He liked the big Irishman's willingness. Shackleton also put a high value on discipline, and Crean, after years in the Royal Navy, regarded an order as something to be obeyed without question. Nor was Crean above giving Shackleton a bit o' the blarney occasionally.

When it came to Charlie Green, the cook, there was a widespread feeling that he was a little 'crackers,' or daft, because of his disorganized and seemingly scatterbrained mannerisms. They called him Chef or Cookie – or sometimes Doughballs because of his high, squeaky voice and because he had in fact lost a testicle in an accident. They poked fun at him on the surface, but underneath there was a fundamental respect, and a fondness, too. Few men were more conscientious. While the others worked only three hours a day, Green was busy in the galley from early morning until long after supper at night.

Green was occasionally the victim of the almost merciless ribbing that all ships' cooks everywhere are subjected to, but he had his jokes, too. Two or three times, when some crewman's birthday was to be celebrated, he produced a cake for the occasion. One proved to be a blown-up toy balloon which he had carefully frosted, and another was a block of wood, daintily covered with icing.

Hudson, the navigator, was a peculiar sort. He meant well, all right, but he was just a little dull. He owed his nickname – Buddha – to a practical joke he had fallen for once while the ship was at South Georgia. The men had convinced him that there was to be a costume party ashore... and any man who had seen South Georgia with his own eyes – its glaciers and rugged mountains, the stink of whale entrails rotting in the harbor – and who could believe it to be the scene of a costume party... but Hudson did. They got him to remove most of his clothing and they dressed him in a bedsheet. Then they tied the lid of a teapot on his head with pieces of ribbon running under his chin. Thus attired, he was rowed to shore, shivering in the icy blasts that howled down off the mountains. A party was held at the home of the whaling factory manager. But when Hudson walked in, he was most assuredly the only one in costume.

In any practical joke such as this, the men knew that the one to look for was Leonard Hussey, the meteorologist. A slightly built little fellow in his early twenties, Hussey was universally liked for his unfailing good humor. He had a sharp, satiric tongue, but he could take a joke against himself without losing his good spirits. It was not always easy to get the best of Hussey in an exchange of wits, though. They liked him, too, because he played the zither banjo and was willing to strike up a tune whenever anybody wanted to sing. Hussey's name was corrupted into a variety of nicknames – Hussbert, Hussbird, and just plain Huss.

A great many of the men looked on McIlroy, one of the surgeons, as a man of the world. He was a handsome, aristocratic-looking individual, slightly older than most of the others, and they immensely enjoyed listening to his tales of past conquests. McIlroy could be bitingly sarcastic, but the others admired him for it. It seemed to go along with his cosmopolitan nature, and there was never any malice in what he said. They called him Mick.

George Marston, the expedition's artist, was a moody fellow, up one day and down the next. He was unique among them in that he worried outwardly about the future, whereas almost everybody else was confident that everything was going to be all right. But Marston, whenever he was feeling downcast, would brood over his wife and children at home. His attitude was not improved by Shackleton's obvious and increasing dislike for him. It was one of those inexplicable things. Perhaps Marston's uneasiness itself was at fault. Shackleton seemed to fear that this attitude would spread to other men. But, apart from his changeable nature and the fact that he was not overly eager to turn out for work, Marston was well liked by most.

Among the forecastle hands, the seamen and firemen, the only outstanding individual was John Vincent, a young, ambitious bully. He was quite short, but ruggedly built and much stronger than any of the other seamen. And he sought to use his superior strength to dominate his shipmates by intimidation. He insisted on being served first at mealtimes so that he could pick the best portion, and when grog was issued, he always managed to get more than his share. The other seamen not only disliked him personally, they had very little respect for his abilities aboard ship. Vincent had been in the Navy, but most of his experience at sea had been aboard trawlers in the North Sea. Unlike How, Bakewell, and McLeod, who had served for years aboard square riggers, Vincent had had no previous experience in sail. Nevertheless, he had his eye on the vacant post of boatswain, and he felt that the best way to get it was to demonstrate a capacity for tyranny. After a time the forecastle hands got fed up, and How, a soft-spoken, agreeable, and extremely competent little chap, went to Shackleton and complained. Shackleton immediately sent for Vincent. Though it is not known what Shackleton told him, Vincent's attitude was considerably less domineering after that.

It was remarkable that there were not more cases of friction among the men, especially after the Antarctic night set in. The gathering darkness and the unpredictable weather limited their activities to an ever-constricting area around the ship. There was very little to occupy them, and they were in closer contact with one another than ever. But instead of getting on each other's nerves, the entire party seemed to become more close-knit.

Early in the winter, George Marston and Frank Wild decided to give each other haircuts. Before they were through, they had shaved off all their hair with the ship's barber clippers. The next evening the fever had spread throughout the crew. Everyone, including Shackleton, had his hair trimmed down to the scalp.

After that, there were many pranks. The following evening, Wild appeared for supper with his face buried in the neck of his jersey, revealing only the top of his shaved head, on which Marston had painted what Greenstreet described as an 'imbecilic looking Johnny.'

And the next night 'Wuzzles' Worsley was put on trial for 'robbing a Presbyterian church of a trouser button out of the oftertory bag and having turned the same to base and ignoble use.' The proceedings were long and disorderly. Wild was the judge, James the prosecuting attorney, and Orde-Lees the defense attorney. Greenstreet and McIlroy gave testimony against the defendant, but when Worsley promised to buy the judge a drink after the trial, Wild charged the jury to find the defendant innocent. Nevertheless, Worsley was found guilty on the first ballot.

Besides these spontaneous affairs, there was a regular series of social occasions. Each Saturday night before the men turned in a ration of grog was issued to all hands, followed by the toast, 'To our sweethearts and wives.' Invariably a chorus of voices added, 'May they never meet.'

On Sunday evenings the men listened to music from the hand-crank phonograph for an hour or two as they lay in their bunks or wrote up their diaries. But the playing of the phonograph was limited because of a shortage of needles. Five thousand had been ordered in England, but Wild, in sending off the requisition, had failed to specify the word 'gramophone.' Only long after the ship had sailed did Orde-Lees, the storekeeper, discover they had five thousand extra sewing needles, and only a small package of the phonograph variety.

Then once each month all hands gathered in the Ritz and Frank Hurley, the photographer, delivered a 'lantern chat' – a slide-illustrated lecture on the places he had visited: Australia, New Zealand, the Mawson expedition. The favorite was one called 'Peeps in Java,' which featured waving palm trees and native maidens.

The Ritz on nights like these was a cozy place. It had been a cargo area just below the main deck and aft of the crew's quarters in the forecastle. Then the stores and men traded places. The supplies were moved to the wardroom area in the deckhouse, and the men took over the hold. The area was about 35 feet long and 25 feet wide, and McNeish had erected partitions to form individual sleeping cubicles for the officers and scientists. In the center was a long table with a paraffin-burning lamp overhead. Here they ate their meals, wrote their diaries, played cards, and read. In one corner was a coal-burning stove which kept the temperature inside quite comfortable. The *Endurance's* thick sides were excellent insulation.

Outside, however, the weather was getting steadily worse. Late in May the temperature dropped below zero, and stayed there. During the first half of June, the average reading was –17 degrees. But the scene from the deck of the *Endurance* was often fantastically beautiful. In clear weather, if the moon was out, it swept in bold, high circles through the starlit skies for days on end, casting a soft, pale light over the floes. At other times, there were breathtaking displays of the *aurora australis*, the Antarctic equivalent of the northern lights. Incredible sunbursts of green and blue and silver shot up from the horizon into the blue-black sky, shimmering, iridescent colors that glinted off the rock-hard ice below. But apart from the increasing cold, the weather remained remarkably stable and free from gales.

Toward the middle of June, at the blackest part of the winter, a chance boast by Frank Hurley that he had the fastest team resulted in a dog derby. Even at high noon when the race was run, it was so dark out that the spectators at this 'Antarctic Sweepstakes' couldn't see the far end of the race course. Wild's team won, but Hurley claimed that he was carrying more weight than Wild, and demanded a rematch. He later won when Shackleton, who was riding as passenger on Wild's sledge, slipped off going around a turn, and Wild was disqualified.

The following night, cagy Dr McIlroy 'brought to light' a pair of dice he happened to find among his things. He first shook with Greenstreet to see who would buy champagne when they got home. Greenstreet lost. By that time several men had gathered around the table in the Ritz, and in subsequent rolls of the dice, an entire evening's entertainment was wagered. Wild got stuck for buying the dinner, McIlroy himself lost the roll for the theater tickets, Hurley the after-theater supper, and parsimonious 'Jock' Wordie, the geologist, was committed to pay for the taxis home.

They held a special celebration on Midwinter's Day, June 22. The Ritz was decked out with bunting and flags, and Hurley built a stage of sorts which was lighted by a row of acetylene gas footlights. Everyone gathered for the festivities at 8 P.M.

Shackleton, as chairman, introduced the participants. Orde-Lees was dressed as a Methodist minister, the 'Rev. Bubbling-Love,' and he exhorted his listeners against the wages of sin. James, as 'Herr Professor van Schopenbaum,' delivered a lengthy lecture on the 'Calorie.' Macklin recited a tropical verse he had written about 'Captain Eno,' the effervescent seafarer, who could have been no one but the effervescent Worsley.

Greenstreet described the evening in his diary: 'I think I laughed most over Kerr who dressed up as a tramp and sang 'Spagoni the Toreador'. He started several keys too high and notwithstanding the accompanist, Hussey, who was vainly whispering "Lower! Lower!" and playing in a much lower key, he kept going until he lost the tune altogether. When he came to the word Spagoni he had forgotten the word so came out with Stuberski the Toreador and had completely forgotten the chorus, so simply saying, "He shall die, he shall die, he shall die!" It was killing and we laughed until the tears ran down our cheeks. McIlroy dressed up as a Spanish girl and a very wicked looking one at that, with very low evening dress and slit skirt showing a bare leg above her stocking tops... gave the Danse Espagnol.'

Marston sang, Wild recited 'The Wreck of the Hesperus,' Hudson was a half-caste girl, Greenstreet was a red-nosed drunk, and Rickinson was a London streetwalker.

The evening ended at midnight with a cold supper and a toast. Then everyone sang, 'God Save the King.'

And so the winter was half done.

MY LIFE AS AN EXPLORER

Roald Amundsen

Roald Amundsen (1872–1928) was a Norwegian polar explorer famous for being the first man to traverse the Northwest Passage and to lead the first verified expedition to reach the South Pole in December of 1911. Amundsen's careful preparation and effective use of dogs and skis led to a relatively smooth and uneventful Antarctic expedition, in contrast to the ill-fated expedition of Robert Scott. Amundsen disappeared while flying on a rescue mission in the Arctic in 1928; his body has never been found.

We were now in a thoroughly dangerous situation. Here we were, 600 miles from civilization, landed upon the ice with airplanes equipped for landing upon water, with the engine of one of the 'planes utterly out of commission, and with provisions adequate for full nourishment for only about three weeks.

The only prospect of salvation lay in transferring the whole party to the *N-25* and making every effort to get that one 'plane into the air again. Under the best possible conditions for rising, this would have been no easy matter, with twice the number of men the *N-25* was supposed to carry. But we did not have the best possible conditions. Instead of rising from the water, as our boats were designed to do, we must rise from the surface of the ice. And that surface was not smooth like a skating pond, but was as rough as hummocks could make it—hummocks of every size that were heaped together in utter confusion.

There was nothing for us to do but to try to flatten out enough of this ice surface to make a runway long enough to permit us to get up sufficient momentum to rise. We worked furiously for twenty-four days at this task. It was truly a race with death, for ultimate starvation could not much longer have been delayed. As it was, we lived on eight ounces of food a day, which is just one half the ration that Peary fed to each of his dogs on his dash to the Pole. Every morning we had a small cake of chocolate dissolved in hot water, making one cup of very weak cocoa, and three crackers, each about the size of a Nabisco wafer. For dinner we had a cup of soup. For supper we had another cup of "near cocoa" and three more biscuits. Of course, we suffered somewhat from unsatisfied appetite, but, on the other hand, it was amazing to see how we maintained our strength on such a meagre diet while performing such strenuous labour.

On the twenty-fourth day we had succeeded in getting a runway more or less level and 500 metres long. The *N-25* theoretically required 1,500 metres of open water in which to get up momentum enough to rise. We could not, however, level off more than 500 metres. At the end of that distance our runway dropped off with an abrupt edge to the surface of a very small pool of water about three feet below and about fifteen feet broad. On the other side was a flat cake of ice about 150 feet in diameter and on the far side of which rose a hummock at least twenty feet high. This, of course, was beyond our strength to break down. As it was, I should estimate that we moved at least 500 tons of ice in the twenty-four days.

Such rest as we took from the heavy work of digging snow was spent partly in sleep (though we got precious little of that) but chiefly in making astronomical observations and depth soundings. These soundings conclusively demonstrated that there was no land near the region where we were. They were made by the use of a new and most ingenious German device which weighed only three pounds. This device utilized the echo principle. Two holes were bored through the ice. Through one of them was dropped a length of wire attached to a very sensitive telephone receiving microphone.

One man listened in on this device while, at the other hole, another man fired a submerged charge of explosive. Synchronized stop watches were used to take the exact time of the explosion and of the "echo." The vibration set up by the explosion was transmitted by the water to the bed of the ocean, whence, by rebound, it returned through the water to the surface where it was perceptible in the ear pieces attached to the submerged wire.

We took two soundings with this device, and in each case five seconds elapsed between the sound of the explosion and the sound of its echo from the ocean bed. This five-seconds interval indicated that the ocean at this point is about 12,000 feet deep. Naturally, no land exists very close to water of such depth.

By June 15th, everything that could be done to get the *N-25* ready had been done. We had transferred from the *N-24* everything useful for the return flight that we could safely attempt to carry in the already overburdened *N-25*. All six of us men stowed ourselves in the crowded cockpit with the motor running and Riiser-Larsen at the "stick." At the signal to go, Riiser-Larsen opened the throttle to the limit and we began moving along our uneven runway of ice.

The most anxious moments of my life were the next few seconds. As we gained speed, the inequalities on the ice multiplied their effect upon us and the fuselage swayed from side to side with such a careening motion that more than once I feared we should be thrown over on one side and have one of our wings crushed. Nearer and faster we approached the end of our runway, but still the bumping motion indicated we had not left the ice. Still gaining momentum but still hugging the ice, we approached the brink of the little drop off into the pool. We reached the pool, jumped over it, dashed down on the flat cake, and then we rose. An enormous sense of relief swept over me, but it lasted only an instant. There, dead ahead of us, and only a few yards away, loomed the twenty-foot hummock on the far side of the little pool. We were headed straight toward it. Five seconds would tell whether we should clear it and at least be in the free air with a fighting chance to return to safe haven, or whether we should crash into it. Should we crash—even if we escaped instant death—we should be faced with the certainty of ultimate death marooned on the Arctic ice. Thoughts and sensations crowd fast at such a moment. The seconds seemed terrible hours. But we did clear it—we could not have had more than an inch to spare. At last we were on the way, after twenty-four days of desperate work and anxiety.

Hour after hour we flew southward. Were we on the right course? Would our fuel last if we were? Lower and lower the fuel gage sank. At length, with only a half-hour's supply of gasoline left, a nighty shout went up from all of us. There, far below us, to the south, lay the familiar peaks of Spitzbergen. Below us appeared a streak of black which meant open water for a landing. Our troubles were not yet over. To reach this water we must make a wide sweep over hummocky ice. This would require some manoeuvring of the 'plane. I had been watching Riiser-Larsen, our pilot, for the last half-hour, and had noticed that when he worked the aileron control it appeared that he had to use unwanted force on the lever. Finally, he had shouted to me that this control was out of order. We had now to come right down, but fortunately we had first reached the open water. If not, we would all have been crushed right there and then.

It was almost as close a call as our experience of the weeks before.

That was the end of our first long flight over the Arctic Ocean to 88° north latitude.

SCOTT'S LAST EXPEDITION
The Journals
Robert Falcon Scott

Captain Robert Falcon Scott (1868–1912) was a British explorer of the Heroic Age of Exploration, whose second Antarctic expedition reached the South Pole on January 17, 1912. Scott was bitterly dismayed to discover a tent left by Roald Amundsen, indicating that the Norwegian explorer had reached the South Pole before him. Scott was therefore unsuccessful in his attempt to be the first man to reach the South Pole and perished on his return journey, but his extraordinary efforts transformed him into a folk hero in his native England and throughout the British Empire. His exploits were memorialized in the form of hundreds of monuments erected in his honour throughout the Empire and beyond.

Sunday, March 18. – Today, lunch, we are 21 miles from the depôt. Ill fortune presses, but better may come. We have had more wind and drift from ahead yesterday; had to stop marching; wind N.W., force 4, temp. -35°. No human being could face it, and we are worn out nearly. My right foot has gone, nearly all the toes – two days ago I was proud possessor of best feet. These are the steps of my downfall. Like an ass I mixed a small spoonful of curry powder with my melted pemmican – it gave me violent indigestion. I lay awake and in pain all night; woke and felt done on the march; foot went and I didn't know it. A very small measure of neglect and have a foot which is not pleasant to contemplate. Bowers takes first place in condition, but there is not much to choose after all. The others are still confident of getting through – or pretend to be – I don't know! We have the last half fill of oil in our primus and a very small quantity of spirit – this alone between us and thirst. The wind is fair for the moment, and that is perhaps a fact to help. The mileage would have seemed ridiculously small on our outward journey.

Monday, March 19. – Lunch. We camped with difficulty last night and were dreadfully cold till after our supper of cold pemmican and biscuit and a half a pannikin of cocoa cooked over the spirit. Then, contrary to expectation, we got warm and all slept well. Today we started in the usual dragging manner. Sledge dreadfully heavy. We are 15½ miles from the depôt and ought to get there in three days. What progress! We have two days' food but barely a day's fuel. All our feet are getting bad – Wilson's best, my right foot worst, left alright. There is no chance to nurse one's feet till we can get hot food into us. Amputation is the least I can hope for now, but will the trouble spread? That is the serious question. The weather doesn't give us a chance – the wind from N. to N.W. and -40° temp. today.

Wednesday, March 21. – Got within 11 miles of depôt Monday night; had to lay up all yesterday in severe blizzard. Today forlorn hope, Wilson and Bowers going to depôt for fuel. Thursday, March 22 and 23. – Blizzard bad as ever – Wilson and Bowers unable to start – tomorrow last chance – no fuel and only one or two of food left – must be near the end. Have decided it shall be natural – we shall march for the depôt with or without our effects and die in our tracks.

Thursday, March 29. – Since the 21st we have had a continuous gale from W.S.W. and S.W. We had fuel to make two cups of tea apiece and bare food for two days on the 20th. Every day we have been ready to start for our depot 11 miles away, but outside the door of the tent it remains a scene of whirling drift. I do not think we can hope for any better things now. We shall stick it out to the end, but we are getting weaker, of course, and the end cannot be far. It seems a pity, but I do not think I can write more.

Surely misfortune could scarcely have exceeded this last blow. We arrived within 11 miles of our old One Ton Camp with fuel for one last meal and food for two days. For four days we have been unable to leave the tent – the gale howling about us. We are weak, writing is difficult, but for my own sake I do not regret this journey, which has shown that Englishmen can endure hardships, help one another, and meet death with as great a fortitude as ever in the past. We took risks, we knew we took them; things have come out against us, and therefore we have no cause for complaint, but bow to the will of Providence, determined still to do our best to the last. But if we have been willing to give our lives to this enterprise, which is for the honour of our country, I appeal to our countrymen to see that those who depend on us are properly cared for. Had we lived, I should have had a tale to tell of the hardihood, endurance, and courage of my companions which would have stirred the heart of every Englishman. These rough notes and our dead bodies must tell the tale, but surely, surely, a great rich country like ours will see that those who are dependent on us are properly provided for. R. SCOTT.

MESSAGE TO THE PUBLIC The causes of the disaster are not due to faulty organisation, but to misfortune in all risks which had to be undertaken.

1. The loss of pony transport in March 1911 obliged me to start later than I had intended, and obliged the limits of stuff transported to be narrowed.

2. The weather throughout the outward journey, and especially the long gale in 83° S., stopped us.

3. The soft snow in lower reaches of glacier again reduced pace. We fought these untoward events with a will and conquered, but it cut into our provision reserve. Every detail of our food supplies, clothing and depôts made on the interior ice-sheet and over that long stretch of 700 miles to the Pole and back, worked out to perfection. The advance party would have returned to the glacier in fine form and with surplus of food, but for the astonishing failure of the man whom we had least expected to fail. Edgar Evans was thought the strongest man of the party. The Beardmore Glacier is not difficult in fine weather, but on our return we did not get a single completely fine day; this with a sick companion enormously increased our anxieties. As I have said elsewhere we got into frightfully rough ice and Edgar Evans received a concussion of the brain – he died a natural death, but left us a shaken party with the season unduly advanced. But all the facts above enumerated were as nothing to the surprise which awaited us on the Barrier. I maintain that our arrangements for returning were quite adequate, and that no one in the world would have expected the temperatures and surfaces which we encountered at this time of the year. On the summit in lat. 85°86° we had -20°, -30°. On the Barrier in lat. 82°, 10,000 feet lower, we had -30° in the day, -47° at night pretty regularly, with continuous head wind during our day marches. It is clear that these circumstances come on very suddenly, and our wreck is certainly due to this sudden advent of severe weather, which does not seem to have any satisfactory cause. I do not think human beings ever came through such a month as we have come through, and we should have got through in spite of the weather but for the sickening of a second companion, Captain Oates, and a shortage of fuel in our depôts for which I cannot account, and finally, but for the storm which has fallen on us within 11 miles of the depôt at which we hoped to secure our final supplies.

LIVING DANGEROUSLY

Ranulph Fiennes

Sir Ranulph Fiennes (born 1944) is a British explorer, army officer and adventurer. Fiennes holds a number of world records relating to exploration, most notably leading the only team to circumnavigate the planet on its polar axis in 1982 and, in 1992–3, becoming the first person to cross the length of the Antarctic continent on foot. Fiennes also famously cut off the tops of his frostbitten fingers and thumb himself with a saw and vice. The 1984 Guinness Book of World Records declared Fiennes the world's greatest living explorer and few would contest that claim almost thirty years later.

On the day we set out from England, Prince Charles commented, 'Transglobe is one of the most ambitious undertakings of its kind ever attempted, the scope of its requirements monumental.' *The New York Times* editorial column, under the heading 'Glory', stated, 'the British aren't so weary as they're sometimes said to be. The Transglobe Expedition, seven years in the planning, leaves England on a journey of such daring that it makes me wonder how the sun ever set on the Empire.'

Our initial plans were mundane enough. With three Land Rovers we would cross Europe, the Sahara and West Africa. The ship would take us from Spain to Algeria and again from the Ivory Coast to Cape Town. Departure from Cape Town had to be timed precisely in order to enter Antarctic waters in mid-summer, when the ice-pack should be at its loosest. This was the reason we left England in early September and why we had to leave South Africa by late December.

Anton Bowring's crew of volunteers were a wonderful bunch, professional at their posts but fairly wild when off duty. They included Quaker, Buddhist, Jew, Christian and atheist, black, white and Asian. They came from Austria, America, both ends of Ireland, South Africa, India, Denmark, Britain, Canada, Fiji and New Zealand. Most of the crew were Merchant Navy men who gave up promising careers, at a time of growing unemployment, to job a three-year voyage with no wage packet.

Ginnie was not the only girl aboard, for Aton had selected an attractive redhead, Jill McNicol, as ships' cook. She soon gained a number of ardent admirers from among the crew.

In Algiers the port officials showed ominous interest in our three-year supply of sponsored spirits and cigarettes. When the first wave of officers departed, promising a second visit in an hour, our skipper decided to make a run for it and quickly unloaded our vehicles and gear. As the crew waved us goodbye, the *Benjamin Bowring*'s thirty-year-old variable pitch control jammed itself in reverse. So the vessel retreated out of the harbour and out of sight steaming backwards.

We drove through Algeria to the sand-dunes of El Golea, a sticky-hot hell-hole dubbed 'El Gonorrhoea' by Oliver. We were pleased to leave the sweltering sands and head south to the Hoggar Mountains. At 8000 feet we reached the Pass of Asekrem, haunt of French monks. We savoured views of vast mountain ranges disappearing to Chad and the centre of Africa. From Tamanrasset we rattled down ever-worsening trails to lonely Tit and thence over trackless miles of sand and scrub to Tim-Missao and the Touareg lands of the Adar des Iforas. Wide starry nights and wind-blown dunes brought back memories of the Dhofari Nejd and desert days long past.

From the Forest of Tombouctou to Goundam on the Niger we roared through veins of 30-foot-high sand in low gear to the cheer of barefoot donkeymen in sampan hats driving cavalcades of pint-size mules along the trail.

At Niafounke we learned that extensive flooding barred our planned route to the Ivory Coast, but a 700-kilometre westerly detour took us at length to Loulouni and the Ivorian border at Ouangolodougou. For a week we camped in thick jungle beside the Bandama Rouge River, an excellent collection point for bilharzia-bearing water-snails, then south to the lush and hilly coastline. In Abidjan harbour, we were met by the *Benjamin Bowring*. Simon and two members of the crew were weak from malarial fever but we pressed on and sailed across the Equator close by the Greenwich Meridian.

The Benguela Current coincided with a Force Seven storm. The electrical system of the stern refrigerator room broke down and a ton of sponsored mackerel turned putrescent. Brave hands volunteered to go below and clean up the mess, slipping about, as the ship rolled and heaved, in a soup of bloody fish bits. They fought their way up the ladders and tossed the sloppy bundles of rotten fish overboard. Then a forklift truck burst its lashings, crushed valuable gear and spattered battery acid about the cargo hold. Nowhere on the bucking ship, from cabins to fo'c'sle, could we escape the fumes of acid and mackerel. We were in the tropics and there was no air conditioning system. With each roll to port our cabin porthole spat jets of sea-water on to our bunks. All night the clash of steel on steel and the groan of hemp under stress sounded from the cargo holds.

The old ship struggled on at a stately eight knots and delivered us, slightly dazed, to Saldanha Bay, near Cape Town, on 3 December. Jackass penguins squawked from the rocky shoreline, hundreds of thousands of them, as we anchored to take on fresh water.

We stayed two weeks in Cape Town to mount an export sales-orientated exhibition of our equipment, one of eight such events to be held during the course of the voyage. On a free evening I drove to Constantia with Ginnie and visited our old house, built by Granny Florrie twenty-five years before. Nothing was as I remembered it. The valley was no longer the wild and wonderful place of my dreams. Residential expansion had tamed and suburbanised the woods and vineyards. The vlei where I had roamed with Archie and the gang was now a row of neat bungalows for foreign embassy staff. Our own house, Broughton, was a transitory post for US Marines on leave and nobody had tended the garden in years. There was no longer a view of the valley, for Granny Florrie's shrubs had flourished unchecked and the four little palm trees named after my sisters and me were now roof-high.

We wandered in silence through the old vegetable patch. No trace remained of the summerhouse, a place of bewitching memories. My mother's rockery, tended with so much care and love, had run amok. Up the valley, beside a building lot where caterpillar trucks were at work, we called on my cousin Googi Marais, whose rifle bullets passing over the roof at night had once caused our English nanny to pack her bags and flee. Googi was now crippled with arthritis, but he and his wife gave us tea and filled me in on the past quarter century of happenings in 'our valley'.

Over the next ten days I was reunited with twenty-two other cousins, one of whom, an alderman of Cape Town, showed me a family tree which proved he and I shared a close relationship with Karl Marx. I regret not having taken away a copy to shock the family back in England.

We left South Africa three days before Christmas, passing by Cape Agulhas, the last land for 2400 miles, and set a course south to Antarctica.

Bothie, our Jack Russell, appointed himself as ship's mascot and, respecting no privacy, left calling-cards in all cabins. Ginnie tried hard to remove all his indiscretions before they were discovered by the crew but it was an uphill struggle.

The Observer, covering the expedition, sent Bryn Campbell to Antarctica with us to record the ship's arrival. A four-man film-team was also on board – provided by millionaire octogenarian Dr Armand Hammer, a friend of Prince Charles – intending to film the entire three-year journey by joining the team in the more accessible areas. Richard Burton was to provide a background commentary.

Swelling the ship's complement still further were a number of oceanographers. At certain stages of the voyage there were eight scientists on board and at all times our own resident boffins, an Irishman and a Cape Town University girl, worked with their bathythermographs and nets to study current patterns and the interaction of water bodies at sub-tropical and Antarctic convergences.

Christmas spirits were dampened by a Force Eight storm which tore away the Christmas tree lashed to our mast and made eating rich festive fare a risky business. Giant Southern Ocean rollers forced the ship to list 47° both ways and, fearing the worst, I asked Anton if the ship could cope. His eyes glinted evilly and he proceeded to tell me a number of Antarctic horror stories.

As the old year slid away, we entered pack-ice and the sea slowly settled. The skipper sent a look-out up to the crow's nest to shout directions to the fo'c'sle through the antiquated intercom. The loose pack-ice yielded to our steel-clad bows and only once did we need to halt and slowly ram a new path through the floes.

On 4 January 1980, we sighted the ice-cliffs of Antarctica and the same day gouged ourselves out a nest, in the thin bay-ice of Polarbjorn Bite, where we could climb straight down a ladder on to the ice. The unloading of over a 100 tons of mixed cargo, including 2000 numbered boxes and the 1600 fuel drums, began at once.

To ensure that the ship could leave before the pack solidified and cut off her escape, we raced to carry every item two miles inland to more solid ice. Everyone helped, even the film-team. Our only tow-machine apart from skidoos was a Groundhog rescued from Greenland but, in only eleven days, the complex operation was complete.

Halfway through the unloading a storm broke up the ice and I watched from the bridge as eight drums of precious aircraft fuel sailed north on a floe. It was as if the bay-ice were a completed jigsaw puzzle dropped onto concrete and shattered.

Our steward, Dave Hicks, drank too much whisky and fell off the gangplank. He floated belly up between the hull and the ice-edge, in imminent danger of being crushed, until someone hauled him out on a boathook.

One of the engineers slipped and broke three ribs and Bothie lost a fight with an angry Chinstrap penguin. The chief engineer, who had hoisted his Honda motorbike on to the ice for a quick ride, lost the machine for ever during the storm. Otherwise the unloading went well and the three of us waved goodbye to the ship and our friends on 15 January. All being well, they would pick us up again in a year or so, two thousand miles away on the far side of the frozen continent.

Ten days later, in conditions of nil visibility, Ollie, Charlie and I took our skidoos and laden sledges on a 370-kilometre journey inland to set up a base in which to spend the next eight months of polar winter and the long disappearance of the sun. Navigation was tricky, for although there were numerous mountain features spearing the overall ice-mantle, our route through them was confined to those slopes not riven by heavy crevassing. I came close to plummeting 200 feet down a crack at a time when I was not roped up to my skidoo.

We passed through a band of especially-difficult crevasse fields known as the Hinge Zone. From there I followed a series of carefully selected compass bearings which wiggled us between the peaks of Draaipunt, Valken and Dassiekop to the 6000-foot-high Borga Massif.

As we inched south, the Twin Otter crew – after a fifteen-day flight from England – flew Ginnie, Simon and 100,000 pounds of gear to an ice-field below Ryvingen Mountain. There they erected four huts made of cardboard which Ginnie had designed. Once snow drifts covered the huts' outer shells and insulated them they would be proof against the worst Antarctic winter.

Antarctica's 5,500,000 frozen square miles dwarf the United States, yet no more than 800 humans live in this continent. Only about fourteen of these polar denizens are women, so it is a very peaceful part of the earth.

The site of our winter base was chosen because of its height above sea-level and its distance inland, the furthest which the Twin Otter could be expected to reach carrying a 2000-pound cargo load. We hoped, by wintering at 6000 feet, to become acclimatised to the bitter conditions before the main crossing journey the following summer, when the average height would be 10,000 feet above sea-level. Until we reached our winter site we would be simple travellers. But once south of our camp, we would become true explorers of one of earth's last untrodden regions. For 900 miles we would pass through terrain neither seen nor touched by mankind.

We reached Ryvingen without mishap and began a race to prepare our camp for the long winter. The Twin Otter crew completed their seventy-eighth ferry flight from Sanae to Ryvingen and we wished them a safe flight back to Britain via the Falkland Islands and Argentina. Ginnie was to stay with us through the winter.

The weather clamped down on the ice-fields, travel for any distance became impossible and we were soon cut off from the outside world. For eight months we must survive through our own common sense. Should anyone be hurt or sick there could be no evacuation and no medical assistance. We must daily handle heavy batteries, generator power lines and heavy steel drums, but avoid the hazards of acid in the eye, serious tooth trouble, appendicitis, cold burns, fuel burns or deep electrical burns. Temperatures would plunge to –50°C and below. Winds would exceed ninety knots and the chill factor would reach –84°C. For 240 days and nights we must live cautiously and mostly without sunlight.

After a few days the huts disappeared under snow drifts, and all exits from the huts were blocked. So I dug tunnels under the snow and stored all our equipment inside them. In two months I completed a 200-yard network of spacious tunnels with side corridors, a loo alcove, a thirty-foot-deep slop-pit and a garage with pillars and archways of ice.

Since our huts were made of cardboard with wooden struts and bunks and our heaters burned kerosene we were apprehensive of fire. A neighbouring Russian base, 500 miles away, had been burnt out the previous winter and all eight occupants were found asphyxiated in an escape tunnel with blocked hatches.

Katabatic winds blasted our camp and snuffed out our fires through chimney back-blow. We experimented with valves, flaps and crooked chimneys, but the stronger gusts confounded all our efforts. The drip-feed pipe of a heater would continue to deliver fuel after a blow-out and this could cause a flash-fire unless great care was taken on re-lighting the heater.

Generator exhaust pipes tended to melt out sub-snow caverns which spread sideways and downwards. One day Oliver discovered a fifteen-foot-deep cave underneath the floor of his generator hut. He moved his exhaust system time and again but nearly died of carbon monoxide poisoning three times despite being alert for the symptoms.

Without power from the generators, Ginnie's radios would not work, so our weather reports, which Oliver must send out every six hours, could not be fed into the World Meteorological System, nor could the complex, very low frequency recording experiments for Sheffield University and the British Antarctic Survey be undertaken. So we fought hard against snow accumulation in certain key areas of the camp.

At night in the main living hut we turned the heater low to save fuel. We slept on wooden slats in the apex of the hut roof and in the mornings there was a difference of 14° between bed level and floor temperature, the latter averaging –15°C.

Ginnie and I slept together on a single slat at one end of the hut, Oliver and Charlie occupied bachelor slats down the other end and Bothie slept in a cavity behind the heater. The long black nights with the roar of the wind so close and the linger of tallow in the dark are now a memory which Ginnie and I treasure.

There was of course friction between us. Forced togetherness breeds dissension and even hatred between individuals and groups. After four years at work together our person-to-person chemistry was still undergoing constant change. Some days, without a word being spoken. I knew that I disliked one or both of the other men and that the feeling was mutual. At other times, without actually going so far as to admit affection, I felt distinctly warm towards them. When I felt positive antagonism towards the others I could let off steam with Ginnie, who would listen patiently. Or else I could spit out vituperative prose in my diary. Diaries on expeditions are often minefields of over-reaction.

Each of us nursed apprehensions about the future. The thought of leaving the security of our cardboard huts for the huge unknown that stretched away behind Ryvingen Mountain was not something on which I allowed my thoughts to dwell.

On May 2nd the wind chill factor dropped to -110°F. Windstorms from the polar plateau blasted through the camp with no warning. Visiting Ginnie's VLF recording hut one morning I was knocked flat by a gust, although a second earlier there had not been a zephyr of wind. A minute later, picking myself up, I was struck on the back by the plastic windshield ripped off my parked skidoo.

Even in thirty-knot wind conditions we found it dangerous to move outside our huts and tunnels except by way of the staked safety lines which we had positioned all around the camp and from hut to hut. Charlie and I both lost the safety-lines one day when an eighty-knot wind hurled ice-needles horizontally through the white-out. We groped separately and blindly in circles until we blundered into a safety line and so found our way to the nearest hut's hatchway. That same day the parachute canopy covering the sunken area of Ginnie's antennae-tuning units was ripped off, and by nightfall two tons of snow had filled in her entire work area.

Before the sun disappeared for twenty-four hours a day we trained daily at skiing. I taught Oliver and Charlie the rudiments of *langlauf* skiing, for I knew, if we succeeded in crossing Antarctica, that we would meet many regions in the Arctic where only skis would be practical.

On a fine autumn day we left camp with laden sledges for a quick sixteen-kilometre trip to a nearby *nunatak*, a lonely rock outcrop. A storm caught us on the return journey and, within minutes, the wind had burned every patch of exposed skin and frostnipped our fingers inside their light ski-mitts. At the time we were each moving at our separate speeds with a mile or so between us. This was not acceptable by current mountain safety practices but it was a system encouraged on all SAS training courses. The fewer individuals in a group, the faster the majority will reach their goal. This presupposes that the weakest link can look after himself. Oliver was last to arrive back at camp, his face and neck bloated with frostnip. For a week he took antibiotics until his sores stopped weeping.

Ginnie's problems were mostly connected with her radio work. Conducting high-frequency experiments with faraway Cove Radio Station, she was wont to work a 1.5 kva generator in the foyer to her hut. A freak wind once blew carbon monoxide fumes under her door. By chance I called her on a walkie-talkie and, receiving no reply, rushed along the tunnels and down to her shack. I found her puce-faced and staggering about dazed. I dragged her out into the fresh –49°C air and told the Cove Radio operator what had happened. Next day we received a worried message from their controller, Squadron-Leader Jack Willis.

> Even in a comfortable environment the operation of electrical radio components is a hazardous operation. In your location the dangers are considerably increased… Remember that as little as thirty mA can kill. Never wear rings or watches when apparatus is live. Beware of snow from boots melting on the floor… Your one kilowatt transmitter can produce very serious radio frequency burns… Static charges will build up to several thousand volts in an aerial. Toxic berillium is employed in some of the components…

Only four days later, with all her sets switched off and no mains power, Ginnie touched the co-ax cable leading to her forty-watt set. She was stunned by a violent shock that travelled up her right arm and, as she put it later that morning, 'felt like an explosion in my lungs'. The cause was static, built up by wind-blown snow.

Not being a technical expert, Ginnie used common sense to repair her sets, replace tiny diodes and solder cold-damaged flex. When a one-kilowatt resistor blew and she had no spares, she thought of cutting a boiling ring out of our Baby Belling cooker. She then wired the cannibalised coil to the innards of her stricken radio and soon had everything working again, the home-made resistor glowing red-hot on an asbestos mat on the floor. The chief back at Cove Radio described her as 'an amazing communicator'.

To keep our little community happy, Ginnie listened to the BBC World Service and produced *The Ryvingen Observer*. On 6 May we learned that the bodies of US airmen were being flown out of Iran following an abortive attempt to rescue American hostages, that Tito's funeral was imminent, that SAS soldiers had killed some terrorists at the Iranian Embassy in London and that food in British motorway cafes had been summed up in a government report as greasy and tasteless.

Bothie spent his days following Ginnie from hut to hut. She kept old bones for him in each shack and dressed him in a modified pullover when the winds were high. I fought a non-stop battle with the terrier for eight months, trying to teach him that 'outside' meant right outside the tunnels as well as the hut. I lost the struggle.

Charlie, in charge of our food, rationed all goodies with iron discipline. Nobody could steal from the food tunnels without his say-so, a law that was obeyed by all but Bothie. Our eggs, originally sponsored in London, were some eight or nine months old by mid-winter and, although frozen much of that time, they had passed through the tropics *en route*. To an outsider they tasted bad – indeed evil – but we had grown used to them over the months and Bothie was addicted. However hard Charlie tried to conceal his egg store, Bothie invariably outwitted him and stole an egg a day, sometimes more.

The months of June, July and August saw an increasing workload in the camp, due both to the scientific programme and our preparations for departure in October. The sun first came back to Ryvingen, for four minutes only, on 5 August, a miserably cold day. Down at our Sanae base hut, Simon, overwintering with one other Transglober, recorded a windspeed in excess of 100 miles per hour. July 30th was our coldest day. With the wind steady at forty-two knots and a temperature of –42°C, our prevailing chill factor was –131°C, at which temperature any exposed flesh freezes in under fifteen seconds.

In August Ginnie discovered the oscillation unit in her VLF time-code generator had failed due to the cold. To complete the VLF experiment without the instrument meant manually pressing a recording button every four minutes for unbroken twenty-four-hour periods. Wrapped in blankets in the isolated VLF hut she kept awake night after night with flasks of black coffee. By October she was dog-tired and hallucinating but determined to complete the three-month experiment.

As the sunlit hours grew longer, the ice-fields reacted. Explosions sounded in the valleys, rebounding as echoes from the peaks all about us. Avalanches or imploding snow-bridges? There was no way of knowing.

With departure imminent, I realised how much I had grown to love the simplicity of our life at Ryvingen, a crude but peaceful existence during which, imperceptibly, Ginnie and I had grown closer together than during the bustle of our normal London lives. Now I felt pangs of regret that it was ending. Also tremors of apprehension. As the days slipped by my stomach tightened with that long-dormant feeling of dread – once so familiar during school holidays as the next termtime approached.

'I wish you weren't leaving,' Ginnie said.

TERRA INCOGNITA
Travels in Antarctica

Sara Wheeler

Sara Wheeler (born 1961) is a British author known for her extensive travels to the polar regions and the accounts of her experiences there. She became the first female writer-in-residence at the South Pole and spent several months living there, after which she wrote *Terra Incognita*. In addition to several books written about her own journeys, Wheeler is also an accomplished biographer who has chronicled the exploits of explorers and travellers like Apsley Cherry-Garrard, the youngest member of Captain Scott's expedition to reach the South Pole, Fanny Trollope and Harriet Martineau.

WE JUMPED over the soft cracks in the snow to reach Scott's hut, and Lucia opened the door with the heavy metal key. It was our last full day at Wooville.

'Remember—' she began, grinning mischievously.

'I think we've been through that enough times,' I said when I knew she was about to refer to our first visit to the hut, when I had struggled to pull the wooden bar back from the door before unlocking it. To free up my hands, without thinking I had put the key in my mouth, where it had instantly frozen to my lips. Lucia had been obliged to exhale energetically over my face to unstick the key without the loss of too much of the skin on my lips. No lasting damage had been done, but the image of me parading around Wooville with the key to Scott's hut glued to my mouth had kept Lucia amused throughout our tenure at Cape Evans.

As sunlight poured through the door, crusts of snow gleamed on the shovels hanging in the small vestibule. For no real reason, we wandered through to the stables at the back of the hut. They were under the same roof as the living quarters but separated by internal wooden walls. The first of two openings on the left in the small snowbound vestibule led to a storage area and then the stables. In the second opening they had hung a sturdy wooden door which opened on to the living quarters.

The stables consisted of a row of eight horse stalls of conventional design. Each horse's name had been stencilled at the end of its stall.

'Abdul,' Lucia read. 'Is that a common name for an English horse?'

She had an endearing habit of assuming that everything Scott had done, or indeed everything that she observed me doing in our camp, was indicative of activities in which all English people were permanently engaged.

'No: I said firmly. 'I've never heard of a horse called Abdul before.'

She was standing alongside a window at the end of the stable, concentrating on a lightning sketch of the horse stalls. Next to her I noticed the blubber stove where Oates had cooked up bran mashes for the horses. I narrowed my eyes and imagined Oates there – I had seen him standing exactly where she was, in one of Ponting's photographs. He used to sleep in the stables sometimes, to be near the sick ponies during the night. He was much taller than Lucia, but there was something similar in the chiselled nose and high cheekbones.

I waited till she had put her sketchbook back in her pocket. On our way into the main part of the hut, I stopped in the storage area at the large pile of glistening seal blubber, slabbed like peat and stored by Scott's men for winter fuel, and bent down to touch it.

'It's amazing that it's still tacky,' I said, 'even after the iron freeze of an Antarctic winter.'

'It must have smelt gross when the hut was heated,' Lucia said. 'I don't know how they stood it. Look at those hockey sticks hanging on the wall. I've not noticed them before. That's an idea. We could have had a game of hockey.'

'Do you think two people can play on their own?' I said, fingering the spokes of a crumbling bicycle hooked on the wall of the passageway between the stable and the main part of the hut.

'Sure they can!' she said. 'At Wooville they can, anyway.'

'Yikes,' I said.

Standing in the kitchen in the main part of the hut, I realised how much the place had come to seem like home. We could have shut our eyes and reeled off the contents of the shelves, from the red Dutch cheeses that looked like cannonballs, their metal shells corroded, up to the set of tiny fluted metal pastry moulds.

'You know, I've always wondered', Lucia said, pointing up to a shelf above the officers' bunks, 'why they brought that trilby hat down here.'

'Maybe it was for dressing up – I mean for plays or little cabaret sketches,' I offered. 'They used to dress up a lot.'

'Do English men often dress up in costume, then?' she asked, looking puzzled.

'It's a class thing, I suppose – it was a kind of male upper-class ritual.' I immediately regretted saying this, as Lucia was fascinated by the class system and had often interrogated me on the subject. I had tried to paint her a comprehensible picture, and was irritably aware of my inability to do so. Sure enough, I had reminded her of unfinished business.

'Now,' she said, 'during the night I was thinking of what you said about "tea" meaning two different things according to which class you come from. You said that "tea" means the evening meal in working-class circles, and a mid-afternoon cup of tea and cucumber sandwiches with the crusts off or one of those things you toast–'

'A crumpet,' I interrupted.

'That's it, a mid-afternoon cup of tea, dainty sandwiches and a crumpet in upper-class circles. When I come to see you in England' – visiting one another in our respective countries was now a popular topic – 'if I meet someone on the sidewalk who invites me for tea, how will I know the time to go and what I'll be fed? Will I have to say, "Excuse me, are you upper-class or working-class?" in order to find out?'

'Well no, you can tell by the accent...' I began, wishing I'd never introduced her to the concept.

'Well, I can't, can I? What's your accent?'

'Sort of middle,' I said miserably.

She threw me a vexed look and then, sensing that I couldn't be bothered to talk about it any more, she began getting out her pastels. I realised how much l was going to miss her.

It was too dark in the hut to paint properly, so once she had laid out her tools on Scott's desk, Lucia went outside and shovelled the snow away from the windows. The hut sprang to life like a mosaic sluiced with water. Scott's bunk was tucked in the lefthand corner at the end furthest from the door, shielded by a seven-foot-high wooden partition separating his quarters from those of the officers. The non-commissioned men slept at the other end of the hut, nearest the door. Half a wall of wooden boxes stencilled with brand names extended between the officers and the men; originally this wall stretched right across the hut, broken only by a narrow gap for traffic, but it had been demolished by the members of the Ross Sea party who camped in the hut. Presumably they were desperate to find anything still languishing in the boxes. Now, except for Ponting's darkroom, which was next to Scott's cubby-hole and directly opposite the entrance at the other end, the rest of the living area consisted of one large open space. It was about fifty feet by twenty-five, and narrow bunks were positioned around the edge while the middle was occupied by a long table and, at Scott's end, a black metal coal stove. The comer on the right at the far end, opposite Scott, functioned as a laboratory for the expedition scientists, and in it three or four benches and tables were piled with a plethora of vials and test tubes. The kitchen, still overflowing with supplies, was at the other end, on the right near the door, and here they had installed another stove.

Scott's den was about eight feet by six, and besides the bunk and a few bookshelves on the partition walls it contained only his desk. Two other bunks were also tucked away in there, a few feet from Scott's on the other side of the desk and underneath the medicine shelves. These were occupied by Bill Wilson, the chief scientist, and Lieutenant Teddy Evans, who went home with scurvy and returned in command of the *Terra Nova*. So Scott had never really been alone – at least outwardly.

At the foot of Scott's bunk the light revealed a hot-water bottle we had not seen before. We felt as if we were massaging the hut back to life. It never heated up though. It was colder inside than out, like a reverse greenhouse.

They had been very happy in the hut, as we had been in ours. Who wouldn't have been, in that place? This is what Cherry wrote about its position. He began by saying that he had seen a lot of volcanoes.

> But give me Erebus for my friend. Whoever made Erebus knew all the charm of horizontal lines, and the lines of Erebus are for the most part nearer the horizontal than the vertical. And so he is the most restful mountain in the world, and I was glad when I knew that our hut would lie at his feet. And always there floated from his crater the lazy banner of his cloud of steam.

Lucia had decided to paint the freshly illuminated bunk, and positioned her stool so that it faced the hot-water bottle. When I shone my torch into the corner behind her, the beam lit up a small brown glass bottle on a shelf full of medicines. A neat, printed label announced simply '*Poison*', and near the base of the bottle another label read '*Harrods*'.

'Isn't that a store for fancy people?' Lucia asked, following the beam of the torch. Oh God, I thought, here we go again. I nipped sharply into Ponting's darkroom.

'Look at this!' I called. Lucia came in. 'It's a glass negative – it shows a man having his hair cut here in the hut!' The image was fogged, but there, indubitably, was a gnarly old explorer sitting on a stool next to Scott's desk, a pipe protruding from his mouth at a jaunty angle and a pair of scissors held to his already cropped scalp by another man. Both were wearing baggy trousers. I had an idea.

'I could cut your hair,' I said, 'and we could set up an automatic-exposure photograph in here – I mean, to be exactly like them.'

This didn't go down at all well.

'What are you going to cut it with?' she asked. 'That Swiss Army knife you carry around? It's too blunt to cut the muffins.' It seemed only fair that she should let me do it after I had submitted like a lamb to the acupuncture needles. As I was thinking about how I might persuade her, she walked out of the darkroom and into the main part of the hut.

'Do you think they missed Cape Evans when they were back at home?' she called. 'The ones who made it home, I mean.'

'They all said they did.'

'Do you think we will?'

I looked at her standing there, a stick of chalk poised in mid-air, and I realised how rich our lives had been at Wooville. As Frank Hurley wrote, 'We had learned to find fullness and contentment in a life which had stripped us of all the distinctions, baubles and trappings of civilisation.'

'I'm sure I'll never stop missing it,' I said. 'I'll think of it every day, sitting in my flat, looking out at the traffic. I know nothing will ever be like this again. I'll never feel quite so separated from my anxieties. It's as though God has given me a gift, once in my life, to step off the planet for two months and listen to a different music.'

'Doesn't it make you unbearably sad – I mean, that it's over?' I had to think about that.

'In a strange way, it doesn't. I sort of feel I'm taking it with me – in my heart, if that doesn't sound naff.'

'What does naff mean?'

'Drippy...' That didn't help. 'Mawkishly sentimental.'

She said no more, and I wandered aimlessly around as she painted, imagining them in this place or that, cosied up in their hut.

I wondered why in the world the bunks were so infernally small, and what had induced them to bring down a blue-and-white Chinese porcelain decorative bowl. I remembered, too, reading about a piano they had brought in from the ship (despite the fact that no one could play it) – but this had vanished.

After some time – it might have been ten minutes – Lucia said, 'No, it doesn't.' I knew immediately she was answering my question about whether what I had said sounded naff. We often had conversations which included long pauses. We had learnt the rhythm of one another's thoughts.

She too, I knew, was preoccupied with the notion that it was all over. She called out –

'What was that quotation you were telling me last night, you know, about being restored to a natural state – it was by one of those guys you call Beards?'

Many of our speech idioms had rubbed off – hers on me and mine on her

– but I had failed to introduce 'Beards' as a term one could use generically, without qualification.

'It was by Reinhold Messner,' I said. 'The greatest mountaineer alive. When he was down here, slogging across the plateau, he wrote, "It seemed to me as if I were restored to that time and that state when nature alone was God." '

I went down to Scott's quarters to look over her shoulder at the pastel drawing.

'Is it okay?' she asked.

'The megaphone's good,' I said. They had presumably brought this object so they could converse from the ship to the sea ice. It had ended up hanging on a hook above Scott's desk. I had tried to think of a use for it at Wooville, but alas! I was too attached to my VHF.

'There's something wrong with the way this light falls here,' I said, pointing to a corner of the painting. We had grown accustomed to being frank about the other's work – false politeness seemed absurdly out of place at Wooville. It would have been like wearing couturier parkas.

'How about a touch of Naples yellow there?' I said, pointing again. She looked at the sketch, and then at me, wide-eyed.

'I've taught you too much,' she said.

Before I could go home, there was one thing I still had to do.

In the hut, a single shaft of light from the midnight sun cut above the mound of snow piled against the window and shone on to the long wooden table, casting distorted shadows against the far wall. It was the table loaded with bottles that Ponting photographed while roistering English voices rang around the hut. Lucia was sleeping peacefully in our own quiet hut, a hundred yards away. The wind reverberated in the small entrance hall like the sound of a train in a tunnel, though the body of the hut was almost silent.

I lay awake for many hours, my head on his pillow, as he, weighed down by his heavy responsibilities, must often have done. How very different the end had been for him. 'Here, then, tonight,' he had written in his diary, 'we have reached the end of our tether.'

The distended shadows shifted along the old wooden walls as the sun wheeled across the sky. I was thinking about my first day in Antarctica and the view from the top of the snowhill as the vulcanologist tap-tapped snow into his specimen tin. I could remember it as if it were yesterday. A great deal had happened since then. I had travelled thousands of miles, lost a lot of body heat, watched hundreds of beards ice up, realised how little I had seen, or knew. It was more of a *terra incognita* than ever. Byrd used the image of a beach and a tide to convey the changing of the seasons in Antarctica: the polar day was the beach, and the night was the tide. I had seen it come in, and I had seen it go out. It had all happened so fast. But I still felt the same about Antarctica.

It was the great thrill of my life – on top of the snowhill, on Scott's bunk, in what was about to become my future. It had allowed me to believe in paradise, and that, surely, is a gift without price.

Then I laid my head on his pillow, and went to sleep.

THE LOST CITY OF Z

David Grann

Percy Harrison Fawcett (1867–disappeared 1925) was a British Army officer, explorer, cartographer and archaeologist who arranged several expeditions to find a mythical lost city in the jungles of Brazil. Despite being nearly fifty years old at the outbreak of the First World War, Fawcett volunteered for service in Flanders and emerged from the conflict a highly decorated veteran. During his expeditions to South America, Fawcett developed a theory positing the existence of a highly advanced ancient civilization he called the Lost City of Z. Despite being ridiculed by his peers in the Royal Geographical Society, Fawcett continued to lead expeditions into the Mato Grosso region of Brazil. Faced with tropical illness and impassable geography, he was forced to abandon many of these missions, but nevertheless persisted in his search for Z. In 1925, during the last of these expeditions, he disappeared along with his son Jack and was never heard from again. In the years following Fawcett's disappearance, archaeologists and explorers have discovered a large archaeological site known as Kuhikugu that is believed to be a potential candidate for Fawcett's Lost City and a possible vindication of his lifelong enterprise.

David Grann (born 1967) is a *New York Times* bestselling author and award-winning journalist, who has been a staff writer for *The New Yorker* since 2003. He is also the author of *The White Darkness* and several of his books, including *The Lost City of Z*, have been adapted into major motion pictures.

Fawcett described the area as "the tickiest place in the world"; the insects swarmed over everything, like black rain. Several bit Raleigh on his foot, and the irritated flesh became infected—"poisoned," in Jack's phrase. As they pressed on the next day, Raleigh grew more and more gloomy. "It is a saying that one only knows a man well when in the wilds with him," Fawcett told Nina. "Raleigh in place of being gay and energetic, is sleepy and silent."

Jack, in contrast, was gaining in ardor. Nina was right: he seemed to have inherited Fawcett's freakish constitution. Jack wrote that he had packed on several pounds of muscle, "in spite of far less food. Raleigh has lost more than I gained, and it is he who seems to feel most the effects of the journey."

Upon hearing about Jack from her husband, Nina told Large, "I think you will rejoice with me in the knowledge that Jack is turning out so capable, and keeping strong and well. I can see his father is very pleased with him, and needless to say *so am I*!"

Because of Raleigh's condition and the weakened animals, Fawcett, who was more careful not to get too far ahead again, stopped for several days at a cattle-breeding ranch owned by Hermenegildo Galvão, one of the most ruthless farmers in Mato Grosso. Galvão had pushed farther into the frontier than most Brazilians and reportedly had a posse of *bugueiros*, "savage hunters," who were charged with killing Indians who threatened his feudal empire. Galvão was not accustomed to visitors, but he welcomed the explorers into his large red-brick home. "It was quite obvious from his manners that Colonel Fawcett was a gentleman and a man of engaging personality," Galvão later told a reporter.

For several days, the explorers remained there, eating and resting. Galvão was curious about what had lured the Englishmen into such wilderness. As Fawcett described his vision of Z, he removed from his belongings a strange object covered in cloth. He carefully unwrapped it, revealing the stone idol Haggard had given to him. He carried it with him like a talisman.

The three Englishmen were soon on their way again, heading east, toward Bakairí Post, where in 1920 the Brazilian government had set up a garrison—"the last point of civilization," as the settlers referred to it. Occasionally, the forest opened up, and they could see the blinding sun and blue-tinged mountains in the distance. The trail became more difficult, and the men descended steep, mud-slicked gorges and traversed rockstrewn rapids. One river was too dangerous for the animals to swim across with the cargo. Fawcett noticed a canoe, abandoned, on the opposite bank and said that the expedition could use it to transport the gear, but that someone would need to swim over and get it—a feat involving, as Fawcett put it, "considerable danger, being made worse by a sudden violent thunderstorm."

Jack volunteered and began to strip. Though he later admitted that he was "scared stiff," he checked his body for cuts that might attract piranhas and dived in, thrashing his arms and legs as the currents tossed him about. When he emerged on the opposite bank, he climbed in the canoe and paddled back across—his father greeting him proudly.

A month after the explorers left Cuiaba, and after what Fawcett described as "a test of patience and endurance for the greater trials" ahead, the men arrived at Bakairí Post. The settlement consisted of about twenty ramshackle huts, cordoned off by barbed wire, to protect against aggressive tribes. (Three years later, another explorer described the outpost as "a pinprick on the map: isolated, desolate, primitive and God forsaken.") The Bakairí tribe was one of the first in the region that the government had tried to "acculturate," and Fawcett was appalled by what he called "the Brazilian methods of civilizing the Indian tribes." In a letter to one of his sponsors in the United States, he noted, "The Bakairís have been dying out ever since they became civilized. There are only about 150 of them." He went on, "They have in part been brought here to plant rice, manioc ... which is sent to Cuiabá, where it fetches, at present, high prices. The Bakairís are not paid, are raggedly clothed, mainly in khaki govt. uniforms, and there is a general squalor and lack of hygiene which is making the whole of them sick."

Fawcett was informed that a Bakairí girl had recently fallen ill. He often tried to treat the natives with his medical kit, but, unlike Dr. Rice, his knowledge was limited, and there was nothing he could do to save her. "They say the Bacairys are dying off on account of fetish [witchcraft], for there is a fetish man in the village who hates them," Jack wrote. "Only yesterday a little girl died—of fetish, they say!"

The Brazilian in charge of the post, Valdemira, put the explorers up in the newly constructed schoolhouse. The men soaked themselves in the river, washing away the grime and sweat. "We have all clipped our beards, and feel better without them," Jack said.

Members of other remote tribes occasionally visited Bakairí Post to obtain goods, and Jack and Raleigh soon saw something that astonished them: "about eight wild Indians, absolutely stark naked," as Jack wrote to his mother. The Indians carried seven-foot-long bows with six-foot arrows. "To Jack's great delight we have seen the first of the wild Indians here—naked savages from the Xingu," Fawcett wrote Nina.

Jack and Raleigh hurried out to meet them. "We gave them some guava cheese," Jack wrote, and "they liked it immensely."

Jack tried to conduct a rudimentary *autopsis*. "They are small people, about five feet two inches in height, and very well built," he wrote of the Indians. "They eat only fish and vegetables—never meat. One woman had a very fine necklace of tiny discs cut from snail shells, which must have required tremendous patience to make."

Raleigh, whom Fawcett had designated as the expedition's photographer, set up a camera and took pictures of the Indians. In one shot, Jack stood beside them, to demonstrate "the comparative sizes"; the Indians came up to his shoulders.

In the evening, the three explorers went to the mud hut where the Indians were staying. The only light inside was from a fire, and the air was filled with smoke. Fawcett unpacked a ukulele and Jack took out a piccolo that they had brought from England. (Fawcett told Nina that "music was a great comfort 'in the wilds,' and might even save a solitary man from insanity.") As the Indians gathered around them, Jack and Fawcett played a concert late into the night, the sounds wafting through the village.

On May 19, a fresh, cool day, Jack woke up exhilarated—it was his twenty-second birthday. "I have never felt so well," he wrote to his mother. For the occasion, Fawcett dropped his prohibition against liquor, and the three explorers celebrated with a bottle of Brazilian-made alcohol. The next morning, they prepared the equipment and the pack animals. To the north of the post, the men could see several imposing mountains and the jungle. It was, Jack wrote, "absolutely unexplored country."

The expedition headed straight for terra incognita. Before them were no clear paths, and little light filtered through the canopy. They struggled to see not just in front of them but above them, where no predators lurked. The men's feet sank in mud holes. Their hands burned from wielding machetes. Their skin bled from mosquitoes. Even Fawcett confessed to Nina, "Years tell, in spite of the spirit of enthusiasm."

Although Raleigh's foot had healed, his other one became infected, and when he removed his sock a large patch of skin peeled off. He seemed to be unraveling; he had already suffered from jaundice, his arm was swollen, and he felt, as he put it, "bilious."

Like his father, Jack was prone to contempt for others' frailty, and complained to his mother that his friend was unable to share his burden of work—he rode on a horse, with his shoe off—and that he was always scared and sullen.

The jungle widened the fissures that had been evident since Raleigh's romance on the boat. Raleigh, overwhelmed by the insects, the heat, and the pain in his foot, lost interest in "the Quest." He no longer thought about returning as a hero: all he wanted, he muttered, was to open a small business and to settle down with a family. ("The Fawcetts can have all my share of the notoriety and be welcome to it!" he wrote his brother.) When Jack talked of the archaeological importance of Z, Raleigh shrugged and said, "That's too deep for me."

"I wish [Raleigh] had more brains, as I cannot discuss any of this with him as he knows nothing of anything," Jack wrote. "We can only converse about Los Angeles or Seaton. What he will do during a year at 'Z' I don't know."

"I wish to *hell* you were here," Raleigh told his brother, adding, "You know there is a saying which I believe is true: 'Two's company—three's none.' It shows itself quite often with me now!" Jack and Fawcett, he said, maintained a "sense of inferiority for others. Consequently at times I feel very 'out of everything.' Of course I do not outwardly show it ... but still, as I have said before, I feel 'awful lonesome' for real friendship."

After nine days, the explorers hacked their way to Dead Horse camp, where the men could still see the "white bones" from Fawcett's old pack animal. The men were approaching the territory of the warlike Suyás and Kayapós. An Indian once described to a reporter a Kayapó ambush of his tribe. He and a few other villagers, the reporter wrote, fled across a river and "witnessed throughout the night the macabre dance of their enemies around their slaughtered brothers." For three days, the invaders remained, playing wooden flutes and dancing among the corpses. After the Kayapós finally departed, the few villagers who had escaped across the river rushed back to their settlement: not a single person was alive. "The women, who they thought would have been spared, lay face up, their lifeless bodies in an advanced state of putrefaction, their legs spread apart by wooden struts forced between the knees." In a dispatch, Fawcett described the Kayapós as an aggressive "lot of stick-throwers who cut off and kill wandering individuals ... Their only weapon is a short club like a policeman's billy"—which, he added, they deploy very skillfully.

After passing through the territory of the Suyás and Kayapós, the expedition would tum eastward and confront the Xavante, who were perhaps even more formidable. In the late eighteenth century, many in the tribe had been contacted by the Portuguese and moved into villages, where they received mass baptisms. Devastated by epidemics and brutalized by Brazilian soldiers, they eventually fled back into the jungle near the River of Death. A nineteenth-century German traveler wrote that "from that time onwards [the Xavante] no longer trusted any white man ... These abused people have therefore changed from compatriots into the most dangerous and determined enemies. They generally kill anyone they can easily catch." Several years after Fawcett's journey, members of the Indian Protection Service tried to make contact with the Xavante, only to return to their base camp and discover the naked corpses of four of their colleagues. One was still clutching in his hand gifts for the Indians.

In spite of the risks, Fawcett was confident—after all, he had always succeeded where others failed. "It is obviously dangerous to penetrate large hordes of lndians traditionally hostile," he wrote, "but I believe in my mission and in its purpose. The rest does not worry me, for I have seen a good deal of lndians and know what to do and what not to do." He added "I believe our little party of three white men will make friends with them all."

The guides, who were already feverish, were reluctant to go any farther, and Fawcett decided that the time had come to send them back. He selected half a dozen or so of the strongest animals to keep for a few more days. Then the explorers would have to proceed with their few provisions on their backs.

Fawcett pulled Raleigh aside and encouraged him to return with the guides. As Fawcett had written to Nina, "I suspect constitutional weakness, and fear that we shall be handicapped by him." After this point, Fawcett explained, there would be no way to carry him out. Raleigh insisted that he would see it through. Perhaps he remained loyal to Jack, in spite of everything. Perhaps he didn't want to be seen as a coward. Or perhaps he was simply afraid to turn back without them.

Fawcett finished his last letters and dispatches. He wrote that he would try to get out other communiqués in the coming year or so, but added that it was unlikely. As he noted in one of his final articles, "By the time this dispatch is printed, we shall have long since disappeared into the unknown."

After folding up his missives, Fawcett gave them to the guides. Raleigh had earlier written to his "dearest Mother" and family. "I shall look forward to seeing you again in old Cal when I return," he said. And he told his brother bravely, "Keep cheerful and things will tum up alright as they have for me."

The explorers gave a final wave to the Brazilians, then turned and headed deeper into the jungle. In his last words to his wife, Fawcett wrote, "You need have no fear of any failure."

THE POWER AND THE GLORY

Graham Greene

Graham Greene (1904–1991) was one of the most distinguished English authors of the twentieth century. His most acclaimed work, *The Power and the Glory*, also stands out as his most poignant. The novel is set in the Mexican state of Tabasco in the 1930s, during the government's anticlerical crusade against the Catholic Church that resulted in the bloody Cristero Rebellion. It tells the story of a renegade Catholic 'whisky priest' – the term was coined by the author to describe a cleric of dubious moral character who nonetheless claimed a higher spiritual authority and displayed courage and resolve on a broader level. Throughout the novel, the unnamed priest struggles with his own moral failures in the face of performing his clerical duties and evading detection by federal militias. He is pursued by a police lieutenant, who thinks that all members of the clergy are fundamentally evil, but who also struggles with his own moral duties.

A voice said, 'Well, have you finished now?'

The priest got up and made a small scared gesture of assent. He recognized the police officer who had given him money at the prison, a dark smart figure in the doorway with the stormlight glinting on his leggings. He had one hand on his revolver and he frowned sourly in at the dead gunman. 'You didn't expect to see me,' he said.

'Oh, but I did,' the priest said. 'I must thank you.'

'Thank me, what for?'

'For letting me stay alone with him.'

'I am not a barbarian,' the officer said. 'Will you come out now, please? It's no use at all your trying to escape. You can see that,' he added, as the priest emerged and looked round at the dozen armed men who surrounded the hut.

'I've had enough of escaping,' he said. The half-caste was no longer in sight; the heavy clouds were piling up the sky: they made the real mountains look like little bright toys below them. He sighed and giggled nervously. 'What a lot of trouble I had getting across those mountains, and now ... here I am ...'

'I never believed you would return.'

'Oh well, lieutenant, you know how it is. Even a coward has a sense of duty.' The cool fresh wind which sometimes blows across before a storm breaks touched his skin. He said with badly-affected ease, 'Are you going to shoot me now?'

The lieutenant said again sharply, 'I am not a barbarian. You will be tried ... properly.'

'What for?'

'For treason.'

'I have to go all the way back there?'

'Yes. Unless you try to escape.' He kept his hand on his gun as if he didn't trust the priest a yard. He said, 'I could swear that somewhere ... '

'Oh yes,' the priest said. 'You have seen me twice. When you took a hostage from my village ... you asked my child: "Who is he?" She said: "My father," and you let me go.' Suddenly the mountains ceased to exist: it was as if somebody had dashed a handful of water into their faces.

'Quick,' the lieutenant said, 'into that hut.' He called out to one of the men. 'Bring us some boxes so that we can sit.'

The two of them joined the dead man in the hut as the storm came up all round them; A soldier dripping with rain carried in two packing-cases. 'A candle,' the lieutenant said. He sat down on one of the cases and took out his revolver. He said, 'Sit down, there, away from the door, where I can see you.' The soldier lit a candle and stuck it in its own wax on the hard earth floor, and the priest sat down, dose to the American; huddled up in his attempt to get at his knife he gave an effect of wanting to reach his companion, to have a word or two in private. They looked two of a kind, dirty and unshaved: the lieutenant seemed to belong to a different class altogether. He said with contempt, 'So you have a child?'

'Yes,' the priest said.

'You – a priest?'

'You mustn't think they are all like me.' He watched the candlelight blink on the bright buttons. He said, 'There are good priests and bad priests. It is just that I am a bad priest.'

'Then perhaps we will be doing your Church a service ...'

'Yes.'

The lieutenant looked sharply up as if he thought he was being mocked. He said, 'You told me twice. That I had seen you twice.'

'Yes, I was in prison. And you gave me money.'

'I remember.' He said furiously, 'What an appalling mockery. To have had you and then to let you go. Why, we lost two men looking for you. They'd be alive today .. .' The candle sizzled as the drops of rain came through the roof.

'This American wasn't worth two lives. He did no real harm.'

The rain poured ceaselessly down. They sat in silence. Suddenly the lieutenant said, 'Keep your hand away from your pocket.'

'I was only feeling for a pack of cards. I thought perhaps it would help to pass the time … '

'I don't play cards,' the lieutenant said harshly.

'No, no. Not a game. Just a few tricks I can show you. May I?'

'All right. If you wish to.'

Mr Lehr had given him an old pack of cards. The priest said, 'Here, you see, are three cards. The ace, the king, and the jack. Now,' he spread them fanwise out on the floor, 'tell me which is the ace.'

'This, of course,' the lieutenant said grudgingly, showing no interest.

'But you are wrong,' the priest said, turning it up. 'That is the jack.'

The lieutenant said contemptuously, 'A game for gamblers – or children.'

'There is another trick,' the priest said, 'called Fly-away Jack. I cut the pack into three – so. And I take this Jack of Hearts and I put it into the centre pack – so. Now I tap the three packs.' His face lit up as he spoke – it was such a long time since he had handled cards – he forgot the storm, the dead man and the stubborn unfriendly face opposite him. 'I say Fly-away Jack' – he cut the left-hand pack in half and disclosed the jack – 'and there he is.'

'Of course there are two jacks.'

'See for yourself.' Unwillingly the lieutenant leant forward and inspected the centre pack. He said, 'I suppose you tell the Indians that that is a miracle of God.'

'Oh no,' the priest giggled, 'I learnt it from an Indian. He was the richest man in the village. Do you wonder? with such a hand. No, I used to show the tricks at any entertainments we had in the parish – for the Guilds, you know.'

A look of physical disgust crossed the lieutenant's face. He said, 'I remember those Guilds.'

'When you were a boy?'

'I was old enough to know …'

'Yes?'

'The trickery.' He broke out furiously with one hand on his gun, as though it had crossed his mind that it would be better to eliminate this beast, now, at this instant, for ever. 'What an excuse it all was, what a fake. Sell all and give to the poor – that was the lesson, wasn't it? and Señora So-and-so, the druggist's wife, would say the family wasn't really deserving of charity, and Señor This, That and the Other would say that if they starved, what else did they deserve, they were Socialists anyway, and the priest – you – would notice who had done his Easter duty and paid his Easter offering.' His voice rose – a policeman looked into the hut anxiously and withdrew again through the lashing rain. 'The Church was poor, the priest was poor, therefore everyone should sell all and give to the Church.'

The priest said, 'You are so right.' He added quickly, 'Wrong too, of course.'

'How do you mean?' the lieutenant asked savagely. 'Right? Won't you even defend ... ?'

'I felt at once that you were a good man when you gave me money at the prison.'

The lieutenant said, 'I only listen to you because you have no hope. No hope at all. Nothing you say will make any difference.'

'No.'

He had no intention of angering the police officer, but he had had very little practice the last eight years in talking to any but a few peasants and Indians. Now something in his tone infuriated the lieutenant. He said, 'You're a danger. That's why we kill you. I have nothing against you, you understand, as a man.'

'Of course not. It's God you're against. I'm the sort of man you shut up every day – and give money to.'

'No, I don't fight against a fiction.'

'But I'm not worth fighting, am I? You've said so. A liar, a drunkard. That man's worth a bullet more than I am.'

'It's your ideas.' The lieutenant sweated a little in the hot steamy air. He said, 'You are so cunning, you people. But tell me this – what have you ever done in Mexico for *us*? Have you ever told a landlord he shouldn't beat his peon – oh yes, I know, in the confessional perhaps, and it's your duty, isn't it, to forget it at once. You come out and have dinner with him and it's your duty not to know that he has murdered a peasant. That's all finished. He's left it behind in your box.'

'Go on,' the priest said. He sat on the packing-case with his hands on his knees and his head bent; he couldn't, though he tried, keep his mind on what the lieutenant was saying. He was thinking – forty-eight hours to the capital. Today is Sunday. Perhaps on Wednesday I shall be dead. He felt it as a treachery that he was more afraid of the pain of bullets than of what came after.

'Well, we have ideas too,' the lieutenant was saying. 'No more money for saying prayers, no more money for building places to say prayers in. We'll give people food instead, teach them to read, give them books. We'll see they don't suffer.'

'But if they want to suffer ...'

'A man may want to rape a woman. Are we to allow it because he wants to? Suffering is wrong.'

'And you suffer all the time,' the priest commented, watching the sour

Indian face behind the candle-flame. He said, 'It sounds fine, doesn't it? Does the jefe feel like that too?'

'Oh, we have our bad men.'

'And what happens afterwards? I mean after everybody has got enough to eat and can read the right books – the books you let them read?'

'Nothing. Death's a fact. We don't try to alter facts.'

'We agree about a lot of things,' the priest said, idly dealing out his cards. 'We have facts, too, we don't try to alter – that the world's unhappy whether you are rich or poor – unless you are a saint, and there aren't many of those. It's not worth bothering too much about a little pain here. There's one belief we both of us have – that we'll all be dead in a hundred years.' He fumbled, trying to shuffle, and bent the cards: his hands were not steady.

'All the same, you're worried now about a little pain,' the lieutenant said maliciously, watching his fingers.

'But I'm not a saint,' the priest said. 'I'm not even a brave man.' He looked up apprehensively: light was coming back: the candle was no longer necessary. It would soon be clear enough to start the long journey back. He felt a desire to go on talking, to delay even by a few minutes the decision to start. He said, 'That's another difference between us. It's no good your working for your end unless you're a good man yourself. And there won't always be good men in your party. Then you'll have all the old starvation, beating, get-rich-anyhow. But it doesn't matter so much my being a coward – and all the rest. I can put God into a man's mouth just the same – and I can give him God's pardon. It wouldn't make any difference to that if every priest in the Church was like me.'

'That's another thing I don't understand,' the lieutenant said, 'why you – of all people – should have stayed when the others ran.'

'They didn't all run,' the priest said.

'But why did you stay?'

'Once,' the priest said, 'I asked myself that. The fact is, a man isn't presented suddenly with two courses to follow: one good and one bad. He gets caught up. The first year – well, I didn't believe there was really any cause to run. Churches have been burnt before now. You know how often. It doesn't mean much. I thought I'd stay till next month, say, and see if things were better. Then – oh, you don't know how time can slip by.' It was quite light again now: the afternoon rain was over: life had to go on. A policeman passed the entrance of the hut and looked in curiously at the pair of them. 'Do you know I suddenly realized that I was the only priest left for miles around? The law which made priests marry finished them. They went: they were quite right to go. There was one priest in particular – he had always disapproved of me. I have a tongue, you know, and it used to wag. He said – quite rightly – that

I wasn't a firm character. He escaped. It felt – you'll laugh at this – just as it did at school when a bully I had been afraid of – for years – got too old for any more teaching and was turned out. You see, I didn't have to think about anybody's opinion any more. The people – they didn't worry me. They liked me.' He gave a weak smile, sideways, towards the humped Yankee.

'Go on,' the lieutenant said moodily.

'You'll know all there is to know about me at this rate,' the priest said, with a nervous giggle, 'by the time I get to, well, prison.'

'It's just as well. To know an enemy, I mean.'

'That other priest was right. It was when he left I began to go to pieces. One thing went after another. I got careless about my duties. I began to drink. It would have been much better, I think, if I had gone too. Because pride was at work all the time. Not love of God.' He sat bowed on the packing-case, a small plump man in Mr Lehr's cast-off clothes. He said, 'Pride was what made the angels fall. Pride's the worst thing of all. I thought I was a fine fellow to have stayed when the others had gone. And then I thought I was so grand I could make my own rules. I gave up fasting, daily Mass. I neglected my prayers – and one day because I was drunk and lonely – well, you know how it was, I got a child. It was all pride. Just pride because I'd stayed. I wasn't any use, but I stayed. At least, not much use. I'd got so that I didn't have a hundred communicants a month. If I'd gone I'd have given God to twelve times that number. It's a mistake one makes – to think just because a thing is difficult or dangerous ... ' He made a flapping motion with his hands.

The lieutenant said in a tone of fury, 'Well, you're going to be a martyr – you've got that satisfaction.'

'Oh no. Martyrs are not like me. They don't think all the time – if I had drunk more brandy I shouldn't be so afraid.'

The lieutenant said sharply to a man in the entrance, 'Well, what is it? What are you hanging round for?'

'The storm's over, lieutenant. We wondered when we were to start?'

'We start immediately.'

He got up and put back the pistol in his holster. He said, 'Get a horse ready for the prisoner. And have some men dig a grave quickly for the Yankee.'

The priest put the cards in his pocket and stood up. He said, 'You have listened very patiently ...'

'I am not afraid,' the lieutenant said, 'of other people's ideas.'

WALKING THE AMAZON
860 Days. One Step at a Time
Ed Stafford

Ed Stafford (born 1975) is a British explorer and Guinness World Record holder notable for being the first man to walk the full length of the Amazon river. A former British Army officer, Stafford completed his Amazon trek in 2010 after a gruelling series of events including being held at gunpoint and wrongfully arrested for drug smuggling and even murder. Stafford was considered an early innovator in the crowdfunding space, relying on online donations from his YouTube followers in order to complete his record-making journey. He has since headlined a number of survival and exploration shows made for British television and remains the only person to have completed the entirety of the Amazonian trek.

On day five of this extreme rationing we came across a beautiful small brook at about two in the afternoon and we both saw the opportunity to try to fish, despite it not being the end of the walking day. I laid the gill net across the three-metre-wide stream and within seconds I had caught six decent-sized fish. We could not pass up this dream river and so decided to stop for the day and make camp. Cho used a rod and line while I just harvested with the net. Both of us brought in big catches that were all for the pot. We had eaten three pots of fish broth by the time we were full. The sensation was phenomenal; the fats in the fish floated on the surface of the broth and as the food entered our stomachs they started to groan, as if they were coming back to life. The fats were absorbed immediately and our brains started to function again. We had highs purely from eating food. We were so hungry that we didn't even consider smoking any of these fish; we just wanted a full recharge. That said, we did save some fish for the morning and slept soundly with full bellies.

On the eighth and last day of rationing we ate all our farine at breakfast knowing that we needed to walk a long twelve-kilometre day to reach Marúą. The jungle had become swampy and gnarled again as it was the flooded forest of the upcoming River Juruá. This meant the going should have been slow but as we had no food in our packs now, and because the end was in sight, we crashed through the stunted black trees at a phenomenal pace.

At 7 p.m. we'd gone straight through the coordinates that the map had given us and we were a full kilometre beyond. Our heads dropped as we realised there was no settlement, and we'd punished ourselves all day to arrive before nightfall. It was now completely dark and we hadn't seen water for hours so we just decided to put up our hammocks and sleep. Without even clearing a space in the undergrowth we each found two trees quickly and hung our hammocks. Unfed and unwashed, we went to bed. The grimy layer of sweat and dirt on our skin made the night unpleasant as well as dispiriting. We took tiny sips from our almost empty water bottles and tried to sleep.

In the morning, with sleep our mood had lifted and nearby we found some aguaje fruits, a nut with a soft, orange flesh with a vague smell of vomit, to complement the last of a single dried piranha. We sipped the dregs of our water, which was brown as it had come from a puddle rather than a river, and had no option but to continue. All we had by way of edible provisions was a half-kilo of salt.

To us, unwashed, exhausted and starving, that morning represented everything that, deep down, I wanted from the expedition. We were 150 kilometres from the main channel of the Solimões (Amazon), about 25 kilometres away from the next big tributary, the Jurua. We had had a deficit of over 3,000 calories a day for the past eight days and we had no option but to put the facts to one side and continue as normal. No words would make any difference, no blame, no analysis. We just had to go on and deep down we expected to be OK.

There were no rivers so we ate only palm hearts all day. Our first trial at sourcing all our carbohydrates from the jungle worked but you have to eat a lot of these salad vegetables to fill you up. *Paimitos*, or heart of palm, in fact became our salvation. Normally we wouldn't have cut these down because, in order to get to the soft, white palm heart in the centre at the top of the tree, you cannot avoid killing the whole palm. In our feeble state, each palm was quite an effort to fell with our machetes but the white flesh inside was the best salad vegetable I have ever tasted. These patches of *palmitos* were sporadic at best, however, and so we kept our eyes peeled for the tops of the trees, looking for the distinctive red stems.

Cho looked like a featherweight boxer now and I, too, had never lost so much weight. My normal weight is about 92 kilograms. I had dropped to about 88 kilograms before I set out on this leg of the expedition and by the time I arrived at the banks of the Solimões again I would be 81 kilograms. As we walked we stumbled frequently and snapped in and out of blood sugar crashes as we impaled ourselves on spiky vegetation in our half-aware state.

The following day everything came good. Cho walked straight into a huge tortoise weighing in at around 10 kilograms. It was morning and so we couldn't lose time by stopping and preparing the animal. We would just have to carry it. We took turns to pack the lead-weight live animal into the tops of our packs.

Eventually we came to a large river, the first we'd seen in a week or so, and I set about cutting up the tortoise. I'd watched Boruga and the Asheninka men do this before and so I knew how to do it; but I never expected it to be quite such a horrific task. If you are not used to killing animals a tortoise is not the best to start with. You have to turn it on its back, hack at the exoskeleton shell between the foot holes until the bottom is loose and then peel it back like the lid of a tin of beans. Except that the bottom is clearly attached to the tortoise still inside and needs to be sliced off with the machete. The underside of the shell has to be off before you can kill the animal and so I grabbed the now defenceless head and cut it off to kill the creature as quickly as I could. The body then kept twitching for the whole time that it took to cut out the rest of the meat and remove the intestines. I washed everything and used the upturned shell as a bowl and cut the tortoise up into small strips which I then salted. Cho made a drying rack and we had a huge amount of meat cured and smoking over the fire. Our morale was flooding back and we were elated by the prospect of food.

I realise that some people might be shocked and distressed about the killing of tortoises, but I think it has to be put into the context of where we were and how long we had travelled with so little food. In our natural state we humans are designed to be omnivores and the jungle is a place where we could survive if we took advantage of what nature had to offer. Although the physical process of killing the animal was quite an ordeal I won't pretend I was sad – this was a natural way of living and the tortoise was part of our food chain. I had begun to see animals in the forest as the locals did – rather than exotic beasts that needed to be preserved, I saw food.

In the morning we crossed the river simply by walking through it. It was perhaps 40 metres wide, but it was shallow and the small part that we had to cross was easy. We strode out on the far side and could immediately tell that people had been in the area. Small paths turned into what appeared to be a dirt logging road which we followed in the hope that it would lead us to people. We ate only our tortoise-meat jerky throughout the morning and had not eaten any carbohydrates, save the limp palm hearts, for over three days.

At about 1 p.m. we saw a wooden shack with a tin roof on top of a hill and made straight for it. As we approached, a woman came to the door and I explained what we were doing. I have no idea what we must have looked like after thirty-seven straight days of jungle from Amatura. The woman called her husband who had been making farine and he came and spoke to us. They were amazed when we told them where we had come from; they said that, to their knowledge, no one had ever made that journey before. They were about to have lunch and invited us to eat.

We dumped our packs outside in the blistering dry heat of the cleared hill and climbed a ladder to enter the cooler wooden hut on stilts. Inside there was no furniture, just a huge pan of fish broth in the middle of the floor, a plastic tub of fresh farine that was still warm, having just been made, and a stack of glass plates. The woman dished us out a plate of soup each as we sat on the floor among the family's children. They watched us wolf down the first plate, then the second, then the third. I know we had eaten tortoise jerky earlier that morning, but the cumulative carbohydrate rationing and overall calorie deficit meant that our bodies had still felt starved and we ate and ate this glut of farine. Looking back, I doubt the farine was any different from farine elsewhere in Brazil, but at the time Cho and I could not stop eating it. It had the most wonderful warm texture and when eaten with the broth it was the best meal I had ever tasted. It is certainly true that the best way to appreciate food is to be truly hungry before you eat. I will never forget that meal as long as I live.

The family waved goodbye to us at about 3 p.m. and pointed us in the direction of Juruá. We had 30 kilometres to cover and expected it to take four days. We bought farine from the family to last us over this time and we also bought coffee, milk powder and sugar, luxuries we hadn't had for weeks.

The next few days saw the worst jungle of the entire expedition: low, tangled rainforest, with a canopy no higher than six metres, with gnarled, black branches blocking our path. Every soggy step gave way and sank our feet up to our thighs; every branch we clung to was covered either in spines or ants. It was the height of dry season now and I dreaded to think what the forest would have been like at any other time of year; completely impassable I suspect. It bore out my decision to cut across the meander from Amaturá to Tefé.

Our progress was painfully slow. One morning we advanced no more than 400 metres. After the false dawn of a house on the hill, I had thought we were home and dry and had pretty much reached Juruá City. I couldn't have been more wrong.

That distance took us six of the toughest days I could remember. I hated every step of it. I was no longer thriving on the thrill of adventure and no longer in survival mode. I had had enough and I just allowed myself to be miserable and pissed off all day long.

Eventually, after hearing motor boats for two days, we could see daylight ahead. The Juruá River itself was vast compared to anything we'd seen since leaving the Solimões and carved out an impressive gorge through the forest, ripping palms and hardwoods from the ground ruthlessly as it constantly altered its course.

Despite our lack of money, I asked for the best hotel in Juruá City. It wasn't luxurious by Western standards but the fact that Cho and I had a double bed each and air conditioning meant that we were in a palace after more than forty days of walking through what must have been some of the most difficult rainforest anywhere in the world.

The Juruá River marked the halfway point to Tefé. Juruá 'City' was a humid, sweaty jungle town with wood-built shops that sat perched on a rare mound of high ground overlooking the low, green sprawl of the Amazon Basin. If a man in a Stetson with low-slung sixshooters had trotted into town on a horse named Silver he would have fitted in perfectly. As long as he spoke a bit of Portuguese.

The contrast between stepping out of the famine of our expedition into the excess of civilisation was remarkable. I spotted one local girl who was not overweight but the rest of the town seemed like personifications of sloth and greed.

We indulged in both those sins. My inbuilt regulator that should have stopped me eating had broken down. I was riding a rollercoaster that was flipping me back and forth between hunger and sickening overeating. Our bodies wanted to build up some fat stores again as we ploughed through cream cakes and egg sandwiches as if we had just been let out of a concentration camp.

CROSSING THE CONTINENT, 1527–1540

The Story of the First African-American Explorer of the American South

Robert Goodwin

Esteban De Dorantes (1500–1539), also known as **Mustafa Azemmouri**, was an African slave who became America's first great explorer and adventurer. As a young man he was sold into slavery in Portuguese Morocco, before joining an expedition led by Spanish explorer Pánfilo de Narváez to establish a permanent settlement in Florida. The group was shipwrecked near contemporary Galveston, Texas and enslaved by a local Native American tribe. After escaping captivity, he and his companions travelled west across Texas and parts of Northern Mexico, making Esteban the first African to traverse the American West. In all, his extraordinary journey of survival, over eight years, brought him from the coast of Spanish Florida to Mexico City. Esteban led a final expedition into the North American interior in search of the legendary Native American city of Cibola, rumoured to contain great riches. He disappeared on this journey and his fate remains a matter of intense speculation.

Dr. Robert Goodwin was born and educated in London and was awarded his PhD by the University of London for his thesis on Golden Age Spain. He is currently a full-time writer and historian and is a Research Associate at University College London.

Cebreros and his men were dumbfounded, struck silent by such a strange and improbable meeting.

These two men were Esteban, an African slave; and Alvar Núñez

Cabeza de Vaca, a Spanish nobleman. One or two days' journey back along the trail were their two Spanish companions: Andrés Dorantes, Esteban's legal owner; and Alonso del Castillo Maldonado, a doctor's son. These four men were the only survivors of the disastrous expedition of 300 would-be conquistadors who had landed at Tampa Bay eight years before, filled with confidence that they could conquer Florida for Spain.

At this meeting with the four Spanish cavalrymen, Esteban and Cabeza de Vaca brought to a conclusion a remarkable odyssey. It is one of the most symbolic and momentous events in American history. They had completed the first crossing of North America made by non-Indians, the first crossing of North America in history.

These four strangers had walked from Texas. They had followed the Rio Grande to El Paso, rounded the Sierra Madre, and crossed through New Mexico and Arizona. From there, they descended into the Sonora valley of northwest Mexico, four godlike shamans who "came naked and barefoot" from the east. With them were a crowd of Native Americans, who flocked to these alien medicine men in search of hope in troubled times.

Cabeza de Vaca later recalled that as they had walked along the ancient trails of the southwest, they had "come across a great many diverse languages, through which God showed them his favor, for they always understood those languages and the Indians understood them. And thus, they conversed through signs so well that it was as if they spoke one another's languages." And, as they wandered among these myriad nations of many tongues, Esteban had "always spoken to the Indians, gathering all the other information they needed. He found out about the trails that they wanted to follow and what nations, tribes and settlements might be thereabouts." Esteban was always the ambassador, a spy and a scout, the advance guard, the diplomat who dealt with the Indians while the Spaniards were mostly silent.

This, the Spaniards claimed, gave them great authority and gravitas, and as a result they commanded much respect. But although their mysterious silence may have created an illusion of supernatural power, it was clearly Esteban who controlled that power by communicating with Indians.

Even as the Spaniards claimed for themselves the credit for their survival, their accounts betrayed the simple truth that Esteban had been their savior. Brave and resourceful, Esteban negotiated with their Indian hosts, always arranging the progress from one tribe to the next that led the four men across a continent and into Mexico. But Esteban needed his Spanish companions too: without them, he would have been a wandering alien; with them, he was a herald of the gods. All four men were locked together by a tight, triangular dynamic: Indian, African, and European.

Cebreros and his men remained silent for a long time. They looked at the strangers. They slowly realized who these peculiar apparitions were. They dismounted. Cebreros tenderly embraced Cabeza de Vaca, official treasurer of the long-lost Florida expedition. The little group of newly reunited Christians now made their way back to the riverside camp.

Esteban no doubt noticed the chained Indian prisoners destined for slavery. He was back among Spaniards, a slave once more.

As the twenty or so amazed Spanish cavalrymen examined the newcomers, Cabeza de Vaca and Esteban introduced themselves, still fearful of the slavers' possible response. But Alcaraz had little time to worry about how they came to be there. He was a man of action and he was in trouble. He confessed that for many days he had seen no sign of Indians, that he was lost, and that his men were going hungry.

Cabeza de Vaca was able to reassure Alcaraz that the necessary sustenance was at hand. He explained that ten leagues back along the trail Dorantes and Castillo were waiting for news. They had sent word into the surrounding lands for the Indian population to join them, and the Indians would have food.

This was welcome news to the desperate Alcaraz, who immediately ordered Esteban to lead three cavalrymen and fifty of his Mexican soldiers to search for Dorantes, Castillo, and their allies. They set out at once.

On the bewildering trails, Esteban held the upper hand. At any time he might have disappeared into the bush to escape or organize an ambush. The Spanish riders and their Mexican foot soldiers were vulnerable, exposed to attack, and blinded by the dense vegetation. But instead, Esteban chose to lead this party of potential enemies into the heart of the four survivors' massed crowd of followers. Why?

Dorantes and Castillo soon heard from their scouts that Esteban was on his way with a posse of foreigners. At first, they must have been overjoyed. They must have begun to think of home and imagine what life would be like among Spaniards once more. But they would have quickly realized that they were far from safe. They had already seen the consequences of Alcaraz's brutality and knew well that such slavers could see them as rivals and might easily murder them. They had to gain the upper hand in any confrontation, and no doubt they now prepared a warlike welcome, calculated to intimidate the Spanish slavers.

Esteban, ever the advance guard and ambassador, must have approached the camp first, arguing in favor of caution. He quickly summarized the slavers' plight and pointed out the power the four survivors had in being able to control the Indians and therefore the food supply. Esteban, it seems, had now decided to throw in his lot with the three Spanish companions who had accompanied him from the Atlantic to the South Sea and once again submit to a Spanish world in which he had always been a slave.

Meanwhile, back at the slavers' camp, Cabeza de Vaca began to tell his history. As the last living royal official of the Narváez expedition, he requested that Alcaraz give him formal testimony of the year, month, and date of their meeting. He insisted that this be accompanied by a description of the manner of their arrival. Alcaraz complied with these bureaucratic requirements and then responded in kind, demanding that Cabeza de Vaca account for his bizzare attire and how he came to be there. In due course, he would interrogate all the survivors, insisting that they explain how and why so many Indians followed them in peace and harmony.

For all that Cabeza de Vaca's storytelling had already begun, the Spanish slavers can have only half-believed that half-recounted tale. How could he claim to have traveled through so many different lands where so many different languages were spoken by so many different peoples? They must have wondered what had really happened to him, and they must have doubted the strange black African would ever return.

But five days later, Esteban, Castillo, and Dorantes appeared at the head of the peaceful Indian army of 600 men and women that they had raised. There could be little doubt that something strange and yet quite wonderful had happened. The force of Indians now amassed before the little band of Spanish slavers with their Mexican servants and their Indian slaves. It may have been an army that seemed to come in peace, but it far outnumbered them and was also clearly a horde equally ready for war. These Indians had fled their homes in the face of Alcaraz's brutality. Their revenge could prove terrible. There was no place for faint hearts, for weakness might be fatal.

Alcaraz was audacious and brave, as well as brutal, and where others might see an army, he also saw potential slaves. He asked them to send word to the people who had lived in the settlements along the riverbanks, but were now in hiding in the bush, that they should bring food to the camp.

The four survivors did as he asked and sent their messengers to take word into the hinterland. Soon, a further 600 Indians arrived—men and women, children and infants. They brought all the corn they could manage, carrying it in cooking pots, which they had used to bury their staples to hide them from the slavers.

Then, as soon as Alcaraz and his men had eaten their fill and looked after their horses, a great argument broke out.

"We quarreled bitterly with the Spaniards because they wanted to enslave the Indians we brought with us," Cabeza de Vaca claimed. "We were so angered by this that when we left we forgot to take many of our bows and arrows with us and also some bags containing five emeralds, which we therefore lost forever. We gave the Spaniards plenty of hides and other things and we worked hard to persuade the Indians they could return home safely and sow the maize crop. But they wanted to go with us until they could leave us with another tribe of Indians, which is what they usually did. They were afraid that if they went home without doing so they would die and that if they went with us instead then they had no need to fear the Christians with their weapons."

This annoyed Alcaraz, who spoke to the Indians through his interpreters. "These four men are of the same nation and race as I am," he explained; "they are Christians and Spaniards, but they have been lost for a long time and are unfortunate people of little importance. I am the master of this land and you must obey me and serve me."

"But," Cabeza de Vaca went on, "the Indians gave little or no credence to all this. Instead, they began to talk amongst themselves, saying that the Christians must be liars, because we four survivors had appeared from where the sun rises, while the slavers came from where it sets. And, while we cured the sick, they murdered the healthy. While we came naked and barefoot, they wore clothes, rode horses, and had weapons. While we coveted nothing, but instead gave away everything that they gave to us so that we were left with nothing, these others seemed to have no other purpose in life than to steal everything they could lay their hands on and never gave anything to anyone."

Cabeza de Vaca is without doubt responsible for embellishing this dialogue between Alcaraz and the Indians, polishing the rhetorical language for the sake of contrast and editing out much extraneous material. Even he admits that he had to rely on interpreters himself to understand what was going on. But the gist of that exchange is entirely plausible because it is ostensibly true.

The four survivors were almost certainly in a position to order their Indian army to confront Alcaraz's weakened force. Instead, they thanked the Indians for all that they had done and advised them to return to their former way of life, to go back to their homes and villages without fear. But according to Cabeza de Vaca, no sooner had they sent those Indians away in peace than Alcaraz ordered custody for the four survivors and sent them with Cebreros to Culiacán. There can be little doubt that the four men at least acquiesced or perhaps colluded in this decision, for their following of Indians was formidable indeed and had they chosen to engage Alcaraz, their army should surely have prevailed.

20 HOURS, 40 MINUTES

Amelia Earhart

Amelia Earhart (1897–disappeared 1937) was a pioneering aviator who, in 1932, became the first woman in history to complete a non-stop solo flight across the Atlantic Ocean. America's 'Queen of the Air' was an international celebrity, her legend secured after she disappeared in 1937 during an attempted circumnavigational flight around the globe. Despite the barriers to women entering the field of aviation, Earhart started flying lessons in her early twenties and worked multiple jobs to pay for her training and so achieve her dream. Earhart's courage and determination, along with her lifetime advocacy for women's rights, made her a feminist icon who inspired a whole generation of young women aviators, many of whom assisted the Allied war effort throughout the Second World War. The precise circumstances of her disappearance have been the subject of extensive debate, and gave rise to a myriad of conspiracy theories and speculative hypotheses regarding her fate.

"Dad, you know, I think I'd like to fly."

Heretofore we had been milling about behind the ropes which lined the field. At my suggestion we invited ourselves into the arena and looked about. I saw a man tagged "official" and asked my father to talk with him about instruction. I felt suddenly shy about making inquiries myself, lest the idea of a woman's being interested in trying to fly be too hilarious a thought for the official.

My father was game; he even went so far as to make an appointment for me to have a trial hop at what was then Rogers Airport. I am sure he thought one ride would be enough for me, and he might as well act to cure me promptly.

Next day was characteristically fair and we arrived early on the field. There was no crowd, but several planes stood ready to go.

A pilot came forward and shook hands.

"A good day to go up," he said, pleasantly.

My father raised an inexperienced eye to the sky and agreed. Agreeing verbally is as far as he went, or has ever gone, for he has not yet found a day good enough for a first flight.

The pilot nodded to another flyer. "He'll go up with us."

"Why?" I asked.

The pair exchanged grins. Then I understood. I was a girl – a "nervous lady." I might jump out. There had to be somebody on hand to grab my ankle as I went over. It was no use to explain I had seen aeroplanes before and wasn't excitable. I was not to be permitted to go alone in the front cockpit.

The familiar "contact" was spoken and the motor came to life. I suppose there must be emotion with all new experiences, but I can't remember any but a feeling of interest on this occasion. The noise of the motor seemed very loud – I think it seems so to most people on their first flight.

The plane rose quickly over some nearby oil derricks which are part of the flora in Southern California. I was surprised to be able to see the sea after a few moments of climbing. At 2,000 feet the pilot idled the motor and called out the altitude for me. The sensation of speed is of course absent, and I had no idea of the duration of the hop. When descent was made I know the field looked totally unfamiliar. I could not have picked it out from among the hundreds of little squares into which populated areas are divided. One of the senses which must be developed in flying is an acuteness in recognizing characteristics of the terrain, a sense seldom possessed by a novice.

Lessons in flying cost twice as much in 1920 as they do now. Five hundred dollars was the price for ten or twelve hours instruction, and that was just half what had been charged a few years before.

When I came down I was ready to sign up at any price to have a try at the air myself. Two things deterred me at that moment. One was the tuition fee to be wrung from my father, and the other the determination to look up a woman flyer who, I had heard, had just come to another field. I felt I should be less selfconscious taking lessons with her, than with the men who overwhelmed me with their capabilities. Neta Snook, the first woman to be graduated from the Curtiss School of Aviation, had a Canuck – an easier plane to fly than a Jenny, whose Canadian sister it was. Neta was good enough to take payments for time in the air, when I could make them, so in a few days I began hopping about on credit with her. I had failed to convince my father of the necessity of my flying, so my economic status itself remained a bit in the air

I had opportunity to get a fair amount of information about details of flying despite my erratic finances. In Northampton, where I had stayed a while after the war, I had taken a course in automobile repair with a group of girls from Smith College. To me the motor was as interesting as flying itself, and I welcomed a chance to help in the frequent pulling down and putting together which it required.

New students were instructed in planes with dual controls; the rudder and stick in the front cockpit are connected with those in the rear so that any false move the student makes can be corrected by the instructor. Every move is duplicated and can be felt by both flyers. One lands, takes off, turns, all with an experienced companion in command. When passengers are carried these controls are removed for safety's sake with little trouble. If there is a telephone connection, communication and explanation are much easier than by any methods of signs or shouting. This telephone equipment, by the way, seems to be more usual in England than here.

I am glad I didn't start flying in the days of the "grass cutters", which exemplified an earlier method of flying instruction. One of the amusing sights of the war training period was that of the novices hopping about the countryside in these penguin planes. They could fly only a few feet from the ground and had to be forced off to do that. The theory had been that such activity offered maximum practice in taking off and landing. In addition it was a sort of Roman holiday for the instructors – they had nothing much to do but, so to speak, wind up their playthings and start them off. And nothing very serious could happen one way or the other.

It was really necessary for a woman to wear breeks and leather coats in these old days of aviation. The fields were dirty and planes hard to enter. People dressed the part in a semi-military khaki outfit, and in order to be as inconspicuous as possible I fell into the same styles. A leather coat I had then, I wore across the Atlantic, eight years later.

Neta sold her plane and I bought one and changed instructors after a few hours' work. John Montijo, an ex-army instructor, took charge of me and soloed me after some strenuous times together. I refused to fly alone until I knew some stunting. It seemed foolhardy to try to go up alone without the ability to recognize and recover quickly from any position the plane might assume, a reaction only possible with practice. In short, to become thoroughly at home in the air, stunting is as necessary as, and comparable to, the ability to drive an automobile in traffic. I was then introduced to aerobatics and felt not a bit afraid when sent "upstairs" alone for the first time.

Usually a student takes off nonchalantly enough but doesn't dare land until his gas supply fails. Any field is familiar with the sight of beginners circling about overhead, staying up solely because they can't bear to come down. The thought of landing without their instructors to help them, if need be, becomes torture, which is only terminated by the force of gravity.

In soloing – as in other activities – it is far easier to start something than it is to finish it. Almost every beginner hops off with a whoop of joy, though he is likely to end his flight with something akin to D. T.'s.

I reversed the process. In taking off for the first time alone, one of the shock absorbers broke, causing the left wing to sag just as I was leaving the ground. I didn't know just what had happened, but I did know something was wrong and wondered what I had done. The mental agony of starting the plane had just been gone through and I was suddenly faced with the agony of stopping it. It was all in a matter of seconds, of course, and somehow I contrived to do the proper thing. My brief "penguin" flight came to a prompt conclusion without further mishap.

When the damage had been repaired, I took courage to try again, this time climbing about 5,000 feet, playing around a little, and returning to make a thoroughly rotten landing. At once I had my picture taken by a gentleman from Iowa who happened to be touring California and wanted a few rare sights for the album back home.

THE BUDDHIST ON DEATH ROW

David Sheff

Jarvis Jay Masters (born 1962) is known for his transformation into one of America's best known Buddhist practitioners while locked in a cell on Death Row. He grew up in California in a home filled with drugs, alcohol and violence. Masters was sent to foster care at the age of five but fell into a life of crime and was imprisoned for armed robbery at age nineteen. In 1990, he was convicted of murdering a prison officer – a charge he denies – sentenced to death and placed in solitary confinement for twenty-one years. His appeal case continues. *The Buddhist on Death Row* is written by the best-selling author **David Sheff.** While not a Buddhist himself, Sheff said he learned many profound lessons from spending time with Masters.

Jarvis tried to follow the lama's instructions, but his despair only worsened over the following months. His friends and lawyers visited and tried to bolster him. The lawyers said that the trial had been a travesty, and they assured him he'd win on appeal. Kelly Hayden, a legal assistant who'd become a friend, visited and commiserated with him. She believed that his conviction and sentence were racist, and she said so. She said, "Don't take it personally." They exchanged horrified looks and burst out laughing. It was a brief moment of levity.

Those first days, Jarvis obsessed about the death sentence and became preoccupied with the message inherent in the judge's words before she'd condemned him: "If people don't want children, they shouldn't have them. Apparently, his mother didn't know how not to have them." He turned those words over in his mind: *If my mother shouldn't have had me, I should never have been born; the world would have been a better place without me.*

Those words affirmed his worst feelings about himself, a message reinforced since he was a child. The judge had seen into his soul. He had been born useless. Those who saw him as evil were right.

Carlette came, but there was little either of them could say. She sobbed and left.

He tried to reread *Life in Relation to Death*, but he couldn't bear it.

In a letter to Chagdud Tulku, Jarvis admitted that he was falling into "the darkest place" and didn't know if he'd ever be able to pull himself out. The lama had said that he and Lisa were there for him if he needed help. Jarvis needed help badly now, and he asked for it.

In her response, Lisa suggested that they talk in person, and he readily agreed. A month later, when her application was approved, she came to San Quentin.

Jarvis had expected a Tibetan like the lama in the photograph, but Lisa looked like a flower child or gypsy. She wasn't a cold and detached Buddhist scholar; she was open and kind and funny.

After some small talk, Lisa said that she and Rinpoche understood his despair and the difficulty of meditating in that state. "Yes, it's hard, but it can save you. Meditation is hardest when we're most afraid, because it forces us to face our fears when all we want to do is run from them. But it's the only way out of our misery.

"It's hard to see that where you are," she acknowledged, "but it's like walking from one mountain to another. If you think about how far you have to go, you'll freeze up and never take the first step. Just take the step."

"I would," Jarvis said. "I can't. I try."

He broached something that had been gnawing at him: "In these books I see pictures of Buddhists sitting in robes on mountain peaks, chanting in these gardens with white flowers and blue skies. Maybe that works there, but how in the world am I going to sit in this hellhole praying to some stone fat man? I live with rapists and killers. Everyone talks about enlightenment, living in the light. But I live in hell."

Lisa responded with a parable about a woman named Kisa Gotami who had lived at the time of the Buddha. Her son died, and she was overcome with grief. Carrying his lifeless body, she set off in search of the Buddha. After many days, she found him and pled with him to bring her child back to life. The Buddha said he'd make a medicine that would revive him, but it required a special ingredient, mustard seeds that came from a home that hadn't been touched by suffering. He sent Kisa Gotami to find some.

She went from village to village and house to house, knocking on door after door. People pitied her and were happy to help, but the seeds they offered were useless, because every person she met had suffered.

She went to more villages and visited more homes, but none had escaped suffering. She was desperate when she reached the thousandth door. She knocked, and a woman answered. Once again she begged for mustard seeds. The woman had some, but then Kisa Gotami asked if she'd experienced suffering in her life. The woman looked up at her. Her life had been filled with suffering.

Kisa Gotami wept. She wasn't crying for herself but for everyone she'd met. She understood at last what the Buddha had wanted her to see, that no one escapes suffering and no one escapes death. She had experienced what she needed to in order to get past her grief. She felt compassion for others. In Lisa's telling, the instant Kisa Gotami felt that, she grasped the universality of suffering and the impermanence of life. She understood that her son had joined the vast pool of souls who have lived and died. She understood that in her suffering she was like all humans. She accepted her son's death, and she was freed from her pain. She became awakened and attained the state of enlightenment as a person who grasped the true nature of existence.

Jarvis was quiet. He'd heard and read other Buddhist stories, but this one touched him differently for some reason.

A moment later, a guard rapped on the door, ending the visit.

That night, Jarvis lay on his cot listening to the prison's unceasing noise: sobs like crying wind, coughing, hacking, and the footsteps of guards. After having been unable to meditate for months, he moved to the floor, crossed his legs, and sat erect. He inhaled as deeply as he ever had.

In a dreamlike state, he saw a man sitting in meditation, his body engulfed in flames. He focused on the meditator and recognized him. The man was himself. As if he were watching through a telephoto lens, he panned back from inside his cell to the tier. Somehow he saw inside the other cells, each containing a man who was also on fire. He zoomed out more and saw San Quentin from the sky. From that vantage, he saw several thousand burning bodies. Still higher, from the clouds, he saw houses across the Bay Area burning. Then he saw California, which was also engulfed in flames. Higher. The country. Then the continent, and then the Western Hemisphere. Next he was watching from space. From that height he saw the whole planet floating in blackness. The water was blue. The landmasses were brown and green. On those expanses, wherever there were humans, fires burned.

Jarvis returned to the prison, to the thousands of men in cages unfit for animals. He thought of Kisa Gotami and realized that suffering was all around him—everywhere humans were. When he opened his eyes, he was shaking, and tears were streaming down his cheeks.

A BLACK EXPLORER AT THE NORTH POLE

Matthew Henson

Matthew Henson (1866–1955) was an African-American explorer who, in 1909, planted the American flag at what he believed to be the North Pole. This was one of seven voyages to the Arctic on which he accompanied American explorer Robert Peary. Born to black sharecroppers in the American South in the immediate aftermath of the Civil War, Henson's family were the target of attacks by the Ku Klux Klan and other white supremacist groups. Henson overcame a variety of obstacles and proved himself a capable sailor and explorer, forming a fast friendship with Robert Peary. In 1937, he became the first African-American to be admitted into the prestigious Explorer's Club.

Captain Bartlett and his two boys had commenced their return journey, and the main column, depleted to its final strength, started northward. We were six: Peary, the commander, the Esquimos, Ootah, Egingwah, Seegloo and Ooqueah, and myself.

Day and night were the same. My thoughts were on the going and getting forward, and on nothing else. The wind was from the southeast, and seemed to push us on, and the sun was at our backs, a ball of livid fire, rolling his way above the horizon in never-ending day.

The Captain had gone, Commander Peary and I were alone (save for the four Esquimos), the same as we had been so often in the past years, and as we looked at each other we realized our position and we knew without speaking that the time had come for us to demonstrate that we were the men who, it had been ordained, should unlock the door which held the mystery of the Arctic. Without an instant's hesitation, the order to push on was given, and we started off in the trail made by the Captain to cover the Farthest North he had made and to push on over one hundred and thirty miles to our final destination.

The Captain had had rough going, but, owing to the fact that his trail was our track for a short time, and that we came to good going shortly after leaving his turning point, we made excellent distance without any trouble, and only stopped when we came to a lead barely frozen over, a full twenty-five miles beyond. We camped and waited for the strong southeast wind to force the sides of the lead together. The Esquimos had eaten a meal of stewed dog, cooked over a fire of wood from a discarded sledge, and, owing to their wonderful powers of recuperation, were in good condition; Commander Peary and myself, rested and invigorated by our thirty hours in the last camp, waiting for the return and departure of Captain Bartlett, were also in fine fettle, and accordingly the accomplishment of twenty-five miles of northward progress was not exceptional. With my proven ability in gauging distances, Commander Peary was ready to take the reckoning as I made it and he did not resort to solar observations until we were within a hand's grasp of the Pole.

The memory of those last five marches, from the Farthest North of Captain Bartlett to the arrival of our party at the Pole, is a memory of toil, fatigue, and exhaustion, but we were urged on and encouraged by our relentless commander, who was himself being scourged by the final lashings of the dominating influence that had controlled his life. From the land to 87° 48' north, Commander Peary had had the best of the going, for he had brought up the rear and had utilized the trail made by the preceding parties, and thus he had kept himself in the best of condition for the time when he made the spurt that brought him to the end of the race. From 87° 48' north, he kept in the lead and did his work in such a way as to convince me that he was still as good a man as he had ever been. We marched and marched, falling down in our tracks repeatedly, until it was impossible to go on. We were forced to camp, in spite of the impatience of the Commander, who found himself unable to rest, and who only waited long enough for us to relax into sound sleep, when he would wake us up and start us off again. I do not believe that he slept for one hour from April 2 until after he had loaded us up and ordered us to go back over our old trail, and I often think that from the instant when the order to return was given until the land was again sighted, he was in a continual daze.

Onward we forced our weary way. Commander Peary took his sights from the time our chronometer-watches gave, and I, knowing that we had kept on going in practically a straight line, was sure that we had more than covered the necessary distance to insure our arrival at the top of the earth.

It was during the march of the 3d of April that I endured an instant of hideous horror. We were crossing a lane of moving ice. Commander Peary was in the lead setting the pace, and a half hour later the four boys and myself followed in single file. They had all gone before, and I was standing and pushing at the upstanders of my sledge, when the block of ice I was using as a support slipped from underneath my feet, and before I knew it the sledge was out of my grasp, and I was floundering in the water of the lead. I did the best I could. I tore my hood from off my head and struggled frantically. My hands were gloved and I could not take hold of the ice, but before I could give the "Grand Hailing Sigh of Distress," faithful old Ootah had grabbed me by the nape of the neck, the same as he would have grabbed a dog, and with one hand he pulled me out of the water, and with the other hurried the team across.

He had saved my life, but I did not tell him so, for such occurrences are taken as part of the day's work, and the sledge he safeguarded was of much more importance, for it held, as part of its load, the Commander's sextant, the mercury, and the coils of piano-wire that were the essential portion of the scientific part of the expedition. My kamiks (boots of sealskin) were stripped off, and the congealed water was beaten out of my bearskin trousers, and with a dry pair of kamiks, we hurried on to overtake the column. When we caught up, we found the boys gathered around the Commander, doing their best to relieve him of his discomfort, for he had fallen into the water also, and while he was not complaining, I was sure that his bath had not been any more voluntary than mine had been.

When we halted on April 6, 1909, and started to build the igloos, the dogs and sledges having been secured, I noticed Commander Peary at work unloading his sledge and unpacking several bundles of equipment. He pulled out from under his *kooletah* (thick, fur outer-garment) a small folded package and unfolded it. I recognized his old silk flag, and realized that this was to be a camp of importance. Our different camps had been known as Camp Number One, Number Two, etc., but after the turning back of Captain Bartlett, the camps had been given names such as Camp Nansen, Camp Cagni, etc., and I asked what the name of this camp was to be—"Camp Peary"? "This, my boy, is to be Camp Morris K. Jesup, the last and most northerly camp on the earth." He fastened the flag to a staff and planted it firmly on the top of his igloo. For a few minutes it hung limp and lifeless in the dead calm of the haze, and then a slight breeze, increasing in strength, caused the folds to straighten out, and soon it was rippling out in sparkling color. The stars and stripes were "nailed to the Pole."

A thrill of patriotism ran through me and I raised my voice to cheer the starry emblem of my native land. The Esquimos gathered around and, taking the time from Commander Peary, three hearty cheers rang out on the still, frosty air, our dumb dogs looking on in puzzled surprise. As prospects for getting a sight of the sun were not good, we turned in and slept, leaving the flag proudly floating above us.

This was a thin silk flag that Commander Peary had carried on all of his Arctic journeys, and he had always flown it at his last camps. It was as glorious and as inspiring a banner as any battle-scarred, blood-stained standard of the world—and this badge of honor and courage was also blood-stained and battle-scarred, for at several places there were blank squares marking the spots where pieces had been cut out at each of the "Farthests" of its brave bearer, and left with the records in the cairns, as mute but eloquent witnesses of his achievements. At the North Pole a diagonal strip running from the upper left to the lower right corner was cut and this precious strip, together with a brief record, was placed in an empty tin, sealed up and buried in the ice, as a record for all time.

Commander Peary also had another American flag, sewn on a white ground, and it was the emblem of the "Daughters of the Revolution Peace Society"; he also had and flew the emblem of the Navy League, and the emblems of a couple of college fraternities of which he was a member.

It was about ten or ten-thirty A. M., on the 7th of April, 1909, that the Commander gave the order to build a snow-shield to protect him from the flying drift of the surface-snow. I knew that he was about to take an observation, and while we worked I was nervously apprehensive, for I felt that the end of our journey had come. When we handed him the pan of mercury the hour was within a very few minutes of noon. Laying flat on his stomach, he took the elevation and made the notes on a piece of tissue-paper at his head. With sun-blinded eyes, he snapped shut the *vernier* (a graduated scale that subdivides the smallest divisions on the sector of the circular scale of the sextant) and with the resolute squaring of his jaws, I was sure that he was satisfied, and I was confident that the journey had ended. Feeling that the time had come, I ungloved my right hand and went forward to congratulate him on the success of our eighteen years of effort, but a gust of wind blew something into his eye, or else the burning pain caused by his prolonged look at the reflection of the limb of the sun forced him to turn aside; and with both hands covering his eyes, he gave us orders to not let him sleep for more than four hours, for six hours later he purposed to take another sight about four miles beyond, and that he wanted at least two hours to make the trip and get everything in readiness.

I unloaded a sledge, and reloaded it with a couple of skins, the instruments, and a cooker with enough alcohol and food for one meal for three, and then I turned in to the igloo where my boys were already sound asleep. The thermometer registered 29° below zero. I fell into a dreamless sleep and slept for about a minute, so I thought, when I was awakened by the clatter and noise made by the return of Peary and his boys.

The Commander gave the word, "We will plant the stars and stripes—*at the North Pole!*" and it was done; on the peak of a huge paleocrystic floe-berg the glorious banner was unfurled to the breeze, and as it snapped and crackled with the wind, I felt a savage joy and exultation. Another world's accomplishment was done and finished, and as in the past, from the beginning of history, wherever the world's work was done by a white man, he had been accompanied by a colored man. From the building of the pyramids and the journey to the Cross, to the discovery of the new world and the discovery of the North Pole, the Negro had been the faithful and constant companion of the Caucasian, and I felt all that it was possible for me to feel, that it was I, a lowly member of my race, who had been chosen by fate to represent it, at this, almost the last of the world's great *work*.

The four Esquimos who stood with Commander Peary at the North Pole, were the brothers, Ootah and Egingwah, the old campaigner, Seegloo, and the sturdy, boyish Ooqueah. Four devoted companions, blindly confident in the leader, they worked only that he might succeed and for the promise of reward that had been made before they had left the ship, which promise they were sure would be kept. Together with the faithful dogs, these men had insured the success of the master. They had all of the characteristics of the dogs, including the dogs' fidelity. Within their breasts lingered the same infatuations that Commander Peary seemed to inspire in all who were with him, and though frequently complaining and constantly requiring to be urged to do their utmost, they worked faithfully and willingly. Ootah, of my party, was the oldest, a married man, of about thirty-four years, and regarded as the best all around member of the tribe, a great hunter, a kind father, and a good provider. Owing to his strong character and the fact that he was more easily managed by me than by any of the others, he had been a member of my party from the time we left the ship. Without exaggeration, I can say that we had both saved each other's lives more than once, but it had all gone in as part of the day's work, and neither of us dwelt on our obligations to the other.

My other boy, Ooqueah, was a young man of about nineteen or twenty, very sturdy and stocky of build, and with an open, honest countenance, a smile that was "child-like and bland," and a character that *was* child-like and bland. It was alleged that the efforts of young Ooqueah were spurred on by the shafts of love, and that it was in the hopes of winning the hand of the demure Miss Anadore, the charming daughter of Ikwah, the first Esquimo of Commander Peary's acquaintance, that he worked so valiantly. His efforts were of an ardent character, but it was not due to the ardor of love, as far as I could see, but to his desire to please and his anxiety to win the promised rewards that would raise him to the grade of a millionaire, according to Esquimo standards.

Commander Peary's boy, Egingwah, was the brother of my boy Ootah, also married and of good report in his community, and it was he who drove the Morris K. Jesup sledge.

If there was any sentiment among the Esquimos in regard to the success of the venture, Ootah and Seegloo by their unswerving loyalty and fidelity expressed it. They had been members of the "Farthest North party" in 1906, the party that was almost lost beyond and in the "Big Lead," and only reached the land again in a state of almost complete collapse. They were the ones who, on bidding Commander Peary farewell in 1906, when he was returning, a saddened and discouraged man, told him to be of good cheer and that when he came back again Ootah and Seegloo would go along, and stay until Commander Peary had succeeded, and they did. The cowardice of their fellow Esquimos at the "Big Lead" on this journey did not in the least demoralize them, and when they were absolutely alone on the trail, with every chance to turn back and return to comfort, wife, and family, they remained steadfast and true, and ever northward guided their sledges.

THE NORTH POLE

Robert Peary

Robert Peary (1856–1920) was an American naval officer and explorer who directed several expeditions to the Arctic regions, including one in 1908–9 that claimed to reach the North Pole. Seven of these expeditions he carried out alongside his friend and fellow explorer Matthew Henson. The claim that his expedition reached the North Pole has been disputed, but was accepted by the National Geographic Society at the time of the voyage. It has since been cast into doubt by a more recent inquest into the travel logs of the expedition.

Only One Day From the Pole

With every passing day even the Eskimos were becoming more eager and interested, notwithstanding the fatigue of the long marches. As we stopped to make camp, they would climb to some pinnacle of ice and strain their eyes to the north, wondering if the Pole was in sight, for they were now certain that we should get there this time.

We slept only a few hours the next night, hitting the trail again a little before midnight between the 3d and 4th of April. The weather and the going were even better than the day before. The surface of the ice, except as interrupted by infrequent pressure ridges, was as level as the glacial fringe from Hecla to Cape Columbia, and harder. I rejoiced at the thought that if the weather held good I should be able to get in my five marches before noon of the 6th.

Again we traveled for ten hours straight ahead, the dogs often on the trot and occasionally on the run, and in those ten hours we reeled off at least twenty-five miles. I had a slight accident that day, a sledge runner having passed over the side of my right foot as I stumbled while running beside a team; but the hurt was not severe enough to keep me from traveling.

Near the end of the day we crossed a lead about one hundred yards wide, on young ice so thin that, as I ran ahead to guide the dogs, I was obliged to slide my feet and travel wide, bear style, in order to distribute my weight, while the men let the sledges and dogs come over by themselves, gliding across where they could. The last two men came over on all fours.

I watched them from the other side with my heart in my mouth—watched the ice bending under the weight of the sledges and the men. As one of the sledges neared the north side, a runner cut clear through the ice, and I expected every moment that the whole thing, dogs and all, would go through the ice and down to the bottom. But it did not.

This dash reminded me of that day, nearly three years before, when in order to save our lives we had taken desperate chances in recrossing the "Big Lead" on ice similar to this—ice that buckled under us and through which my toe cut several times as I slid my long snowshoes over it. A man who should wait for the ice to be really safe would stand small chance of getting far in these latitudes. Traveling on the polar ice, one takes all kinds of chances. Often a man has the choice between the possibility of drowning by going on or starving to death by standing still, and challenges fate with the briefer and less painful chance.

That night we were all pretty tired, but satisfied with our progress so far. We were almost inside of the 89th parallel, and I wrote in my diary: "Give me three more days of this weather!" The temperature at the beginning of the march had been minus 40°. That night I put all the poorest dogs in one team and began to eliminate and feed them to the others, as it became necessary.

We stopped for only a short sleep, and early in the evening of the same day, the 4th, we struck on again. The temperature was then minus 35°, the going was the same, but the sledges always haul more easily when the temperature rises, and the dogs were on the trot much of the time. Toward the end of the march we came upon a lead running north and south, and as the young ice was thick enough to support the teams, we traveled on it for two hours, the dogs galloping along and reeling off the miles in a way that delighted my heart. The light air which had blown from the south during the first few hours of the march veered to the east and grew keener as the hours wore on.

I had not dared to hope for such progress as we were making. Still the biting cold would have been impossible to face by anyone not fortified by an inflexible purpose. The bitter wind burned our faces so that they cracked, and long after we got into camp each day they pained us so that we could hardly go to sleep. The Eskimos complained much, and at every camp fixed their fur clothing about their faces, waists, knees, and wrists. They also complained of their noses, which I had never known them to do before. The air was as keen and bitter as frozen steel.

At the next camp I had another of the dogs killed. It was now exactly six weeks since we left the *Roosevelt*, and I felt as if the goal were in sight. I intended the next day, weather and ice permitting, to make a long march, "boil the kettle" midway, and then go on again without sleep, trying to make up the five miles which we had lost on the 3d of April.

During the daily march my mind and body were too busy with the problem of covering as many miles of distance as possible to permit me to enjoy the beauty of the frozen wilderness through which we tramped. But at the end of the day's march, while the igloos were being built, I usually had a few minutes in which to look about me and to realize the picturesqueness of our situation—we, the only living things in a trackless, colorless, inhospitable desert of ice. Nothing but the hostile ice, and far more hostile icy water, lay between our remote place on the world's map and the utmost tips of the lands of Mother Earth.

I knew of course that there was always a *possibility* that we might still end our lives up there, and that our conquest of the unknown spaces and silences of the polar void might remain forever unknown to the world which we had left behind. But it was hard to realize this. That hope which is said to spring eternal in the human breast always buoyed me up with the belief that, as a matter of course, we should be able to return along the white road by which we had come.

Sometimes I would climb to the top of a pinnacle of ice to the north of our camp and strain my eyes into the whiteness which lay beyond, trying to imagine myself already at the Pole. We had come so far, and the capricious ice had placed so few obstructions in our path, that now I dared to loose my fancy, to entertain the image which my will had heretofore forbidden to my imagination—the image of ourselves at the goal.

We had been very fortunate with the leads so far, but I was in constant and increasing dread lest we should encounter an impassable one toward the very end. With every successive march, my fear of such impassable leads had increased. At every pressure ridge I found myself hurrying breathlessly forward, fearing there might be a lead just beyond it, and when I arrived at the summit I would catch my breath with relief—only to find myself hurrying on in the same way at the next ridge.

At our camp on the 5th of April I gave the party a little more sleep than at the previous ones, as we were all pretty well played out and in need of rest. I took a latitude sight, and this indicated our position to be 89° 25′, or thirty-five miles from the Pole; but I determined to make the next camp in time for a noon observation, if the sun should be visible.

Before midnight on the 5th we were again on the trail. The weather was overcast, and there was the same gray and shadowless light as on the march after Marvin had turned back. The sky was a colorless pall gradually deepening to almost black at the horizon, and the ice was a ghastly and chalky white, like that of the Greenland ice-cap—just the colors which an imaginative artist would paint as a polar ice-scape. How different it seemed from the glittering fields, canopied with blue and lit by the sun and full moon, over which we had been traveling for the last four days.

The going was even better than before. There was hardly any snow on the hard granular surface of the old floes, and the sapphire blue lakes were larger than ever. The temperature had risen to minus 15°, which, reducing the friction of the sledges, gave the dogs the appearance of having caught the high spirits of the party. Some of them even tossed their heads and barked and yelped as they traveled.

Notwithstanding the grayness of the day, and the melancholy aspect of the surrounding world, by some strange shift of feeling the fear of the leads had fallen from me completely. I now felt that success was certain, and, notwithstanding the physical exhaustion of the forced marches of the last five days, I went tirelessly on and on, the Eskimos following almost automatically, though I knew that they must feel the weariness which my excited brain made me incapable of feeling.

When we had covered, as I estimated, a good fifteen miles, we halted, made tea, ate lunch, and rested the dogs. Then we went on for another estimated fifteen miles. In twelve hours' actual traveling time we made thirty miles. Many laymen have wondered why we were able to travel faster after the sending back of each of the supporting parties, especially after the last one. To any man experienced in the handling of troops this will need no explanation. The larger the party and the greater the number of sledges, the greater is the chance of breakages or delay for one reason or another. A large party cannot be forced as rapidly as a small party.

Take a regiment, for instance. The regiment could not make as good an average daily march for a number of forced marches as could a picked company of that regiment. The picked company could not make as good an average march for a number of forced marches as could a picked file of men from that particular company; and this file could not make the same average for a certain number of forced marches that the fastest traveler in the whole regiment could make.

So that, with my party reduced to five picked men, every man, dog, and sledge under my individual eye, myself in the lead, and all recognizing that the moment had now come to let ourselves out for all there was in us, we naturally bettered our previous speed.

When Bartlett left us the sledges had been practically rebuilt, all the best dogs were in our pack, and we all understood that we must attain our object and get back as quickly as we possibly could. The weather was in our favor. The average march for the whole journey from the land to the Pole was over fifteen miles. We had repeatedly made marches of twenty miles. Our average for five marches from the point where the last supporting party turned back was about twenty-six miles.

We Reach the Pole

The last march northward ended at ten o'clock on the forenoon of April 6. I had now made the five marches planned from the point at which Bartlett turned back, and my reckoning showed that we were in the immediate neighborhood of the goal of all our striving. After the usual arrangements for going into camp, at approximate local noon, of the Columbia meridian, I made the first observation at our polar camp. It indicated our position as 89° 57′.

We were now at the end of the last long march of the upward journey. Yet with the Pole actually in sight I was too weary to take the last few steps. The accumulated weariness of all those days and nights of forced marches and insufficient sleep, constant peril and anxiety, seemed to roll across me all at once. I was actually too exhausted to realize at the moment that my life's purpose had been achieved. As soon as our igloos had been completed and we had eaten our dinner and double-rationed the dogs, I turned in for a few hours of absolutely necessary sleep, Henson and the Eskimos having unloaded the sledges and got them in readiness for such repairs as were necessary. But, weary though I was, I could not sleep long. It was, therefore, only a few hours later when I woke. The first thing I did after awaking was to write these words in my diary: "The Pole at last. The prize of three centuries. My dream and goal for twenty years. Mine at last! I cannot bring myself to realize it. It seems all so simple and commonplace."

Everything was in readiness for an observation at 6 P.M., Columbia meridian time, in case the sky should be clear, but at that hour it was, unfortunately, still overcast. But as there were indications that it would clear before long, two of the Eskimos and myself made ready a light sledge carrying only the instruments, a tin of pemmican, and one or two skins; and drawn by a double team of dogs, we pushed on an estimated distance of ten miles. While we traveled, the sky cleared, and at the end of the journey, I was able to get a satisfactory series of observations at Columbia meridian midnight. These observations indicated that our position was then beyond the Pole.

Nearly everything in the circumstances which then surrounded us seemed too strange to be thoroughly realized; but one of the strangest of those circumstances seemed to me to be the fact that, in a march of only a few hours, I had passed from the western to the eastern hemisphere and had verified my position at the summit of the world. It was hard to realize that, in the first miles of this brief march, we had been traveling due north, while, on the last few miles of the same march, we had been traveling south, although we had all the time been traveling precisely in the same direction. It would be difficult to imagine a better illustration of the fact that most things are relative. Again, please consider the uncommon circumstance that, in order to return to our camp, it now became necessary to turn and go north again for a few miles and then to go directly south, all the time traveling in the same direction.

As we passed back along that trail which none had ever seen before or would ever see again, certain reflections intruded themselves which, I think, may fairly be called unique. East, west, and north had disappeared for us. Only one direction remained and that was south. Every breeze which could possibly blow upon us, no matter from what point of the horizon, must be a south wind. Where we were, one day and one night constituted a year, a hundred such days and nights constituted a century. Had we stood in that spot during the six months of the arctic winter night, we should have seen every star of the northern hemisphere circling the sky at the same distance from the horizon, with Polaris (the North Star) practically in the zenith.

All during our march back to camp the sun was swinging around in its ever-moving circle. At six o'clock on the morning of April 7, having again arrived at Camp Jesup, I took another series of observations. These indicated our position as being four or five miles from the Pole, towards Bering Strait. Therefore, with a double team of dogs and a light sledge, I traveled directly towards the sun an estimated distance of eight miles. Again I returned to the camp in time for a final and completely satisfactory series of observations on April 7 at noon, Columbia meridian time. These observations gave results essentially the same as those made at the same spot twenty-four hours before.

I had now taken in all thirteen single, or six and one-half double, altitudes of the sun, at two different stations, in three different directions, at four different times. All were under satisfactory conditions, except for the first single altitude on the sixth. The temperature during these observations had been from minus 11° Fahrenheit to minus 30° Fahrenheit, with clear sky and calm weather (except as already noted for the single observation on the sixth). I give here a facsimile of a typical set of these observations. (See the two following pages.)

In traversing the ice in these various directions as I had done, I had allowed approximately ten miles for possible errors in my observations, and at some moment during these marches and countermarches, I had passed over or very near the point where north and south and east and west blend into one.

Of course there were some more or less informal ceremonies connected with our arrival at our difficult destination, but they were not of a very elaborate character. We planted five flags at the top of the world. The first one was a silk American flag which Mrs. Peary gave me fifteen years ago. That flag has done more traveling in high latitudes than any other ever made. I carried it wrapped about my body on every one of my expeditions northward after it came into my possession, and I left a fragment of it at each of my successive "farthest norths": Cape Morris K. Jesup, the northernmost point of land in the known world; Cape Thomas Hubbard, the northernmost known point of Jesup Land, west of Grant Land; Cape Columbia, the northernmost point of North American lands; and my farthest north in 1906, latitude 87° 6′ in the ice of the polar sea. By the time it actually reached the Pole, therefore, it was somewhat worn and discolored.

A broad diagonal section of this ensign would now mark the farthest goal of earth—the place where I and my dusky companions stood.

It was also considered appropriate to raise the colors of the Delta Kappa Epsilon fraternity, in which I was initiated a member while an undergraduate student at Bowdoin College, the "World's Ensign of Liberty and Peace," with its red, white, and blue in a field of white, the Navy League flag, and the Red Cross flag.

After I had planted the American flag in the ice, I told Henson to time the Eskimos for three rousing cheers, which they gave with the greatest enthusiasm. Thereupon, I shook hands with each member of the party—surely a sufficiently unceremonious affair to meet with the approval of the most democratic. The Eskimos were childishly delighted with our success. While, of course, they did not realize its importance fully, or its world-wide significance, they did understand that it meant the final achievement of a task upon which they had seen me engaged for many years.

Then, in a space between the ice blocks of a pressure ridge, I deposited a glass bottle containing a diagonal strip of my flag and records of which the following is a copy:

90 N. Lat., North Pole,
April 6, 1909.

Arrived here to-day, 27 marches from C. Columbia.

I have with me 5 men, Matthew Henson, colored, Ootah, Egingwah, Seegloo, and Ookeah, Eskimos; 5 sledges and 38 dogs. My ship, the S. S. *Roosevelt*, is in winter quarters at C. Sheridan, 90 miles east of Columbia.

The expedition under my command which has succeeded in reaching the Pole is under the auspices of the Peary Arctic Club of New York City, and has been fitted out and sent north by the members and friends of the club for the purpose of securing this geographical prize, if possible, for the honor and prestige of the United States of America.

The officers of the club are Thomas H. Hubbard, of New York, President; Zenas Crane, of Mass., Vice-president; Herbert L. Bridgman, of New York, Secretary and Treasurer.

I start back for Cape Columbia to-morrow.

ROBERT E. PEARY,
United States Navy.

90 N. LAT., NORTH POLE,
April 6, 1909.

I have to-day hoisted the national ensign of the United States of America at this place, which my observations indicate to be the North Polar axis of the earth, and have formally taken possession of the entire region, and adjacent, for and in the name of the President of the United States of America.

I leave this record and United States flag in possession.

ROBERT E. PEARY,
United States Navy.

If it were possible for a man to arrive at 90° north latitude without being utterly exhausted, body and brain, he would doubtless enjoy a series of unique sensations and reflections. But the attainment of the Pole was the culmination of days and weeks of forced marches, physical discomfort, insufficient sleep, and racking anxiety. It is a wise provision of nature that the human consciousness can grasp only such degree of intense feeling as the brain can endure, and the grim guardians of earth's remotest spot will accept no man as guest until he has been tried and tested by the severest ordeal.

Perhaps it ought not to have been so, but when I knew for a certainty that we had reached the goal, there was not a thing in the world I wanted but sleep. But after I had a few hours of it, there succeeded a condition of mental exaltation which made further rest impossible. For more than a score of years that point on the earth's surface had been the object of my every effort. To its attainment my whole being, physical, mental, and moral, had been dedicated. Many times my own life and the lives of those with me had been risked. My own material and forces and those of my friends had been devoted to this object. This journey was my eighth into the arctic wilderness. In that wilderness I had spent nearly twelve years out of the twenty-three between my thirtieth and my fifty-third year, and the intervening time spent in civilized communities during that period had been mainly occupied with preparations for returning to the wilderness. The determination to reach the Pole had become so much a part of my being that, strange as it may seem, I long ago ceased to think of myself save as an instrument for the attainment of that end. To the layman this may seem strange, but an inventor can understand it, or an artist, or anyone who has devoted himself for years upon years to the service of an idea.

But though my mind was busy at intervals during those thirty hours spent at the Pole with the exhilarating thought that my dream had come true, there was one recollection of other times that, now and then, intruded itself with startling distinctness. It was the recollection of a day three years before, April 21, 1906, when after making a fight with ice, open water, and storms, the expedition which I commanded had been forced to turn back from 87° 6′ north latitude because our supply of food would carry us no further. And the contrast between the terrible depression of that day and the exaltation of the present moment was not the least pleasant feature of our brief stay at the Pole. During the dark moments of that return journey in 1906, I had told myself that I was only one in a long list of arctic explorers, dating back through the centuries, all the way from Henry Hudson to the Duke of the Abruzzi, and including Franklin, Kane, and Melville—a long list of valiant men who had striven and failed. I told myself that I had only succeeded, at the price of the best years of my life, in adding a few links to the chain that led from the parallels of civilization towards the polar center, but that, after all, at the end the only word I had to write was failure.

But now, while quartering the ice in various directions from our camp, I tried to realize that, after twenty-three years of struggles and discouragement, I had at last succeeded in placing the flag of my country at the goal of the world's desire. It is not easy to write about such a thing, but I knew that we were going back to civilization with the last of the great adventure stories—a story the world had been waiting to hear for nearly four hundred years, a story which was to be told at last under the folds of the Stars and Stripes, the flag that during a lonely and isolated life had come to be for me the symbol of home and everything I loved—and might never see again.

The thirty hours at the Pole, what with my marchings and countermarchings, together with the observations and records, were pretty well crowded. I found time, however, to write to Mrs. Peary on a United States postal card which I had found on the ship during the winter. It had been my custom at various important stages of the journey northward to write such a note in order that, if anything serious happened to me, these brief communications might ultimately reach her at the hands of survivors. This was the card, which later reached Mrs. Peary at Sydney:—

"90 North Latitude, April 7th.

"*My dear Jo*,

"I have won out at last. Have been here a day. I start for home and you in an hour. Love to the "kidsies."

"Bert."

In the afternoon of the 7th, after flying our flags and taking our photographs, we went into our igloos and tried to sleep a little, before starting south again.

I could not sleep and my two Eskimos, Seegloo and Egingwah, who occupied the igloo with me, seemed equally restless. They turned from side to side, and when they were quiet I could tell from their uneven breathing that they were not asleep. Though they had not been specially excited the day before when I told them that we had reached the goal, yet they also seemed to be under the same exhilarating influence which made sleep impossible for me.

Finally I rose, and telling my men and the three men in the other igloo, who were equally wakeful, that we would try to make our last camp, some thirty miles to the south, before we slept, I gave orders to hitch up the dogs and be off. It seemed unwise to waste such perfect traveling weather in tossing about on the sleeping platforms of our igloos.

Neither Henson nor the Eskimos required any urging to take to the trail again. They were naturally anxious to get back to the land as soon as possible—now that our work was done. And about four o'clock on the afternoon of the 7th of April we turned our backs upon the camp at the North Pole.

Though intensely conscious of what I was leaving, I did not wait for any lingering farewell of my life's goal. The event of human beings standing at the hitherto inaccessible summit of the earth was accomplished, and my work now lay to the south, where four hundred and thirteen nautical miles of ice-floes and possibly open leads still lay between us and the north coast of Grant Land. One backward glance I gave—then turned my face toward the south and toward the future.

PROMETHEUS

Lord Byron

Lord Byron (1788–1824) is considered one of the greatest English poets and was a leading figure of the Romantic movement. A Scottish peer, he is particularly famous for the brooding hero of *Childe Harold's Pilgrimage* and his more satirical work *Don Juan*, as well as his many love affairs. Byron travelled extensively throughout Europe and, in 1816, left England permanently for a life in Switzerland, Italy and Greece. He was fighting in the Greek War of Independence when he died of a fever aged 36, and he is still considered a hero in Greece today. This poem follows the Greek myth of the Titan Prometheus, who steals fire from the Gods on Olympus and gives it to humanity. For this act, Zeus sentences Prometheus to eternal torment – to be tied to a rock whilst his liver is pecked out by vultures, only for it to regrow and be eaten again the next day – but in Prometheus Byron salutes the nobility of his sacrifice and rebellion against tyranny.

Titan! to whose immortal eyes
        The sufferings of mortality,
        Seen in their sad reality,
Were not as things that gods despise;
What was thy pity's recompense?
A silent suffering, and intense;
The rock, the vulture, and the chain,
All that the proud can feel of pain,
The agony they do not show,
The suffocating sense of woe,
        Which speaks but in its loneliness,
And then is jealous lest the sky
Should have a listener, nor will sigh
        Until its voice is echoless.

Titan! to thee the strife was given
 Between the suffering and the will,
 Which torture where they cannot kill;
And the inexorable Heaven,
And the deaf tyranny of Fate,
The ruling principle of Hate,
Which for its pleasure doth create
The things it may annihilate,
Refus'd thee even the boon to die:
The wretched gift Eternity
Was thine—and thou hast borne it well.
All that the Thunderer wrung from thee
Was but the menace which flung back
On him the torments of thy rack;
The fate thou didst so well foresee,
But would not to appease him tell;
And in thy Silence was his Sentence,
And in his Soul a vain repentance,
And evil dread so ill dissembled,
That in his hand the lightnings trembled.

Thy Godlike crime was to be kind,
 To render with thy precepts less
 The sum of human wretchedness,
And strengthen Man with his own mind;
But baffled as thou wert from high,
Still in thy patient energy,
In the endurance, and repulse
 Of thine impenetrable Spirit,
Which Earth and Heaven could not convulse,
 A mighty lesson we inherit:
Thou art a symbol and a sign
 To Mortals of their fate and force;
Like thee, Man is in part divine,
 A troubled stream from a pure source;
And Man in portions can foresee
His own funereal destiny;
His wretchedness, and his resistance,
And his sad unallied existence:

To which his Spirit may oppose
Itself—and equal to all woes,
        And a firm will, and a deep sense,
Which even in torture can descry
        Its own concenter'd recompense,
Triumphant where it dares defy,
And making Death a Victory.

THE CRUEL WAY

Ella Maillart

Ella Maillart (1903–1997) was a Swiss photographer and adventurer. An avid sailor and skier, she represented Switzerland in the Olympics in 1924 and 1931. Maillart travelled extensively throughout the predominately Muslim countries of the newly established Union of Soviet Socialist Republics in the 1930s. She wrote about her travels through China under Japanese occupation and voyages by train, automobile, horseback, camelback and foot through the Himalayan and steppe regions of Central Asia. In *The Cruel Way,* she describes her journey from Geneva to Kabul in a Ford car with her friend Annemarie Schwarzenbach, whom she calls 'Christina'. Maillart details the dangers the pair navigated alongside her struggles to help Christina with her battle with drug addiction.

THE first mistake had been our reaching Sofia too late. At the border-post I had caused an exasperating delay. Of the six visas we needed, I had forgotten the Bulgarian one. When at last that little matter was settled, we found the main road under repair and had to crawl at a snail's pace along a track in a small valley. There, against the red of the earth and the green of the great tobacco leaves, the black aprons of the women looked surprisingly severe.

When we reached the beautiful plain of Sofia the lingering snowfields of Mount Vitosha were already powdered pink by the declining sun. Tacitly we went on: with no tent or meal to think of, we could continue till we found a hotel.

Christina was so tired that she went to bed as soon as we arrived. I took the car to a garage, driving for the first time through the maze-like lanes of a capital where crowds stroll in the middle of the roads without fear or remorse. The success of my performance excited me so much that it wiped away my tiredness. Back at the hotel I found Christina dressed: she had rung up Marjorie and soon a taxi took us to Boulevard Osvoboditel.

In the wake of Marjorie, we lunched and dined out: I came across smart Silianoff, the only Bulgar I knew. Gloomily we spoke of the League of Nations at Geneva where we had last met. We chatted about our friends who had quitted the big building before it sank—cheerful Hilary St. George Saunders the gourmet, Clarence Streit the elegant Utopian carrying about the manuscript of his *Union Now* which was to cause a great stir in the States. Meanwhile, in a nearby public garden, I was watching a group of school-children gathered around their national flag: they were taught patriotism and that their country had been freed from the Turks seventy years ago. There was a general prayer at the end of which the children looked as if they were catching flies on their noses; but no: they were hurriedly crossing themselves before the lowering of the flag at sunset.

Next day Christina was seedy. She vomited and slept throughout the afternoon. Even when she opened her eyes, looked at me from the depth of her helplessness and said she felt like death, I failed to grasp what was happening. Her words rang so true that I should have been frightened had I not decided to believe that all was due to her extreme lassitude.

But the evidence flashed when I saw on the floor of the bathroom the brittle glass of an empty ampoule. She had succumbed once more. She had disregarded our compact; she had done what she pretended to abhor. My presence, my confidence in her, the fear of displeasing me had had no effect.

What did it mean? What were we to do?

Which was the true Christina? Grey like wood-pulp, her face looked dead, the eyelids brown, the lips bluish, the stiff chin frozen by an inner contraction. That could not be the true Christina. She was lost in a world to which I had no access. I did not know what to say: my words might hurt or exasperate. And indeed I had nothing to say!

Leaving Sofia behind, we drove in silence. Having put us on the right road, Marjorie had wished us "bon voyage" with all the power of her kind heart. Last evening, delicately, she had tried to help Christina by mentioning a deep experience of her own. One day when was utterly drowned in woe she had the revolution of how little she mattered: since then, it was impossible for her to dramatize her difficulties.

In silence we followed the "Valley of Roses": attar is made with the flowers of every crop. In the villages, the men wore small turbans and drank coffee out of tiny Turkish cups; the women were wrapped in black shawls: and old 'uns sitting on a bench held up their distaffs like so many queer weapons.

At the edge of a wide field we bought a small crate of strawberries for ninepence. We heard that German students had volunteered to help the gatherers. They told them that but for Hitler who bought their eight thousand tons of strawberries as well as their tobacco and essence of flowers, there would be all-round starvation. Some of the women pickers wore ample musulman breeches which we photographed perfunctorily. Back in the car, an invisible heaviness seemed to rest on our shoulders.

She drove by my side with her usual nonchalance. Silence was tangible, a viscosity for which I had no solvent. How soon would she cease to identify herself with that living corpse, how soon begin to speak about what mattered?

She was probably suffering agonies ... Even so I should be firm and treat her like a man, showing no emotion or tender weakness that she might use towards her ruin. I wondered if those who helped her before had not been too fond of her, too worried by her misery, too ready to let her have her own way?

I hoped to succeed where they had failed because I was different from them, not loving her as they did: this was probably why I had a slight hold on her. In our ordinary life I sometimes felt so distant from her that I could not bring myself to respond to her "thou". Such reserve is most unusual with me.

Nevertheless, I think I loved her profoundly. I loved the generous courage she showed in her campaigning against injustice, the honesty with which she judged herself, the way she bore her solitariness, her conviction that love is a mystery we have to penetrate. But a jealous fatality made her strangle what she embraced: she asked too much from love.

I had been moved to see that in her distress last winter it was to me she wrote from Neuchâtel: "This nursing-home drives me crazy. I should like to talk with you if you could spare the time." I felt strong, coming down from the mountains where I had "managed" our ski-ing team. We met at the station where she came in her car. She was harassed by her fight with the doctors: they would not understand that writing was life and food to her, that the regenerative cure of enforced rest applied to dyspeptics and hysterics could not suit her. Around us, all was white, covered by new snow. It was exhilarating to walk near the lake through the crisp, frosty, sparkling air. A mass of thick long bristles, Palazi, my Eskimo dog, was playing at dragging Christina along the deserted road. The strong animal had crossed Greenland as the leader of Robert's pack. One look at the sincerity of that kingly creature was enough to cheer you, gave the proof that goodness and nobility exist. In the presence of Christina also, narrow or mean thoughts were banished. The gist of our dialogue had been that if she was mad I was mad too: I was unwilling to let myself be strangled by that prudent life that everybody advocated. I also was convinced that—whether we succeed or not—it is our job to search for the significance of life.

That night outside Sofia, by a small stream under a big elm, she spoke at last, sullenly:

"If you decide to keep me with you I suppose I shall have to submit to a constant control. You won't trust me any more?"

"I will do it if you want. I have not ceased to hope: I am sure that when you know yourself better this misery of your own making will vanish like a harmless mist." I wanted to say much more but I knew she was not willing to listen.

Next day she looked like an ordinary convalescent.

South of Philippopoli the country turned out to be miserable: bare, with nothing but a few huts built haphazardly. Desultorily we switched on our wireless. A suggestive *czarda* submerged us in its passionate waves. And I entered the no man's land of reverie. This was my fifth journey to the East. I knew the way as far as Herat in Afghanistan, so till then I could not be entranced by the feeling of discovery. Besides, the chief object of this journey was to rescue Christina from a negative atmosphere, to find conditions such that she would of her own accord live normally. I felt sure I could reach the Hackins if we followed the road I knew: but it barred any escapade into difficult (and really interesting!) country. Briefly stated, my main aims were to acquire self-mastery and to save my friend from herself.

The second aim depended upon the first. Only clear knowledge of myself would allow me to help Christina in the fundamental problem she was raising. Of course that mastery of myself should bring me nearer to reality, and ever since I began roughing it among sailors and nomads I was in search of a "real" life. Towards achieving my aim I could so far only visualise a whitewashed room in a Pamirian village where I might learn to think: this training was to take place in the morning before I went abroad in search of heads for my calipers!

Lucien Fabre had said that an Islamic country could not be helpful. But I did not want help. Away from a shaky and feverish Europe I simply wanted to look within. My search for an Edenic mountain tribe was merely the excuse for breaking away from the helplessness prevalent in Europe. I knew that self-knowledge can be acquired wherever one is but I was too weak or too silly to insulate myself from the revolt, panic, militarism and febrile planning that was sweeping Europe off her feet. Distance would help, I felt sure. In the West, anyway, everyone seemed as lost as I was: why not try the East?

As for my second aim, if at the start there was an elan of unselfishness in my desire to help Christina, it was mixed by now with pride and vanity: I, Kini, could not bear to be beaten—could not, therefore, fail. But my doubts of success were as sincere, now, as the confidence I had felt in London when I picked up the gauntlet thrown by Irene.

I repeated to Christina the sombre predictions I had listened to then. But to counterbalance them I added that Blaise Cendarars had felt her sterling quality and was impatient to meet her again. On her side, Christina had been advised not to travel with me because I was an immoral cynic: had I not written in a book that I started for Central Asia with salvarsan in my luggage?—proof that I had made up my mind to live licentiously. Christina had calmly answered: "Kini had a great desire to live with nomads. They are known to be plagued with venereal diseases. Therefore she bought a remedy in case she should be contaminated through drinking from their filthy vessels."

A pungent smell of burning alarmed us. The car was stopped and we searched for the causes. Approaching Christina I found it was due to her first Bulgarian cigarette, just lighted!

We were advancing through the plain of the Maritza, wide, treeless, dotted with storks patiently standing in the high grass. With their oppressed skulls, flattened horns, tired and swollen eyes, shiny skins, wallowing for protection in the mud under the culverts, buffaloes were typical of hot mosquito-lands.

The car reached the barrier at the Turkish border. In spite of our explanations, the clerk in the office got muddled. He had taken down that Madame Silvaplana *née* Francis, was Legation Secretary. "No, that is her husband's profession."

"Then what shall I call her? Professor?" At that moment, discovering that he was handling a diplomatic passport, the man became afraid: rubber stamps one after another thudded on our documents. Hurrying as he did, the Turk probably forgot to warn us that we were entering a military zone.

The map told us we were nearing Edirne (Adrianople) and soon we reached its vegetable gardens. But we were not prepared for what we saw when we rounded a small hill. Lit by the horizontal rays of the evening sun, a vision from the *Thousand and One Nights* floated above the lilac mist. Many minarets arose, slender and precious in the level golden light; incredibly tall beside huge round domes powdered with purple shadows, they looked like the masts of a marmoreal ship sailing through iridescent skies. Fading out, the glorious vision left only a few outlines, livid after that shining splendour.

Curious to see how the morning sun would play on such a setting, we pitched camp on the spot. But so many children pestered us that we packed up again, all the time answering that we were not German: "*Aleman iok*—no, no!"—a denial we were to repeat day after day from then on.

Having traversed the town and gained the country, we looked for a by-lane where we might spend the night. But the east side of Adrianople was deserted: abandoned steles showed that it was a huge cemetery, sure sign that the town had once been a Muslim capital. Giving up our search, we rigged the tent in the waste-land. (That night, the ground as hard as rock, Christina decided to buy a circular air-cushion for her hip!)

Daylight showed the town far behind. We wanted to reach Istanbul that Saturday evening before closing-time. So from our camp, we took leave of its monuments. One of them, built by the great Sinan in the sixteenth century, is said to be his masterpiece—the mosque of Sultan Selim II, as big as Santa Sophia.

We were right to have started early: the foundations of the "international road" were being laid and in three atrocious hours we covered only eighteen miles. In deep ruts made by high-bodied lorries, the Ford was either stuck or grating against stones. On reaching the good road at Chorlu, in the enthusiasm of relief we filmed our first stork's nest topping a minaret.

Handling a camera is a risky business in Turkey and the danger increases as you approach Persia, Russia or Japan. I knew this very well but had never expected that our first arrest would be so early. Our storks were unluckily within twenty yards of a Poste de Gendarmerie. We very soon found ourselves sitting inside: it was one of those old wooden houses I was so keen to visit!

I could not think how to propitiate the Inspector who sat behind our five cameras. Their number looked most suspicious, they were not sealed and we had used them in military territory. "Why were we feigning ignorance? Would we give an account of every hour we had spent in Turkey? Where had we slept last night? Why did not our passports bear the obligatory stamp of the hotel? Slept on the road outside Adrianople! You want me to believe that? I shall release you only when your films have been developed."

It took a long time to grasp that: the Inspector spoke nothing but Turkish and in that tongue we could only say *Aleman iok*. Unhappily the country swarmed with German spies saying all the same thing. The village photographer was sent for and a schoolmistress who spoke French. In the presence of authority, they shook with fear. Christina's patience was exhausted: she was thinking what kind of threat would work on the Inspector. She made me tremble; I used all my power of persuasion to explain to her that in newly reformed countries it pays to pretend one is much impressed by officials.

With thick fingers tanned by acids the photographer was in vain trying to open our small-film cameras. At last he said he could not handle or develop such material. To our immense surprise, we were at once released. The Inspector, no doubt being human, was hungry: we had spent more than four hours in his office.

Fields of yellow wheat waved as we speeded onward, while the motionless dark-blue line of the Sea of Marmora barred the south.

At sunset we approached another capital. Dark foliage covered an undulating strip of land. Above it, plunging and rising in noble style, looking like a Wall of China astray among clusters of trees, towered the grand walls of Constantinople. In 1453 the last emperor of Byzantium fought and died on this wall, defeated by Muhammad II. Powerful enough to bring a world to its end, the world of the Holy Roman Empire whose magnificence had endured for a thousand years, this terrible Turk was only twenty-one when he led his one hundred and sixty thousand men into battle. Before the final attack he said to them: "There are three conditions for winning a war—to want victory, to be ashamed of defeat and to obey your chief."

By the Gate of Adrianople that Saturday evening there were only a few men wearing worker's caps, bored, taking an airing.

Somewhere in the huge town spread at our feet there would have to be an explanation between Christina and myself.

ON SLEDGE AND HORSEBACK TO OUTCAST SIBERIAN LEPERS

Kate Marsden

Kate Marsden (1859–1931) was a British nurse and missionary who, in 1890, travelled to Siberia in search of a cure for leprosy. Her journeys never resulted in a cure, but she founded a hospital for lepers in Siberia some years later and recalled her travails in her memoir *On Sledge and Horseback to Outcast Siberian Lepers*. While she achieved acclaim in England, aspersions were cast on the authenticity of her stories, perhaps out of spite, given rumours that Marsden was a lesbian. She founded a charity in support of lepers that exists to this day.

More bogs and marshes for several miles; and then I grew so sleepy and sick that I begged for rest, not withstanding our position on semi-marshy ground, which had not as yet dried from the heat of the summer sun. I was asleep in five minutes, lying on the damp ground with only a fan to shelter me from the sun.

On again for a few more miles; but I began to feel the effects of this sort of travelling—in a word, I felt utterly worn out. It was as much as I could do to hold on to the horse, and I nearly tumbled off several times in the effort. The cramp in my body and lower limbs was indescribable, and I had to discard the cushion under me, because it became soaked through and through with the rain, and rode on the broad, bare, wooden saddle. What feelings of relief arose when the time of rest came, and the pitching of tents, and the brewing of tea! Often I slept quite soundly till morning, awaking to find that the mosquitoes had been hard at work in my slumbers, in spite of veil and gloves, leaving great itching lumps, that turned me sick. Once we saw two calves that had died from exhaustion from the bites of these pests, and the white hair of our poor horses was generally covered with clots of blood, due partly to mosquitoes and partly to prodigious horseflies. But those lepers—they suffered far more than I suffered, and that was the one thought, added to the strength which God supplied, that kept me from collapsing entirely.

Sometimes we rested all day, and travelled at night to avoid the intense heat. We passed forests, where hundreds of trees had fallen. According to popular superstition, the Yakut witches quarrelled, and met in the forest to fight out the dispute. But the spirit of the forest became so angry at this conduct that he let loose a band of inferior spirits; and then, in a moment, a tempest began and rushed through the forest, tearing up the trees and causing them to fall in the direction of those disputants who were in the right. But the true meaning of those fallen trees is yet more interesting and singular than the superstitious one. Underneath the upper soil of these forests combustion goes on, beginning in the winter. The thaw of summer and the deluge of rain seem to have little effect upon the fire, for it still works its way unsubdued. When the tempest comes the trees drop by hundreds, having but slight power of resistance. I brought home with me some of this burnt earth, intending to send it to the British Museum, should no specimen be already there.

My second thunderstorm was far worse than the first. The forest seemed on fire, and the rain dashed in our faces with almost blinding force. My horse plunged and reared, flew first to one side, and then to the other, dragging me amongst bushes and trees, so that I was in danger of being caught by the branches and hurled to the ground. After this storm one of the horses, carrying stores and other things, sank into a bog nearly to its neck; and the help of all the men was required to get it out. Those stores! I may say now, that long before the journey was accomplished they were nearly all spoilt or gone, thus adding to our already accumulating difficulties too numerous to particularise. Hard bread became flour, and we had to make a kind of cold pudding with it.

Soon after the storm we were camping and drinking tea, when I noticed that all the men were eagerly talking together and gesticulating. I asked the tchinovnick what it all meant, and was told that a large bear was supposed to be in the neighbourhood, according to a report from a post-station close at hand. There was a general priming of fire-arms, except in my case, for I did not know how to use my revolver, so thought I had better pass it on to some one else, lest I might shoot a man in mistake for a bear. We mounted again and went on. The usual chattering this time was exchanged for a dead silence, this being our first bear experience; but we grew wiser as we proceeded, and substituted noise for silence. We hurried on, as fast as possible, to get through the miles of forests and bogs. I found it best not to look about me, because, when I did so, every large stump of a fallen tree took the shape of a bear. When my horse stumbled over the roots of a tree, or shied at some object unseen by me, my heart began to gallop.

A WOMAN IN ARABIA

Gertrude Bell, edited by Georgina Howell

Gertrude Bell (1868–1926) was a British archaeologist, explorer and travel writer. Along with T. E. Lawrence, Bell was instrumental in supporting the Hashemite Kingdoms that shaped the Middle East in the vacuum created by the fall of the Ottoman Empire. Her knowledge of local customs and culture was significant in the British diplomatic and war effort during and immediately after the First World War, and won her more affection from the local populations than was common among British officials. Her warnings leading up to the creation of the state of Iraq were poignant, given the next century of conflict emerging from misguided European policies in the region.

Georgina Howell (1942–2016) wrote the acclamed biography *Gertrude Bell: Queen of the Desert, Shaper of Nations* and was a journalist for *Vanity Fair*, *Vogue* and the *Sunday Times*.

On the eve of his departure to Albania, Gertrude wrote to Doughty-Wylie of her plans to undertake an epic journey in one of the most dangerous parts of the desert. Dick was as worried as she could wish. He wrote back, "I am nervous about you ... south of Maan and from there to Hayil is surely a colossal trek. For your palaces your road your Baghdad your Persia I do not feel so nervous – but Hayil from Maan – Inshallah!"

Her destination would be Hayyil, the almost mythical city described by Charles M. Doughty, Dick's uncle, in his famous *Arabia Deserta*, the book that had accompanied her on all expeditions. She proposed to travel sixteen hundred miles by camel, taking a circular route south from Damascus to central Arabia, then east across the interior and the shifting sands of the Nefud, becoming the first Westerner to cross that angle of the desert. She would make her way to the Misma Mountains, a coal-black landscape with flint pinnacles as high as ten-story buildings. She would then descend into the plateau of granite and basalt at the heart of which the snow-white medieval city of Hayyil floated like a mirage. Her return journey would be north to Baghdad and west across the vast Syrian Desert, back to Damascus. Much of the journey would be through unmapped territory and areas where her caravan was likely to come under tribal attack.

It was as daunting a prospect politically as geographically. Britain was supplying arms and money to the chieftain Ibn Saud, allied to the puritan Wahhabi sect of Islam. The Ottoman government supported the opposing dynasty of Ibn Rashid of the Shammar federation, perhaps the cruelest, most violent tribe of Arabia, centered on Hayyil. The trip would allow her to provide the Foreign Office with detailed new information at a critical moment, with both sides poised to strike to take control of the Arabian Peninsula.

She had already warned the British government that Ibn Saud was better as a friend than as an enemy. She set out with the ultimate intention of reaching Ibn Saud and making contact with him in his stronghold of Riyadh.

Among the distinguished men who warned her against the journey were Sir Louis Mallet, future ambassador to Constantinople; her old friend David Hogarth; and the Indian government's resident in the Persian Gulf, Lieutenant Colonel Percy Cox, a name that would come to mean much to her later. Defiantly, Gertrude decided that she would go nevertheless.

Her first weeks in Damascus were taken up with organizing the most elaborate caravan she had ever undertaken, hiring her crew, buying gifts for sheikhs in the bazaars, and choosing seventeen camels. She wired her father for an extra £400 ($55,000 RPI adjusted), a not inconsiderable sum. She also visited the Rashid's sinister agent in the city, to whom she paid £200 ($27,500), getting in return a promissory note that she intended to cash in Hayyil to fund her return journey.

She had started keeping parallel diary entries. The first would be a cursory memorandum written daily while the memory was fresh. Reading these factual, ill-organized jottings, full of Arabic words and phrases, gives a vivid picture of Gertrude, tired and dirty after a day's march, her hair falling out of its pins, scribbling away at her folding desk while Fattuh put up her bedroom tent, unpacked, and arranged her possessions. These notes contained positions of water holes and Turkish barracks, routes through unmapped areas and other information that she would pass on to the Foreign Office. They were also the raw material for her upbeat letters home.

The second diary, with entries written a few days apart, was a thoughtful and polished account of her journey and feelings, kept solely for Dick – with the proviso that his wife might read them – and portraying her as a shade less robust in her attitude to dangers and setbacks. She bundled up these diary entries and sent them to Dick when she arrived at an outpost or town big enough to have a post office. As she now had to avoid the Turkish soldiers who were looking out for her, she often had to carry her papers with her for weeks until she could send them.

She traveled on through all kinds of danger and difficulty but once again fell under the spell of the desert, terrifying and beautiful, with its roaring silence and jeweled nights.

The caravan left Damascus on December 16, 1913. The journey was marred by torrential rain and bitter winds. Not a week later came a tribal attack in which shots were fired, and they were nearly robbed of all their rifles and possessions. Not long afterward, her camp was invaded by Turkish soldiers who demanded permits she did not have. They arrested her faithful servant, Fattuh, and her guide, Muhammad al-Ma'rawi. She managed to persuade the local governor, the qaimaqam, to get them released but was ordered to telegraph for permits to travel before the caravan could leave. Unfortunately, the British ambassador in Constantinople, Sir Louis Mallet, was the very man who had warned her against making the journey. He told her that His Majesty's Government would disclaim all responsibility for her if she went farther. It was no more than she had expected, and she wrote in her diary that night: "Decided to run away." Before she left she was obliged to write a letter absolving the Turks from responsibility for her welfare.

The following extracts are taken from Gertrude's diary, which was written expressly for Dick. Since he spoke Arabic, Gertrude did not bother to translate every word.

January 16, 1914

I have cut the thread.... Louis Mallet has informed me that if I go on towards Nejd my own government washes its hands of me, and I have given a categorical acquittal to the Ottoman Government, saying that I go on at my own risk. ... We turn towards Nejd, inshallah, renounced by all the powers that be, and the only thread which is not cut though is that which runs through this little book, which is the diary of my way kept for you.

I am an outlaw!

February 11, 1914

Yesterday ... we began to see landmarks; but the country through which we rode was very barren. In the afternoon we came to a big valley, the Wadi Niyyal, with good herbage for the camels and there we camped. And just at sunset the full moon rose in glory and we had the two fold splendour of heaven to comfort us for the niggardliness of the earth. She was indeed niggardly this morning. We rode for 4 hours over a barren pebbly flat entirely devoid of all herbage. They call such regions *jellad*. In front of us were the first great sand hills of the Nefud [al-Nafud]. And turning a little to the west we came down into a wide bleak *khabra* wherein we found water pools under low heaps of sand. The place looked so unpromising that I was prepared to find the water exhausted which would have meant a further westerly march to a well some hours away and far from our true road. We watered our camels and filled the water skins in half an hour and turned east into the Nefud. We have come so far south (the khabra was but a day's journey from Taimah) in order to avoid the wild sand mountains (*tu'us* they are called in Arabic) of the heart of the Nefud and our way lies now within its southern border. This great region of sand is not desert. It is full of herbage of every kind, at this time of year springing into green, a paradise for the tribes that camp in it and for our own camels. We marched through it for an hour or two and camped in deep pale gold sand with abundance of pasturage all about us, through the beneficence of God. We carry water for 3 days and then drink at the wells of Haizan [Bir Hayzan]. The *Amir*, it seems is not at Hayyil, but camping to the north with his camel herds. I fear this may be tiresome for me; I would rather have dealt with him than with his *wakil*. Also report says that he informed all men of my corning but whether to forward me or to stop me I do not know. Neither do I know whether the report is true.

February 13, 1914

We have marched for 2 days in the Nefud, and are still camping within its sands. It is very slow going, up and down in deep soft sand, but I have liked it; the plants are interesting and the sand hills are interesting. The wind driving through it hollows out profound cavities, *ga'r* they are called. You come suddenly to the brink and look down over an almost precipitous wall of sand. And from time to time there rises over the *ga'r* a head of pale driven sand, crested like a snow ridge and devoid of vegetation. These are the *tu'us*. At midday yesterday we came to a very high *ta's* up which I struggled – it is no small labour – and saw from the top the first of the Nejd mountains, Iman, and to the W. the hills above Tairnah and all round me a wilderness of sandbanks and *ti'as*. When I came down I learnt that one of my camels had been seized with a malady and had sat down some 10 minutes away. Muhammad and the negro boy, Fellah, and I went back to see what could be done for her but when we reached her we found her in the death throes. "She is gone" said Muhammad. "Shall we sacrifice her?" "It were best" said I. He drew his knife out, "*Bismillah allaha akbar*!" and cut her throat...

We have a wonderfully peaceful camp tonight in a great horseshoe of sand, with steep banks enclosing us. It is cloudy and mild – last night it froze like the devil – and I feel as if I had been born and bred in the Nefud and had known no other world. Is there any other?

We came yesterday to a well, one of the rare wells on the edge of the Nefud, and I rode down to see the watering. Haizan is a profound depression surrounded by steep sand hills and the well itself is very deep – our well rope was 48 paces long. They say it is a work of the... first forefathers, and certainly no Beduin of today would cut down into the rock and build the dry walling of the upper parts – but who can tell how old it is? There are no certain traces of age, only sand and the deep well hole. We found a number of Arabs watering their camels, the 'Anazeh clan of the Awaji who were camped near us. The men worked half naked with the passionate energy which the Arabs will put into their job for an hour or two – no more. I watched and photographed and they left me unmolested, though none had seen a European of any kind before. One or two protested at first against the photography, but the Shammar with me reassured them and I went on in peace. We go two days more through the Nefud because it is said to be the safest road and I am filled with a desire not to be stopped now, so near Hayyil. My bearings are onto Jebel Misma, which is but a few day's journey from Ibn al Rashid. I want to bring this adventure to a prosperous conclusion since we have come so far *salinum* – in the security of God.

February 16, 1914

I am suffering from a severe fit of depression today – will it be any good if I put it into words, or shall I be more depressed than ever afterwards? It springs, the depression, from a profound doubt as to whether the adventure is after all worth the candle. Not because of the danger – I don't mind that; but I am beginning to wonder what profit I shall get out of it all. A compass traverse over country which was more or less known, a few names added to the map – names of stony mountains and barren plains and of a couple of deep desert wells (for we have been watering at another today) – and probably that is all. I don't know what *tete* [offer] the Rashid people will make to me when I arrive, and even if they were inspired by the best will in the world, I doubt whether they could do more than give me a free passage to Baghdad, for their power is not so great nowadays as it once was. And the road to Baghdad has been travelled many times before. It is nothing, the journey to Nejd, so far as any real advantage goes, or any real addition to knowledge, but I am beginning to see pretty clearly that it is all that I can do. There are two ways of profitable travel in Arabia. One is the *Arabia Deserta* way, to live with the people and to live like them for months and years. You can learn something thereby, as he [Charles Doughty] did; though you may not be able to tell it again as he could. It's clear I can't take that way; the fact of being a woman bars me from it. And the other is Leachman's way – to ride swiftly through the country with your compass in your hand, for the map's sake and for nothing else. And there is some profit in that too. I might be able to do that over a limited space of time, but I am not sure. Anyway it is not what I am doing now. The net result is that I think I should be more usefully employed in more civilized countries where I know what to look for and how to record it. Here, if there is anything to record the probability is that you can't find it or reach it, because a hostile tribe bars your way, or the road is waterless, or something of that kind, and that which has chanced to lie upon my path for the last 10 days is not worth mentioning – two wells, as I said before, and really I can think of nothing else. So you see the cause of my depression. I fear when I come to the end I shall not look back and say: That was worth doing; but more likely when I look back I shall say: It was a waste of time .... That's my thought tonight, and I fear it is perilously near the truth. I almost wish that something would happen – something exciting, a raid, or a battle! And yet that's not my job either. What do ineffective archaeologists want with battles? They would only serve to pass the time and leave as little profit as before. There is such a long way between me and letters, or between me and anything and I don't feel at all like the daughter of kings, which I am supposed here to be. It's a bore being a woman when you are in Arabia.

February 17, 1914

We were held up today by rain. It began, most annoyingly, just after we had struck camp – at least I don't know that it was so very annoying, for we put in a couple of hours' march. But the custom of the country was too strong for me. You do not march in the rain. It was, I must admit, torrential. It came sweeping upon us from behind and passing on blotted out the landscape in front, till my *rafiq* said that he should lose his way, there were no landmarks to be seen. "No Arabs move camp today" said he "they fear to be lost in the Nefud." And as he trudged on through the wet sand, his cotton clothes clinging to his drenched body, he rejoiced and gave thanks for the rain. "Please God it goes over all the world" he said and "The camels will pasture here for 3 months time." The clouds lifted a little but when a second flood overtook us I gave way. We pitched the men's tent and lighted a great fire at which we dried ourselves – I was wet too. In a moment's sunshine we pitched the other tents, and then came thunder and hail and rain so heavy that the pools stood twinkling in the thirsty sand. I sat in my tent and read *Hamlet* from beginning to end and, as I read, the world swung back into focus. Princes and powers of Arabia stepped down into their true place and there rose up above them the human soul conscious and answerable to itself, made with such large discourse, looking before and after –. Before sunset I stood on the top of the sand hills and saw the wings of the rain sweeping round 'Irnan and leaving Misma' light-bathed – Then the hurrying clouds marched over the sand and once more we were wrapped in rain. No fear now of drought ahead of us.

February 20, 1914

God is merciful and we have done with the Nefud. The day after the rain – oh but the wet sand smelt good and there was a twittering of small birds to gladden the heart! – we came in the afternoon to some tents of the Shammar and pitching our camp not far off we were visited by the old shaikh, Mhailam, who brought us a goat and some butter. Him we induced to come with us as *rafiq*. He is old and lean, gray haired and toothless, and ragged beyond belief; he has not even an '*agal* to bind the kerchief on his head and we have given him a piece of rope. But he is an excellent *rafiq* – I have not had a better. He knows the country and he is anxious to serve us well. And next day we rode over sand to the northern point of Jebel Misma'. Then Mhailam importuned me to camp saying there was no pasturage in the *jellad*, the flat plain below; and Muhammad al-Ma'rawi backed him for he feared that we might fall in with Hetaim raiders if we left the Nefod. But I held firm. Raiders and hunger were as nothing to the possibility of a hard straight road. For you understand that travelling in the Nefud is like travelling in the Labyrinth. You are forever skirting round a deep horseshoe pit of sand, perhaps half a mile wide, and climbing up the opposite slope, and skirting round the next horseshoe. If we made a mile an hour as the crow flies we did well. Even after I had delivered the ultimatum, my two old parties were constantly heading off to the Nefud and I had to keep a watchful eye on them and herd them back every half hour. It was bitter cold; the temperature had fallen to 27° [-2.8°C] in the night and there was a tempestuous north wind. And so we came to the last sand crest and I looked down between the black rocks of Misma' and saw Nejd. It was a landscape terrifying in its desolation. Misma' drops to the east in precipices of sandstone, weathered to a rusty black; at its feet are gathered endless companies of sandstone pinnacles, black too, shouldering one over the other. They look like the skeleton of a vast city planted on a sandstone and sand-strewn floor. And beyond and beyond more pallid lifeless plain and more great crags of sandstone mountains rising abruptly out of it. Over it all the bitter wind whipped the cloud shadows. "*Subhan Allah!*" said one of my Damascenes, "we have come to Jehannum." Down into it we went and camped on the skirts of the Nefud with a sufficiency of pasturage. And today the sun shone and the world smiled and we marched off gaily and found the floor of Hell to be a very pleasant place after all. For the rain has filled all the sandstone hollows with clear water, and the pasturage is abundant, and the going, over the flat rocky floor, is all the heart could desire. In the afternoon we passed between the rocks of Jebel Habran, marching over a sandy floor with black pinnacled precipices on either hand, and camped on the east, in a bay of rock with *khabras* of rain water below and pasturage all round us in the sand. We have for neighbours about a mile away a small *ferij* of Shammar tents, and lest there should be anyone so evil minded as to dream of stealing a camel from us, Mhailam has just now stepped out into the night and shouted: "Ho! Anyone who watches! come in to supper!... Let anyone who is hungry come and eat!" And having thus invited the universe to our bowl, we sleep, I trust, in peace.

February 24, 1914

We are within sight of Hayyil and I might have ridden in today but I thought it better to announce my auspicious coming! So I sent in two men early this morning. Muhammad and 'Ali, and have myself camped a couple of hours outside. We had ... a most delicious camp in the top of a mountain, Jebel Rakham. I climbed the rocks and found flowers in the crevices – not a great bounty, but in this barren land a feast to the eyes.... Yesterday we passed by two more villages and in one there were plum trees flowering – oh the gracious sight! And today we have come through the wild granite crags of Jebel 'Ajja and are camped in the Hayyil plain. From a little rock above my tent I have spied out the land and seen the towers and gardens of Hayyil, and Swaifly lying in the plain beyond, and all is made memorable by *Arabia Deserta*. I feel as if I were on a sort of pilgrimage, visiting sacred sites. And the more I see of this land the more I realize what an achievement that journey was. But isn't it amazing that we should have walked down into Nejd with as much ease as if we had been strolling along Piccadilly!

March 2, 1914

What did I tell you as to the quality most needed for travel among the Arabs? Patience if you remember; that is what one needs. Now listen to the tale of the week we have spent here. I was received with the utmost courtesy. Their slaves, *'abds*, slave is too servile and yet that is what they are – came riding out to meet me and assured me that Ibrahim, the *Amir's wakil* was much gratified by my visit. We rode round the walls of the town and entered in by the south gate – the walls are of quite recent construction, towered, all round the town – and there, just within the gate I was lodged in a spacious house which Muhammad ibn Rashid had built for his summer dwelling. My tents were pitched in the wide court below. Within our enclosure there is an immense area of what was once gardens and cornfields but it is now left unwatered and uncultivated. The Persian Hajj used to lodge here in the old days. As soon as I was established in the *Roshan*, the great columned reception room, and when the men had all gone off to see to the tents and camels, two women appeared. One was an old widow, Lu.lu.ah, who is caretaker in the house; she lives here with her slave woman and the latter's boy. The other was a merry lady, Turkiyyeh, a Circassian who had belonged to Muhammed al Rashid and had been a great favourite of his. She had been sent down from the *qasr* to receive me and amuse me and the latter duty she was most successful in performing. In the afternoon came Ibrahim, in state and all smiles. He is an intelligent and well educated man – for Arabi – with a quick nervous manner and a restless eye. He stayed till the afternoon prayer. As he went out he told Muhammad al-Ma'rawi that there was some discontent among the *'ulema* at my coming and that etc etc in short, I was not to come further into the town till I was invited. Next day I sent my camels back to the Nefud borders to pasture. There is no pasture here in the granite grit plain of Hayyil and moreover they badly needed rest. I sold 6, for more than they were worth, for they were in wretched condition; but camels are fortunately dear here at this moment, with the *Amir* away and all available animals with him. And that done I sat still and waited on events. But there were no events. Nothing whatever happened, except that two little Rashid princes came to see me, 2 of the 6 male descendants who are all that remain of all the Rashid stock, so relentlessly have they slaughtered one another. Next day I sent to Ibrahim and said I should like to return his call. He invited me to come after dark and sent a mare for me and a couple of slaves. I rode through the dark and empty streets and was received in the big *Roshan* of the *qasr*, a very splendid place with great stone columns supporting an immensely lofty roof, the walls white washed, the floor of white juss, beaten hard and shining as if it were polished. There was a large company. We sat all round the wall on carpets and cushions, I on Ibrahim's right hand, and talked mostly of the history of the Shammar in general and of the Rashids in particular. Ibrahim is well versed in it and I was much interested. As we talked slave boys served us with tea and then coffee and finally they brought lighted censors and swung the sweet smelling '*ud*' before each of us three times. This is the signal that the reception is over and I rose and left them. And then followed day after weary day with nothing whatever to do. One day Ibrahim sent me a mare and I rode round the town and visited one of his gardens – a paradise of blossoming fruit trees in the bare wilderness. And the Circassian, Turkiyyeh, has spent another day with me; and my own slaves (for I have 2 of my own to keep my gate for me) sit and tell me tales of raid and foray in the stirring days of 'Abd al Aziz, Muhammad's nephew; and my men come in and tell me the gossip of the town. Finally I have sent for my camels – I should have done so days ago if they had not been so much in need of rest. I can give them no more time to recover for I am penniless. I brought with me a letter of credit on the Rashid's from their agent in Damascus – Ibrahim refuses to honour it in the absence of the *Amir* and if I had not sold some of my camels I should not have had enough money to get away. As it is I have only the barest minimum. The gossip is that the hand which has pulled the strings in all this business is that of the *Amir*'s grandmother, Fatima, of whom Ibrahim stands in deadly fear. In Hayyil murder is like the spilling of milk and not one of the shaikhs but feels his head sitting unsteadily upon his shoulders. I have asked to be allowed to see Fatima and have received no answer. She holds the purse strings in the *Amir*'s absence and she rules. It may be that she is at the bottom of it all. I will not conceal from you that there have been hours of considerable anxiety. War is all round us. The *Amir* is raiding Jof to the north and Ibn Sa'ud is gathering up his powers to the south – presumably to raid the *Amir.* If Ibrahim chose to stop my departure till the *Amir*'s return (which is what I feared) it would have been very uncomfortable. I spent a long night contriving in my head schemes of escape if things went wrong. I have however two powerful friends in Hayyil, shaikhs of 'Anezah, with whose help the Rashids hope to recapture that town [Jof]. I have not seen them – they dare not visit me – but they have protested vigorously against the treatment which has been accorded me. I owe their assistance to the fact that I have their nephew with me, 'Ali the postman who came with me 3 years ago across the Hamad.

Yesterday I demanded a private audience of Ibrahim and was received, again at night, in an upper hall of the *qasr*. I told him that I would stay here no longer, that the withholding of the money due to me had caused me great inconvenience and that I must now ask of him a *rafiq* to go with me to the 'Anazeh borders. He was very civil and assured me that the *rafiq* was ready.

It does not look as if they intended to put any difficulties in my way. My plan is to choose out the best of my camels and taking with me Fattuh, 'Ali and the negro boy Fellah, to ride to Nejef [al-Najaf]. The Damascenes I send back to Damascus. They will wait a few days more to give the other camels longer rest and then join a caravan which is going to Medina [al-Madinah] – 10 days' journey. Thence by train. Since I have no money I can do nothing but push on to Baghdad, but it is at least consoling to think that I could not this year have done more. I could not have gone south from here; the tribes are up and the road is barred. Ibn Sa'ud has – so we hear – taken the Hasa, and driven out the Turkish troops. I think it highly probable that he intends to turn against Hayyil and if by any chance the *Amir* should not be successful in his raid on (Jof], the future of the Shammar would look dark indeed. The Turkish Govt. are sending them arms... but I think that Ibn Saud's star is in the ascendant and if he combines with Ibn Sha'lan (the Ruwalla 'Anazeh) they will have Ibn Rashid between the hammer and the anvil. I feel as if I had lived through a chapter of the *Arabian Nights* during this last week. The Circassian woman and the slaves, the doubt and the anxiety, Fatima weaving her plots behind the *qasr* walls, Ibrahim with his smiling lips and restless shifting eyes – and the whole town waiting to hear the fate of the army which has gone up with the *Amir* against Jof. And to the spiritual sense the place smells one of blood. Twice since Khalil was here have the Rashids put one another to the sword – the tales round my camp fire are all of murder and the air whispers murder. It gets upon your nerves when you sit day after day between high mud walls and I thank heaven that my nerves are not very responsive. They have kept me awake only one night out of seven! And good, please God! Please God nothing but good.

March 6, 1914

We have at last reached the end of the comedy – for a comedy it has after all proved to be. And what has been the underlying reason of it all I cannot tell, for who can look into their dark minds? On March 3 there appeared in the morning a certain eunuch slave Sa'id, who is a person of great importance and with him another, and informed me that I could not travel, neither could they give me any money, until a messenger had arrived from the *Amir*. I sent messages at once to 'Ali's uncles and the negotiations were taken up again with renewed vigour. Next day came word from the *Amir*'s mother, Mudi, inviting me to visit them that evening. I went (riding solemnly through the silent moonlit streets of this strange place), and passed two hours taken straight from the *Arabian Nights* with the women of the palace. I imagine that there are few places left wherein you can see the unadulterated East in its habit as it has lived for centuries and centuries – of those few Hayyil is one. There they were, those women – wrapped in Indian brocades, hung with jewels, served by slaves and there was not one single thing about them which betrayed the base existence of Europe or Europeans – except me! I was the blot. Some of the women of the shaikhly house were very beautiful. They pass from hand to hand – the victor takes them, with her power and the glory, and think of it! his hands are red with the blood of their husband and children. Mudi herself – she is still a young woman and very charming – has been the wife of 3 A*mirs* in turn. Well, some day I will tell you what it is all like, but truly I still feel bewildered by it. I passed the next day in solitary confinement – I have been a prisoner, you understand, in the big house they gave me. Today came an invitation from two boys, cousins of the *Amir*'s to visit them in their garden. I went after the midday prayers and stayed till the *'asr*. Again it was fantastically oriental and medieval. There were 5 very small children, all cousins, dressed in long gold embroidered robes, solemn and silent, staring at me with their painted eyes. And my hosts, who may have been 13 or I4 years old – one had a merry face like a real boy, the other was grave and impassive. But both were most hospitable. We sat in a garden house on carpets – like the drawings in Persian picture books. Slaves and eunuchs served us with tea and coffee and fruits. Then we walked about the garden, the boys carefully telling me the names of all the trees. And then we sat again and drank more tea and coffee. Sa'id the eunuch was of the party and again I expressed my desire to depart from Hayyil and again was met by the same negative – Not till the *Amir*'s messenger has come. Not I nor anyone knows when the messenger will come, neither did I know whether there were more behind their answer. Sa'id came to us after the *'asr* and I spoke to him with much vigour and ended the interview abruptly by rising and leaving him. I thought indeed that I had been too abrupt, but to tell you the truth I was bothered. An hour later came in my camels and after dark Sa'id again with a bag of gold and full permission to go where I liked and when I liked. And why they have now given way, or why they did not give way before, I cannot guess. But anyhow I am free and my heart is at rest – it is widened.

March 6, 1914

I have not written any of my tale for these ten days, because of the deadly fatigue of the way. But today, as I will tell you, I have had a short day and I will profit by it. I did not leave Hayyil till March 8. I asked and obtained leave to see the town and the *qasr* by daylight – which I had never been allowed to do – and to photograph. They gave me full permission to photograph – to my surprise and pleasure, and I went out next day, was shown the *modif* and the great kitchen of the *qasr* and took many pictures. Every one was smiling and affable – and I thought all the time of Khalil, coming in there for his coffee and his pittance of *taman*. It is extraordinarily picturesque and I make no doubt that it preserves the aspect of every Arabian palace that has ever been since the Days of Ignorance. Some day, *inshallah*, you shall see my pictures. Then I photographed the *meshab* and the outside of the mosque and as I went through the streets I photographed them too. As I was going home there came a message from my Circassian friend, Turkiyyeh, inviting me to tea at her house. I went, and photographed Hayyil from her roof and took an affectionate farewell of her. She and I are now, I imagine, parted for ever, except in remembrance. As I walked home all the people crowded out to see me, but they seemed to take nothing but a benevolent interest in my doings. And finally the halt, the maim and the blind gathered round my door and I flung out a bag of copper coins among them.

And thus it was that my strange visit to Hayyil ended, after 11 days' imprisonment, in a sort of apotheosis!

ARABIAN SANDS

Wilfred Thesiger

Wilfred Thesiger (1910–2003) was a British explorer and writer who travelled extensively in Arabia. Born to a high-ranking British diplomat in Ethiopia, Thesiger developed an early fascination with non-Western traditions and cultures, which he cultivated through his lengthy travels in the Middle East. He gained notoriety in the West for his two expeditions through the vast Rub' al Khali desert, known as the Empty Quarter, in the south of the Arabian Peninsula. His chronicles of lost cultures and unexplored regions inspired my own journeys in the region, especially across the Empty Quarter and what remained of the Mesopotamian marshes of Southern Iraq.

Next morning Hamad, Jadid, and bin Kabina went to the settlements in Liwa to buy food. They took three camels with them, and I told bin Kabina to buy flour, sugar, tea, coffee, butter, dates, and rice if he could get any, and above all to bring back a goat. Our flour was finished, but that evening Musallim produced from his saddle-bags a few handfuls of maize, which we roasted and ate. It was to be the last food we had until the others returned from Liwa three days later. They were three interminable nights and days.

I had almost persuaded myself that I was conditioned to starvation, indifferent to it. After all, I had been hungry for weeks and even when we had had flour I had had little inclination to eat the charred or sodden lumps which Musallim had cooked. I used to swallow my portion with even less satisfaction than that with which I eventually voided it. Certainly I thought and talked incessantly of food, but as a prisoner talks of freedom, for I realized that the joints of meat, the piles of rice, and the bowls of steaming gravy which tantalized me could have no reality outside my mind. I had never thought then that I should dream of the crusts which I was rejecting.

For the first day my hunger was only a more insistent feeling of familiar emptiness; something which, like a toothache, I could partly overcome by an effort of will. I woke in the grey dawn craving for food, but by lying on my stomach and pressing down I could achieve a semblance of relief. At least I was warm. Later, as the sun rose, the heat forced me out of my sleeping-bag.

I threw my cloak over a bush and lay in the shade and tried to sleep again. I dozed and dreamt of food; I woke and thought of food. I tried to read, but it was difficult to concentrate. A moment's slackness and I was thinking once more of food. I filled myself with water, and the bitter water, which I did not want, made me feel sick. Eventually it was evening and we gathered round the fire, repeating, 'Tomorrow they will be back'; and thought of the supplies of food which bin Kabina would bring with him, and of the goat which we should eat. But the next day dragged out till sunset, and they did not come.

I faced another night, and the nights were worse than the days. Now I was cold and could not even sleep, except in snatches. I watched the stars; some of them – Orion, the Pleiades, and the Bear – I knew by name, others only by sight. Slowly they swung overhead and dipped down towards the west, while the bitter wind keened among the dunes. I remembered how I had once awakened with hunger during my first term at school and cried, remembering some chocolate cake which I had been too gorged to eat when my mother had taken me out to tea two days before. Now I was maddened by the thought of the crusts which I had given away in the Uruq al Shaiba. Why had I been such a fool? I could picture the colour and texture, even the shape, of the fragments which I had left.

In the morning I watched Mabkhaut tum the camels out to graze, and as they shuffled off, spared for a while from the toil which we imposed upon them, I found that I could only think of them as food. I was glad when they were out of sight. Al Auf came over and lay down near me, covering himself with his cloak; I don't think we spoke. I lay with my eyes shut, insisting to myself, 'If I were in London I would give anything to be here.' Then I thought of the jeeps and lorries with which the Locust Officers in the Najd were equipped. So vivid were my thoughts that I could hear the engines, smell the stink of petrol fumes. No, I would rather be here starving as I was than sitting in a chair, replete with food, listening to the wireless, and dependent upon cars to take me through Arabia. I clung desperately to this conviction. It seemed infinitely important. Even to doubt it was to admit defeat, to forswear everything to which I held.

I dozed and heard a camel roaring. I jerked awake, thinking, 'They have come at last', but it was only Mabkhaut moving our camels. The shadows lengthened among the sand-hills; the sun had set and we had given up hope when they returned. I saw at once that they had no goat with them. My dream of a large hot stew vanished. We exchanged the formal greetings and asked the formal questions about the news. Then we helped them with the only camel which was loaded. Bin Kabina said wearily, 'We got nothing. There is nothing to be had in Liwa. We have two packages of bad dates and a little wheat. They would not take our *riyals* – they wanted rupees. At last they took them at the same valuation as rupees, God's curse on them!' He had run a long palm-splinter into his foot and was limping. I tried to get it out but it was already too dark to see.

We opened a package of dates and ate. They were of poor quality and coated with sand, but there were plenty of them. Later we made porridge from the wheat, squeezing some dates into it to give it a flavour. After we had fed, al Auf said, 'If this is all we are going to have we shall soon be too weak to get on our camels.' We were a depressed and ill-tempered party that evening.

The past three days had been an ordeal, worse for the others than for me, since, but for me, they could have ridden to the nearest tents and fed. However, we had not suffered the final agony of doubt. We had known that the others would return and bring us food. We had thought of this food, talked of this food, dreamt of this food. A feast of rich and savoury meat, the reward of our endurance. Now all we had was this. Some wizened dates, coated with sand, and a mess of boiled grain. There was not even enough of it. We had to get back across Arabia, travelling secretly, and we had enough food for ten days if we were economical. I had eaten tonight, but I was starving. I wondered how much longer I should be able to face this fare. We must get more food. Al Auf said, 'We must get hold of a camel and eat that', and I thought of living for a month on sun-dried camel's meat and nothing else. Hamad suggested that we should lie up near Ibri in the Wadi el Ain, and send a party into lbri to buy food. He said, 'It is one of the biggest towns in Oman. You will get everything you want there.' With difficulty I refrained from pointing out that he had said this of Liwa.

Musallim interrupted and said that we could not possibly go into the Duru country; the Duru had heard about my visit to Mughshin last year and had warned the Bait Kathie not to bring any Christians into their territory. Al Auf asked him impatiently where in that case he did propose to go. They started to wrangle. I joined in and reminded Musallim that we had always planned to return through the Duru country. Excitedly he turned towards me and, flogging the ground with his camel-stick to give emphasis to his words, shouted: 'Go through it? Yes, if we must, quickly and secretly, but through the uninhabited country near the sands. We never agreed to hang about in the Duru country, nor to go near Ibri. By God, it is madness! Don't you know that there is one of the Imam's governors there. He is the Riqaishi. Have you never heard of the Riqaishi? What do you suppose he will do if he hears there is a Christian in his country? He hates all infidels. I have been there. Listen, Umbarak, I know him. God help you, Umbarak, if he gets hold of you. Don't think that Oman is like the desert here. It is a settled country – villages and towns, and the Imam rules it all through his governors, and the worst of them all is the Riqaishi. The Duru, yes; Bedu like our selves; our enemies, but we might smuggle you quickly through their land. But hang about there – no; and to go near Ibri would be madness. Do you hear? The first people who saw you, Umbarak, would go straight off and tell the Riqaishi.'

Al Auf asked him quietly, 'What do you want to do?' and Musallim stormed, 'God, I don't know. I only know I am not going near lbri.' I asked him if he wanted to return to Salala by the way we had come, and added, 'It will be great fun, with worn-out camels and no food.' He shouted back that it would not be worse than going to lbri. Exasperated by this stupidity, al Auf turned away muttering 'There is no god but God', while Musallim and I continued to wrangle until Mabkhaut and Hamad intervened to calm us.

Eventually we agreed that we must get food from lbri and that meanwhile we would buy a camel from the Rashid who were ahead of us in the Rabadh, so that we should have an extra camel with us to eat if we were in trouble. Hamad said, 'You must conceal the fact that Umbarak is a Christian.' Mabkhaut suggested that I should pretend that I was a *saiyid* from the Hadhramaut, since no one would ever mistake me for a Bedu. I protested, 'That is no good; as a *saiyid* I should get involved in religious discussions. I should certainly be expected to pray, which I don't know how to do; they would probably even expect me to lead their prayers. A nice mess I should make of that.' The others laughed and agreed that this suggestion would not work. I said, 'While we are in the sands here I had better be an Aden townsman who has been living with the tribes and is now on his way to Abu Dhabi. When we get to Oman I will say I am a Syrian who has been visiting Riyadh and that I am now on my way to Salala.' Bin Kabina asked, 'What is a Syrian?' and I said, 'If you don't know what a Syrian is I don't suppose the Duru will either. Certainly they will never have seen one.'

I AM MALALA

Malala Yousafzai

Malala Yousafzai (born 1997) is a Pakistani activist for female education and the youngest ever person to receive a Nobel Peace Prize. From a young age she campaigned for girls to be allowed to receive an education after the Taliban took control of the town where she lived in Pakistan's Swat Valley and banned them from attending school. In 2012, aged just fifteen, she was shot in the head by a Taliban gunman while on a bus. She survived, and the attempt on her life received worldwide media coverage and sparked international support for her cause. Now based in Birmingham, she co-founded the non-for-profit Malala Fund offering education and training to girls and also recently graduated from Oxford University. *I am Malala* is an international bestseller.

IT WAS DURING one of those dark days that my father received a call from his friend Abdul Hai Kakar, a BBC radio correspondent based in Peshawar. He was looking for a female teacher or a schoolgirl to write a diary about life under the Taliban. He wanted to show the human side of the catastrophe in Swat. Initially Madam Maryam's younger sister Ayesha agreed, but her father found out and refused his permission saying it was too risky.

When I overheard my father talking about this, I said, 'Why not me?' I wanted people to know what was happening. Education is our right, I said. Just as it is our right to sing and play. Islam has given us this right and says that every girl and boy should go to school. The Quran says we should seek knowledge, study hard and learn the mysteries of our world.

I had never written a diary before and didn't know how to begin. Although we had a computer, there were frequent power cuts and few places had Internet access. So Hai Kakar would call me in the evening on my mother's mobile. He used his wife's phone to protect us as he said his own phone was bugged by the intelligence services. He would guide me, asking me questions about my day, and asking me to tell him small anecdotes or talk about my dreams. We would speak for half an hour or forty-five minutes in Urdu, even though we are both Pashtun, as the blog was to appear in Urdu and he wanted the voice to be as authentic as possible. Then he wrote up my words and once a week they would appear on the BBC Urdu website. He told me about Anne Frank, a thirteen-year-old Jewish girl who hid from the Nazis with her family in Amsterdam during the war. He told me she kept a diary about their lives all cramped together, about how they spent their days and about her own feelings. It was very sad as in the end the family was betrayed and arrested and Anne died in a concentration camp when she was only fifteen. Later her diary was published and is a very powerful record.

Hai Kakar told me it could be dangerous to use my real name and gave me the pseudonym Gul Makai, which means 'cornflower' and is the name of the heroine in a Pashtun folk story. It's a kind of *Romeo and Juliet* story in which Gul Makai and Musa Khan meet at school and fall in love. But they are from different tribes so their love causes a war. However, unlike Shakespeare's play their story doesn't end in tragedy. Gul Makai uses the Holy Quran to teach her elders that war is bad and they eventually stop fighting and allow the lovers to unite.

My first diary entry appeared on 3 January 2009 under the heading I AM AFRAID: 'I had a terrible dream last night filled with military helicopters and Taliban. I have had such dreams since the launch of the military operation in Swat.' I wrote about being afraid to go to school because of the Taliban edict and looking over my shoulder all the time. I also described something that happened on my way home from school: 'I heard a man behind me saying, "I will kill you." I quickened my pace and after a while I looked back to see if he was following me. To my huge relief I saw he was speaking on his phone, he must have been talking to someone else.'

It was thrilling to see my words on the website. I was a bit shy to start with but after a while I got to know the kind of things Hai Kakar wanted me to talk about and became more confident. He liked personal feelings and what he called my 'pungent sentences' and also the mix of everyday family life with the terror of the Taliban.

I wrote a lot about school as that was at the centre of our lives. I loved my royal-blue school uniform but we were advised to wear plain clothes instead and hide our books under our shawls. One extract was called DO NOT WEAR COLOURFUL CLOTHES. In it I wrote, 'I was getting ready for school one day and was about to put on my uniform when I remembered the advice of our principal, so that day I decided to wear my favourite pink dress.'

I also wrote about the burqa. When you're very young, you love the burqa because it's great for dressing up. But when you are made to wear it, that's a different matter. Also it makes walking difficult! One of my diary entries was about an incident that happened when I was out shopping with my mother and cousin in the Cheena Bazaar: 'There we heard gossip that one day a woman was wearing a shuttlecock burqa and fell over. When a man tried to help her she refused and said. "Don't help me, brother, as this will bring immense pleasure to Fazlullah." When we entered the shop we were going to, the shopkeeper laughed and told us he got scared thinking we might be suicide bombers as many suicide bombers wore the burqa.'

At school people started talking about the diary. One girl even printed it out and brought it in to show my father.

'It's very good,' he said with a knowing smile.

I wanted to tell people it was me, but the BBC correspondent had told me not to as it could be dangerous. I didn't see why as I was just a child and who would attack a child? But some of my friends recognised incidents in it. And I almost gave the game away in one entry when I said, 'My mother liked my pen name Gul Makai and joked to my father we should change my name ... I also like the name because my real name means "grief-stricken".'

The diary of Gul Makai received attention further afield. Some newspapers printed extracts. The BBC even made a recording of it using another girl's voice, and I began to see that the pen and the words that come from it can be much more powerful than machine guns, tanks or helicopters. We were learning how to struggle. And we were learning how powerful we are when we speak.

MY LIFE AS AN EXPLORER

Sven Hedin

Sven Hedin (1865–1952) was a Swedish explorer whose travels through Central Asia and the Himalayas provided a wealth of geographical and cultural information about Central Asian cultures and languages to Western audiences. His early admiration of the German Third Reich brought him into contact with prominent Nazi leaders, whose association with Hedin has permanently marred his scientific and moral reputation. Although initially sympathetic to the cause of National Socialism, Hedin later became fiercely critical of the regime, particularly its persecution of Jews. Before and throughout the Second World War, Hedin worked tirelessly to rescue prominent Jewish scholars and scientists from the system of concentration camps set up across Europe. Wooed and courted by the Nazi high command, Hedin was often blackmailed into making public overtures of support for the regime in exchange for the release of high-profile Jewish prisoners. His troubled legacy raises lasting questions about the moral implications of associating with the purveyors of evil in order to accomplish a perceived ultimate good.

By my orders, Cherdon, Islam Bai, Turdu Bai, and a few more of my men moved our head-quarters to the small town of Charkhlik, there to await my arrival in the following spring.

I was accompanied by the Cossack Shagdur, the Mohammedans Faizullah, Tokta Ahun, Mollah, Kuchuk, Khodai Kullu, Khodai Verdi, Ahmed, and another Tokta Ahun, a Chinese-speaking huntsman whom we called Li Loye, in order not to confuse the two like-named men. We had eleven camels, eleven horses, and Yoldash, Malenki, and Malchik, the dogs. All the animals were thoroughly rested and in excellent form. It was my plan to march two hundred and forty miles between the parallel ranges of the Astin-tagh to Anambaruin-ula (a mountain mass in the east), then northward through the Gobi Desert, thence westward to Altmish-bulak, and finally south-westward to the ancient city and by way of the Lop-nor to Charkhlik.

We left on December 12. In the beginning we had some troublesome days, pushing our way through the narrow valleys of the Akato-tagh, with their soft slate-clay. Nobody had ever been there before, and not even the natives knew the glen which we hoped would lead to a pass across the range. The lateral mountains were perpendicular, and several hundred yards high.

The bottom of the valley was dry as tinder and absolutely barren. The bronze bells echoed wonderfully in the yellow passage. There had been landslides in various places, but the rocks did not stop us. We were, however, always in danger of being buried by new landslides. The valley became narrower. Eventually the packs scraped the walls on both sides; and the camels, squeezing their way through, made the dust fly. I hurried ahead to reconnoitre, and found that the valley shrank to two feet, and that at the very end there was only a vertical crack, which not even a cat could have squeezed through.

There was nothing to do but turn back. We hoped there had not been a landslide meantime; for in that event we might very well have been entrapped like so many mice.

After a thorough reconnaissance we finally succeeded in surmounting the ranges; and thereafter we walked east and north-east over good terrain.

New Year's Eve, the last night of the century, was cold and clear, and the moon shone like an arc-light. I read the texts that were being listened to in every church in Sweden that evening. Alone in my tent I awaited the approach of the new century. There were no bells here other than those of the camels, no organ music but the roar of the continual storms.

On January 1, 1901, we encamped in the Valley of Anam-baruin-gol, and I decided to encircle the entire mountainous bulk of that name, a stretch of a hundred and eighty-six miles. On one occasion we surprised twelve beautiful wild sheep climbing the steeps of an almost perpendicular mountain wall with the agility of monkeys. They eyed us steadily, the while Shagdur managed to steal in below them. A shot rang out, and a dignified ram tumbled two hundred feet down the precipice and received a death-blow on the round pads of his twisted horns.

A week later we were at the Lake of Bulungir-gol and visited some yurts of the Mongolian Sartang tribe on the surrounding steppe. The road back to Anambaruin-gol took us north of the mountain group, and we had to cross its deep valleys, which stretched towards the Gobi Desert. There were countless springs and ice-cakes. The pasturage was good and we encamped under old willow trees. It did not matter much that the cold dropped to –27°, for fuel was plentiful. Partridges were abundant and they gave pleasant variety to my dinners. Two old Mongols, of whom we inquired our way, sold us grain for our camels and horses. And at last we camped on the Anambaruin-gol, on the same spot as before.

From there I sent Tokta Ahun and Li Loye to head-quarters in Charkhlik, with six tired horses and the specimens I had collected thus far. They also took a written order to Islam Bai to send a relief party to the northern shore of the Lop-nor (or Kara-koshun), to establish a base of supplies there, and to light a bonfire every morning and evening, from March 13 on, for at about that time we would be on our way from the ancient city through the desert.

The rest of us, carrying six bags full of ice, started north into the desolate Gobi Desert. We walked through stretches of high sand-dunes, across small weather-worn granite mountains, across a clay desert and a steppe, and came out on a very ancient road, identifiable only by the heaps of stones that had withstood Time. Wild camels, antelopes and wolves appeared now and then. We dug a well in a welcome depression. It yielded potable water and the camels and horses quenched their thirst.

With enough ice to last men and horses ten days, we marched northward through an unknown desert. Wild-camel prints were now exceedingly frequent. The desert was as smooth as a lake. After a while the terrain rose, and we crossed some small weather-worn ridges. There was not a drop of water; and it would have been useless to dig for any. We accordingly turned south-west and west, and I made for Altmish-bulak by compass.

We made long marches for the next week or so. Our friend Abd-ur Rahim, who had shown us the way to Altmish-bulak the year before, had mentioned three salt springs situated east of that place. The camels had had no water now for ten days, and only a few mouthfuls of snow from a crevice. On

February 17 our situation began to be critical, and it became imperative to find one of Abd-ur Rahim's three springs. All day long and the entire next day we searched in vain for water. The terrain, too, was now against us. We reached those parts of the clay desert where the wind had ploughed furrows, twenty feet deep and thirty-five feet wide, between long, perpendicular clay ridges. They ran from north to south and we had to explore endlessly before we got past them. There was not a stick of firewood at the camp that evening, and so we sacrificed our tent-pole.

By February 19 the camels had not drunk for twelve days. They would soon die of thirst if water was not found. I moved on in advance. My horse followed me like a dog. Yoldash I was with me. A small mountain ridge made me swerve to the south-west. I walked in a dry bed, in the sandy bottom of which I discerned the fresh trail of about thirty wild camels. A small glen opened on the right. All the camel tracks radiated from there like a fan. There must be a well there. I walked up the valley and soon found a cake of ice, forty feet in diameter and three inches thick. Thus the camels were saved. When they got into the valley, we broke the ice-cake into pieces and fed them to the animals. They crunched the ice like so much sugar.

During the following days we discovered the other two springs also. They were surrounded by reed-fields. Eighteen wild camels were grazing near the last spring. Shagdur stole upon them, but he shot at too great a range, and the camels vanished like the wind.

We were scheduled to be twenty-eight kilometres from Altmish-bulak on February 24. The little oasis should be in latitude S. 60° W. Consequently, in the morning, I promised my men that before evening came we would pitch our tent among the tamarisks and reed-thickets of "The Sixty Springs."

There was a strong north-east wind, which helped us along. But an enveloping haze rested on the waste; and what would become of us if we should inadvertently pass the little oasis? I was headed for a certain point in the desert, but the dust-haze obstructed my view.

I had already covered twenty-eight kilometres and began to fear that the oasis was behind me. But what was this? Something straw-yellow gleamed right in front of me. It was reeds. And I caught sight of fourteen wild camels. I stopped, while Shagdur stole upon them. He succeeded in bringing down a young female, who was still on her feet when we got to her; also an older specimen, a male, whose skeleton we prepared during the following days and now reposes in the Zoological Museum of the Stockholm High School.

According to my calculations, our distance from the spring should have been twenty-eight kilometres; but it proved to be thirty-one. This miscalculation – three kilometres in 1,450, or two-tenths per cent. – was not great.

We indulged in a thorough rest after these forced marches. Then I left one man, the horses and some tired camels in pasturage, and went south with the rest of the caravan. We took all our luggage and nine bags of ice.

On March 3 we camped at the base of a clay tower, twenty-nine feet high. We stowed our ice in the shadow of a clay ridge and sent a man back to the spring with all the camels. These were to return to us again in six days with a further supply of ice. We promised to have a beacon fire burning on the sixth day.

We were now cut off from the world. I felt like a king in his own country, in his own capital. No one else on earth knew of the existence of this place. But I had to make good use of my time. First I located the place astronomically. Then I drew plans of the nineteen houses near our camp. I offered a tempting reward to the first man who discovered human writing in any form. But they found only scraps of blankets, pieces of red cloth, brown human hair, boot-soles, fragments of skeletons of domestic animals, pieces of rope, an ear-ring, Chinese coins, chips of earthenware and other odds and ends.

Nearly all the houses had been built of wood, the walls of bunched osiers or clay-covered wicker. In three places the door frames still remained upright. One door actually stood wide open, just as it must have been left by the last inhabitant of this ancient city, more than fifteen hundred years ago.

Shagdur succeeded in finding the place which Ordek had discovered the year before, when he went back for the shovel. There we came upon the remains of a Buddhist temple. This, in its day, must have presented a charming sight. Originally the town was situated on the old Lop-nor, which, because of the altered course of the Kuruk-daria, had since moved south. No doubt the temple stood in a park, with wide waters extending to the south. Houses, towers, walls, gardens, roads, caravans and pedestrians were then to be seen everywhere. Now it was the habitat of Death and Silence.

Our excavations yielded the frame of a standing image of Buddha, three and a half feet high; horizontal friezes with seated Buddhas, and vertical wooden posts with standing Buddhas artistically carved thereon; lotus and other flower ornaments; also sections of breast-works, all carved in wood and very well preserved. It was Shagdur who finally found a small wooden board with inscriptions (Karoshti, India), and won the prize. A similar amount was promised for the next discovery. My men worked as long as there was a trace of daylight on the waste land.

The days went by. Dawn found us already at work. We made excavations in every house. At last there remained only one house, of sun-dried clay, in the shape of a stable, with three cribs opening outwards. Mollah found a slip of paper with Chinese ideographs in the crib on the extreme right; and he got the reward. The paper lay two feet deep under sand and dust. We dug deeper and sifted the sand and dust between our fingers. One piece of paper after another was brought to light, thirty-six in all, every one of which bore writing. We also discovered a hundred and twenty-one small wooden staves covered with inscriptions. Apart from these ancient documents we found only some rags, fish-bones, a few grains of wheat and rice, and a small fragment of rug with a swastika design and colours still quite clear. For all I knew, it may have been the oldest rug in the world. The whole collection looked like a rubbish heap. Yet I had a feeling that those leaves contained a slight contribution to the history of the world. We found nothing in the other two cribs. March 9, our last day, was due. I completed the plans and measurements of the houses, and examined a clay tower, finding it solid. We found two hair-pencils such as the Chinese write with to this day; an unbroken earthenware pot, two and a third feet high; a smaller pot; and a great number of coins and small objects of various kinds. The tallest post still standing in a house measured i4‘i feet.

At dusk the two men returned from the spring with all the camels, and ten bags and six goatskins filled with water. The sun sank, and our work in the ancient city came to an end.

TIGER OF THE SNOWS

The Autobiography of Tenzing of Everest

Tenzing Norgay and James Ramsey Ullman

Tenzing Norgay (*c.*1914–1986) was one of the first two humans known to have summited Mount Everest, a feat he accomplished alongside Sir Edmund Hillary on 29 May 1953. Norgay was born to Tibetan yak herders somewhere in the Himalayan regions of Nepal or Tibet, though accounts differ on his actual birthplace. After resolving not to become a monk, Norgay served as a Sherpa guide for European mountaineers in the 1930s and 1940s, developing a reputation for himself as a capable and reliable mountaineer. After saving Edmund Hillary's life following a nearly disastrous fall into a crevasse, Norgay was selected by Hillary as his climbing partner of choice for all future expeditions. Norgay's incredible journey from the son of a rural yak herdsman to one of TIME Magazine's *100 Most Influential People of the 20th Century* ultimately solidified his place among the canon of the world's greatest explorers and adventurers, and he remains a symbol of regional pride for the people of Nepal, Tibet and India.

James Ramsey Ullmann (1907–1971) was an American writer and mountaineer who worked with Tenzing Norgay on *The Tiger of the Snows*. Travel, and particularly mountaineering, provided the physical background and spiritual core of most of Ullman's books.

MAY 28.... It had been on the twenty-eighth that Lambert and I had made our final effort, struggling up as far as we could above our high camp on the ridge. Now we were a day's climb lower; a day later. A year later.

When it first grew light it was still blowing, but by eight o'clock the wind had dropped. We looked at each other and nodded. We would make our try.

But a bad thing had happened during the night: Pemba had been sick. And it was clear that he could not go higher. The day before we had lost Ang

Tempa, who had been one of those supposed to go up to Camp Nine, and now, with Pemba out, only Ang Nyima was left of the original Sherpa team of three. This meant that the rest of us would all have to carry heavier loads, which would make our going slower and harder; but there was nothing we could do about it. A little before nine Lowe, Gregory and Ang Nyima started off, each of them carrying more than forty pounds and breathing oxygen, and about an hour later Hillary and I followed, with fifty pounds apiece. The idea of this was that our support party would do the slow hard work of cutting steps in the ice, and then we would be able to follow at our own pace, without tiring ourselves.... Or maybe I should say without tiring ourselves *too much*.

We crossed the frozen rocks of the col. Then we went up the snow-slope beyond and a long couloir, or gully, leading toward the southeast ridge. As had been planned, the fine steps cut by the others made the going easier for us, and by the time they reached the foot of the ridge – about noon – we had caught up with them. A little above us here, and off to one side, were some bare poles and a few shreds of canvas that had once been the highest camp for Lambert and me; and they brought back many memories. Then slowly we passed by and went on up the ridge. It was quite steep, but not too narrow, with rock that sloped upward and gave a good foothold, if you were careful about the loose snow that lay over it. About 150 feet above the old Swiss tent we came to the highest point that Colonel Hunt and Da Namgyal had reached two days before, and there in the snow were the tent, food and oxygen tanks which they had left for us. These now we had to add to our own loads, and from there on we were carrying weights of up to sixty pounds.

The ridge grew steeper, and our pace was now very slow. Then the snow became thicker, covering the rocks deeply, and it was necessary to cut steps again. Most of the time Lowe did this, leading the way with his swinging ax, while the rest of us followed. But by two in the afternoon all of us, with our great loads, were beginning to get tired, and it was agreed that we must soon find a camping place. I remembered a spot that Lambert and I had noticed the year before – in fact, that we had decided would be our highest campsite if we had another chance at the top; but it was still hidden above us, and on the stretch between there was no place that could possibly have held a tent. So on we went, with myself now leading: first still along the ridge, then off to the left, across steep snow, toward the place I was looking for.

"Hey, where are you leading us to?" asked Lowe and Gregory. "We have to go down."

"It can't be far now," I said. "Only five minutes."

But still we climbed, still we didn't get there. And I kept saying, "Only five minutes.... Only five minutes."

"Yes, but how many five minutes are there?" Ang Nyima asked in disgust.

Then at last we got there. It was a partly level spot in the snow, down a little from the exposed ridge and in the shelter of a rocky cliff; and there we dropped our loads. With a quick "Good-by – good luck," Lowe, Gregory and Ang Nyima started down for the col, and Hillary and I were left alone. It was then the middle of the afternoon, and we were at a height of about 27,900 feet. The summit of Lhotse, the fourth-highest peak in the world, at which we had looked up every day during the long expedition, was now below us. Over to the southeast, Makalu was below us. Everything we could see for hundreds of miles was below us, except only the top of Kangchenjunga, far to the east – and the white ridge climbing on above us into the sky.

We started pitching the highest camp that has ever been made. And it took us almost until dark. First we chopped away at the ice to try to make our sleeping place a little more level. Then we struggled with frozen ropes and canvas and tied the ropes around oxygen cylinders to hold them down. Everything took five times as long as it would have in a place where there was enough air to breathe; but at last we got the tent up, and when we crawled in it was not too bad. There was only a light wind, and inside it was not too cold to take off our gloves. Hillary checked the oxygen sets, while I got our little stove going and made warm coffee and lemon juice. Our thirst was terrible, and we drank them down like two camels. Later we had some soup, sardines, biscuits and canned fruit, but the fruit was frozen so hard we had first to thaw it out over the stove.

We had managed to flatten out the rocks and ice under the tent, but not all at one level. Half the floor was about a foot higher than the other half, and now Hillary spread his sleeping bag on the upper shelf and I put mine on the lower. When we were in them each of us rolled over close against the canvas, so that the weight of our bodies would help hold it in place. Mostly the wind was still not too bad, but sometimes great gusts would come out of nowhere and the tent would seem ready to fly away. Lying in the dark, we talked of our plans for the next day. Then, breathing the "night oxygen," we tried to sleep. Even in our eiderdown bags we both wore all our clothes, and I kept on my Swiss reindeer boots. At night most climbers take off their boots, because they believe this helps the circulation in the feet; but at high altitudes I myself prefer to keep them on. Hillary, on the other hand, took his off and laid them next to his sleeping bag.

The hours passed. I dozed and woke, dozed and woke. And each time I woke, I listened. By midnight there was no wind at all. "God is good to us," I thought. "Chomoltingma is good to us," The only sound was that of our own breathing as we sucked at our oxygen.

May 29.... On the twenty-ninth Lambert and I had descended in defeat from the col to the cwm. Down – down – down....

At about three-thirty in the morning we began to stir. I got the stove going and boiled snow for lemon juice and coffee, and we ate a little of the food left over from the night before. There was still no wind. When, a while later, we opened the tent-flap, everything was clear and quiet in the early morning light. It was then that I pointed down and showed Hillary the tiny dot that was the Thyangboche Monastery, 16,000 feet below. "God of my father and mother," I prayed in my heart, "be good to me now—today."

But the first thing that happened was a bad thing. Hillary's boots, lying all night outside his sleeping bag, had frozen, and now they were like two lumps of black iron. For a whole hour we had to hold them over the stove, pulling and kneading them, until the tent was full of the smell of scorched leather and we were both panting as if we were already climbing the peak. Hillary was very upset, both at the delay and at the danger to his feet. "I'm afraid I may get frostbitten, like Lambert," he said. But at last the boots were soft enough for him to put on, and then we prepared the rest of our gear. For this last day's climbing I was dressed in all sorts of clothes that came from many places. My boots, as I have said, were Swiss; my windjacket and various other items had been issued by the British. But the socks I was wearing had been knitted by Ang Lahmu. My sweater had been given me by Mrs. Henderson of the Himalayan Club. My wool helmet was the old one that had been left to me by Earl Denman. And, most important of all, the red scarf around my neck was Raymond Lambert's. At the end of the fall expedition he had given it to me and smiled and said, "Here, maybe you can use it sometime." And ever since, I had known exactly what that use must be.

At six-thirty, when we crawled from the tent, it was still clear and windless. We had pulled three pairs of gloves onto our hands—silk, wool and windproof; and now we fastened our crampons to our boots and onto our backs slung the forty pounds of oxygen apparatus that would be the whole load for each of us during the climb. Around my ax were still the four flags, tightly wrapped. And in the pocket of my jacket was a: small red-and-blue pencil.

"All ready?"

"*Ah chah*. Ready."

And off we went.

Hillary's boots were still stiff, and his feet cold, so he asked me to take the lead. And for a while that is how we went on the rope – up from the campsite to the southeast ridge and then on along the ridge toward the south summit. Sometimes we found the footprints of Bourdillon and Evans and were able to use them; but mostly they had been wiped away by the winds of the two days before and I had to kick or chop our own steps. After a while we came to a place I recognized: the point where Lambert and I had stopped and turned back. I pointed it out to Hillary and tried to explain through my oxygen mask, and as we moved on I thought of how different it was these two times – of the wind and the cold then and the bright sunshine now—and how lucky we were on this day of our great effort. By now Hillary's feet were feeling better, so we changed places on the rope; and we kept doing this from then on, with first one of us leading the way and then the other, in order to share the work of kicking and chopping. As we drew near to the south summit we came upon something we had been looking for: two bottles of oxygen that had been left for us by Bourdillon and Evans. We scraped the ice off the dials and were happy to see that they were still quite full. For this meant that they could be used later for our downward trip to the col, and meanwhile we could breathe in a bigger amount of what we were carrying with us.

We left the two bottles where they were and climbed on. Up until now the climbing – if not the weather – had been much the same as I remembered from the year before: along the steep broken ridge, with a rock precipice on the left and snow cornices hiding another precipice on the right. But now, just below the south summit, the ridge broadened out into a sort of snow face, so that the steepness was not so much to the sides as straight behind us, and we were climbing up an almost vertical white wall. The worst part of it was that the snow was not firm, but kept sliding down, sliding down – and we with it – until I thought, "Next time it will keep sliding, and we will go all the way to the bottom of the mountain." For me this was the one really bad place on the whole climb, because it was not only a matter of what you yourself did, but what the snow under you did, and this you could not control. It was one of the most dangerous places I had ever been on a mountain. Even now, when I think of it, I can still feel as I felt then, and the hair almost stands up on the back of my hands.

At last we got up it, though, and at nine o'clock we were on the south summit. This was the highest point that Bourdillon and Evans had reached, and for ten minutes we rested there, looking up at what was still ahead. There was not much farther to go – only about 300 feet of ridge – but it was narrower and steeper than it had been below, and, though not impossible-looking, would certainly not be easy. On the left, as before, was the precipice falling away to the Western Cwm, 8,000 feet below, where we could now see the tiny dots that were the tents of Camp Four. And on the right were still the snow cornices, hanging out over a 10,000-foot drop to the Kangshung Glacier. If we were to get to the top it would have to be along a narrow, twisting line between precipice and cornices: never too far to the left, never too far to the right – or it would be the end of us.

One thing we had eagerly been waiting for happened on the south summit.

Almost at the same moment we each came to the end of the first of our two bottles of oxygen, and now we were able to dump them here, which reduced the weight we were carrying from forty to only twenty pounds. Also, as we left the south summit, another good thing happened. We found that the snow beyond it was firm and sound. This could make all the difference on the stretch that we still had to go.

"Everything all right?"

"*Ah chah*. All right."

From the south summit we first had to go down a little.

Then up, up, up. All the time the danger was that the snow would slip, or that we would get too far out on a cornice that would then break away; so we moved just one at a time, taking turns going ahead, while the second one wrapped the rope around his ax and fixed the ax in the snow as an anchor. The weather was still fine. We were not too tired. But every so often, as had happened all the way, we would have trouble breathing and have to stop and clear away the ice that kept forming in the tubes of our oxygen sets. In regard to this, I must say in all honesty that I do not think Hillary is quite fair in the story he later told, indicating that I had more trouble than he with breathing and that without his help I might have been in serious difficulty. In my opinion our difficulties were about the same – and luckily never too great – and we each helped and were helped by the other in equal measure.

Anyhow, after each short stop we kept going, twisting always higher along the ridge between the cornices and the precipices. And at last we came to what might be the last big obstacle below the top. This was a cliff of rock rising straight up out of the ridge and blocking it off, and we had already known about it from aerial photographs and from seeing it through binoculars from Thyangboche. Now it was a question of how to get over or around it, and we could find only one possible way. This was along a steep, narrow gap between one side of the rock and the inner side of an adjoining cornice, and Hillary, now going first, worked his way up it, slowly and carefully, to a sort of platform above. While climbing, he had to press backwards with his feet against the cornice, and I belayed him from below as strongly as I could, for there was great danger of the ice giving way. Luckily, however, it did not. Hillary got up safely to the top of the rock and then held the rope while I came after.

Here again I must be honest and say that I do not feel his account, as told in *The Conquest of Everest*, is wholly accurate. For one thing, he has written that this gap up the rock wall was about forty feet high, but in my judgment it was little more than fifteen. Also, he gives the impression that it was only he who really climbed it on his own, and that he then practically pulled me, so that I "finally collapsed exhausted at the top, like a giant fish when it has just been hauled from the sea after a terrible struggle." Since then I have heard plenty about that "fish," and I admit I do not like it. For it is the plain truth that no one pulled or hauled me up the gap. I climbed it myself, just as Hillary had done; and if he was protecting me with the rope while I was doing it, this was no more than I had done for him. In speaking of this I must make one thing very plain. Hillary is my friend. He is a fine climber and a fine man, and I am proud to have gone with him to the top of Everest. But I do feel that in his story of our final climb he is not quite fair to me; that all the way through he indicates that when things went well it was his doing and when things went badly it was mine. For this is simply not true. Now here do I make the suggestion that I could have climbed Everest by myself; and I do not think Hillary should suggest that he could have, or that I could not have done it without his help. All the way up and down we helped, and were helped by, each other – and that was the way it should be. But we were not leader and led. We were partners.

On top of the rock cliff we rested again. Certainly, after the climb up the gap we were both a bit breathless, but after some slow pulls at the oxygen I am feeling fine. I look up; the top is very close now; and my heart thumps with excitement and joy. Then we are on our way again. Climbing again. There are still the cornices on our right and the precipice on our left, but the ridge is now less steep. It is only a row of snowy humps, one beyond the other, one higher than the other. But we are still afraid of the cornices and, instead of following the ridge all the way, cut over to the left, where there is now a long snow slope above the precipice. About a hundred feet below the top we come to the highest bare rocks. There is enough almost level space here for two tents, and I wonder if men will ever camp in this place, so near the summit of the earth. I pick up two small stones and put them in my pocket to bring back to the world below. Then the rocks, too, are beneath us. We are back among the snowy humps. They are curving off to the right, and each time we pass one I wonder, "Is the next the last one? Is the next the last?" Finally we reach a place where we can see past the humps, and beyond them is the great open sky and brown plains. We are looking down the far side of the mountain upon Tibet. Ahead of us now is only one more hump – the last hump. It is not a pinnacle. The way to it is an easy snow slope, wide enough for two men to go side by side. About thirty feet away we stop for a minute and look up. Then we go on....

I have thought much about what I will say now: of how Hillary and I reached the summit of Everest. Later, when we came down from the mountain, there was much foolish talk about who got there first. Some said it was I, some Hillary. Some that only one of us got there – or neither. Still others that one of us had to drag the other up. All this was nonsense. And in Katmandu, to put a stop to such talk, Hillary and I signed a statement in which we said, "we reached the summit almost together." We hoped this would be the end of it. But it was not the end. People kept on asking questions and making up stories. They pointed to the "almost" and said, "What does that mean?" Mountaineers understand that there is no sense to such a question; that when two men are on the same rope they are *together*, and that is all there is to it. But other people did not understand. In India and Nepal, I am sorry to say, there has been great pressure on me to say that I reached the summit before Hillary. And all over the world I am asked, "Who got there first? Who got there first?"

Again I say: it is a foolish question. The answer means nothing. And yet it is a question that has been asked so often—that has caused so much talk and doubt and misunderstanding—that I feel, after long thought, that the answer should be given. As will be clear, it is not for my own sake that I give it. Nor is it for Hillary's. It is for the sake of Everest—the prestige of Everest—and for the generations who will come after us. "Why," they will say, "should there be a mystery to this thing? Is there something to be ashamed of? To be hidden? Why can we not know the truth?".... Very well: now they will know the truth. Everest is too great, too precious, for anything but the truth.

A little below the summit Hillary and I stopped. We looked up. Then we went on. The rope that joined us was thirty feet long, but I held most of it in loops in my hand, so that there was only about six feet between us. I was not thinking of "first" and "second." I did not say to myself, "There is a golden apple up there. I will push Hillary aside and run for it." We went on slowly, steadily. And then we were there. Hillary stepped on top first. And I stepped up after him.

So there it is: the answer to the "great mystery." And if, after all the talk and argument, the answer seems quiet and simple, I can only say that that is as it should be. Many of my own people, I know, will be disappointed at it. They have given a great and false importance to the idea that it must be I who was "first." These people have been good and wonderful to me, and I owe them much. But I owe more to Everest – and to the truth. If it is a discredit to me that I was a step behind Hillary, then I must live with that discredit. But I do not think it was that. Nor do I think that, in the end, it will bring discredit on me that I tell the story. Over and over again I have asked myself, "What will future generations think of us if we allow the facts of our achievement to stay shrouded in mystery? Will they not feel ashamed of us – two comrades in life and death – who have something to hide from the world?" And each time I asked it the answer was the same: "Only the truth is good enough for the future. Only the truth is good enough for Everest."

Now the truth is told. And I am ready to be judged by it.

*

We stepped up. We were there. The dream had come true....

What we did first was what all climbers do when they reach the top of their mountain. We shook hands. But this was not enough for Everest. I waved my arms in the air and then threw them around Hillary, and we thumped each other on the back until, even with the oxygen, we were almost breathless. Then we looked around. It was eleven-thirty in the morning, the sun was shining, and the sky was the deepest blue I have ever seen. Only a gentle breeze was blowing, coming from the direction of Tibet, and the plume of snow that always blows from Everest's summit was very small. Looking down the far side of the mountain, I could see all the familiar landmarks from the earlier expeditions: the Rongbuk Monastery, the town of Shekar Dzong, the Kharta Valley, the Rongbuk and East Rongbuk Glaciers, the North Col, the place near the northeast ridge where we had made Camp Six in 1938. Then, turning, I looked down the long way we ourselves had come: past the south summit, the long ridge, the South Col; onto the Western Cwm, the icefall, the Khumbu Glacier; all the way down to Thyangboche and on to the valleys and hills of my homeland.

Beyond them, and around us on every side, were the great Himalayas, stretching away through Nepal and Tibet. For the closer peaks – giants like Lhotse, Nuptse and Makalu – you now had to look sharply downward to see their summits. And farther away, the whole sweep of the greatest range on earth – even Kangchenjunga itself – seemed only like little bumps under the spreading sky. It was such a sight as I had never seen before and would never see again: wild, wonderful and terrible. But terror was not what I felt. I loved the mountains too well for that. I loved Everest too well. At that great moment for which I had waited all my life my mountain did not seem to me a lifeless thing of rock and ice, but warm and friendly and living. She was a mother hen, and the other mountains were chicks under her wings. I too, I felt, had only to spread my own wings to cover and shelter the brood that I loved.

We turned off our oxygen. Even there on top of the world it was possible to live without it, so long as we were not exerting ourselves. We cleared away the ice that had formed on our masks, and I popped a bit of sweet into my mouth. Then we replaced the masks. But we did not turn on the oxygen again until we were ready to leave the top. Hillary took out his camera, which he had been carrying under his clothing to keep it from freezing, and I unwound the four flags from around my ax. They were tied together on a string, which was fastened to the blade of the ax, and now I held the ax up and Hillary took my picture. Actually he took three, and I think it was lucky, in those difficult conditions, that one came out so well. The order of the flags from top to bottom was United Nations, British, Nepalese, Indian; and the same sort of people who have made trouble in other ways have tried to find political meaning in this too. All I can say is that on Everest I was not thinking about politics. If I had been, I suppose I would have put the Indian or Nepalese flag highest-though that in itself would have been a bad problem for me. As it is, I am glad that the U.N. flag was on top. For I like to think that our victory was not only for ourselves – not only for our own nations – but for all men everywhere.

I motioned to Hillary that I would now take his picture. But for some reason he shook his head; he did not want it. Instead, he began taking more pictures himself, around and down on all sides of the peak, and meanwhile I did another thing that had to be done on the top of our mountain. From my pocket I took the package of sweets I had been carrying. I took the little red-and-blue pencil that my daughter Nima had given me. And scraping a hollow in the snow, I laid them there. Seeing what I was doing, Hillary handed me a small cloth cat, black and with white eyes, that Hunt had given him as a mascot, and I put this beside them. In his story of our climb Hillary says it was a crucifix that Hunt gave him and that he left on top; but if this was so I did not see it. He gave me only the cloth cat. All I laid in the snow was the cat, the pencil and the sweets. "At home," I thought, "we offer sweets to those who are near and dear to us. Everest has always been dear to me, and now it is near too." As I covered up the offerings I said a silent prayer. And I gave my thanks. Seven times I had come to the mountain of my dream, and on this, the seventh, with God's help, the dream had come true.

"*Thuji chey, Chomolungma*. I am grateful.... "

We had now been on top almost fifteen minutes. It was time to go. Needing my ax for the descent, I could not leave it there with the flags; so I untied the string that held them, spread the flags across the summit, and buried the ends of the string as deeply as I could in the snow. A few days later planes of the Indian Air Force flew around the peak, making photographs, but the fliers reported they could see nothing that had been left there. Perhaps they were too far off. Or perhaps the wind had blown the flags away. I do not know.

Before starting down we looked around once more. Had Mallory and Irvine reached the top before they died? Could there be any sign of them? We looked, but we could see nothing. Still they were in my thoughts, and I am sure in Hillary's too. All those who had gone before us were in my thoughts – sahibs and Sherpas, English and Swiss – all the great climbers, the brave men, who for thirty-three years had dreamed and challenged, fought and failed on this mountain, and whose efforts and knowledge and experience had made our victory possible. Our companions below were in my thoughts, for without them, too – without their help and sacrifice – we could never have been where we were that day. And closest of all was one figure, one companion: Lambert. He was so near, so real to me, that he did not seem to be in my thoughts at all, but actually standing there beside me. Any moment now I would turn and see his big bear face grinning at me. I would hear his voice saying, "*Ça va bien*, Tenzing. *Ça va bien!*"

Well, at least his red scarf was there. I pulled it more tightly around my throat. "When I get back home," I told myself, "I will send it to him." And I did.

Since the climbing of Everest all sorts of questions have been put to me, and not all of them have been political. From the people of the East there have been many that have to do with religion and the supernatural. "Was the Lord Buddha on the top?" I have been asked. Or, "Did you see the Lord Siva?" From many sides, among the devout and orthodox, there has been great pressure upon me to say that I had some vision or revelation. But here again – even though it may be disappointing to many – I can tell only the truth; and this is no, that on the top of Everest I did not see anything supernatural or feel anything superhuman. What I felt was a great closeness to God, and that was enough for me. In my deepest heart I thanked God. And as we turned to leave the summit I prayed to Him for something very real and very practical: that, having given us our victory, he would get us down off the mountain alive.

A LADY'S SECOND JOURNEY AROUND THE WORLD

Ida Pfeiffer

Ida Pfeiffer (1797–1858) was an Austrian ethnographer, adventurer and traveller who became one of the first preeminent female travel writers. She began her travelling career already well into middle age, completing two circumnavigational trips and recording her observations in her bestselling journals. As a woman, she was forbidden membership of the prestigious Royal Geographical Society but was admitted to the geographical societies of both Paris and Berlin. Pfeiffer voraciously collected plant, animal and mineral specimens on her various travels, many of which she later sold to the Berlin Natural History Museum and the British Museum. She is considered an early forerunner of the independent female travellers of the late nineteenth and early twentieth centuries, and her notable ethnographical contributions and honours secured her a lasting reputation as an accomplished writer and amateur natural scientist.

ON the 6th of August I set out on my rather dangerous journey, and traveled twenty miles to Sipirok. All was forest or Jungle-grass, and from the top of a small chain of hills I looked over the wide undulating valley of Lawas, one of the largest in Sumatra. I had now passed through a great part of the island, and found it equal, or perhaps superior even, to Java in the beauty of its natural scenery. What a glorious country it might become! but, as yet, it is to a great extent unpeopled and uncultivated. A few plantations there are, but the vast forests of the interior are inhabited only by wild beasts, and the blood-thirsty tiger has his lair in the scorched jungle-grass.

Yet a great part of Sumatra must certainly be a splendid country for European emigrants. Near as it lies to the equator, the climate on the great table-lands, of which there are many, is extremely moderate, and the exuberant richness of vegetation in the thick forests, and the height of the jungle-grass, affords sufficient evidence of the fertility of the soil. Where nature unassisted is so bountiful, she would certainly perform wonders with the aid of cultivation. But the Dutch government does not encourage the settlement of Europeans—not even of its own subjects; and it declares, alas! with perfect truth, that the example of the whites tends to corrupt the natives. There is, nevertheless, I believe, another motive secretly influential in the case; namely, the fear that the whites might in time become too powerful for the little mother country to hold in check, and even perhaps unite with the natives to render themselves independent.

Sipirok lies in a small smooth valley. Here is the last coffee plantation under the superintendence of a native Writer. I happened to arrive just as the coffee was being delivered, and this gave me an opportunity of seeing a great number of the people. Their appearance was not prepossessing; their faces were of the same type as the Malays, but actually uglier, and the females excessively diminutive. In the elegant art, too, of filing and blackening their teeth they have obtained such proficiency, that they are able to do what might seem impossible—to render themselves more frightful than nature has made them. They were very scantily attired, extremely dirty, and all had their cheeks puffed out with the siri they were chewing, and were spitting right and left among, or at least close to, the coffee-beans that lay spread out. By way of amusing their leisure, too, they were occupying themselves with hunting for vermin on the heads and clothes of their children, who, covered as they were with horrid cutaneous eruptions, were playing and pelting each other with the coffee-beans.

After the coffee had been examined, put into sacks, and placed in the magazine, and when the people had received their money, the open space or square was transformed into a bazaar. From the apartments of the Writer various kinds of goods were brought out; and traders who had been for hours impatiently waiting for the clearing away of the coffee, began to unpack and exhibit colored stuffs, glass beads, brass rings, eatables, and so forth. With eager glances did the fortunate possessors of money survey all these tempting wares, only distracted, poor creatures, by the difficulty of choosing among so many seductive articles, when their means of purchasing were so small. In the course of an hour the bazaar was over; that is to say, the people had spent all their money.

At Sipirok my traveling on horseback was to cease, and I should again, as in Borneo, have to renounce for a time all the conveniences of life, and recommence my wanderings on foot.

August 6th. Twelve miles to Donan. The way led through thick woods, and over steep hills and mountains, by terribly slippery paths. When I reached Donan, I was shown to a dilapidated hut, which contained two sleeping-places; and from this time I was, of course, at every Battaker village, or *Utta*, as they are called, surrounded by a crowd of curious gazers. Even at Muara Sipongie I had begun to find myself an attractive spectacle, as no European woman had ever before made her appearance in that country; but here I *drew*, as actors say, such very full houses—that is to say, my hut was so thronged—that I could not tell for a long time who were to be my fellow-lodgers; but at length I had the pleasure of discovering that the other inmates were a nearly dying man and—a murderer! The latter had killed one of his neighbors in a fit of jealousy, and was to be beheaded for it in two days in the bazaar. He lay on the ground naked, and fastened to a post, his feet being drawn through a block; and he was behaving almost like a madman—laughing, weeping, and screaming alternately, and dashing himself wildly from one side to the other. It was a horrible sight! The sick person was a youth of eighteen years of age, who also lay on the ground without mat or covering, though he appeared to be in the last stage of consumption, and had dreadful fits of coughing. Unfortunately I could do nothing for the poor fellow, having neither medicine nor any thing else with me.

I could not help observing on this occasion how much more sympathy appeared to be felt with the murderer than with the sick man. The women prepared his siri for him, brought him rice and dried fish for his meals, fed him, as his hands were bound, like a little child, fanned the flies off him, etc.; and the men even unbound him and carried him to a neighboring river to take a bath. The poor fellow dying of consumption they took no notice of, but let him cough and groan, and gave him neither food nor drink; they could not have done less for him if he had been already dead. I could give him only rice and water, for this was all I had myself.

Diseases of the lungs are very prevalent in the highlands of Sumatra, and I heard many of the people coughing fearfully. Although the heat is great in the daytime, the nights are almost cold; it rains much; and the people go as scantily clothed as in regions of unvaried sunshine, and have nothing to cover themselves at night when they sleep, so the fact is not so very surprising.

As I was determined I would not, if I could help it, pass the night in the room with the murderer, I sent to beg the Rajah to find me some other place of shelter, and he had the complaisance to order both the criminal and the sick man to be removed; but the people were not to be hindered from thronging in to gaze at me; and even during the night I was not left for a moment alone. Until midnight the fires were kept burning, and they gossiped incessantly; then most of them lay down; pulled their sarangs closely about them, and were soon snoring one against the other.

The next day, also, I was obliged to pass in Donan, for the Rajah, who was nominally under the protection of the Dutch government, assured me that without his escort I could not venture into the country of the wild Battakers, now only a few miles off. He would go with me, he said, and use his personal influence with the Rajah, with whom he was acquainted, to secure my safety. In pursuance of this friendly resolution, he first slaughtered a buffalo calf in my honor, to secure the patronage and favor of the evil spirits—who, if they were offended, might oppose insurmountable obstacles to our undertaking—to induce them to refrain from increasing the perils of our journey.

Early in the morning he paid me a visit, attended by some dozens of women and girls, mostly his relatives. They defiled before me in a profoundly humble attitude, bending down, and shading their faces with their hands—the mode of salutation, I was told, for inferiors toward persons of rank. Then they sat down on the ground at the back of the hut, and took out of some pretty plaited baskets a quantity of siri that was intended for my delectation.

The girls wore from ten to fifteen leaden rings in their ears, and had also the upper part of the cartilage pierced and decorated with a button, or a string of glass beads; but when they marry I was told they have to lay aside all these trinkets. The girls have their bosoms covered, the married women mostly bare, and both women and girls twist their hair up into a knot, putting a straw cushion under it to increase its apparent mass. What is rather perplexing, too, the gentlemen wear it just as long, and twist it up in the same manner as the ladies; they have no beards, and they wear the sarang, their only garment, fastened round them mostly in the same manner. Fortunately they, for the most part, stick on a straw cap, or twist a handkerchief round their heads, and by this sign one may recognize the superior sex.

Many of the girls were of considerable *embonpoint*, and were only young ladies by courtesy, as they had, in fact, passed their youth, although, as it appeared, without entering the conjugal state; a circumstance to be accounted for from wives being purchasable articles here.

The purpose of the Rajah's visit was to invite me to the solemn slaughtering of the buffalo-calf, and I soon accompanied him and his ladies to his hut. The ceremony began with a wild dance, performed by the Rajah's son, a youth of eighteen; and as every one desired to witness the *pas seul* of the young chief, the hut soon became so full that there was no moving. It was whispered about, probably to flatter the young man, that he was possessed by an evil spirit; and, as if he wished to justify the opinion, his dancing became even more and more fast and furious, until at last he fairly fell down exhausted. Then another took his place; but this was an inferior performer, who did not enjoy the advantage of demoniacal possession, and he soon retired, for the Rajah's son sprung up again, and recommenced his mad exhibition to the accompaniment, as before, of a kind of uproarious music. A bowl filled with unboiled rice was then presented to him, and he raised it several times above hie head, as if he wished to offer the contents to the spirits, or beg their blessing upon it. Then he took a small portion and flung it into the air, and after that he rushed out of the hut, scattering the rice as he went, and at last poured the remainder over the buflalo-calf, which lay on a sort of scaffold, bound, and ready to be slaughtered.

After this the prince retumed to the hut and continued his extraordinary *ballet d'action* until he could no longer stand, and fell exhausted into the arms of the much edified spectators. Thereupon the calf was slain, cut into many little bits, and for the most part distributed among the people; and the liver was politely put aside for me, and in the evening presented to me; but, unluckily, it had been roasted till it was as hard as a stone, and quite uneatable, so that I had again to content myself with rice and salt, although the calf had been killed expressly to do me honor.

August 8th. I left Donan with a suit of more than twenty persons, of whom, however, the greater part did not proceed farther than the frontier, about three miles off. At parting they took my hand and wished me a safe return, but accompanied their good wishes with a pantomine more expressive than agreeable—pointing to my throat, and giving me to understand by signs that the wild men would cut my head off, and eat me up. This was not very encouraging, but it never once entered my head to desist from my undertaking. My party consisted of the Rajah and five of his people, my guide, and two coolies—one for the guide and one for me.

The way led now into the untrodden wilderness, through all but impenetrable woods and jungle-grass six feet high. We saw no habitation, nor any trace of man, though many of wild beasts, especially tigers. We came to a river, and found that the only way of crossing it was by climbing a tree and scrambling along the branches that stretched over the water about twenty feet above it, and then on to those of a corresponding tree on the opposite bank, which extended its leafy arms across, so as to form, between the two, a kind of natural bridge.

From time to time we came to openings in the forest which afforded glimpses of wide, lonely valleys, watered by the innumerable windings of the river Padang-Toru, and of a small lake glittering in the sunlight on one of the hills. The Padang-Toru, to which we often came quite close, is a fine broad stream, and its surface is as yet unruffled even by the smallest canoe: whichever way we looked, all was one vast solitude; it seemed as if we were the only inhabitants of the earth.

At this time of year it rains in Sumatra almost regularly every afternoon, and, unluckily, the rain always caught us on our road; for here, as in Borneo, it was impossible to get the people to start early. The bad weather was particularly disagreeable to me, as I had no opportunity of changing clothes or linen; in the first place, because the people never left me for a sufficient time, day or night; and, secondly, because I very often could not get at my small luggage, even had it been ever so necessary.

My guide, like the one I had at Sarawak, did just what he pleased, and always demanded the services of the first cooly that was to be got for his own benefit, leaving my carpet-bag to take its chance of being brought by any helping hand that might turn up, and often enough leaving it behind, with merely an order to send it on. This day the rain was beyond measure annoying, and we had, besides, to pass the night in the woods. The men put up, indeed, a little roof of large leaves, and spread some also on the ground; but when we arrived I was already drenched through and through, and, having had to walk through a morass, was up to my knees in mud.

I went down to the small stream near which we had encamped, and washed off the mud, and then came back dripping wet and shivering with cold (the evenings and mornings were very cold), and crept to the fire; but as we had no dry wood, this rather glimmered than burned. The people collected wood for the night, and caught some small fish in the river; they also brought some perfectly green bamboo canes, the use of which I could not at all understand, but presently I found they were to serve as cooking utensils. The men put some rice and water upon *bisang* leaves, made them up into long rolls, and pushed them into the canes. The fish they prepared in the same way; then laid the canes across the fire, and left them till they began to brown, which was a considerable time, as they were extremely moist. They then split the canes, and took out what they had put in. The larger fish were done in a different way, being stuck on small pieces of wood, and roasted before the fire.

The meal thus prepared was neither very tempting nor very clean, for the rice had not been washed nor the fish gutted; but I had eaten nothing the whole day, and had also walked eighteen miles so I was not inclined to be very critical, and was well content with it. Before we lay down for the night, I begged my attendants to make up a good fire, in order to keep the tigers from us; but the wood was too wet to burn well, and the men were soon fast asleep, and not to be waked by any calling of mine. We were involved in the thickest darkness, and I found it impossible to sleep for a moment, for I was in constant dread of an attack of wild beasts, or of men little less savage.

As often as I saw the gleam of a diamond beetle I thought it was the glare of a tiger's eye, and every leaf that rattled appeared to me to indicate the approach of a serpent. Altogether it was a terrible night!

NOMAD
The Diaries of Isabelle Eberhardt

Isabelle Eberhardt

Isabelle Eberhardt (1877–1904) was a Swiss author and traveller who wrote extensively about French colonial rule in Algeria. A convert to Islam who wrote under a male pseudonym and wore men's clothing, Eberhardt was seen as a pariah among polite European society in Algeria and was targeted by the French administration as a subversive, spy and traitor. Her writings depicted the odious and violent nature of colonial rule and she is considered an early advocate of decolonization. Her unique legacy has inspired a number of films, books, and even an opera about her.

Isabelle discusses the wrongs of colonialism, and her own battles with oppression.

11 December 7.15 in the evening
'It is one thing to know that somewhere in some distant place certain people are busy torturing others and subjecting them to every possible form of suffering and humiliation, and something else again to be present for three months during such torture and see such suffering and humiliation being inflicted every day.'
Resurrection, Leo Tolstoy

Algiers, 25 December 1902, half-past noon
The past and all its Christmas seasons are way behind me. My yearning for the past no longer extends beyond the Souf now.

The hardest thing to do, perhaps, is to *free oneself*, and what is more, to adopt a *lifestyle* that is free. I am increasingly resentful of mankind for its intolerance of unusual people, and for bending to slavery only to impose it on others. Where is that hermitage where I would be beyond imbecility and free of physical desires?

The same day, eleven o'clock in the evening
My discontent with people grows by leaps and bounds... dissatisfaction with myself as well, for I have not managed to find a suitable *modus vivendi*, and fear that none is possible with my temperament.

There is only one thing that can help me get through the years I have on earth, and that is writing. It has the huge advantage of giving the will a free hand and letting one project oneself without having to cope with the world outside. That is its essential value, whatever else it may yield in the way of a career or gain, especially as I am more and more convinced that life as such is hostile and a dead end. I trust that I can bring myself to be content with living quietly as a writer.

As of now, I will probably go to Medeah and Bou Saada once the five days of Ramadan are over.

Once again, my soul is caught in a period of transition, undergoing changes and, no doubt, growing darker still and more oppressed. If this foray of mine into the world of darkness does not stop, what will be its terrifying outcome?

Yet, I do believe there is a remedy, but in all heartfelt humanity ∪ *it lies in the realm of the Islamic religion.* That is where I shall find peace at last, and solace for my heart. The impure and, so to speak, hybrid atmosphere I now live in does me no good. My soul is withering and turns inward for its distressing observations.

As agreed, I set off for the Dahra on the evening of Thursday 11 December by the light of the Ramadan moon. The night was clear and cool. There was total silence over that desert town, as rider Muhammad and I slipped through like shadows. That man is so much a Bedouin and so close to nature that he is my favourite companion, for he is in total harmony with the landscape and the people... not to mention my own frame of mind. He is not aware of it, but he is as preoccupied as am I with the puzzle and enigma of the senses.

At Montenotte and Cavaignac we went to the Moorish cafés there. We crossed *oueds*, went up slopes and down ravines, past graveyards...

In a desert full of diss and doum, above a grim-looking shelf rather like those in the Sahara with shrubs perched high up on mounds, we dismounted to eat and get some rest. The place felt so unsafe that we started at the slightest noise. I spotted a vague, white silhouette against one of the shrubs down below. The horses snorted restlessly... who was it? The shadow vanished, and when we went by that spot, the horses were uneasy.

Our path led through a narrow valley intersected by many *oueds*. Jackals howled nearby. Farther down we came to the *mechta* of Kaddour-bel-Korchi, the *caïd* of the Talassa. The *caïd* was not there and we had to go farther still, until we found him in the *mechta* of a certain Abdel Kader ben Aissa, a pleasant, hospitable man. We had our meal there and once the moon had set, we went off for Baach by paths riddled with holes and full of mud and rolling stones. At dawn, the *borj* of Baach, the most beautiful in the area, came into sight high up on a pointed hill, looking very similar to a *borj* in the Sahara.

Algiers, 29 December 1902, 2.30 in the morning
How curious and dreamlike is my impression – is it a pleasant one? I can't tell! – of life in Algiers, with all the weariness that goes with the end of Ramadan!

Ramadan! We spent its first few days over in Ténès, in the soothing climate of family life the way it is at that time of year. What a curious family we are, made up of people who have drifted together by accident, Sliméne and I, Bel Hadj from Bou Saada, and Muhammad, who has one foot in the unforgettable Souf and another on those poetic slopes overlooking the blue bay and road to Mostaganem…

Isabelle is contemplative and ponders transience and her blood family's disintegration: the deaths of her mother, the man she calls her father, Vava, the suicide of her eldest brother Vladimir (Volodia), and the collapse of her relationship with her favourite brother, Augustin. She has been cut out of inheriting her mother's money by her half brother and sister, Nicholas and Nathalie, legitimate heirs to the de Moerder estate. In a final irony, all she is left is debt and lawyers' bills.

Algiers, 31 December 1902, midnight
Another year has slipped by… One year less to live… And I love life, out of sheer curiosity for nature and its mysteries.

Whatever became of my old dreams, those dreams of my youth, gazing at the snow-clad Jura mountaintops and great oak forests? Where are my beloved who have departed from this world?

Even when I was tiny, I used to think with terror of the time when those older than me would have to die. It seemed impossible they would! Five years have now gone by since Mama was laid to rest in a Muslim graveyard on Islamic soil. It will soon be four years ago that Vava was buried in Vernier over in the land of exile, next to Volodia whose death has never been explained. All around Mother's grave at Bône, Algerian winter flowers are now in bloom, and those two graves over there must be covered with snow.

And everything else is gone. That fateful, hapless house has passed into other hands. Augustin has vanished from the horizon of my life where he used to loom so large for so many years. Everything that once existed has now been demolished and destroyed for good. And for these past four years I have been roaming in anguish by myself, my only companion the man I found over in the pristine Souf, long may he stay at my side and bring solace, ∪ *please Allah.*

What does this next year hold in store for us? What new hopes and what new disillusions? Despite so many changes, it is good to have a loving heart to call one's own, and friendly arms in which to rest.

Algiers, Sunday 9 January 1903, midnight
It would be nice to die in Algiers, over there on Mustapha's hill, facing that sensuous yet melancholy panorama, facing the harmony of that vast bay with the jagged profile of the Kabyle mountains in the distance. It would be nice to die there, slowly, on a sunny autumn day, to be aware of dying while taking in the soft strains of music and inhaling fragrances as ethereal as our souls, which we would then breathe out together, in a slow, infinitely smooth and sensual act of renunciation, free of torment and regret.

Everthing about my present life is temporary and uncertain. Everything is hazy, and the strange fact is that I no longer mind.

Who knows how long my stay in Algiers will last, who knows how it will end? Who knows where I will be tomorrow? Another journey southwards, in the direction of the desert, that blessed land where the sun is fiery and the palm trees' shadows blue upon the soil.

What I would like right now is to live over in Ténès, lead a quiet life there free of shackles, and keep going off on horseback in pursuit of my dream, from tribe to tribe.

After a period of brief and fragmented trips, Isabelle journeys to Bou Saada to visit the *maraboute* Lella Zeyneb. Her reward is a spiritual one.

Bou Saada, Wednesday 28 January 1903, half-past noon
Left Algiers at six o'clock on Monday 26th in clear weather. Reached Bou Saada at 7.30 in the evening, stayed at the Moorish bath. Never have I been so keenly aware here of the vaguely ominous weight that seems to hang over all the occupied territories; it is something one cannot put one's finger on, there are so many ambiguities and innuendoes, so many mysteries...

In spite of my fatigue, and lack of sleep and food, this has been a good journey. The Ziar are kind and simple people, and sang their saint's *medha* to the accompaniment of *gasba*, *zorna* and *bendar*, each in turn, while the train wended its way in the sunshine I was so happy to have found again.

Chellal is a cheerless village built of mud; a handful of wretched cottages set in a hollow full of water. An acrid smell of iodine and saltpetre hangs in the air. The native population is made up of Ouled–Madhi and of Hachem, who are not very congenial. The *maghreb* was superb, with the mountains standing out in bluish-black against the red-gold of the sky.

I visited the Arab Bureau this morning, and by about one o'clock I went for a stroll in the Arab part of town, and in the *oued*, where Arab washerwomen stood out in blue and red dots of an incredibly warm intensity.

Tomorrow I will go to El Ijamel. Once I have had some rest tomorrow night, I will do a better job of writing down my observations. The physical fatigue and lack of food I have suffered until tonight have worn me out. The ride to El Hamel will be good training for the long journey to Sahari and Boghar.

It looks as if I am no longer being persecuted. They tell me there had been no advance word of my arrival, yet they have been most pleasant, even the commanding officer… how shadowy and mysterious these people are!

El Hamel, Thursday 29 January 1903, about four in the afternoon

There is a heavy silence all around and the only sound to break it is the occasional noise coming from the village or the *zawyia*, the distant sound of dogs barking and the raucous growls of camels.

El Hamel! How appropriate that name is for this corner of old Islam, so lost in these barren mountains and so veiled in unfathomable mystery.

The same evening, about ten o'clock

I am sitting on my bed, near the fireplace in the vaulted main room. The fire and my bed right on the floor make the room look so much more cheerful and cosy than it did earlier in the day.

The 'hotel', a large square edifice, boasts a deep and desolate-looking inner courtyard full of bricks and stones. It leads to the upper floor which is divided into two rooms, a small and a large one, both of them with semicircular vaults, like well-to-do houses in the Souf. One of the windows looks out over the cemeteries in a south-westerly direction, the three others give out on to the east. There are three French beds, an oval table, chairs, all of it set on very thick rugs. With a few more authentically Arab touches, the room would look truly grand. I wish I could arrange it myself and do it justice. On the western side stand the tall buildings made of *toub* where the *maraboute* lives. To the north is the new mosque with its great, round cupola surrounded by smaller ones, and inside it stands the tomb of Sidi Muhammad Belkassem.

I am going to lie down and rest, for tomorrow I must rise early to go and see the *maraboute*. No doubt I will return to Bou Saada tomorrow afternoon, and will try to be there by the *maghreb*. After that I will have a week for a good look, and that is a time I must not waste.

Bou Saada, Saturday 31 January, one o'clock in the afternoon
We arrived here from El Hamel yesterday at three in the afternoon.

Every time I see Lella Zeyneb I feel rejuvenated, happy for no tangible reason and reassured. I saw her twice yesterday in the course of the morning. She was very good and very kind to me, and was happy to see me again.

Visited the tomb of Sidi Muhammad Belkassem, small and simple in that large mosque, and which will be very beautiful by the time it is finished. I then went on to pray on the hillside facing the grave of El Hamel's pilgrim founders.

I did some galloping along the road, together with Si Bel Abbès, under the paternal gaze of Si Ahmed Mokrani. Some women from the brothel were on their way back from El Hamel. Painted and bedecked, they were rather pretty, and came to have a cigarette with us. Did *fantasias* in their honour all along the way. Laughed a lot...

The legend of El Hamel's pilgrims appeals to my imagination. It must be one of Algeria's most biblical stories...

I began this diary over in that hated land of exile, during one of the blackest and most painfully uncertain periods in my life, a time fraught with suffering of every sort. Today it is coming to an end.

Everything is radically different now, myself included.

For a year now I have been on the blessed soil of Africa, which I never want to leave again. In spite of my poverty, I have still been able to travel and explore unknown regions of my adoptive country. My Ouïha is alive and we are relatively happy materially.

This diary, begun a year and a half ago in horrible Marseilles, comes to an end today, while the weather is grey and transparent, soft and almost dream-like here in Bou Saada, another Southern spot I used to yearn for over there!

I am getting used to this tiny room of mine at the Moorish bath; it is so much like me and the way I live. I will be staying here for a few more days before setting off on my journey to Boghar, through areas I have never seen; living in this poorly whitewashed rectangle, a tiny window giving out on the mountains and the street, two mats on the floor, a line on which to hang my laundry, and the small torn mattress I am sitting on as I write. In one corner lie straw baskets; in the opposite one is the fireplace; my papers lie scattered about... And that is all. For me that will do.

There is no more than a vague echo in these pages of all that has happened these last eighteen months; I have filled them at random, whenever I have felt the need to *articulate*... For the uninitiated reader, these pages would hardly make much sense. For myself they are a vestige of my earlier cult of the past. The day may come, perhaps, when I will no longer record the odd thought and impression in order to make them last a while. For the moment, I sometimes find great solace in rereading these words about days gone by.

I shall start another diary. What shall I record there, and where shall I be, the day in the distant future when I close it, the way I am closing this one today?

U *Allah knows what is hidden and the measure of people's sincerity!*

Edited by Elizabeth Kershaw

LUCY

The Beginnings of Humankind

Donald Johanson

Donald Johanson (born 1943) is a renowned American paleoanthropologist who, in November of 1974, discovered the remains of a bipedal hominin in the Afar region of Ethiopia. This hominin would become known as 'Lucy', the first – and by far the most famous – member of the hominin species *Australopithecus afarensis*. The discovery of Lucy offered key insights into a missing chapter of human evolution and made Johanson famous. In 1981, he established the Institute of Human Origins, a multidisciplinary research organization dedicated to bridging the academic approaches of various scientific fields to answer questions relating to the origins of humanity.

Some people are good at finding fossils. Others are hopelessly bad at it. It's a matter of practice, of training your eye to see what you need to see. I will never be as good as some of the Afar people. They spend all their time wandering around in the rocks and sand. They have to be sharp-eyed; their lives depend on it. Anything the least bit unusual they notice. One quick educated look at all those stones and pebbles, and they'll spot a couple of things a person not acquainted with the desert would miss.

Tom and I surveyed for a couple of hours. It was now close to noon, and the temperature was approaching 110. We hadn't found much: a few teeth of the small extinct horse *Hipparion*; part of the skull of an extinct pig; some antelope molars; a bit of a monkey jaw. We had large collections of all these things already, but Tom insisted on taking these also as added pieces in the overall jigsaw puzzle of what went where.

"I've had it," said Tom. "When do we head back to camp?"

"Right now. But let's go back this way and survey the bottom of that little gully over there."

The gully in question was just over the crest of the rise where we had been working all morning. It had been thoroughly checked out at least twice before by other workers, who had found nothing interesting. Nevertheless, conscious of the "lucky" feeling that had been with me since I woke, I decided to make that small final detour. There was virtually no bone in the gully. But as we turned to leave, I noticed something lying on the ground partway up the slope.

"That's a bit of a hominid arm," I said.

"Can't be. It's too small. Has to be a monkey of some kind."

We knelt to examine it.

"Much too small," said Gray again.

I shook my head. "Hominid."

"What makes you so sure?" he said.

"That piece right next to your hand. That's hominid too."

"Jesus Christ," said Gray. He picked it up. It was the back of a small skull. A few feet away was part of a femur: a thighbone. "Jesus Christ," he said again. We stood up, and began to see other bits of bone on the slope: a couple of vertebrae, part of a pelvis – all of them hominid. An unbelievable, impermissible thought flickered through my mind. Suppose all these fitted together? Could they be parts of a single, extremely primitive skeleton? No such skeleton had ever been found – anywhere.

"Look at that," said Gray. "Ribs."

A single individual?

"I can't believe it," I said. "I just can't believe it."

"By God, you'd better believe it!" shouted Gray. "Here it is. Right here!" His voice went up into a howl. I joined him. In that 110-degree heat we began jumping up and down. With nobody to share our feelings, we hugged each other, sweaty and smelly, howling and hugging in the heat-shimmering gravel, the small brown remains of what now seemed almost certain to be parts of a single hominid skeleton lying all around us.

"We've got to stop jumping around," I finally said. "We may step on something. Also, we've got to make sure."

"Aren't you sure, for Christ's sake?"

"I mean, suppose we find two left legs. There may be several individuals here, all mixed up. Let's play it cool until we can come back and make absolutely sure that it all fits together."

We collected a couple of pieces of jaw, marked the spot exactly and got into the blistering Land-Rover for the run back to camp. On the way we picked up two expedition geologists who were loaded down with rock samples they had been gathering.

"Something big," Gray kept saying to them. "Something big. Something *big*."

"Cool it," I said.

But about a quarter of a mile from camp, Gray could not cool it. He pressed his thumb on the Land-Rover's horn, and the long blast brought a scurry of scientists who had been bathing in the river. “We've got it,” he yelled. “Oh, Jesus, we've got it. We've got The Whole Thing!”

That afternoon everyone in camp was at the gully, sectioning off the site and preparing for a massive collecting job that ultimately took three weeks. When it was done, we had recovered several hundred pieces of bone (many of them fragments) representing about forty percent of the skeleton of a single individual. Tom's and my original hunch had been right. There was no bone duplication.

But a single individual of what? On preliminary examination it was very hard to say, for nothing quite like it had ever been discovered. The camp was rocking with excitement. That first night we never went to bed at all. We talked and talked. We drank beer after beer. There was a tape recorder in the camp, and a tape of the Beatles song “Lucy in the Sky with Diamonds” went belting out into the night sky, and was played at full volume over and over again out of sheer exuberance. At some point during that unforgettable evening – I no longer remember exactly when – the new fossil picked up the name of Lucy, and has been so known ever since, although its proper name – its acquisition number in the Hadar collection – is AL 288-1.

For five years I kept Lucy in a safe in my office in the Cleveland Museum of Natural History. I had filled a wide shallow box with yellow foam padding, and had cut depressions in the foam so that each of her bones fitted into its own tailor-made nest. *Everybody* who came to the Museum – it seemed to me – wanted to see Lucy. What surprised people most was her small size.

Her head, on the evidence of the bits of her skull that had been recovered, was not much larger than a softball. Lucy herself stood only three and one-half feet tall, although she was fully grown. That could be deduced from her wisdom teeth, which were fully erupted and had been exposed to several years of wear. My best guess was that she was between twenty-five and thirty years old when she died. She had already begun to show the onset of arthritis or some other bone ailment, on the evidence of deformation of her vertebrae. If she had lived much longer, it probably would have begun to bother her.

Her surprisingly good condition – her completeness – came from the fact that she had died quietly. There were no tooth marks on her bones. They had not been crunched and splintered, as they would have been if she had been killed by a lion or a saber-toothed cat. Her head had not been carried off in one direction and her legs in another, as hyenas might have done with her. She had simply settled down in one piece right where she was, in the sand of a long-vanished lake edge or stream – and died. Whether from illness or accidental drowning, it was impossible to say. The important thing was that she had not been found by a predator just after death and eaten. Her carcass had remained inviolate, slowly covered by sand or mud, buried deeper and deeper, the sand hardening into rock under the weight of subsequent depositions. She had lain silently in her adamantine grave for millennium after millennium until the rains at Hadar had brought her to light again.

That was where I was unbelievably lucky. If I had not followed a hunch that morning with Tom Gray, Lucy might never have been found. Why the other people who looked there did not see her, I do not know. Perhaps they were looking in another direction. Perhaps the light was different. Sometimes one person sees things that another misses, even though he may be looking directly at them. If I had not gone to Locality 162 that morning, nobody might have bothered to go back for a year, maybe five years. Hadar is a big place, and there is a tremendous amount to do. If I had waited another few years, the next rains might have washed many of her bones down the gully. They would have been lost, or at least badly scattered; it would not have been possible to establish that they belonged together. What was utterly fantastic was that she had come to the surface so recently, probably in the last year or two. Five years earlier, she still would have been buried. Five years later, she would have been gone.

PUSHING THE LIMITS

BEYOND POSSIBLE

Nimsdai Purja

Nimsdai Purja (born 1983) is a mountaineer and holder of numerous mountaineering world records. Born in Nepal, Purja joined the British army at the age of eighteen. He spent six years as a Ghurka before becoming the first Gurkha to join the UK Special Forces (SBS), where he served for ten years. Whilst he was in the army, Purja developed an interest in mountaineering and discovered he had a natural physiological advantage that enabled him to acclimatise more quickly than other climbers at high altitude. In 2019, when he left the armed forces, he set himself the challenge, called Project Possible, of climbing the world's fourteen highest peaks – the 8,000-ers – in seven months. Later that year, Purja and his team of Nepalese mountaineers completed Project Possible in six months and six days, smashing the previous record of eight years and setting multiple other historic firsts, including completing the first-ever winter ascent of K2.

On occasions, my attitude towards climbing a mountain was not dissimilar to the psychological position I'd adopted for military operations. In both circumstances I was never sure if I was going to come back alive, and once a mission was finished, it was often the best course of action to leave safely and efficiently. Of course, I was always supported by information: before military jobs we worked from the gathered intelligence regarding the enemy, their location and their capabilities. On climbing expeditions, we relied heavily on our knowledge of the mountain, weather reports, and our equipment; but some factors were forever beyond our determination in both scenarios.

In war, unexpected hostiles might be lurking nearby. During climbs, a rock fall might explode from nowhere. Once a team stepped into action, the unpredictable became a dangerous opponent. The best approach to neutralising it was to work steadily, and methodically, without emotion.

I wasn't a robot, though. There were some moments when my work on the mountains felt overwhelming. On some occasions during the mission I even prayed for death (though I never worried about it, or feared it), in moments when the very thought of putting one foot in front of the other felt too draining to contemplate.

These events, though rare, usually happened during trailblazing or line-fixing efforts, where I'd become so exhausted after climbing for twenty hours or more, that to close my eyes, even for a second or two, caused me to slip into sleep. As my body dropped to the ground, my brain would often jolt, shocking me awake like one of those falling dreams that kick in whenever a person drifts off too quickly, their muscles having relaxed all at once; triggering a hypnagogic jerk, where the brain imagines the body as tumbling from bed, or tripping over a step. On some peaks, fighting off the urge to snooze fitfully became a never-ending battle.

Sometimes the elements seemed capable of overwhelming me, too. Climbing through extreme weather conditions with sudden drops in temperature caused my bones and extremities to burn. Hurricane winds whipped up spindrifts; they ricocheted off my summit suit, and weirdly, banner-day conditions were sometimes equally demoralising. On Kanchenjunga, under clear, still skies, every muscle in my body had trembled with pain as I struggled to bring those casualties to safety.

Each step had felt tortuous as I heaved their weight towards Camp 4, and every now and then I briefly imagined the sweet release of an avalanche collapsing above me. Picturing my fall in the whiteout, I felt the eruption sucking me down deeper, a rock or chunk of ice knocking me unconscious. Out cold, I'd suffocate quickly, blissfully free of pain; the suffering would come to an end. Thankfully, those thoughts were only ever fleeting.

I've never been someone that grumbles about pain in front of others, or opens up about any emotional hurt I might be experiencing – not too much anyway. I feel weirdly exposed even writing these ideas down: it's a vulnerable process. But a lot of that strength had to do with being a soldier, where an alpha-male culture encouraged individuals within it to suffer silently.

The lads I served with rarely grumbled about discomfort, or discussed any psychological hurdles they might be overcoming, and I followed suit, managing my issues alone during Selection. Sometimes Suchi would ask me what it had been like and I'd mumble some vague and limited description. I suppose to admit that pain was part of my job would be to accept its reality, and to do that would increase my chances of becoming crushed by it. What UK Special Forces required were people that could grin and bear it. I sucked up the agony thrown my way and laughed as much as I could.

My combat mentality later helped me to overcome the turbulence of the Death Zone. For one, I was able to use discomfort in positive ways by turning it into a motivational fuel. During rare moments of weakness, where I'd briefly envision turning around, I'd think: *Yeah, but what happens if I give up now?* Sure, quitting would have brought some much-needed respite, but the relief would prove temporary – the long-lasting pain of giving up would be bloody miserable.

My biggest concern throughout the mission was not finishing, either through weakness or dying, so I used the potential consequences of failure as a way of *not quitting*. I pictured the disappointed people who had once looked to my project for inspiration, or the joking doubters that would inevitably make comments in interviews and call me out online. Their faces fired me up. Most of all it felt important that I complete the fourteen expeditions in one piece. I needed my story to be told truthfully and in full because my success was not a coincidence.

Then I remembered the financial risks I'd taken. I visualised my parents living together in the not-too-distant future, once my mission was completed, and the love I had for them was enough to inspire positive action. My heart and intentions were pure. I didn't want failure to sully them, so at my lowest ebbs, such as in the middle of the day-long trailblaze, I forgot about aching muscles in my legs and back simply by imagining the burn of humiliation. I was soon able to make another step through heavy snow, or along the rope, until one step became two, two steps became ten, ten steps became a hundred.

No way was I allowing myself to quit, so I also recalled my undefeated record. *I have reached all my objectives, from the Gurkhas to the Special Forces and then at high altitude. Now is not the time to break down.*

In many ways, this was the echo of an old mind trick I had used in war. When trying to negotiate pain, I often worked to create a bigger, more controllable hurt, one that would shut out the first – *replacing an agony that was beyond my control*. If my Bergen felt too heavy, I'd run harder. Any backache I'd been experiencing was soon overshadowed by the jabbing pain in my knees. On K2, I moved faster to forget the cramps in my bowels, but I also once climbed with a grinding, pounding toothache, the result of a condition called barodontalgia, where the barometric pressure trapped inside a cavity or filling changes with the high altitude. (Some people have complained of fillings popping out during a mountain expedition.)

On that occasion there was no option to turn around; I had a group of clients to lead, so instead I worked towards locating a second, more uncomfortable pain, one that I could turn off if necessary. That day, I climbed non-stop, working for a full twenty-four hours at a speed that left me fighting for breath. My lungs were tight, my whole body was in turmoil, but by the time I reached the summit, the throbbing in my gums had been forgotten.

Suffering sometimes created a weird sense of satisfaction for me. The psychological power of always giving 100 per cent, where simply *knowing* I was delivering my all, was enough to drive me on a little bit further: it created a sense of pride when seeing a job through to the end.

NOTHING VENTURE, NOTHING WIN

Sir Edmund Hillary

Sir Edmund Percival Hillary (1919–2008) was an accomplished mountaineer who became the first man to set foot on the summit of Mount Everest as a part of the Hunt Expedition in 1953. He would never claim this honour, instead emphasizing the role of his friend and fellow climber Tenzing Norgay, who later disclosed that Hillary was the first man to actually summit the highest peak in the world. Over the course of the next decade, he participated in many expeditions and became the first man in history to summit Everest and reach both the North and South Poles. He devoted the rest of his life to constructing hospitals and schools for the Sherpa people of Eastern Nepal through his Himalayan Trust and advocating for progressive political causes in his native New Zealand. A profoundly humble and conscientious figure in life, he remains a cultural icon and hero in Nepal and across the Commonwealth of Nations to this day.

May 26.

The whole group of 11 of us got away promptly in the morning for the South Col. I led all the way and Tenzing and I went pretty quickly. Things seemed to be going all right behind so we carried on fairly smartly. At 9.30 we saw our first view of two ropes on the South-east ridge above the couloir. John had left at 7 a.m. with Da Namgyl – John on 4 litres and Da Namgyl on 2 litres. Tom and Charles had a lot of trouble with their closed circuit oxygen and didn't get away until 7.50 a.m. We got a great thrill out of watching their progress. Tom and Charles surged ahead but John soon stopped and dumped loads about 150 feet above the Swiss ridge camp – at about 27,350 feet.

Tenzing and I reached the South Col in two and three-quarter hours. We saw John and Da Namgyl descending and went up to meet them and assisted them back to camp. John was very exhausted. The rest of our party duly arrived including the five Sherpas who were only carrying to the South Col – unfortunately the three Sherpas who were meant to carry loads up the ridge for us trailed the field by miles. At 1 p.m. Tom and Charles disappeared over the South Summit – what an achievement!

A great achievement it was indeed! They were higher than men had ever climbed before; but to my shame the delight I felt at their success was tempered with secret thoughts of envy and fear. Would they go on towards the top, I wondered? They had already done so much better than I had expected – perhaps they would have the strength to continue? Tenzing was more obviously concerned than I – he felt sure they would now reach the summit and felt some bitterness that a Sherpa had not been given the opportunity to get there with the first party. The South Summit was shrouded in cloud – we had no way of knowing what the men were doing.

Our unworthy thoughts were wasted effort. At three thirty p.m. Evans and Bourdillon appeared again out of the mist on their way down the ridge – they had turned back from the South Summit 28,700 feet. Absolutely exhausted, their descent of the ridge was hazardous in the extreme. They had a number of tumbles in tricky places and then fell from top to bottom of the great couloir – it was a miracle they survived. They reached the South Col more exhausted than any men I have ever seen. Tom was still bitterly disappointed they hadn't tried for the top – but Charles knew they would never have returned. Too tired almost to talk, they painted a gloomy picture of the ridge running on towards the summit; and expressed their doubts about us making it. By mid-afternoon it was blowing furiously and life on the Col was an extreme of misery.

May 27.

One of the worst nights I have ever experienced… very strong wind and very cold, –25 centigrade… and particularly uncomfortable. Tenzing, George, Greg and I were in the pyramid tent and breathing a little oxygen. John, Tom and Charles were crammed into a Meade and the three Sherpas in the small dome. It was a very windy morning indeed and I couldn't get warm. Tom and Charles were completely exhausted. They finally decided they must get away. Angtember had been sick all night so he too was to go down. John, after some rather pointed discussion decided to descend too. It was a pathetic sight to see the bunch climb the slope above camp. Tom was on his knees on numerous occasions and we had to give him an oxygen bottle. John, too, was very tired and the determined Charles seemed the only rational member of the party. They finally disappeared and had a most difficult descent to Camp VII where Mike Ward was fortunately in residence and was able to give them help.

May 28.

A fine but windy morning, –25 centigrade. The first blow was that Pember was sick leaving us only one Sherpa... Ang Nima. We decided we'd have to carry up the camp ourselves. George took 3 oxygen cylinders which made up a load of about 41 lbs; Greg carried oxygen plus a primus and food ... about 40 lbs; Ang Nima had 3 oxygen bottles... 41 lbs; Tenzing had two oxygen bottles and so did I... plus all our personal and camping gear, camera and food ... at least 49 or 50 lbs each. The other three departed at 8.45 and had made a lot of height up the couloir by the time Tenzing and I left at 10 a.m. Our loads slowed us down but we were going very well. The couloir was hard windpacked snow and had to be cut for many hundreds of feet and George did most of this work.

Tenzing and I caught up to the others on the ridge by the Swiss tent at 27,200 feet approximately. After another 150 feet we came to the dump made by John and Da Namgyl of a tent, fuel, food and oxygen. After some discussion we decided to push on carrying all the gear. I took on the tent making my load to over 60 lbs and the others divided the rest and probably had over 50 lbs. We continued on up the ridge with George doing most of the leading and plugging. The ridge was steep with a little snow over the rocks but the upward sloping strata gave easy going. We continued for some time but found no sign of a camping site. Oxygen was running low and we had to switch over to assault supplies. The position was getting a bit desperate when Tenzing did a lead out over deep unstable snow to the left and finally we found a somewhat more flattish spot beneath a rock bluff. We decided to camp here at 27,900 feet and gave the others a little oxygen and sent them down. It was 2.30 p.m.

Tenzing and I took off our oxygen and started making a camp site – a frightful job. We chopped out frozen rubble with our ice axes and tried to level a suitable area. By 5 p.m. we had cleared a site large enough for a tent but on two levels. We decided it would have to do so pitched the tent on it. We had no effective means of tying down the tent so I hitched some ropes to comers of rocks and to oxygen bottles sunk in the snow and hoped for the best. At 6 p.m. we moved into the tent. Tenzing had his Lilo along the bottom level overhanging the slope while I sat on the top level with my feet on the bottom ledge and was able to brace the whole tent against the quarter hourly fierce gusts of wind.

The primus worked like a charm and we consumed large amounts of very sweet lemon water, soup and coffee and ate with relish sardines on biscuits, a tin of apricots, dates, and biscuits with jam. I had made an inventory of our oxygen supplies which were inevitably low due to the reduced porter lift and found that we only had one and three-quarters light alloy cylinders each for the assault. By relying on the two one-third full bottles left by Tom and Charles about 500 feet below the South summit I thought we could make an attack using about 3 litres a minute (I had adjustments for this on my set and fortunately Tenzing's set was faulty and on 4 litres a minute it was really only delivering a little over three litres).

We also had a little excess oxygen in three nearly empty bottles and this could give us about 4 hours of sleeping at 1 litre a minute. The thermometer was registering –27 degrees C. but it wasn't unpleasantly cold as the wind was confined to casual strong gusts. I spread the oxygen into two two-hour periods and although I was sitting up I dozed reasonably well. Between oxygen sessions we brewed up and had lemon juice and biscuits. It was very noticeable that although we had used no oxygen from our arrival at the camp at 2.30 p.m. until we went to sleep at 9 p.m. (six and a half hours) that we were only slightly breathless and could work quite hard… certainly harder than I expected at 27,900 feet.

May 29.

At 4 a.m. the weather looked perfect and the view superb. Tenzing pointed out Thyangboche Monastery far below us. We commenced making drinks and food and thawing out frozen boots over the primus stove. I got the oxygen sets into the tent and tested them out.

At 6.30 a.m. we moved off and taking turns plugged up the ridge above camp. The ridge narrowed considerably and the breakable crust made plugging tedious and balance difficult. We soon reached the oxygen bottles and were greatly relieved to find about 1100 lbs pressure in each. The narrow ridge led up to the very impressive steep snow face running to the South Summit. Evans and Bourdillon had ascended the rocks on the left and then descended the snow on their return. Their tracks were only faintly visible and we liked neither route. We discussed the matter and decided for the snow. We commenced plugging up in foot deep steps with a thin wind crust on top and precious little belay for the ice axe. It was altogether most unsatisfactory and whenever I felt feelings of fear regarding it I'd say to myself, 'Forget it! This is Everest and you've got to take a few risks.' Tenzing expressed his extreme dislike but made no suggestions regarding turning back. Taking turns we made slow speed up this vast slope. After several hundred feet the angle eased a little and the slope was broken by more rock outcrops and the tension eased. At 9.00 a.m. we cramponed up onto the fine peak of the South Summit.

We looked with some eagerness on the ridge ahead as this was the crux of the climb. Both Tom and Charles had expressed comments on the difficulties of the ridge ahead and I was now feeling particularly hopeful. The sight ahead, in fact, was impressive but not disheartening. On the right long cornices like fingers hung over the Kangshung face. From these cornices a steep snow slope ran down to the left to the top of the rocks which drop 8,000 feet to the Western Cwm. I thought I saw a middle route by cutting steps along the snow above the rocks and sufficiently far down from the crest to be out of danger from the cornices.

Our first three-quarter bottles were finished so we discarded them and set off with a light 19 lb apparatus – one full bottle of oxygen which we were using at 3 litres a minute. We dropped off the South Summit and keeping low on the left I commenced cutting steps in excellent firm frozen snow. It was first class going and as I was feeling very well we made steady progress. Some of the cornice bumps proved tricky but I was able to turn them by dropping right down onto the rocks and scrambling by that way. Tenzing had me on a tight rope all the time and we moved throughout one at a time. After an hour or so we came to a vertical rock step in the ridge. This appeared quite a problem. However the step was bounded on its right by a vertical snow cliff and I was able to work my way up this 40 foot crack and finally get over the top. I was rather surprised and pleased that I was capable of such effort at this height. I brought Tenzing up and noticed he was proving a little sluggish, but an excellent and safe companion for all that. I really felt now that we were going to get to the top and that nothing would stop us. I kept frequent watch on our oxygen consumption and was encouraged to find it at a steady rate.

I continued on, cutting steadily and surmounting bump after bump and cornice after cornice looking eagerly for the summit. It seemed impossible to pick it and time was running out. Finally I cut around the back of an extra large hump and then on a tight rope from Tenzing I climbed up a gentle snow ridge to its top. Immediately it was obvious that we had reached our objective. It was 11.30 a.m. and we were on top of Everest!

To the north an impressive corniced ridge ran down to the East Rongbuk glacier. We could see nothing of the old North route but were looking down on the North Col and Changtse. The West ridge dropped away in broad sweeps and we had a great view of the Khumbu and Pumori far below us. Makalu, Kangchenjunga and Lhotse were dominant to the east looking considerably less impressive than I had ever seen them. Tensing and I shook hands and then Tenzing threw his arms around my shoulders. It was a great moment! I took off my oxygen and for ten minutes I photographed Tenzing holding flags, the various ridges of Everest and the general view. I left a crucifix on top for John Hunt and Tenzing made a little hole in the snow and put in it some food offerings – lollies, biscuits and chocolate. We ate a Mint Cake and then put our oxygen back on. I was a little worried by the time factor so after 15 minutes on top we turned back at 11.45.

The steps along the ridge made progress relatively easy and the only problem was the rock step which demanded another jamming session. At 12.45 we were back on the South Summit both now rather fatigued. Wasting no time (our oxygen was getting low) we set off down the great slope in considerable trepidation about its safeness. This was quite a mental strain and as I was coming down first I repacked every step with great care. Tenzing was a tower of strength and his very fine ability to keep a tight rope was most encouraging. After what seemed a lifetime the angle eased off and we were soon leading down onto the narrow snow ridge and finally to the dump of oxygen bottles. We loaded these on and then, rather tired, wended our way down our tracks and collapsed into our camp at 2 p.m. Our original bottles were now exhausted. They had given us four and three-quarter hours and allowing 800 litres in these very full bottles our consumption had averaged two and five-sixth litres per minute.

At the ridge camp we had a brew of lemon and sugar and then picked up all our personal gear and connected up our last bottles – one-third full. At 3 p.m. we left the ridge camp and although we were tired we made good time down the ridge to the Swiss camp and the couloir. The snow in the couloir was firm and we had to recut all the steps. We kicked down the lower portions and then cramponed very weakly down to meet George...

George met us with a mug of soup just above camp, and seeing his stalwart frame and cheerful face reminded me how fond of him I was. My comment was not specially prepared for public consumption but for George... 'Well, we knocked the bastard off!' I told him and he nodded with pleasure... 'Thought you must have!'

Wilf Noyce and Pasang Puta were also in Camp and they looked after us with patience and kindness. I felt a moment of sympathy for Wilf – he was the only one left with the strength to try for the summit; but now he wouldn't get the chance. It was another foul night with strong wind and very cold temperatures but Tenzing and I were given oxygen and the time passed amazingly quickly. Reaction was starting to set in next day and I felt very lethargic as we traversed the long slopes of the Lhotse face and then tramped slowly down the Cwm towards Camp IV. The party drifted out of camp towards us, not knowing if we had been successful or not. When George exuberantly gave the thumbs up signal of success they rushed towards us and soon we were embracing them all, and shaking hands, and thumping each other on the back. It was a touching and unforgettable moment; and yet somehow a sad one too.

INDIA AND TIBET

A history of the relations which have subsisted between the two countries from the time of Warren Hastings to 1910; with a particular account of the mission to Lhasa of 1904

Francis Younghusband

Francis Younghusband (1863–1942) was a British army officer, explorer and devoted spiritualist who, in 1903, led an expedition into Tibet to settle a border dispute. This excursion amounted to a de facto invasion of the region and resulted in the deaths of several hundred Tibetans – mostly monks – following a disastrous confrontation. Wracked with guilt regarding the invasion of Tibet and having experienced a mystical vision on his journey through the territory, Younghusband abandoned his Christian beliefs for a deep spiritualism. His transformed views would resemble elements of the New Age movement that would follow some decades later.

On January 7 we encamped at the foot of the pass, the thermometer that night falling 18° below zero. As I looked out of my tent at the first streak of dawn the next morning there was a clear cutting feel in the atmosphere, such as is only experienced at great altitudes. The stars were darting out their rays with almost supernatural brilliance. The sky was of a steely clearness, into which one could look unfathomable depths. Behind the great sentinel peak of Chumalhari, which guards the entrance to Tibet, the first streaks of dawn were just appearing. Not a breath of air stirred, but all was gripped tight in the frost which turned buckets of water left out overnight into solid ice, and made the remains of last night's stew as hard as a rock. Under such conditions we prepared for our advance over the pass, and as the troops were formed on parade, preparatory to starting, it was found that many of the rifles and one of the Maxims would not work, on account of the oil having frozen.

The rise to the pass was very gradual, and the pass itself, 15,200 feet above sea-level, was so wide and level that we could have advanced across it in line. But soon now the wind got up, and swept along the pass with terrific force. At this altitude, and clad in such heavy clothing, we could advance but slowly, and the march seemed interminable. The clearness of the atmosphere made the little hamlet of Tuna appear quite near; but hour after hour we plodded wearily over the plateau, and it was late in the afternoon before we reached it, and even then, for the sake of water, we had to go a mile or more beyond, and encamp in the open.

Column Crossing the Tang-La, January 1904

A Tibetan force was near at hand, and as they were credited with a habit of attacking at night, General Macdonald took special precautions against such an eventuality; but as darkness set in and the cold increased in intensity, we felt we should be pretty helpless in an open camp, and there were some thoughts of retiring again across the pass, for the military risks were very great. But, on the whole, we thought it would be better to face it now we were there; and as, next morning, we examined the hamlet of Tuna, and found it could be turned into a good defensible post, and had a well within the walls, we decided that the Mission should remain there, with an escort of four companies of the 23rd Pioneers, Lieutenant Hadow's Maxim-gun detachment, and a 7-pounder—the whole under Colonel Hogge; while General Macdonald, with the flying column, returned to Chumbi to complete his arrangements.

The immediate surroundings in which we now found ourselves were miserable in the extreme. Tuna was nearly 15,000 feet above the sea, and was the filthiest place I have ever seen. We tried to live in the houses, but after a few days preferred our tents, in spite of the cold, which was intense, and against which we could not have the comfort and cheer of a fire, for only sufficient fuel for cooking could be obtained, most of it being yak-dung, and much having to be brought from Chumbi. The saving feature was the grand natural scenery, which was a joy of which I never tired. Immediately before us was an almost level and perfectly smooth gravel plain ten or twelve miles in width, and on the far side of this rose the great snowy range, which forms the main axis of the Himalayas, and here separates Tibet from Bhutan. Snow seldom fell. The sky was generally clear, and the sunshine brilliant, and well wrapped up, away from the dirty hamlet and sheltered from the terrific wind, there was pleasure to be had out of even Tuna. And the sight of the serene and mighty Chumalhari, rising proudly above all the storms below and spotless in its purity, was a never-ending solace in our sordid winter post.

Chapter XII
Tuna

The first event of importance after our arrival at Tuna was the receipt, on January 12, of a message from the Lhasa officials, saying that they wished for an interview. At noon, the time I had appointed, several hundreds of men appeared on the plain below the village. They halted there, and asked that I should come out and meet them halfway. Perhaps unnecessarily, I refused this request. It was bitingly cold in the open plain, and I thought the Tibetan leaders might have come into my camp, where I had said I would receive them, and where a guard of honour was ready. However, I sent out the indispensable and ever-ready Captain O'Connor to hear what they had to say, and on his return he replied that they once more urged us to return to Yatung, but afterwards stated that they were prepared to discuss matters there, at Tuna.

This constituted a distinct improvement on the attitude adopted by them at Phari, and their general demeanour was much more cordial, according to Captain O'Connor. But they told him that if we advanced and they were defeated, they would fall back upon another Power, and that things would then be bad for us. In conversation with the Munshi they said that they would prevent us from advancing beyond our present position, and they repudiated our treaty with the Chinese, saying they were tired of the Chinese, and could conclude a treaty by themselves.

Chumalhari

Encouraged by the fact that they showed some little signs of a desire to discuss matters, I determined now to make a bold move to get to close quarters with them. I was heartily tired of this fencing about at a distance; I wanted to get in under their reserve. And I thought that if we could meet and could tell them in an uncontentious and unceremonious manner what all the pother was about, we might at any rate get a start—get what the Americans call a "move on." It was worth while, it seemed to me, to make a supreme effort to get this intrinsically small matter settled by peaceful means, even if a very considerable risk was incurred in the process; and I wished particularly to see them, and to judge of them, in their own natural surroundings. I was constantly being called upon by Government to give my opinion upon the probable action of the Tibetans, but so far I had only seen them in our own camps, and they had steadily refused to admit me into theirs. I therefore determined on the following morning, without any formality, without any previous announcement, and without any escort, to ride over to their camp, about ten miles distant, at Guru, and talk over the general situation—not as British Commissioner, with a list of grievances for which he had to demand redress, but as one who wished to understand them, and by friendly means to effect a settlement. I was only too well aware that such an attempt was likely to be taken by the Tibetans as a sign of weakness; still, when I saw these people so steeped in ignorance of what opposing the might of the British Empire really meant, I felt it my duty to reason with them up to the latest moment, to save them from the results of their ignorance.

Captain O'Connor and Captain Sawyer, of the 23rd Pioneers, who was learning Tibetan, accompanied me, but we did not take with us even a single sepoy as escort. On our way we were met by messengers, who had come to say that the Tibetan chiefs would not come to see me at Tuna, and I was all the more pleased that I had left Tuna before the message arrived.

On reaching Guru, a small village under a hill, we found numbers of Tibetan soldiers out collecting yak-dung in the surrounding plain; but there was no military precaution whatever taken, and we rode straight into the village. About 600 soldiers were huddled up in the cattle-yards of the houses. They were only armed with spears and matchlocks, and had no breech-loaders. As we rode through the village they all crowded out to look at us, and not with any scowls, but laughing to each other, as if we were an excellent entertainment. They were not very different in appearance from the ordinary Bhutia dandy-bearers of Darjiling or the yak-drivers we had with us in camp.

We asked for the General, and on reaching the principal house I was received at the head of the stairs by a polite, well-dressed, and well-mannered man, who was the Tibetan leader, and who was most cordial in his greeting. Other Generals stood behind him, and smiled and shook hands also. I was then conducted into a room in which the three Lhasa monks were seated, and here the difference was at once observable. They made no attempt to rise, and only made a barely civil salutation from their cushions. One object of my visit had already been attained: I could from this in itself see how the land lay, and where the real obstruction came from.

The Lhasa General and the Shigatse Generals—we had become accustomed to calling them Generals, though the English reader must not imagine they at all resembled Napoleon—took their seats on cushions at the head of the room and opposite to the monks. We were given three cushions on the right, and two Shigatse Generals and another Shigatse representative had seats on the left. Tea was served, and the Lhasa General, as the spokesman of the assembly, asked after my health.

After I had made the usual polite replies and inquiries after their own welfare, I said I had not come to them now on a formal visit as British Commissioner, or with any idea of officially discussing the various points of difference between us; but I was anxious to see them and know them, and to have an opportunity of freely discussing the general situation in a friendly, informal manner. So I had ridden over, without ceremony and without escort, to talk matters over, and see if there was no means of arriving at a settlement by peaceful means. I said that I had been appointed British Commissioner on account of my general experience in many different countries, that I had no preconceived ideas upon this question and no animus against them; from what I had seen of them, I was convinced there was no people with whom we were more likely to get on, and I hoped now we had really met each other face to face we should find a means of settling our differences and forming a lasting friendship.

The Lhasa General replied that all the people of Tibet had a covenant that no Europeans were ever to be allowed to enter their country, and the reason was that they wished to preserve their religion. The monks here chimed in, saying that their religion must be preserved, and that no European, on any account, must be admitted. The General then went on to say that, if I really wanted to make a friendly settlement, I should go back to Yatung.

I told him that for a century and a half we had remained quietly in India, and made no attempt to force ourselves upon them. Even though we had a treaty right to station an officer at Yatung, we had not exercised that right. But of recent years we had heard from many different sources that they were entering into friendly relations with the Russians, while they were still keeping us at arm's length. One Dorjieff, for instance, had been the bearer of autograph letters from the Dalai Lama to the Czar and Russian officials at the very time when the Lama was refusing letters from the Viceroy of India. We could understand their being friendly with both the Russians and ourselves, or their wishing to have nothing to do with either; but when they were friendly with the Russians and unfriendly with us, they must not be surprised at our now paying closer attention to our treaty rights.

The General assured me that it was untrue that they had any dealings with the Russians, and the monks brusquely intimated that they disliked the Russians just as much as they disliked us; they protested that they had nothing to do with the Russians, that there was no Russian near Lhasa at that time, and that Dorjieff was a Mongolian, and the custom of Mongolians was to make large presents to the monasteries. They asked me, therefore, not to be so suspicious.

I said it was difficult not to be suspicious when they persistently kept us at such a distance. I then addressed them in regard to religion, and asked them if they had ever heard that we interfered with the religions of the people of India. They admitted that we did not interfere, but they maintained, nevertheless, that it was to preserve their religion that they adhered to their determination to keep us out.

As the Buddhist religion nowhere preaches this seclusion, it was evident that what the monks wished to preserve was not their religion, but their priestly influence. This was the crux of the whole situation. And it entirely bore out what Mr. Nolan, the Commissioner of Darjiling, had observed many years before—that it was "the breaking of the beggars' bowl" that was in question, the loss of these presents from Mongolians and others.

So far the conversation, in spite of occasional bursts from the monks, had been maintained with perfect good-humour; but when I made a sign of moving, and said that I must be returning to Tuna, the monks, looking as black as devils, shouted out: "No, you won't; you'll stop here." One of the Generals said, quite politely, that we had broken the rule of the road in coming into their country, and we were nothing but thieves and brigands in occupying Phari Fort. The monks, using forms of speech which Captain O'Connor told me were only used in addressing inferiors, loudly clamoured for us to name a date when we would retire from Tuna before they would let me leave the room. The atmosphere became electric. The faces of all were set. One of the Generals left the room; trumpets outside were sounded, and attendants closed round behind us.

A real crisis was on us, when any false step might be fatal. I told Captain O'Connor, though there was really no necessity to give such a warning to anyone so imperturbable, to keep his voice studiously calm, and to smile as much as he possibly could, and I then said that I had to obey the orders of my Government, just as much as they had to obey the orders of theirs; that I would ask them to report to their Government what I had said, and I would report to my Government what they had told me. That was all that could be done at present; but if the Viceroy, in reply to my reports, ordered me back to India I should personally be only too thankful, as theirs was a cold, barren, and inhospitable country, and I had a wife and child at Darjiling, whom I was anxious to see again as soon as I could.

This eased matters a little. But the monks continued to clamour for me to name a date for withdrawal, and the situation was only relieved when a General suggested that a messenger should return with me to Tuna to receive there the answer from the Viceroy. The other Generals eagerly accepted the suggestion, and the tension was at once removed. Their faces became smiling again, and they conducted me to the outer door with the same geniality and politeness with which they had received us, though the monks remained seated and as surly and evil-looking as men well could look.

We preserved our equanimity of demeanour and the smiles on our faces till we had mounted our ponies and were well outside the camp, and then we galloped off as hard as we could, lest the monks should get the upper hand again and send men after us. It had been a close shave, but it was worth it.

I had sized up the situation, and felt now I knew how I stood. I knew from that moment that nowhere else than in Lhasa, and not until the monkish power had been broken, should we ever make a settlement. But it was still treason to mention the word "Lhasa" in any communication to Government, and I had to keep these conclusions to myself for many months yet, for fear I might frighten people in England who had not yet got accustomed to the idea of our going even as far as Gyantse.

While I perceived that the monks were implacably hostile, that they had the preponderating influence in the State, and were entirely convinced of their power to dictate to us, I perceived also that the lay officials were much less unfriendly, less ignorant of our strength, and more amenable to reason, and that the ordinary people and soldiers, though perhaps liable to be worked on by the monks, had no innate bad feeling against us. Hereon I based my hopes for the security of the eventual settlement.

SILENCE

Shūsaku Endō

Shūsaku Endō (1923–1996) is considered one of twentieth-century Japan's greatest novelists. *Silence* tells the story of two Portuguese Jesuit missionaries who travel to Japan in 1639 to discover the fate of their mentor who is rumoured to have apostatized under pressure from the Japanese authorities. The novel is a commentary on the meaning of faith and endurance in the face of persecution, and the importance of cultural differences in determining religious attitudes. *Silence* proceeds from a position of moral absoluteness to a deep sense of moral ambiguity as the arrival of the two priests results in horrific persecution for the underground community of Christians in Japan. Father Sebastião Rodrigues becomes deeply troubled by the suffering of the Japanese Christians, particularly as it becomes evident that their suffering is needless, and could come to a swift end if only he, a foreigner and interloper, would willingly renounce his faith. The inner turmoil that engulfs Rodrigues acts as a corollary to the outer turmoil his arrival brings upon the Christian community in Japan. In the end, he is faced with the prospect of denying his faith to save his people.

Ferreira stood there motionless, his head hanging down like an old animal. The interpreter, true to type, put his head down to the barely opened door and for a long time peered in at the scene. Waiting and waiting, he heard no sound, and uneasily whispered in a hoarse voice: 'I suppose you're not dead. Oh no! No! It's not lawful for a Christian to put an end to that life given him by God. Sawano! The rest is up to you.' With these words he turned around and disappeared from sight, his footsteps echoing in the darkness.

When the footsteps had completely died out, Ferreira, silent, his head hanging down, made no movement. His body seemed to be floating in air like a ghost; it looked thin like a piece of paper, small like that of a child. One would think that it was impossible even to clasp his hand.

'Eh!' he said putting his face in at the door. 'Eh! Can you hear me?'

There was no answer and Ferreira repeated the same words. 'Somewhere on that wall,' he went on, 'you should be able to find the lettering that I engraved there. "Laudate Eum." Unless they have been cut away, the letters are on the right-hand wall ... Yes, in the middle ... Won't you touch them with your fingers?'

But from inside the cell there came not the faintest sound. Only the pitch darkness where the priest lay huddled up in the cell and through which it seemed impossible to penetrate.

'I was here just like you.' Ferreira uttered the words distinctly, separating the syllables one from another. 'I was imprisoned here, and that night was darker and colder than any night in my life.'

The priest leaned his head heavily against the wooden wall and listened vaguely to the old man's words. Even without the old man's saying so, he knew that that night had been blacker than any before. Indeed, he knew it only too well. The problem was not this; the problem was that he must not be defeated by Ferreira's temptings – the tempting of a Ferreira who had been shut up in the darkness just like himself and was now enticing him to follow the same path.

'I, too, heard those voices. l heard the groaning of men hanging in the pit.' And even- as Ferreira finished speaking, the voices like snoring, now high, now low, were carried to their ears. But now the priest was aware of the truth. It was not snoring. It was the gasping and groaning of helpless men hanging in the pit.

While he had been squatting here in the darkness, someone had been groaning, as the blood dripped from his nose and mouth. He had not even adverted to this; he had uttered no prayer; he had laughed. The very thought bewildered him completely. He had thought the sound of that voice ludicrous, and he had laughed aloud. He had believed in his pride that he alone in this night was sharing in the suffering of that man. But here just beside him were people who were sharing in that suffering much more than he. Why this craziness, murmured a voice that was not his own. And you call yourself a priest! A priest who takes upon himself the sufferings of others! 'Lord, until this moment have you been mocking me?' he cried aloud.

'Laudate Eum! I engraved those letters on the wall,' Ferreira repeated. 'Can't you find them? Look again!'

'I know!' The priest, carried away by anger, shouted louder than ever before. 'Keep quiet!' he said. 'You have no right to speak like this.'

'I have no right? That is certain. I have no right. Listening to those groans all night I was no longer able to give praise to the Lord. I did not apostatize because I was suspended in the pit. For three days, I who stand before you was hung in a pit of foul excrement, but I did not say a single word that might betray my God.' Ferreira raised a voice that was like a growl as he shouted: 'The reason I apostatized . . . are you ready? Listen! I was put in here and heard the voices of those people for whom God did nothing. God did not do a single thing. I prayed with all my strength; but God did nothing.'

'Be quiet!'

'All right. Pray! But those Christians are partaking of a terrible suffering such as you cannot even understand. From yesterday – in the future – now at this very moment. Why must they suffer like this? And while this goes on, you do nothing for them. And God – he does nothing either.'

The priest shook his head wildly, putting both fingers into his ears. But the voice of Ferreira together with the groaning of the Christians broke mercilessly in. Stop! Stop! Lord, it is now that you should break the silence. You must not remain silent. Prove that you are justice, that you are goodness, that you are love. You must say something to show the world that you are the august one.

A great shadow passed over his soul like that of the wings of a bird flying over the mast of a ship. The wings of the bird now brought to his mind the memory of the various ways in which the Christians had died. At that time, too, God had been silent. When the misty rain floated over the sea, he was silent. When the one-eyed man had been killed beneath the blazing rays of the sun, he had said nothing. But at that time, the priest had been able to stand it; or, rather than stand it, he had been able to thrust the terrible doubt far from the threshold of his mind. But now it was different. Why is God continually silent while those groaning voices go on?

'Now they are in that courtyard.' (It was the sorrowful voice of Ferreira that whispered to him.) 'Three unfortunate Christians are hanging. They have been hanging there since you came here.'

The old man was telling no lie. As he strained his ears the groaning that had seemed to be that of a single voice suddenly revealed itself as a double one – one groaning was high (it never became low): the high voice and the low voice were mingled with one another, coming from different persons.

'When I spent that night here five people were suspended in the pit. Five voices were carried to my ears on the wind. The official said: "If you apostatize, those people will immediately be taken out of the pit, their bonds will be loosed, and we will put medicine on their wounds." I answered: "Why do these people not apostatize?" And the official laughed as he answered me: "They have already apostatized many times. But as long as you don't apostatize these peasants cannot be saved."'

'And you . . .' The priest spoke through his tears. 'You should have prayed . . . .'

'I did pray. I kept on praying. But prayer did nothing to alleviate their suffering. Behind their ears a small incision has been made; the blood drips slowly through this incision and through the nose and mouth. I know it well, because I have experienced that same suffering in my own body. Prayer does nothing to alleviate suffering.'

The priest remembered how at Saishoji when first he met Ferreira he had noticed a scar like a burn on his temples. He even remembered the brown colour of the wound, and now the whole scene rose up behind his eyelids. To chase away the imagination he kept banging his head against the wall. 'In return for these earthly sufferings, those people will receive a reward of eternal joy,' he said.

'Don't deceive yourself!' said Ferreira. 'Don't disguise your own weakness with those beautiful words.'

'My weakness?' The priest shook his head; yet he had no self-confidence. 'What do you mean? It's because I believe in the salvation of these people . . .'

'You make yourself more important than them. You are preoccupied with your own salvation. If you say that you will apostatize, those people will be taken out of the pit. They will be saved from suffering. And you refuse to do so. It's because you dread to betray the Church. You dread to be the dregs of the Church, like me.' Until now Ferreira's words had burst out as a single breath of anger, but now his voice gradually weakened as he said: 'Yet I was the same as you. On that cold, black night I, too, was as you are now. And yet is your way of acting love? A priest ought to live in imitation of Christ. If Christ were here . . .'

For a moment Ferreira remained silent; then he suddenly broke out in a strong voice: 'Certainly Christ would have apostatized for them.'

Night gradually gave place to dawn. The cell that until now had been no more than a lump of black darkness began to glimmer in a tiny flicker of whitish light.

'Christ would certainly have apostatized to help men.'

'No, no!' said the priest, covering his face with his hands and wrenching his voice through his fingers. 'No, no!'

'For Love Christ would have apostatized. Even if it meant giving up everything he had.'

'Stop tormenting me! Go away, away!' shouted the priest wildly. But now the bolt was shot and the door opened – and the white light of the morning flooded into the room.

'You are now going to perform the most painful act of love that has ever been performed,' said Ferreira, taking the priest gently by the shoulder.

Swaying as he walked, the priest dragged his feet along the corridor. Step by step he made his way forward, as if his legs were bound by heavy leaden chains – and Ferreira guided him along. In the gentle light of the morning, the corridor seemed endless; but there at the end stood the interpreter and two guards, looking just like three black dolls.

'Sawano, is it over? Shall we get out the *fumie*?' As he spoke the interpreter put on the ground the box he was carrying and, opening it, he took out a large wooden plaque.

'Now you are going to perform the most painful act of love that has ever been performed.' Ferreira repeated his former words gently. 'Your brethren in the Church will judge you as they have judged me. But there is something more important than the Church, more important than missionary work: what you are now about to do.'

The *fumie* is now at his feet.

A simple copper medal is fixed onto a grey plank of dirty wood on which the grains run like little waves. Before him is the ugly face of Christ, crowned with thorns and the thin, outstretched arms. Eyes dimmed and confused the priest silently looks down at the face which he now meets for the first time since coming to this country.

'Ah,' says Ferreira. 'Courage!'

'Lord, since long, long ago, innumerable times I have thought of your face. Especially since coming to this country have I done so tens of times. When I was in hiding in the mountains of Tomogi; when I crossed over in the little ship; when I wandered in the mountains; when I lay in prison at night . . . Whenever I prayed your face appeared before me; when I was alone I thought of your face imparting a blessing; when I was captured your face as it appeared when you carried your cross gave me life. This face is deeply ingrained in my soul – the most beautiful, the most precious thing in the world has been living in my heart. And now with this foot I am going to trample on it.'

The first rays of the dawn appear. The light shines on his long neck stretched out like a chicken and upon the bony shoulders. The priest grasps the *fumie* with both hands, bringing it close to his eyes. He would like to press to his own face that face trampled on by so many feet. With saddened glance he stares intently at the man in the centre of the *fumie*, worn down and hollow with the constant trampling. A tear is about to fall from his eye. 'Ah,' he says trembling, 'the pain!'

'It is only a formality. What do formalities matter?' The interpreter urges him on excitedly. 'Only go through with the exterior form of trampling.'

The priest raises his foot. In it he feels a dull, heavy pain. This is no mere formality. He will now trample on what he has considered the most beautiful thing in his life, on what he has believed most pure, on what is filled with the ideals and the dreams of man. How his foot aches! And then the Christ in bronze speaks to the priest: 'Trample! Trample! I more than anyone know of the pain in your foot. Trample! It was to be trampled on by men that I was born into this world. It was to share men's pain that I carried my cross.'

The priest placed his foot on the *fumie*. Dawn broke. And far in the distance the cock crew.

Translated by William Johnston

SELECTED POLITICAL WRITINGS

Mahatma Gandhi

Mahatma Gandhi (1869–1948) was an Indian civil rights leader, lawyer and activist who successfully led the campaign of non-violent resistance against British colonial rule in India. Gandhi lived to see India gain its independence from the British Empire but was murdered not long after by a radical Hindu nationalist who opposed his interfaith tolerance towards Muslims, Sikhs and Jains. Gandhi's philosophy of non-violence inspired later civil rights leaders such as Nelson Mandela in South Africa and Dr Martin Luther King Jr. in the United States.

NON-VIOLENCE

When a person claims to be non-violent, he is expected not to be angry with one who has injured him. He will not wish him harm; he will wish him well; he will not swear at him; he will not cause him any physical hurt. He will put up with all the injury to which he is subjected by the wrongdoer. Thus non-violence is complete innocence. Complete non-violence is complete absence of ill will against all that lives. It therefore embraces even sub-human life not excluding noxious insects or beasts. They have not been created to feed our destructive propensities. If we only knew the mind of the Creator, we should find their proper place in His creation. Non-violence is therefore, in its active form, goodwill towards all life. It is pure Love. I read it in the Hindu scriptures, in the Bible, in the Koran.

Non-violence is a perfect state. It is a goal towards which all mankind moves naturally though unconsciously. Man does not become divine when he personifies innocence in himself. Only then does he become truly man. In our present state, we are partly men and partly beasts and, in our ignorance and even arrogance, say that we truly fulfill the purpose of our species when we deliver blow for blow and develop the measure of anger required for the purpose. We pretend to believe that retaliation is the law of our being, whereas in every scripture we find that retaliation is nowhere obligatory but only permissible. It is restraint that is obligatory. Retaliation is indulgence requiring elaborate regulating. Restraint is the law of our being. For, highest perfection is unattainable without highest restraint. Suffering is thus the badge of the human tribe.

The goal ever recedes from us. The greater the progress, the greater the recognition of our unworthiness. Satisfaction lies in the effort, not in the attainment. Full effort is full victory.

Therefore, though I realize more than ever how far I am from that goal, for me the Law of complete Love is the law of my being. Each time I fail, my effort shall be all the more determined for my failure ....

A drop of water must yield to the analyst the same results as a lakeful. The nature of my non-violence towards my brother cannot be different from that of my non-violence to the universe, it must still satisfy the same test ....

The political non-violence of the non-co-operator [in the civil disobedience campaign of 1920–22] does not stand this test in the vast majority of cases. Hence the prolongation of the struggle. Let no one blame the unbending English nature. The hardest "fibre" must melt in the fire of love.

I cannot be dislodged from the position because I know it. When British or other nature does not respond, the fire is not strong enough, if it is there at all.

Our non-violence need not be of the strong, but it *has* to be of the truthful. We must not intend harm to the English or to our co-operating countrymen if and whilst we claim to be non-violent. But the majority of us *have* intended harm, and we have refrained from doing it because of our weakness or under the ignorant belief that mere refraining from physical hurt amounted to due fulfillment of our pledge. Our pledge of nonviolence excludes the possibility of future retaliation. Some of us seem, unfortunately, to have merely postponed the date of revenge.

Let me not be misunderstood. I do not say that the policy of non-violence excludes the possibility of revenge when the policy is abandoned. But it does most emphatically exclude the possibility of future revenge after a successful termination of the struggle. Therefore, whilst we are pursuing the policy of non-violence, we are bound to be actively friendly to English administrators and their co-operators ....

Swaraj by non-violent means can therefore never mean an interval of chaos and anarchy. Swaraj by non-violence must be a progressively peaceful revolution such that the transference of power from a close corporation to the people's representatives will be as natural as the dropping of a fully ripe fruit from a well-nurtured tree. I say again that such a thing may be quite impossible of attainment. But I know that nothing less is the implication of non-violence. And if the present workers do not believe in the probability of achieving such comparatively non-violent atmosphere, they should drop the non-violent program and frame another which is wholly different in character. If we approach our program with the mental reservation that, after all, we shall wrest the power from the British by force of arms, then we are untrue to our profession of non-violence. If we believe in our program, we are bound to believe that the British people are not unamenable to the force of affection as they are undoubtedly amenable to force of arms. For the unbelievers, the [alternative is] ... a rapid but bloody revolution probably never witnessed before in the world. I have no desire to take part in such a revolution. I will not be a willing instrument for promoting it.

MY PATH

I am conscious of the fact that the truth for which I stand has not yet been fully accepted by India. It has not yet been fully vindicated. My work in India is still in the experimental stage.

My path is clear. Any attempt to use me for violent purposes is bound to fail. I have no secret methods. I know no diplomacy save that of truth. I have no weapon but non-violence. I may be unconsciously led astray for a while but not for all time. I have therefore well-defined limitations, within which alone I may be used ....

I am yet ignorant of what exactly Bolshevism is. I have not been able to study it. I do not know whether it is for the good of Russia in the long run. But I do know that in so far as it is based on violence and denial of God, it repels me. I do not believe in short-violent-cuts to success. Those Bolshevik friends who are bestowing their attention on me should realize that however much I may sympathize with and admire worthy motives, I am an uncompromising opponent of violent methods even to serve the noblest of causes. There is, therefore, really no meeting ground between the school of violence and myself. But my creed of non-violence not only does not preclude me but compels me even to associate with anarchists and all those who believe in violence. But that association is always with the sole object of weaning them from what appears to me to be their error. For experience convinces me that permanent good can never be the outcome of untruth and violence.

THE PRISON DIARY OF HO CHI MINH

Ho Chi Minh

Ho Chi Minh (1890–1969) was a Vietnamese politician and communist revolutionary. He led the Vietnamese independence movement from 1941, successfully fighting against French imperial control and later the military invasion of Vietnam by the United States. An author and poet, little is known about Ho's early life and various competing accounts exist regarding the details of his birth, childhood and activities before he became involved in the independence movement. In 1942, he was arrested in China by Chiang Kai-shek's nationalist government, probably under suspicion of spying. His *Prison Diary* chronicles his harsh eighteen-month imprisonment, and his hopes and aspirations, through poetry. He is considered the father of modern Vietnam and a revolutionary icon for his determined opposition to Western colonial rule.

Hard is the Road of Life

I

Having travelled over steep mountain and deep ravine,
How could I expect in the plain to meet even greater danger?
In the mountain I suffered no harm from the tiger;
In the plain I met with men and was thrown in jail.

The Flute of a Fellow-Prisoner

Nostalgically a flute wails in the ward.
Sad grows the tone, mournful the melody.
Miles away, beyond passes and streams, in infinite melancholy,
A lonely figure mounts a tower gazing far and wide.

Learning to Play Chess

I

To while the time away we learn to play chess.
Horse and foot are engaged in endless chase.
Move with lightning speed in attack or defence
Talent and nimble feet will give you the upper hand.

Moonlight

In jail there is neither flower nor wine.
What could one do when the night is so exquisite?
To the window I go and look at the moonshine.
Through the bars the moon gazes at the poet.

Transferred to Tianbao on "Double Ten" Day

Every house is decked with lanterns and flowers:
It's national day, the whole country is filled with delight.
But this is the moment I am put in chains for transfer:
Contrary winds continue to hamper the eagle's flight.

Out On the Road

Out on the road we learn to know hardships
One peak hardly climbed, another above us rises.
But once we've struggled to the highest pass,
Our eyes can encompass at one glance ten thousand li.

Arrival at Tianbao Jail

Today I have walked fifty-three kilometres.
My hat and clothes are soaking through, my shoes in tatters.
Without a place to sleep, all through the night
I sit by the edge of the latrine, waiting for light.

Advice to Myself

Without the cold and bleakness of winter
The warmth and splendour of spring could never be.
Misfortunes have steeled and tempered me
And even more strengthened my resolve.

The Bonds

Entwined round my arms and legs is a long dragon:
I look like a foreign officer with braid on the shoulders
But the cords of these officers are woven of gold,
While my decoration is a thick rope of fibre.

Guards Carry a Pig

I

Going with us, guards carry a pig
On their shoulders, while I am rudely dragged along.
A man is treated worse than a pig,
Once he's deprived of his liberty.

Sadness

The whole world is ablaze with the flames of war.
Fighters eagerly ask to be sent to the front.
In jail inaction weighs on the prisoner heavier still:
His noble ambitions are not worth a paltry cent.

The Morning Sun

The morning sun into the prison penetrates:
The smoke clears away, the mist dissipates.
The breath of life suddenly fills the skies,
And the prisoners' faces are now all smiles.

Midnight

In sleep an honest look on all faces is worn;
Only when people wake does good or evil show
Good and evil are not qualities inborn;
More often than not from education they grow.

Sleepless Nights

During the long and sleepless nights in prison,
I've written more than a hundred poems on thraldom.
Often at the end of a quatrain I put down my brush
And through the bars look up at the sky of freedom.

Autumn Night

At the gate guards holding their rifles stand.
Above them shredded clouds are drifting with the moon.
Bed-bugs swarm about like tanks manoeuvring.
Like air squadrons, mosquitoes regroup and disperse.
My heart travels a thousand li to my country.
Sadness twists my dreams into a thousand tangled skeins.
An innocent man, yet I've been a whole year in chains.
With tears dropping on my inkslab, I make another poem on captivity.

After Prison, Practising Mountain-Climbing

The mountains embrace the clouds, the clouds hug the mountains.
The river below shines like a spotless mirror.
On the slopes of the Western Range, my heart beats as I wander,
Looking towards the Southern skies and thinking of old friends.

ONE GIRL ONE DREAM

Laura Dekker

Laura Dekker (born 1995) is a Dutch sailor who, in 2012, became the youngest person to complete a solo circumnavigation of the globe at the age of sixteen. Born in New Zealand to a family of German and Dutch heritage, Dekker was raised a sailor, and often accompanied her father on lengthy expeditions. In August of 2010, she departed from Gibraltar in secret in her 40-foot (12.4 m) sailboat *Guppy* and arrived – 518 days and 5,600 nautical miles later – at the Caribbean island of Sint Maarten. She donated *Guppy* to a Los Angeles non-profit educational organization which wrecked it on a Pacific reef six months later. Dekker started a foundation in her name dedicated to educating young people about long-distance sailing.

DAY 17: ***12 October***

No bird shit today, and no squalls either, but a strong wind and something to go for! It's still cloudy, but the sun breaks through from time to time and that cheers me up. The wind gives me a broad reach and I've boomed out the genoa. The sheets are still getting chafed by the spinnaker pole and I invent a new solution. A sort of safety rope. I make a short loop in the eye of the genoa and fix the spinnaker boom to this. This line is sure to tear, too, but that's not serious. It's holding so far, but then I think of all my other attempts – the duct tape, Rescue Tape, the patches bound around ... But theoretically this should work. *Guppy* is in her element and races through the waves.

I'm too late to see it coming ... A massive wave breaks over the cockpit and soaks me to the bone. I've had my shower, but it leaves me even more salty. When I go below to change into some dry clothes, I feel *Guppy* balancing on the top of a wave and, before I know what's happening, I'm flung through the cabin, along with everything else that's loose. Everything in *Guppy*land is back to normal ... Welcome back, wind!

DAY 18: ***13 October***

In the meantime, the wind has got a little too frisky... Braids of white foam are flying over the water and the seas are mounting. In contrast to the Pacific, the waves are steep and high with a swell that's coming from a different direction to the wind. *Guppy* is being blown forward at a speed of 8 knots while massive waves wash over the deck. The companionway has to stay closed, and I see walls of water chasing past when I look outside. But *Guppy* is handling it well; I'm proud of her and know that she will continue to thunder on until the sea calms down again. All I have to do is keep watch. I've been at sea for 18 days now and this has been my longest crossing so far in terms of time; and I'm not even halfway yet.

Sitting on the chart table with one foot on the cabin steps and the other firmly against the cabin wall, I switch on the SSB. *Guppy* is occasionally surfing off the waves at speeds exceeding 10 knots, and is rolling dangerously from side to side. I have to reduce sail, put a second reef in the mainsail and possibly set the storm jib before night falls, because otherwise it is simply too dangerous. I'm busy thinking about all this when I receive a call from *Sogno d'Oro*. We've been talking for a few minutes when *Guppy* starts to surf faster and faster off a wave.

'Oh, shit!' is all I can say.

A huge breaker crashes over *Guppy* from the side, taking poor *Guppy* down a mountain of white foam to land on her side at the bottom of the trough with a mighty bump. Looking through the Plexiglas door, I see the sea wash into the cockpit. Still holding the microphone in one hand with the other on a handgrip, I'm hanging horizontally to the companionway and am looking at the oncoming water in shock. Slowly, *Guppy* manages to right herself while I survey the chaos inside and the water that is slowly running out of the cockpit.

'I, I, we – *Guppy* has just been knocked down,' I stutter into the radio. 'I'll call you back in half an hour.'

I switch off the SSB, click myself into the harness and wait for the right moment to venture on deck. In the meantime, the windpilot has got everything under control again. Almost everything that was in the cockpit has been swept away. The sprayhood has been totally flattened on one side, and I'm standing up to my knees in water in the cockpit ... I take in the remaining bit of the genoa that's still attached to the spinnaker boom. With water flying over me, and cursing myself, I insert the second reef in the mainsail; something I should have done hours ago. Several lines are trailing in the water behind *Guppy*, and I bring them back on board. Half an hour later, everything looks to be under control again. There doesn't seem to be much damage to the mast or equipment. Cold and soaked to the bone, I get back to Henk and explain what's just happened. *Guppy* is more stable now that she is going slower, and I'm more comfortable about facing the night. We chat about life on board. Things that are so easy to do at home are a real challenge on board. Just going to the toilet is a major task, and you have to wedge yourself into a certain position just in case an unexpected wave launches you through the boat ... But what must be done must be done; including eating liquid food that flies through the cabin the moment you let go of it, and losing stuff you left on deck. Reefing on time, but not too early, in case *Guppy* becomes a toy in the waves — it's all part of it.

It feels as though *Guppy* has been on a rollercoaster all night. I hear the breakers gathering height in the dark, but only see them when they crash over *Guppy* with force. The cockpit is underwater regularly. All the hatches have to stay shut tightly, which makes it very stuffy inside. I'm impressed by the waves here; not only are they really huge, but they are particularly steep. Each big breaker could knock *Guppy* down again, but she's handling it well and is running at 7 knots on a small piece of sail.

DAY 19: ***14 October***

It's already light when the wind starts to drop a little. By noon it's just 25 knots and the waves are becoming longer. The breakers have disappeared. I shake out a reef and unfurl a good bit of the genoa. The situation is improving steadily and I suddenly feel exhausted. I've been on standby all night watching the turbulent conditions from behind the Plexiglas door. Before turning in, I check *Guppy*'s position and see that we've made good progress in the past few days.

DAY 20: ***15 October***

The wind has totally died and we start the umpteenth grey, wet day. I can't even remember the last time I saw the sun. Everything is timeless here. If I didn't make a diary entry every day, I would lose my sense of time altogether. What does it matter if you're at sea for 20 or 25 days? Even though there's a big difference between one and five days. I'm still very tired, miss the sun and sometimes feel like running. At the same time, I'm intensely happy here on *Guppy* on the waves that have calmed down now. There are times when I'd like to be on land, but there are always more moments on land when I wish I was at sea. The sea continues to draw me onwards, and so does my curiosity to experience what lies beyond the horizon.

DAY 21: ***16 October***

I'm woken up by the sun for a change. It warms me and *Guppy* all morning and things get a chance to dry, but now slowly but surely the sun disappears once again and so does the wind that has been blowing *Guppy* along nicely for the past few days. The speed drops to below 3 knots, leaving *Guppy* to float aimlessly. It drizzles from time to time. The sea is an endless dale of grey waves over which *Guppy* glides up and down. But, as always, the wind will return at some stage. It's all or nothing, and it's actually quite pleasant to be able to walk over the deck without a wave washing over me, and to be able to sit in a dry cockpit.

DAY 22: *17 October*

Drizzle and more drizzle. There's still little wind and the grey horizon ends 500 metres further off. Eating has become an important part of my day; something to look forward to. I try to make a special meal out of the large stock of long-life products. This will require a little imagination ... I watch a film from time to time and read a lot. Now and then it rains a little harder, which means I've managed to collect 10 litres of lovely rainwater. I'm happy, even though there's little wind. *Guppy* has just passed the halfway mark and we have another 3000 miles to go before we reach South Africa. When I open one of my laptops to download my email via SailMail, I notice that it hasn't survived the wave crash we had earlier when it flew through the cabin. The screen has a huge crack in it! Fortunately, I have other laptops on board and have installed the SailMail programme on one of them as well.

How is it possible to be at sea for three weeks without having two consecutive days of good wind? *Guppy* floats into the night surrounded by spooky whitish-grey veils of rain. Everything is wet and damp on board. But *Guppy* is still moving forward, and the sails aren't slapping. It may be miserable weather outdoors, but it's really cosy inside.

Triggered by an approaching rain front, the radar alarm goes off in the middle of the night. Wham! A few minutes later, there's suddenly 35 knots of wind and *Guppy* is up to her portholes in water and begging me to reef the sails. All the stars have disappeared behind a big black cloud and the rain pelts down. I furl in the genoa halfway in the streaming rain, reef the mainsail and sail with a broad reach until it gets a little calmer. It's all over an hour later, and I now have a 20-knot wind. I've lost my bucket of lovely fresh rainwater in the squall. A pity, because it was still half-full and good for a nice freshwater shower. Oh well, shit happens ... I should have tied the bucket down or put it away!

DAY 23: *18 October*

At first light, *Guppy* is still sailing with a good broad reach and, although there are some strange cross-waves, she's running well at more than 7 knots.

While I'm making breakfast in the galley, a big wave comes from nowhere and almost throws *Guppy* on her side again. Its twin follows and washes over the entire boat. The plastic flap in front of the entrance hasn't been able to keep the wave out, of course, and a good deal of water washes inside and over me. Dripping water, I try to find what's left of my breakfast ... In the cockpit, the water washes over the seats while the windpilot brings *Guppy* back on course to continue undeterred. While I'm mopping up the water that drips from the steps, I notice that they have worn and decide that I'll sand them down and varnish them in the corning days. The wood of the cabin steps has been wet for days, and this saltwater doesn't do them any good. But there isn't much I can do about it at the moment. I've had countless grey days. I read an entire book, forget to eat, and log in my position. Before I know it, the day has passed.

DAY 24: ***19 October***

A sunny day at last! That's good for the batteries and for me. The wind and waves also offer good sailing conditions. *Guppy* and I enjoy it while it lasts, and are making good progress. I've managed to cover a good many miles on this journey so far, and I enjoy looking back at all the lovely places that *Guppy* and I have visited. I'm feeling great and decide to bake some bread. A little later, the entire cabin is full of the delicious aroma of fresh bread. Yum, this is a real treat at sea! I can open the companionway now that the rain has stopped and the waves seem to have calmed down. It's great to be able to let everything dry and to feel the warmth of the sun.

DAY 25: ***20 October***

The sun is shining on my face when I wake up. I move immediately on deck to discover that the wind is still good, and finally everything on deck gets the chance to dry. I'm glad that the solar panels are doing their best to charge the batteries. I never sail without the radar, mast light and Echomax, so I really need to charge the batteries well after all the rainy days we've had. This is also good for my morale; the sun feels so wonderful! Funny that I should say this now, because I was cursing the sun while I was on the Equator ... All in all, I've made good progress over the past few days and everything is taking its course on board, but sometimes things do break and I repair them; such as a broken fastening, a worn line or the water pump. I've also repaired the jib drum with some improvising, and hope it holds. And so the days pass.

DAY 26: ***21 October***

The wind is easing, but *Guppy* is still making steady progress. After a good start this morning, the sun disappears, alas. There's just a watery sun now and it's fairly cold, but *Guppy* sails on at a good 3.5 knots, so I can't complain. I'm going to use this calm day to check everything and carry out the necessary maintenance. I grease the rudder bearings, cables and discs again. While I do this regularly, I have neglected doing so over the past few days because it wasn't possible with the waves washing over a staggering *Guppy*. My extra lines for the genoa aren't really working as well as I'd hoped; they are wearing faster than I'd anticipated, which is why I've come up with something new. I've attached some stainless-steel rings in the hope that this will work. It's hard work bending the sprayhood frame back into shape, and it takes me a few hours before I can look at my work with satisfaction and get an approving wink from *Guppy*. Eventually, it's back in place and looking pretty good. I then hear boink and a dragging sound behind me on deck. The boom vang has dropped off because of a broken locking pin. This is why the thick pivot pin is also missing. I soon find a spare one and try to repair the troublesome thing. But there's too much tension on it, which means that I need to get the mainsail down in order to get the pin in. I replace the pivot pin and then the split pin, and now Guppy can sail under full sail again. Let's hope the wind doesn't drop entirely so that *Guppy* can continue to make progress.

DAY 27: ***22 October***

During the night the wind changes its mind, and towards morning it's calm again and I've dropped the mainsail to stitch up a hole and do some other mending. I have the feeling that I'll be at sea forever, but it feels good. A feeling of peace has come over me. We are just floating around, but that's just fine.

There's a splendid sunset with the sky looking as though it's had big blobs of white, orange and blue thrown at it. Some are thick and bubbly, and others soft and fluid, and all of it is reflected in the sea which doesn't have a ripple. It's enchanting to look at. Lying in the cockpit, I gaze at the sky until it turns black with pin pricks of light that start to glitter overhead. *Guppy's* speed is about 1.5 knots; not something to write home about, but I don't really mind.

DAY 28: ***23 October***

I'm woken in the middle of the night by the sound of slapping sails and busy myself trying to get as much headway as possible out of the wind, which is constantly changing. But when there is no wind, there is no headway. There's nothing I can do about it, however much I try trimming the sails, and that's really frustrating. A few hours later, I'm treated to a massive downpour. I'm sitting inside and don't feel like going out into the dark, cold night to enjoy the freshwater.

I feel good and am thinking about everything. There are moments when I miss my family and friends and feel like giving them a hug; or going out for an evening with people I know instead of having to make new friends all the time, whom I then have to leave behind just as I'm getting to know them. But that's what I've chosen, and I've really grown since my departure in that respect. (And I'm not talking about my height. At 1.63 metres, I'm not really tall.) I've acquired a better view of the world and a better idea of what I want to do with my life. I have so many plans that I sometimes worry about how I'm going to pack them all into one lifetime. But the most important plan I have at the moment is to return to my country of birth, New Zealand. I like Australia, what I've seen of it so far, and I'd certainly like to return there. People's attitudes there are very different compared with the Netherlands. Australians and New Zealanders have a 'can do' mentality, making the impossible possible and going for your dreams. In the Netherlands, people are put into little boxes and those who say 'can do' are given a hard time. I know one thing for sure, and that's that I don't fit into one of those little Dutch boxes! I want to fulfil my dreams; something that's normal in New Zealand and Australia.

In the morning there's hardly a cloud in the sky and the sun casts a magic glow over the water. I take a nice freshwater shower from the bucket that has filled with rainwater overnight, and realise that it's not only the air but the water, too, that's a lot colder here. Whoa, this is a really cold shower! Brr ... But the sun makes up for it all, and I'm enjoying the space around me. I play music all day long, and check the weather reports on the SSB from time to time. There's not much promise of improvement, alas, but at least *Guppy* is making some headway. Towards nightfall, there's enough wind to keep her on course and make some progress. The stars are awesome tonight; more radiant than they have been for some time. Mostly it's been too overcast to see much of the unknown world above. Now that *Guppy* is sailing without any assistance from me, I can sleep through the night without having to battle wjth the sails every 15 minutes.

SIX YEARS A HOSTAGE

Stephen McGown

Stephen McGown is the longest-held, surviving Al-Qaeda captive in the world. During a once-in-a-lifetime journey riding his motorbike from London to Johannesburg, McGown and two others were attacked at gunpoint and taken hostage. He was held for almost six years in various camps in the Sahara Desert, and went to every length to survive and improve his status with his captors, including learning Arabic and French, and converting to Islam. McGown's story was told to **Tudor Caradoc-Davies**, an author based in Cape Town, South Africa, who wrote *Six Years a Hostage.*

If you worked for the CIA or the NSA's eye-in-the-sky satellite divisions circa 2016, and you happened to be zooming in on a small brigade of what you thought might be AQIM mujahideen camped in the far northwestern reaches of Mali, there's a good chance you saw a man with long hair and a ginger tinge to his beard halfway up a sand dune, arms open, trying to embrace a massive thunderstorm whipping in from the forests of West Africa.

That was me. Music is *haram*, but internally I was singing, my heart was dancing and my mind was surfing the wind, free. There were two songs in my mental jukebox. Somewhat predictably, considering I was standing in the rains in Africa, one was Toto's 'Africa', while the other was 'Break My Stride' by Matthew Wilder.

These thunderstorms were so enormous and so much greater than I was that it felt exhilarating to be alive, so small and insignificant in the face of Earth's power. At that stage I was about 80% through my self-imposed rehab. Standing in a thunderstorm, just so I could feel something, anything, was a legitimate part of my process.

I realised early on in the desert that I had an indefinite prison sentence. I decided that if Al Qaeda did not kill me, then I had to go home a better person, complete in myself. I began to consciously rehabilitate myself.

Of all the things that I have EVER done, this was the most satisfying yet emotionally challenging. For most of my time in the desert, I struggled with depression and anxiety. In fact, these two things had been omnipresent since my early 20s, from my time in Johannesburg right through my London days.

My journey home was supposed to set me back on course and allow me to find myself again. Then I was taken against my will into the desert, and for a long time I was completely overwhelmed.

I knew that I needed to get some kind of control over my situation. I realised I actually had to figure out who the hell I was. My conversion helped me with this; it allowed me a more 'normal' environment with fewer moving parts. Then, as I came to grips with who I was – or at least who I wanted to be again – I had a choice to make: either I could fight what Allah/God/fate had dished me, or accept the situation and heal.

My battles with anxiety and depression were things the other two prisoners knew nothing about. I kept them very close to my chest. My mom and I were very similar. Her depression and anxiety attacks, made worse by going though typical South African crime situations like being held at gunpoint, were so debilitating. I would sometimes be reduced to tears thinking about the pain she had to endure.

I sent out a silent prayer for her on my 40th birthday, up in The Holes in Group Abdul Aziz, on 28 January 2015. We all woke up for the morning prayer, and then everyone went back to their hut, lay down and went back to sleep as per usual. At this stage it was dawn and the horizon was starting to get light. I walked back to my hut and made a small fire in my *virna* burner, which I'd made from a pineapple can with holes punched in the side for ventilation. I boiled some water, made coffee and sat there by myself, next to a small *sabaya* bush, watching the sunrise and thinking about Cath, my mom, my dad and my sister.

My mom's birthday was a few months later on 20 March, when we were in Group Khalid, also up in The Holes. A surveillance plane flew over our camp mid-morning. After it completed two loops above us, I noticed a small silver pinprick behind it. This turned out to be a Mirage fighter jet, so we had to abandon camp. We walked five kilometres northeast, dashing from tree to grass clump to *sabaya*, until Khalid arrived in the Land Cruiser to pick us up so we could reconvene at another camp. We drove about 20 kilometres northeast and happened to land up in the same camp where I had my 40th birthday.

For my mom's birthday, I ended up sitting next to the same bush. At this stage, I had run out of coffee. Again, I boiled some water, and this time I made some Arabic tea. It was quite profound that I spent both my 40th and my mom's birthday sitting in the same camp next to the same bush, looking at the same view. On both occasions, I wasn't interested in talking to anyone else about it; this was for me, sharing thought with my family. It was beautiful watching the sunrise and the sky going pink before the day heated up.

Depression is something you cannot reason with. It moves in and takes over, wiping aside all logic and determination. Without an arm or without a leg you can still function, but when you are not in control of your mind, you feel desperate, full of fear and anxiety every moment of the day. Without a clear mind, it is difficult to say that you are actually alive.

Although my wife would say that I am friendly and fun, and everybody around me thought the same, inside I felt out of balance. Until I found equilibrium inside myself, until I felt complete, it was irrelevant what people saw from the outside. I had to be able to live with myself.

In the first two years of my captivity, I was broken down by fear to a mere skeleton of who I had been. I had those stupid Hollywood sayings running through my head, like 'break them down to build them up', but I really had been broken. I felt that there was nothing to me. I began to stutter again. I truly had nothing left to lose. I saw myself as a blank canvas; I could decide what picture I would paint on it.

I liked the person I used to be, the person God intended me to be. My rehabilitation was about reconnecting with the complete person I was before I sold my soul for money and became somebody that society expected me to be. Me before my depression. When life was about passion. I was going to reprogram myself to be that person again.

I wrote down my goals on a piece of milk carton. I kept this scrap of milk carton inside the top pocket of my *kamees* (shirt). I would bring it out and literally spend hours reading through my list. If I felt I had achieved one of my points, some of which would take a few months, I would cross off that point and add a new word to my list. I reprogrammed myself through reading those cards and living out the words in my everyday actions to create positive habits.

I had zero passion for anything, but I used to love the outdoors, birds and animals, people, so I started to force these things on myself, hoping that I would reignite that flame. A bird would fly past my hut and I would make myself get up and have a look at the bird, to recognise things in it that I used to find beautiful. Often I would look at the bird, shrug and return to my hut – I felt nothing.

The part of me that I had lost was buried very deep. But with time – and it was slow – I began to see the beauty in that bird. It gave me a joy that would literally bubble inside my chest, knowing that I was finding myself again. I could feel the flicker of the passion that I used to feel for everything and anything, no matter the situation. I could begin to see the best in the very mundane, and I knew that I was on the path to re-finding myself.

I'd force myself to stand in thunderstorms, hunt for things in the sand – beads, fossils, and ostrich shells – and learn Saharan 'things', like how to milk a camel, milk a goat or make rope from grass. I figured that if I saturated myself, or even forced myself into things that used to inspire me, my passion would begin to reappear. It took a while, but I began to feel life starting to flow back into my being.

My kidnapping and rehabilitation were too perfect not to be part of God's plan. I was kidnapped with two very different people, who played a big part in my rehabilitation. Sjaak saw the lighter side of life and was quick to be silly and laugh, and this reminded me what it was like to be carefree and fun. Johan was structured and robot-like, and taught me how to stand up for myself. It took him walking over me in the beginning for me to push back against his overbearing nature.

In the beginning, Johan was known by Al Qaeda as the Professor, Sjaak was the Comedian, and I was Britannia, the quiet guy. I knew that I had to throw some personality into the mix to get a name for myself with Al Qaeda; if they liked me, they were less likely to kill me. My rehabilitation was imperative to my survival.

There was another part of my rehabilitation that ran parallel to me re-finding myself: I suffered from panic attacks. It was absolutely irrational and I had no control over it. A grey mist would appear in my head, and I would not be able to think or make conversation. I would sit in the presence of someone, overwhelmed with anxiety. As part of my rehab, I would plan a conversation and then go over to Al Qaeda and speak until I had exhausted my topics. When the grey cloud returned, I would crawl back to my hut.

Sjaak and I had a few fights because I got close to Al Qaeda. I told him that he could handle his situation however he wanted to, but I wanted to go home a better person. I did not want to be a further burden to my family once I was released. He could not relate to this, but he did not know what I was struggling with. Had I spent six years being an angry recluse in my hut, I would have come out of the Sahara a very different person.

Sometimes I felt like I was making progress and the grey cloud was not so thick. Other times my mind would just seize up, I would forget everything I wanted to speak about and I would sit in front of one of the mujahideen in a very awkward silence. Al Qaeda, through Islam, showed good character, and every day was like Groundhog Day with them; today I could embarrass myself, but tomorrow would be a fresh start. I got to try again and again until I got it right.

My growth would be two steps forward, one step back. The reason for this was that we changed groups every two months; by the time I got used to people in one group and found it easier to make conversation, we would change groups and I would begin again from the bottom, where I could feel the grey cloud come in. But the trend of my growth-graph was upwards; I was making progress.

Time truly does heal. I was held captive for five years and eight months, and slowly I believe my chemical balance started to change. I am not a doctor, so it may never have been chemical, but with time I could see I was getting better. At this stage I felt that I had progressed far enough to work on the next stage of my rehabilitation. I needed to learn how to laugh again and make silly conversation... I was too serious. I would make an effort to smile and laugh. In the process I learned that we are all a product of our environment, and beneath the weapons, death and religious fanaticism, we are all the same.

Over the better part of four years, I finally got free from the grey cloud that had always haunted me. Somehow I had managed to reset my mind, become that guy I enjoyed again. After about five years and three months, I felt that I had reached the top of my Saharan schooling and it was time to graduate. Five months later I was released. The whole situation was too perfect. I believe God has a plan; we may not understand it or like it, but there is perfection in it. Everything happens in God's time; we just need to be patient and trust.

LONG WALK TO FREEDOM

The Autobiography of Nelson Mandela

Nelson Mandela

Nelson Mandela (1918–2013) was a revolutionary, activist and author who served as President of South Africa from 1994–9. A committed socialist and African Nationalist, Mandela dedicated his life to dismantling the apartheid regime in South Africa and provoking a national reckoning with the country's racist past. An associate of the banned South African Communist Party in his early years, Mandela was sentenced to life in prison. He served a term of twenty-seven years. During his time in prison, he composed the initial manuscript for his autobiographical *Long Walk to Freedom,* which he published in 1994. Shortly after his release, he was elected President of the African National Congress and after a sustained period of negotiations with the ruling National Party, initiated the first multiracial general election, after which he became the first black president of South Africa. During his presidency, Mandela embarked on a broad campaign of national reconciliation designed to address the grievances of centuries of white minority rule. This campaign, along with his extraordinary record of progressive policymaking, cemented his legacy as a civil rights hero and a champion of democracy both within his native South Africa and in the outside world.

On the day of the inauguration, I was overwhelmed with a sense of history. In the first decade of the twentieth century, a few years after the bitter Anglo-Boer war and before my own birth, the white-skinned peoples of South Africa patched up their differences and erected a system of racial domination against the dark-skinned peoples of their own land. The structure they created formed the basis of one of the harshest, most inhumane, societies the world has ever known. Now, in the last decade of the twentieth century, and my own eighth decade as a man, that system had been overturned forever and replaced by one that recognized the rights and freedoms of all peoples regardless of the colour of their skin.

That day had come about through the unimaginable sacrifices of thousands of my people, people whose suffering and courage can never be counted or repaid. I felt that day, as I have on so many other days, that I was simply the sum of all those African patriots who had gone before me. That long and noble line ended and now began again with me. I was pained that I was not able to thank them and that they were not able to see what their sacrifices had wrought.

The policy of apartheid created a deep and lasting wound in my country and my people. All of us will spend many years, if not generations, recovering from that profound hurt. But the decades of oppression and brutality had another, unintended, effect, and that was that it produced the Oliver Tamboss the Walter Sisuluss the Chief Luthulis, the Yusuf Dadoos, the Bram Fischers, the Robert Sobukwes of our time – men of such extraordinary courage, wisdom and generosity that their like may never be known again. Perhaps it requires such depths of oppression to create such heights of character. My country is rich in the minerals and gems that lie beneath its soil, but I have always known that its greatest wealth is its people, finer and truer than the purest diamonds.

It is from these comrades in the struggle that I learned the meaning of courage. Time and again, I have seen men and women risk and give their lives for an idea. I have seen men stand up to attacks and torture without breaking, showing a strength and resilience that defies the imagination. I learned that courage was not the absence of fear, but the triumph over it. I felt fear myself more times than I can remember, but I hid it behind a mask of boldness. The brave man is not he who does not feel afraid, but he who conquers that fear.

I never lost hope that this great transformation would occur. Not only because of the great heroes I have already cited, but because of the courage of the ordinary men and women of my country. I always knew that deep down in every human heart, there was mercy and generosity. No one is born hating another person because of the colour of his skin, or his background, or his religion. People must learn to hate, and if they can learn to hate, they can be taught to love, for love comes more naturally to the human heart than its opposite. Even in the grimmest times in prison, when my comrades and I were pushed to our limits, I would see a glimmer of humanity in one of the guards, perhaps just for a second, but it was enough to reassure me and keep me going. Man's goodness is a flame that can be hidden but never extinguished.

We took up the struggle with our eyes wide open, under no illusion that the path would be an easy one. As a young man, when I joined the African National Congress, I saw the price my comrades paid for their beliefs, and it was high. For myself, I have never regretted my commitment to the struggle, and I was always prepared to face the hardships that affected me personally. But my family paid a terrible price, perhaps too dear a price, for my commitment.

In life, every man has twin obligations – obligations to his family, to his parents, to his wife and children; and he has an obligation to his people, his community, his country. In a civil and humane society, each man is able to fulfil those obligations according to his own inclinations and abilities. But in a country like South Africa, it was almost impossible for a man of my birth and colour to fulfil both of those obligations. In South Africa, a man of colour who attempted to live as a human being was punished and isolated. In South Africa, a man who tried to fulfil his duty to his people was inevitably ripped from his family and his home and was forced to live a life apart, a twilight existence of secrecy and rebellion. I did not in the beginning choose to place my people above my family, but in attempting to serve my people, I found that I was prevented from fulfilling my obligations as a son, a brother, a father and a husband.

In that way, my commitment to my people, to the millions of South Africans I would never know or meet, was at the expense of the people I knew best and loved most. It was as simple and yet as incomprehensible as the moment a small child asks her father, 'Why can you not be with us?' And the father must utter the terrible words: 'There are other children like you, a great many of them ...' and then one's voice trails off.

I was not born with a hunger to be free. I was born free – free in every way that I could know. Free to run in the fields near my mother's hut, free to swim in the clear stream that ran through my village, free to roast mealies under the stars and ride the broad backs of slow-moving bulls. As long as I obeyed my father and abided by the customs of my tribe, I was not troubled by the laws of man or God.

It was only when I began to learn that my boyhood freedom was an illusion, when I discovered as a young man that my freedom had already been taken from me, that I began to hunger for it. At first, as a student, I wanted freedom only for myself, the transitory freedoms of being able to stay out at night, read what I pleased and go where I chose. Later, as a young man in Johannesburg, I yearned for the basic and honourable freedoms of achieving my potential, of earning my keep, of marrying and having a family – the freedom not to be obstructed in a lawful life.

But then I slowly saw that not only was I not free, but my brothers and sisters were not free. I saw that it was not just my freedom that was curtailed, but the freedom of everyone who looked like I did. That is when I joined the African National Congress, and that is when the hunger for my own freedom became the greater hunger for the freedom of my people. It was this desire for the freedom of my people to live their lives with dignity and self-respect that animated my life, that transformed a frightened young man into a bold one, that drove a law-abiding attorney to become a criminal, that turned a family-loving husband into a man without a home, that forced a life-loving man to live like a monk. I am no more virtuous or self-sacrificing than the next man, but I found that I could not even enjoy the poor and limited freedoms I was allowed when I knew my people were not free. Freedom is indivisible; the chains on any one of my people were the chains on all of them, the chains on all of my people were the chains on me.

It was during those long and lonely years that my hunger for the freedom of my own people became a hunger for the freedom of all people, white and black. I knew as well as I knew anything that the oppressor must be liberated just as surely as the oppressed. A man who takes away another man's freedom is a prisoner of hatred, he is locked behind the bars of prejudice and narrow-mindedness. I am not truly free if I am taking away someone else's freedom, just as surely as I am not free when my freedom is taken from me. The oppressed and the oppressor alike are robbed of their humanity.

When I walked out of prison, that was my mission, to liberate the oppressed and the oppressor both. Some say that has now been achieved. But I know that that is not the case. The truth is that we are not yet free; we have merely achieved the freedom to be free, the right not to be oppressed. We have not taken the final step of our journey, but the first step on a longer and even more difficult road. For to be free is not merely to cast off one's chains, but to live in a way that respects and enhances the freedom of others. The true test of our devotion to freedom is just beginning.

I have walked that long road to freedom. I have tried not to falter; I have made missteps along the way. But I have discovered the secret that after climbing a great hill, one only finds that there are many more hills to climb. I have taken a moment here to rest, to steal a view of the glorious vista that surrounds me, to look back on the distance I have come. But I can rest only for a moment, for with freedom come responsibilities, and I dare not linger, for my long walk is not yet ended.

ALONE

Richard Byrd

Richard Byrd (1888–1957) was an American aviator, naval officer, and polar explorer who served as the leader of several aerial expeditions to the polar regions. A decorated military leader and recipient of the Congressional Medal of Honor, Byrd was appointed to lead Operation Highjump, the largest Antarctic expedition ever undertaken at the time. Highjump represented the first large-scale attempt to geographically map largely unexplored regions of the Antarctic landmass. During his final Antarctic expedition – Operation Deep Freeze – Byrd established a permanent American military presence at the South Pole and McMurdo sound, securing his legacy as one of the most important polar explorers.

As I saw the situation, the necessities were these: To survive I must continue to husband my strength, doing whatever had to be done in the simplest manner possible and without strain. I must sleep and eat and build up strength. To avoid further poisoning from the fumes, I must use the stove sparingly and the gasoline pressure lantern not at all. Giving up the lantern meant surrendering its bright light, which was one of my tew luxuries; but I could do without luxuries for a while. As for the stove, the choice there lay between freezing and inevitable poisoning. Cold I could feel, but carbon monoxide was invisible and tasteless. So I chose the cold, knowing that the sleeping bag provided a retreat. From now on, I decided, I would make a strict rule of doing without the fire for two or three hours every afternoon.

So much for the practical procedure. If I depended on this alone, I should go mad from the hourly reminders of my own futility. Something more – the will and desire to endure these hardships–was necessary. They must come from deep inside me. But how? By taking control of my thought. By extirpating all lugubrious ideas the instant they appeared and dwelling only on those conceptions which would make for peace. A discordant mind, black with confusion and despair, would finish me off as thoroughly as the cold. Discipline of this sort is not easy. Even in April's and May's serenity I had failed to master it entirely.

That evening I made a desperate effort to make these conclusions work for me. Although my stomach was rebellious, I forced down a big bowl of thin soup, plus some vegetables and milk. Then I put the fire out; afterwards, propped up in the sleeping bag, I tried to play Canfield. But the games, I remember, went against me; and this made me profoundly irritable. I tried to read Ben Ames Williams' *All the Brothers Were Valiant;* but, after a page or two, the letters became indistinct; and my eyes ached–in fact, they had never stopped aching. I cursed inwardly, telling myself that the way the cards fell and the state of my eyes were typical of my wretched luck. The truth is that the dim light from the lantern was beginning to get on my nerves. In spite of my earlier resolve to dispense with it, I would have lighted the pressure lantern, except that I wasn't able to pump up the pressure. Only when you've been through something like that do you begin to appreciate how utterly precious light is.

Something persuaded me to take down the shaving mirror from its nail near the shelf. The face that looked back at me was that of an old and feeble man. The cheeks were sunken and scabrous from frostbite, and the bloodshot eyes were those of a man who has been on a prolonged debauch. Something broke inside me then. What was to be gained by struggling? No matter what happened, if I survived at all, I should always be a physical wreck, a burden upon my family. It was a dreadful business. All the fine conceptions of the afternoon dissolved in black despair.

The dark side of a man's mind seems to be a sort of antenna tuned to catch gloomy thoughts from all directions. I found it so with mine. That was an evil night. It was as if all the world's vindictiveness were concentrated upon me as upon a personal enemy. I sank to depths of disillusionment which I had not believed possible. It would be tedious to discuss them. Misery, after all, is the tritest of emotions. All that need be said is that eventually my faith began to make itself felt; and by concentrating on it and reaffirming the truth about the universe as I saw it, I was able again to fill my mind with the fine and comforting things of the world that had seemed irretrievably lost. I surrounded myself with my family and my friends; I projected myself into the sunlight, into the midst of green, growing things. I thought of all the things I would do when I got home; and a thousand matters which had never been more than casual now became surpassingly attractive and important. But time after time I slipped back into despond. Concentration was difficult, and only by the utmost persistence could I bring myself out of it. But ultimately the disorder left my mind; and, when I blew out the candles and the lantern, I was living in the world of the imagination–a simple, uncomplicated world made up of people who wished each other well, who were peaceful and easy-going and kindly.

The aches and pains had not subsided; and it took me several hours to fall asleep; but that night I slept better than on any night since May 31st; and in the morning was better in mind and body both.

The melancholy began to lift, and I was able to do a little more for myself. Wednesday the 6th I succeeded in getting topside for the 8 A.M. weather "ob." Although the morning was clear, drift still blurred the horizon and peppered my face. I sank to my knees in soft snow at every step. It was good to possess the spaciousness of the Barrier after the narrowness of the shack. I threw the beam of my flashlight at the wind vane, and saw that the wind was in the southeast. That means cold, I muttered. Rime covered everything. The breathing slats in the sides of the instrument shelter were thick with drift, but I did not feel up to brushing it off. I was satisfied to read the thermometer, reset the pin, and retreat below.

Later, I crept to the far end of the fuel tunnel and secured a small piece of asbestos, which I cut to fit over the top of the stove. My idea was that it would help to shut off the initial fumes that poured through the chinks while the burner was still cold and smoky. I cut the piece to fit snugly around the stovepipe and fold over the edges of the stove.

In the afternoon I eavesdropped on Little America's weekly broadcast to the United States. One reason was that prudence suggested testing my handiness with the battery-powered emergency set. But the moving reason was a hunger for familiar voices. I missed much of what was said, but I did catch the solemn to-do over the three cows at Little America; how one stood up all the time and refused to lie down; and how another, which was accustomed to lying down every night, lay down with the coming of the winter night and refused to stand up; and how the third, poor thing, couldn't make up her mind just what to do, except roll her eyes at Cox, the carpenter, who haunted the cowbarn. I really chuckled over that and over "Ike" Schlossbach's baritone solo called "Love, You Funny Thing." Other people in Antarctica, I realized, had their problems too.

June II: The Struggle

My scheduled contact with little america which fell on the next day–Thursday the 7th–confirmed what I knew in my heart: that the improvement in my condition was more mental than physical. Though I was less weak, I was at least three hours getting fuel, heating the engine, sweating it into the shack and out, and completing the other preparations. I moved feebly like a very old man.

Once I leaned against the tunnel wall, too far gone to push the engine another inch. You're mad, I whispered to myself. It would be better to stay in the bunk and cut out paper dolls than keep up this damnable nonsense.

That day the cold was worse. The thermograph showed a minimum of 48° below zero. From all indications the "heat wave" was broken. The slick white film of ice on the walls had climbed from the floor halfway to the ceiling. All my resistance to cold seemed to have vanished. My flesh crawled, and my fingers beat an uncontrollable tattoo against everything they touched. It was disheartening to be so much at the mercy of something from which there was no lasting escape. Resting betweenwhiles, I huddled over the stove. The warmth was only superficial. My blood ran cold as ice.

In spite of all my efforts, I was late making contact. Dyer was playing a record, as he sometimes did when he grew tired of repeating the call. I finally recognized what it was "The Pilgrim's Song" from *Tannhäuser*; I waited much stirred, until the record was played out. When I broke in, Charlie Murphy chided me: "Oversleep, Dick?"

"No, busy."

Charlie had little to say. Siple, however, was waiting to read a paper. As well as I can remember, it dealt with the theoretical configuration of the undiscovered coastal reaches of the Pacific Quadrant, where I was supposed to make a flight of exploration in the spring. Interesting, yes; there was no denying the thought that Siple had put into it. And, desperate as I was to close the conversation, I could not help but reflect that this was the most exquisite of ironies: that I should sit there, gripping the table for support, and listen to a theory about a coast I had never seen and now might never see.

If I remember rightly, I said: "Very interesting. Submit to scientific staff." Dyer broke in to ask if I had any further messages before signing off. I asked him if it would be possible to shift the schedules from the morning to the afternoon. He replied: "Wait a minute, please." Although the words were unintelligible, I could hear voices talking in the background. Dyer said that they were perfectly willing to make the change if I wished it, but that to do so would involve shifting their own schedules with the United States, which had been fixed after long testing. "Never mind," I said. And, for the time being, let the subject drop, not wishing to excite suspicion. Directly after the schedule I took to my bunk, and scarcely moved the rest of the day. The pain was back, and with it the bitterness and the discouragement.

Why bother? they argued mockingly. *Why not let things drift? That would be the simple way. Your philosophy tells you to immerse yourself in the universal processes. Well the processes here are in the direction of uninterrupted disintegration. That is the direction of everlasting peace. So why resist?*

From that day on I came to dread the radio schedules with Little America.

The task of getting the engine ready, together with the ever-present fumes, emptied me of whatever strength and resistance I had accumulated meanwhile. It seemed almost better to let the radio go by the board. I tried to think up excuses for stopping the contacts altogether which I could present to Little America over the next schedule, but none made sense. I couldn't very well say that the conversations were beginning to bore me or that the transmitter was on the verge of breakdown and not to be worried if Advance Base went off the air. For one thing, many expedition problems remained to be discussed with Poulter and Murphy; and, in spite of the explicit orders I had left with the officers at Little America and reiterated to the tractor crew just before they departed for Advance Base, I couldn't help but feel that any lasting silence on my part might tempt the camp into some rash move.

Thus, I was caught up in a vicious circle. If I kept up the schedules, the drain upon my strength plus the fumes would almost certainly finish me off; if I failed to keep them up, I had better be finished off anyway. This was the way I looked at the situation–the way I believe that any member of the expedition in my position would have viewed it. Just as any normal man would be, I was doggedly determined to keep them from attempting a possibly disastrous relief excursion.

So I came to dread the schedules for a second reason: the fear that I would betray the condition which I was trying to hide by maintaining them. I knew that Murphy was always studying me in his cynical, penetrating fashion. As time went on, I sent amusing messages to throw him off. (I must say, though, that most of these looked pretty silly afterwards.) But, when a course of action is obvious, you take it for granted that you could not deliberately do otherwise; and, when your conscience is too weak to hold you to that course, you continue along it from momentum. By an ironical twist of circumstances the radio, which should have been the greatest safety factor, had instead become my greatest enemy.

That Thursday night I really grasped how far I had fallen. In the diary I wrote: " ... These early morning schedules are killing. They leave me without strength to go through the day. Tremendously difficult to get even a little sleep. There are strange nagging pains in my arms, legs, shoulders, and lungs .... I'm doing everything possible to hold out. Could I but read, the hours would not seem half so long, the darkness half so oppressive, and my minor misfortunes half so formidable .... "

Across the room, in the shadows beyond the reach of the storm lantern, were rows of books, many of them great books, preserving the distillates of profound lives. But I could not read them. The pain in my eyes would not let me. The phonograph was there, but the energy to crank it had to be saved for the business of living. Every small aspect of the shack bespoke my weakness:

the wavering, smoking flame in the lantern and the limp outlines of the clothing on the walls; the frozen cans of food on the table, the slick patches of ice on the deck, the darker stains of spilt kerosene, and the yellowed places where I had vomited; the overturned chair beside the stove which I hadn't bothered to pick up, and the book–John Marquand's *Lord Timothy Dexter of Newburyport*–which lay face down on the table.

THE WHITE DARKNESS

David Grann

Henry Worsley (1960–2016) was a British explorer and army officer. He perished in 2016 during a solo attempt to cross the Antarctic landmass which, if completed, would have been the first of its kind. An accomplished explorer and soldier, Worsley undertook several Antarctic expeditions retracing the steps of his heroes Roald Amundsen, Robert Falcon Scott, and Ernest Shackleton. On his final expedition, he succumbed to severe dehydration and ill health and, after radioing for help and being airlifted to a base in Chile, passed away at the age of 55.

David Grann (born 1967) is a *New York Times* bestselling author and award-winning journalist, who has been a staff writer for *The New Yorker* since 2003. He is also the author of *The Lost City of Z* and *and other bestselling books.*

Then, as on Shackleton's *Endurance* expedition, everything began to go wrong. On November 22, a little more than a week into his journey, he was engulfed in a whiteout and was pinned down in his tent. "Proper Antarctic storm!" he wrote in his diary, noting that there was no chance of moving forward that day. The next morning, the gusts felt strong enough to hurl a small dog; one of the tent poles broke, and he had to repair it. "A salutary reminder just who is in control around here," he said of the conditions. "Trespassers will be punished."

He emerged on November 24, and found himself plowing through a dust bowl of ice in which all he could see, hour after hour, was his compass strapped to his chest and his skis with their metronomic rhythm—an experience that he described as "miserable, mind-numbing, monochromatic monotony." He was ascending a section of the Transantarctic Mountains, and on November 25 he came upon a steep slope of ice that rose hundreds of feet. He tried to climb with his crampons, but the sled wouldn't move. Again, he tried. Again, it wouldn't budge.

If he didn't keep moving, he would freeze. He decided to lighten his sled, and unloaded most of his bags of food and stored them on a flat part of the ice. Then he began to climb. When he reached the top of the ridge, gasping for air behind his face mask, he deposited what he'd dragged up. After a short rest, he scaled back down to retrieve the rest. He made trip after trip.

Once, in the poor visibility, he failed to notice the scar of a crevasse and his foot broke through the surface. He felt himself slipping into the hole, which was widening around him. He grabbed the edge and clung to it, dangling over an abyss, before he hauled himself up. When he peered into the chasm, he wrote in his diary, he "suddenly felt very alone, vulnerable and scared."

His body was weakening more rapidly than on his previous expeditions. Not only was the sled heavier; he constantly had to break track, and he had to carry out alone the tiresome tasks of making camp each night and packing up in the morning.

On November 30, after trekking for nearly three weeks and traversing a hundred and sixty-five nautical miles, he reported that he had "aching shoulders, lower–back pain, very snotty nose ...and coughing due to breathing in cold air." He developed a rash on his groin. His feet were covered in bruises and blisters, and he took a knife to his boots, hoping to smooth the lining and alleviate the pain. One day, he suffered from a mysterious stomach ache, which was aggravated by the sled harness yanking at his waist.

Though he usually maintained a buoyant tone in his broadcast, he was more despairing in his diary. "It was a real physical battle with fatigue," he wrote, adding, "I was stopping literally every minute or so to catch my breath or just get ready for the next exertion required." Two nights later, after another whiteout, he lamented that he hadn't had enough strength "to pull the sledge through" the storm. His diary entries became a litany of suffering: "hard day"; "a very difficult day"; "a brutal day"; "awful day—floundering around in a complete whiteout"; "another awful day—worse than yesterday"; "swimming against a strong tide"; "still swimming against the tide"; "totally spent and demoralized." Each morning, he unzipped the flap of his tent and peeked out, hoping for clear skies, only to behold what he called "more of the white darkness." At times, he could not even discern the tips of his skis through the murk, which, he wrote, was as "thick as clotted cream."

On December 1, he marched into what he described as "the mother of all storms." Trudging uphill, with his head bowed against a fusillade of ice pellets, he moved at less than a mile per hour. After many hours, he abruptly paused. "I sat huddled on my sledge, down jacket on, wondering whether to go or to stop," he later recalled. It was so windy that he did not know if he could set up his tent, and so he resumed trekking. "My hands took a battering, and often I had to stop to give them some warmth," he said. "And the light was so flat that on two occasions, immediately after stopping, I fell straight over, such is the disorienting effect it has on your senses."

The next day, he blindly skied over a ridge, and the sled overtook him and pulled him down. His head and back and legs slammed against the ice. The sled flipped over twice, dragging him for twenty yards. He lay splattered on the ice, cursing. When he got to his feet, he nervously checked his fuel canisters. One crack and he would be doomed, but there were none, and, conscious of time slipping away, he untangled his harness and set off again.

And incredibly, despite every obstacle and every calamity, he was on track to reach the South Pole around New Year's Day. Nothing seemed to stop him. One morning, he forged on even when the conditions were so awful that he conceded that it was "crazy" to set off. Another time, he wrote in his diary, "I just can't go further—I don't have it in me." And yet he rose the next day and marched onward. On December 18, the thirty-sixth day, he walked more than seventeen nautical miles, a remarkable trek that took him fifteen hours. After another punishing day—which he described as a "combination of eating, bending, driving, tying, pushing, bracing, draining, swearing, pausing and despairing"—he told himself, "I just have to accept it and keep moving."

His existence had been reduced to a single purpose: making his mileage. When approaching sastrugi, he commanded himself to "attack, attack, attack." After one such battle, he wrote proudly in his diary that he had stormed "the ramparts of every piece that was unfortunate enough to get in my way." He added, "The sledge, now a battering ram and not a burden, smashed through all in its path." When he was asked by his radio listeners how he persevered, he said that it was less about physical prowess than about how "strong your mind and will are—hours at the gym cannot prepare you."

Robert Swan, a British adventurer who had trekked to both poles, was monitoring Worsley's journey, and expressed awe at his daily progress. In an audio message posted on Worsley's website, on December 5, Swan said, "His average is fantastic," adding, "He's facing some quite odd conditions, but, being Henry, he's slugging it out." In a second message, posted later that month, Swan described Worsley proceeding as if a traffic light were glowing in front of him: "Very, very rarely in your mind do you ever see the color green, for the simple reason if you're in green you're probably not pushing hard enough .... You're thinking about your feet, your legs, your calves, your hips, your arms, your neck, your shoulders, and you're constantly doing these checks to see whether everything's OK. ... As Henry has said, as he moves towards those last few hours every day, you can feel that he's pushing into the red zone. And the red zone is not a place to stay in, because in the red zone your body is starting to eat itself. You're much more likely to get frostbite. So you live on the edge of the orange, occasionally push into the red, and then, very sensibly, he comes back off the red, back into the orange. And hopefully, when he's into his sleeping bag and speaking to us, you know he's back into the green."

By Christmas Day, Worsley was nearly within a hundred nautical miles of the Pole. Prince William broadcast a message, saying, "We're thinking of you at the Christmas period as you're lugging all your kit up and down the slopes and the hills of the southern Atlantic in the Antarctic." Worsley opened a package that Joanna and the children had given him. Inside were miniature versions of traditional Christmas sweets: a mincemeat pie and a fruitcake. Alicia had written him a note that quoted lyrics from *The Jungle Book:* "Look for the bare necessities / The simple bare necessities/ Forget about your worries and your strife." And Joanna had included a sample of Amouage Journey Man cologne. "I figured his tent would be so smelly by then," she recalled.

On his broadcast, Worsley said, "Packages from home, especially at times like this, no matter where you are in the world, carry special meaning. And none more so for me this morning."

Using his satellite phone, he called Joanna and Alicia, in London, and then Max, in France. Throughout the journey, Worsley had made a record in his diary of virtually every communication that he had had with them. Once, after speaking to Joanna, he wrote, "I do love her so much." Another time, after he received a text from Alicia saying "I am thinking of you constantly, and love you more than ever," he jotted down, "Sweet text from Shrimp"—his nickname for her. And he noted that a conversation one morning with Max had "lifted my spirits." On Christmas Day, he wrote in his diary, "Lovely to hear their voices."

Despite the holiday, Worsley marched twelve nautical miles. As he lay in his tent that night, he lit a cigar, the sweet smoke filling the air, and ate his Christmas treats. It was, he said, like a "little heaven."

Soon, he was almost at nine thousand feet, the elevation of the South Pole. He was so tired that once, while sitting on his sled during a snack break, he nodded off, even though the wind-chill temperature was minus twenty-two. "I may be drained of all power and energy," he reported on a broadcast. "But I still seem to have the will that says, to my heart and nerves and sinews, *Hold on.*" He kept telling himself, "Keep your eyes on the prize."

On January 2, only a day behind schedule, he reached the Pole. He was greeted by a group of well-wishers from the scientific-research station. They were the first people he had seen in fifty-one days. But this was not the climax of his journey—it was only the end of the first phase—and, because he was making his trek unaided, he couldn't go inside the base to receive a hot meal or even to take a bath. "It was weird arriving here and not stopping," he wrote in his diary, adding, "Very tempting to stay at Pole—eat and sleep." But he set up his camp as usual, maintaining his self-imposed exile.

During his broadcast, he told his listeners, "I owe so much to all of you for your support in getting me this far. I cannot emphasize too strongly just how much it has urged me on over the darker days, of which there have been many. But those I have to thank most are Joanna, Max, and Alicia." His voice cracked. "They have been with me every step of the way, each with a warm hand in the small of my back, lifting me when I am down, strengthening me when I am weak, and filling me when I'm empty. I owe them everything." He concluded, "At the southerly point where the world spins on—good night."

In London, Joanna listened to his broadcasts each evening before falling asleep. Shortly before Christmas, she was interviewed by a reporter from the *Daily Express*, and she said, "Henry was away abroad a lot in his army days so we've been used to separations ... but I miss him much more now. I do worry about him because I know how frail he is getting—he does lose a huge amount of weight and he has had a really rough time with the weather." She went on, "He is so determined. In my head I know there's no way he won't succeed, even if he has to walk all day and night. He has enormous mental strength." She was overcome: "He's an amazing man—isn't it wonderful to be married to someone like that?"

Worsley estimated that it would take him about three weeks to complete the rest of his journey, and he hoped that the hardest part was behind him. In his diary, he had written, "Just pray going North is that much easier." Yet, as he climbed the Titan Dome, he found the ascent to be "a killer." He had lost more than forty pounds, and his unwashed clothing hung on him heavily. "Still very weak—legs are stick thin and arms puny," he noted in his diary. His eyes had sunk into shaded hollows. His fingers were becoming numb. His Achilles tendons were swollen. His hips were battered and scraped from the constantly jerking harness. He had broken his front tooth biting into a frozen protein bar, and told ALE that he looked like a pirate. He was dizzy from the altitude, and he had bleeding hemorrhoids.

On January 7, he woke in the middle of the night with another stomach-ache. "I felt pretty awful," he admitted on his broadcast. "The weakest I felt in the entire expedition." The earbuds on his iPod had broken, leaving him in silence. "I feel alone," he confessed on a broadcast, adding, "Occasionally, it would be nice to have somebody to talk to about the day."

He kept thinking that he would soon reach the top of the Titan Dome. "I'll be okay if the promised 'downhill' materializes," he wrote in his diary.

But the peak eluded him—he was trapped in an infinite beyond. On January 11, he told his listeners, "I'm desperate to go down and into air thick enough to breathe."

Listening to the broadcasts, Joanna was increasingly concerned. "I felt in his voice this exhaustion and sadness," she recalled. He had no companion to tell him that he had remained too long in the red zone; nor was he held back by the worry that his actions might jeopardize the lives of others. And he was confident that he could do what he always had done: prevail through unbending will. In his commonplace book, he had once written down a quote, from the cyclist Lance Armstrong, that said, "Losing and dying: it's the same thing."

And so Worsley pressed on, muttering to himself a line from Tennyson's poem "Ulysses": "To strive, to seek, to find, and not to yield." Once, he looked up in the sky and saw, through his frozen goggles, a dazzling sun halo. On the edge of the circle, there were intense bursts of light, as if the sun were being splintered into three fiery balls. He knew that the phenomenon was caused by sunlight being refracted through a scrim of ice particles. Yet, as he stumbled onward through the void, he wondered if the light was actually some guiding spirit, like the "fourth man" that Shackleton had spoken of. Perhaps Worsley, too, had pierced the "veneer of outside things"—or perhaps his mind was simply unravelling. His diary entries had become sparer and darker: "So breathless ...I am fading ...hands/fingers are forever shutting down ...wonder how long they will last."

On January 17, he staggered through a whiteout, pulling his sled for sixteen hours. When he stopped, it was late evening, and he struggled to build his camp again—to plant the tent poles in the ice, to unload his food, to light the cooker, to melt snow for water. "It's now one o'clock in the morning," he said in his broadcast. "In sum, it's been a punishing day." He continued, "What little energy I have left ..." His voice faded in and out.

Joanna panicked upon hearing the broadcast. She called many of Worsley's close friends, asking if someone should ask ALE to dispatch a rescue plane. They thought that Worsley would be OK, given his experience and abilities, and that he should be the one to make such a call. Robert Swan, in one of his earlier broadcasts, had noted that Worsley had on his belt "a wonderful, unbelievable" Iridium satellite telephone, adding, "If he does have a problem, he can hit the button and get some support and rescue very, very quickly."

On January 19, after man-hauling through another storm, Worsley was too tired to give a broadcast, and with his frozen hand he scribbled only a few words in his diary, the writing almost illegible: "Very desperate ...slipping away ...stomach ...took painkillers." He was incontinent, and repeatedly had to venture outside to squat in the freezing cold. His body seemed to be eating itself.

The next day, the sixty-ninth day of his journey, he could drag his sled for only a few hours. He built his tent and collapsed inside. At one point, he called Max on the satellite phone, waking him in the middle of the night in France. All Henry kept saying was "I just want to hear your voice, I just want to hear your voice."

Max told him, "You will always be a polar warrior in my eyes. You just need to pull out and come home."

On the morning of January 21, Joanna spoke to him. He was suffering from, as she put it, "complete shutdown." He couldn't even muster the energy to boil water or to brush his teeth, and she pleaded with him to call ALE and evacuate. "You've absolutely got to call them," she said.

He told her that, though he wasn't going to leave the tent, he needed some time to think through what to do next.

THROUGH GATES OF SPLENDOUR

Elisabeth Elliot

Elisabeth Elliot (1926–2015) was a Protestant missionary and author who, in 1955, embarked on a mission to reach the remote Huaorani people of Ecuador alongside her husband Jim Elliot. After attempting to communicate with local members of the tribe, her husband and his four companions were speared to death by Huaorani tribesmen. Despite this tragedy, Elisabeth remained in Ecuador, went to live among the same indigenous tribe with her young daughter for two years and became one of the first Westerners to engage in extensive contact with the Huaorani people. While some have argued that this contact with Westerners facilitated the decline of this distinct culture and the assimilation of their ethnic group into the developed world, Elisabeth's efforts, in fact, reduced the prevalence of vengeance killings among the tribesmen and helped them stand up to loggers and land developers.

TWO DAYS LATER we widows – already we were adjusting ourselves to the use of the word – sat together at the kitchen table in Shell Mera. Dr. Art Johnston was describing the finding of the bodies. He had just returned with the weary, straggling ground party. When he hesitated, we urged him to give us all the facts.

It was evident that death had been caused by lance wounds. But how had ten Aucas managed to overwhelm five strong men who were armed with guns? Over and over we asked ouselves this question. The only possible answer was an ambush. Somehow, the Aucas must have succeeded in convincing the men of their peaceful intentions. Nate had assured Marj that they would never allow Aucas with spears in their hands to approach them. Perhaps the 'commission of ten' that Nate mentioned on the radio had been a decoy party. Certainly if this party had carried spears Nate would have reported this and the men would not have looked forward so eagerly to their arrival. This group may have walked peacefully on to the beach while a second party, carrying spears, moved up under cover of the jungle foliage to carry out a surprise attack. It seems likely that the missionaries and the unarmed Aucas had been mingling together, as they had on the previous Friday, with friendly words and gestures. And then, at a secret signal....

There was evidence of a struggle on the beach—marks of Ed's leather heels in the sand; one bullet-hole through the windshield of the plane. However, no blood was found. If any Aucas had suffered, it was not apparent. Had the men tried to avoid shooting by backing into the river? A lance was found thrust into the sand in the river bottom near the body of Jim Elliot. The fact that all the bodies were in the water might indicate that they had tried desperately to show the Aucas that they would shoot only as a last resort.

The condition of the Piper showed real malice. Possibly some Auca had punctured the fabric of the plane with a spear, and, finding it vulnerable, had begun to peel it off. Others helped, and soon they had denuded it completely, tossing the strips into the water near by. But someone intended to put this man-carrying bird out of commission once and for all. Some of the framework was bent, and a part of the landing gear, made of tubular steel, was battered in as if by a very heavy object. The propeller and instrument panel, however, were intact. Perhaps to touch the 'soul' of the creature was taboo, but they had torn the stuffings from the seats, as if to disembowel the flying beast.

Why, after the overtures of friendship on Friday, had the Aucas turned with such sudden and destructive anger on their white visitors on Sunday? The answer can only be guessed. Among the most qualified to venture a guess is Frank Drown, whose work with the Jivaros has given him shrewd insight into Indian thinking. He says:

> An Indian, when he first hears or sees something new, will accept it. Perhaps he accepts merely from normal curiosity, but he does accept. But after he has had time to think about the novelty he begins to feel threatened, and that is the time when he may attack. A group of Indians will sit back and discuss a new contrivance or a new way of doing things with some eagerness; but the witch-doctors, who are the real conservatives, can be counted on for rejection. They have a lot of authority and, when they work on their fellow tribesmen to reject an innovation, the people seldom go contrary to their advice. As in any culture, the younger men may be looking for a new way of life, but the older ones hang on to their traditions and maintain the *status quo*. Furthermore, most Indians are basically and understandably sceptical of anything the white man offers him. And don't forget that, after all, this was the first time within memory that the Aucas have had an encounter with the white man which was completely friendly. We can only hope they are pondering that fact right now.

In the kitchen we sat quietly as the reports were finished, fingering the watches and wedding rings that had been brought back, trying for the hundredth time to picture the scene. Which of the men watched the others fall? Which of them had time to think of his wife and children? Had one been covering the others in the tree house, and come down in an attempt to save them? Had they suffered long? The answers to these questions remained a mystery. This much we knew: 'Whosoever shall lose his life for my sake and the gospel's, the same shall save it.' There was no question as to the present state of our loved ones. They were 'with Christ'.

And, once more, ancient words from the Book of Books came to mind:

> All this has come upon us, yet have we not forgotten thee.... Our heart is not turned back, neither have our steps declined from Thy way, though Thou hast sore broken us in the place of dragons, and covered us with the shadow of death.

The quiet trust of the mothers helped the children to know that this was not a tragedy. This was what God had planned. 'I know my daddy is with Jesus, but I miss him, and I wish he would just come down and play with me once in a while,' said three-year-old Stevie McCully. Several weeks later, back in the States, Stevie's little brother, Matthew, was born. One day the baby was crying and Stevie was heard to say, 'Never you mind; when we get to Heaven I'll show you which one is *our* daddy.' Was the price too great?

To the world at large this was a sad waste of five young lives. But God has His plan and purpose in all things. There were those whose lives were changed by what happened on Palm Beach. In Brazil, a group of Indians at a mission station deep in the Mato Grosso, upon hearing the news, dropped to their knees and cried out to God for forgiveness for their own lack of concern for fellow Indians who did not know of Jesus Christ. From Rome, an American official wrote to one of the widows: 'I knew your husband. He was to me the ideal of what a Christian should be.' An Air Force Major stationed in England, with many hours of jet flying, immediately began making plans to join the Missionary Aviation Fellowship. A missionary in Africa wrote: 'Our work will never be the same. We knew two of the men. Their lives have left their mark on ours.'

Off the coast of Italy, an American naval officer was involved in an accident at sea. As he floated alone on a raft, he recalled Jim Elliot's words (which he had read in a news report): 'When it comes time to die, make sure that all you have to do is die.' He prayed that he might be saved, knowing that he had more to do than die. He was not ready. God answered his prayer, and he was rescued. In Des Moines, Iowa, an eighteen-year-old boy prayed for a week in his room, then announced to his parents: 'I'm turning my life over completely to the Lord. I want to try to take the place of one of those five.'

Letters poured in to the five widows – from a college in Japan, 'We are praying for you'; from a group of Eskimo children in a Sunday School in Alaska; from a Chinese church in Houston; from a missionary on the Nile River who had picked up *Time* magazine and seen a photograph of her friend, Ed McCully.

Only eternity will measure the number of prayers which ascended for the widows, their children, and the work in which the five men had been engaged. The prayers of the widows themselves are for the Aucas. We look forward to the day when these savages will join us in Christian praise.

Plans were promptly formulated for continuing the work of the martyrs. The station at Arajuno was manned to be ready in case the Aucas should come out for friendly contact. Gift flights were resumed by Johnny Keenan, so that the Aucas would know, beyond any doubt, that the white man had nothing but the friendliest of motives. Revenge? The thought never crossed the mind of one of the wives or other missionaries.

Barbara Youderian returned to her work among the Jivaros, with the two little children, and I went back to Shandia with ten-month-old Valerie to carry on as much as I could of the work of the Quichua station. Another pilot, Hobey Lowrance, with his family and a new plane, were sent to the mission air base in Shell Mera, while Marj Saint took up a new post in Quito. After the birth of her third son in the United States, a few weeks after the death of her husband, Marilou McCully returned to Ecuador with her boys to work in Quito with Marj. For Olive Fleming, who had spent only two months in the jungle when her husband died, the problem regarding the future has been more difficult. But for her, as for all, one thing is certain: her life belongs to God, as had her husband's, and He will show the way.

In the months since the killing of the five men, Nate Saint's sister Rachel has continued with the study of the Auca language, working with the Auca woman, Dayuma. Many flights have been made over the houses of the Aucas. The first group of houses was found to have been burned, a common Auca practice after a killing, but not far away new houses were discovered, and gifts were dropped to the waiting Indians. When Johnny Keenan swoops over, 'George' appears, jumping and waving the little model plane given him by Nate Saint. 'Delilah' also seems to be there with him. Patches of bright yellow fabric from Nate's plane adorn the roofs of some of the houses.

Thousands of people in all parts of the world pray every day that 'the light of the knowledge of the glory of God' may be carried to the Aucas, a people almost totally unheard of before. How can this be done? God, who led the five, will lead others, in His time and way.

From among the Quichuas with whom Jim, Ed, and Pete worked, several have surrendered their lives to God for His use, to preach to their own people – or even to the Aucas, if He chooses. They have carried on the work begun by the missionaries, speaking to their relatives of Christ, reading the Scriptures that have been translated for them, travelling sometimes in canoes and over muddy trails to teach the Bible to others who do not know its message. A converted Indian, formerly a notorious drinker, came to me one day and said, 'Señora, I lie awake at night chinking of my people. "How will I reach them?" I say. "How will they hear of Jesus?" I cannot get to them all. But they *must know*. I pray to God, asking Him to show me what to do.' In the little prayer meetings the Indians never forget to ask God to bless their enemies: 'O God, You know how those Aucas killed our beloved Señor Eduardo, Señor Jaime, and Señor Pedro. O God, You know that it was only because they didn't know You. They didn't know what a great sin it was. They didn't understand why the white men had come. Send some more messengers, and give the Aucas, instead of fierce hearts, soft hearts. Stick their hearts, Lord, as with a lance. They stuck our friends, but You can stick them with Your Word, so that they will listen, and believe.'

For the wives and relatives of the five men, the mute longing of their hearts was echoed by words found in Jim Elliot's diary:

> I walked out to the hill just now. It is exalting, delicious, to stand embraced by the shadows of a friendly tree with the wind tugging at your coat-tail and the heavens hailing your heart, to gaze and glory and give oneself again to God – what more could a man ask? Oh, the fullness, pleasure, sheer excitement of knowing God on earth! I care not if I never raise my voice again for Him, if only I may love Him, please Him. Mayhap in mercy He shall give me a host of children that I may lead them through the vast star fields to explore His delicacies whose finger-ends set them to burning. But if not, if only I may see Him, touch His garments, and smile into His eyes – ah then, not stars nor children shall matter, only Himself. O Jesus, Master and Centre and End of all, how long before that Glory is Thine which has so long waited Thee? Now there is no thought of Thee among men; then there shall be thought for nothing else. Now other men are praised; then none shall care for any other's merits. Hasten, hasten, Glory of Heaven, take Thy crown, subdue Thy Kingdom, enthral Thy creatures.

THE AUTOBIOGRAPHY OF MARÍA ELENA MOYANO

The Life and Death of a Peruvian Activist

María Elena Moyano

María Elena Moyano (1958–1992) was a Peruvian feminist, activist and schoolteacher. She was one of the founding members of the Popular Federation of Women of Villa El Salvador (FEPOMUVES) and a community organizer who set up training centers for women and soup kitchens for the poor. She believed in non-violent revolution and opposed the violent tactics of the Maoist revolutionary guerrilla group, Shining Path, as well as the austerity measures of the Peruvian government. Her vocal opposition to the Shining Path resulted in her assassination at their hands in 1992. Her murder was a flashpoint for change in the impoverished shantytown community of Villa El Salvador on the outskirts of Lima, and support for the Shining Path then dwindled rapidly. An estimated 300,000 people attended her funeral.

On the Side of Life

I believe it is very difficult to defeat, with fear and terror, what people have built with their own hands, strength, and spirit.

This is what we say to the *compañeras* in the local organizations. And this spirit has been demonstrated by the women of Villa El Salvador. Terror must be defeated—the fear that can be sown among us. Here is our example of how to defuse this fear. On the twenty-sixth [September 26, 1991], the workers in all of the kitchens in the Lima metropolitan area are going to demonstrate, using the example of the women of Villa El Salvador. They are going to march in the streets. Marches have been a traditional form of protest for the people, a way to learn and to teach.

We are marching to protest terrorist acts of intimidation, like the assassination of our leaders. The terrorists are threatening our organizations, and when they touch the people and what the people have built, the people will fight. The people will defend what they have built. I don't think this will happen, however. Organized women in this country have learned to stand up for what they believe in, and now they are doing it in a significant way.

I want to explain that the people haven't reacted until now because they lacked faith [in their institutions]. So many human rights have been violated! Our young people have been murdered; they have been made to disappear. There's the case of the disappearance of the young student in Villa El Salvador. So where are the people going to turn if there is no confidence in this government or in its military forces?

The political parties, without exception, are in total disrepute; the people do not think well of them. In these circumstances, it is the people who must confront Sendero. The first defeat must be a political one. I don't think that there should be a military confrontation. I don't have guns or the arms to challenge Sendero. I believe that there has to be a contest of goals. Sendero has a political agenda for the country, with a strategy and all. We, too, have to come up with our proposal. What is it we want for this country? What is it we want to build?

The country has not yet realized that it's the women's organizations that deal with survival issues and the neighborhood organizations that we've built, with our hands and efforts, that offer the direction for the new Peru, assuming, of course, that the government does its part.

How can we say that we're going to defeat Sendero if, at the same time, the government denies its support to the thousands of survival organizations that help people endure this country's crisis? What has the government given these organizations that help people meet their basic needs? Is there legislation to support the kitchens? The legislation is useless. Is there a budget?

This country's government, political parties, and representatives don't recognize just how much they are damaging what the people, who have lost faith in the political parties, have built. The people have organized and created their own defense mechanisms. Of course, the people are not going to approve of terrorist acts because they are the ones who suffer.

We have responded to Sendero, telling the Senderistas they are mistaken if they think they can change the country in their manner. You don't change a country through the assassination of popular leaders, by attacking the people's organizations, and by killing even the priests who work with the people. To those who think that Sendero is fighting for the country, we reply that it will not be its members who move this country ahead; it will be the people who learn to take charge of their own destinies.

The Senderistas are killing our directors. Today we learned that they killed a director of Vaso de Leche in Callao. Are they only killing police? Are they only killing mayors? Are they only killing congressmen? Today they are killing mothers with families. The general public does not react because Sendero has not yet touched them. But today they touched us, and we must unite with an organized and strong response. We know how to do this. We must very carefully state, to this terrorist group, that it is the Senderistas who oppose the people. Until now, many of the women directors have said, "Yes, they are *compañeras* who fight for the people." Not any longer. False. They fight against the people. They are opposed to our organizations.

We align ourselves with life. Let those with political proposals present them so they may be discussed and debated. Sendero must not threaten our directors, because if it touches just one *compañera* or one director in Villa El Salvador, the people of Villa will rise up. We are not afraid of anyone, and we're prepared to give our lives.

Letter to Emma Hilario

Emma:
I learned of the attempt on your life, after having returned from a trip.
I want to tell you that an attempt to kill you is an attempt to kill the poorest and neediest people. There is no name for what they have done. Nothing remains to us except the efforts of an organized people and the consciousness of the women of our towns. I do not tell you to be careful because many people tell us that, and, in the end, nobody takes care of us or supports us. All that remains, Emma, is the fortitude that comes from the hearts of our oppressed people. The terror will only be defeated with that force.
Your compañera,
Marfa Elena Moyano
January 1992

The People's Movement

Twenty thousand women demonstrated against hunger and terror.

We are suffering economic conditions that are the product of neoliberal politics that oppress and crush the poorest people. Furthermore, the military forces are violating human rights. We cannot forget the thousands of dead. There is also the terrorist group that annihilates the leaders of our people and threatens to impose its terror on all of our country. Today it threatens the communal kitchens, where it finds the poorest people.

In light of these circumstances, we propose that the government change the political and economic situation by taking the following steps: raising salaries; providing equity in taxation; supporting nutrition legislation; and fortifying peasants by giving them machinery and seeds, instead of arms, to combat Sendero.

In contrast to the group that says we should "fight for justice," we propose that the people should govern themselves. We believe in the organizations the people, out of necessity, have generated to deal with the economic situation. And we believe in the people's right of self-defense.

We are convinced that this oppressive situation must change so that the needy have opportunities in all areas – health, education, and also in politics, where the elite have dominated, and the people have not participated.

This change will be achieved through the support supplied by the local organizations and through popular democracy, autonomy, and justice. I don't think the armed forces should deliver food in the barrios, and they shouldn't have a presence there. We don't have confidence in them. There would be indignation. It is too late.

We propose that there be an accord that arises from the coordinated efforts of the people, orchestrated by an entity previously known as the National Popular Assembly (ANP). This assembly would include all of the organizations and would represent the poorest people. This accord must have two fundamental tenets: it must be against hunger; it must be against terror.

The people must democratically decide when to use the mechanism of self-defense, without having it imposed upon them. They must take a position and decide how to defend themselves.

In addition to bearing the effects of the government's crushing economic policies that have worsened the already exhausted condition of our people, the people now suffer attacks from the Senderistas, who are programmed to destroy our organizations and, through threats and terror, to end neighborhood leadership.

It is clear that Sendero Luminoso is concentrating its efforts in Lima and, particularly, in the barrios, in accordance with its plan geared at achieving "strategic equilibrium." But in order for it to advance, it must do away with the Vaso de Leche committees, the communal kitchens, and the various survival and neighborhood organizations. This is a task it will not accomplish, because the women of the barrios want peace for the country, and they reject terrorism, as seen at the mass demonstration of September 26.

They killed Juana Lopez Leon, a Vaso de Leche director. They tried to destroy one of the supply centers of the Women's Federation of Villa El

Salvador. However, the women, who have worked in the communal kitchens since 1978, and in Vaso de Leche since 1984, are standing firm, providing an example, teaching how to construct democracy from the bottom up, demonstrating that they can survive and still generate new jobs, thus contributing to national development and to social transformation.

These are the women who demand that the government pay attention to Law 25307 in its national budget. This law recognizes the grassroots organizations and creates a support program for their current nutritional work. Day after day, the women in these organizations demonstrate their management skills. These are the women who forge unity; they know that by being united, they will provide for their children's well-being.

In the immediate future, it is essential to generate a wide citizen mobilization that results, for example, in district accords that favor peace and democracy and unite those of us who stand for life and not for war and death. Unity will help us overcome our fears. Additionally, as a means of protecting the grassroots organizations, and as an affirmation of democracy, we should use the neighborhood patrols for our own defense in everyday life.

INCIDENTS IN THE LIFE OF A SLAVE GIRL

Harriet Jacobs

Harriet Jacobs (1813 or 1815–1897) was an African-American writer who was born into slavery in the American South and recalled the horrific realities of enslaved life in her groundbreaking *Incidents in the Life of a Slave Girl.* For seven years she hid in a tiny crawlspace in her grandmother's home to avoid the sexual advances of her white master, who threatened to sell her children to a different owner if she did not comply with his desires. In 1842, she escaped from hiding to the Free North and wrote her memoir. Living in the confinement of the crawlspace for so long led to serious health issues which plagued Jacobs till the end of her life. They did not, however, prevent her from organizing relief work before, during, and in the aftermath of the Civil War for freed slaves and soldiers. Her work came to be seen as one of the most important slave narratives and she is remembered as an icon of feminism and Black liberation.

XXI. The Loophole Of Retreat.

A small shed had been added to my grandmother's house years ago. Some boards were laid across the joists at the top, and between these boards and the roof was a very small garret, never occupied by any thing but rats and mice. It was a pent roof, covered with nothing but shingles, according to the southern custom for such buildings. The garret was only nine feet long and seven wide. The highest part was three feet high, and sloped down abruptly to the loose board floor. There was no admission for either light or air. My uncle Phillip, who was a carpenter, had very skilfully made a concealed trap-door, which communicated with the storeroom. He had been doing this while I was waiting in the swamp. The storeroom opened upon a piazza. To this hole I was conveyed as soon as I entered the house. The air was stifling; the darkness total. A bed had been spread on the floor. I could sleep quite comfortably on one side; but the slope was so sudden that I could not turn on my other without hitting the roof. The rats and mice ran over my bed; but I was weary, and I slept such sleep as the wretched may, when a tempest has passed over them. Morning came. I knew it only by the noises I heard; for in my small den day and night were all the same. I suffered for air even more than for light. But I was not comfortless. I heard the voices of my children. There was joy and there was sadness in the sound. It made my tears flow. How I longed to speak to them! I was eager to look on their faces; but there was no hole, no crack, through which I could peep. This continued darkness was oppressive. It seemed horrible to sit or lie in a cramped position day after day, without one gleam of light. Yet I would have chosen this, rather than my lot as a slave, though white people considered it an easy one; and it was so compared with the fate of others. I was never cruelly overworked; I was never lacerated with the whip from head to foot; I was never so beaten and bruised that I could not turn from one side to the other; I never had my heel-strings cut to prevent my running away; I was never chained to a log and forced to drag it about, while I toiled in the fields from morning till night; I was never branded with hot iron, or torn by bloodhounds. On the contrary, I had always been kindly treated, and tenderly cared for, until I came into the hands of Dr. Flint. I had never wished for freedom till then. But though my life in slavery was comparatively devoid of hardships, God pity the woman who is compelled to lead such a life!

My food was passed up to me through the trap-door my uncle had contrived; and my grandmother, my uncle Phillip, and aunt Nancy would seize such opportunities as they could, to mount up there and chat with me at the opening. But of course this was not safe in the daytime. It must all be done in darkness. It was impossible for me to move in an erect position, but I crawled about my den for exercise. One day I hit my head against something, and found it was a gimlet. My uncle had left it sticking there when he made the trap-door. I was as rejoiced as Robinson Crusoe could have been at finding such a treasure. It put a lucky thought into my head. I said to myself, "Now I will have some light. Now I will see my children." I did not dare to begin my work during the daytime, for fear of attracting attention. But I groped round; and having found the side next the street, where I could frequently see my children, I stuck the gimlet in and waited for evening. I bored three rows of holes, one above another; then I bored out the interstices between. I thus succeeded in making one hole about an inch long and an inch broad. I sat by it till late into the night, to enjoy the little whiff of air that floated in. In the morning I watched for my children. The first person I saw in the street was Dr. Flint. I had a shuddering, superstitious feeling that it was a bad omen. Several familiar faces passed by. At last I heard the merry laugh of children, and presently two sweet little faces were looking up at me, as though they knew I was there, and were conscious of the joy they imparted. How I longed to *tell* them I was there!

My condition was now a little improved. But for weeks I was tormented by hundreds of little red insects, fine as a needle's point, that pierced through my skin, and produced an intolerable burning. The good grandmother gave me herb teas and cooling medicines, and finally I got rid of them. The heat of my den was intense, for nothing but thin shingles protected me from the scorching summer's sun. But I had my consolations. Through my peeping-hole I could watch the children, and when they were near enough, I could hear their talk. Aunt Nancy brought me all the news she could hear at Dr. Flint's. From her I learned that the doctor had written to New York to a colored woman, who had been born and raised in our neighborhood, and had breathed his contaminating atmosphere. He offered her a reward if she could find out any thing about me. I know not what was the nature of her reply; but he soon after started for New York in haste, saying to his family that he had business of importance to transact. I peeped at him as he passed on his way to the steamboat. It was a satisfaction to have miles of land and water between us, even for a little while; and it was a still greater satisfaction to know that he believed me to be in the Free States. My little den seemed less dreary than it had done. He returned, as he did from his former journey to New York, without obtaining any satisfactory information. When he passed our house next morning, Benny was standing at the gate. He had heard them say that he had gone to find me, and he called out, "Dr. Flint, did you bring my mother home? I want to see her." The doctor stamped his foot at him in a rage, and exclaimed, "Get out of the way, you little damned rascal! If you don't, I'll cut off your head."

Benny ran terrified into the house, saying, "You can't put me in jail again. I don't belong to you now." It was well that the wind carried the words away from the doctor's ear. I told my grandmother of it, when we had our next conference at the trap-door, and begged of her not to allow the children to be impertinent to the irascible old man.

Autumn came, with a pleasant abatement of heat. My eyes had become accustomed to the dim light, and by holding my book or work in a certain position near the aperture I contrived to read and sew. That was a great relief to the tedious monotony of my life. But when winter came, the cold penetrated through the thin shingle roof, and I was dreadfully chilled. The winters there are not so long, or so severe, as in northern latitudes; but the houses are not built to shelter from cold, and my little den was peculiarly comfortless. The kind grandmother brought me bedclothes and warm drinks. Often I was obliged to lie in bed all day to keep comfortable; but with all my precautions, my shoulders and feet were frostbitten. O, those long, gloomy days, with no object for my eye to rest upon, and no thoughts to occupy my mind, except the dreary past and the uncertain future! I was thankful when there came a day sufficiently mild for me to wrap myself up and sit at the loophole to watch the passers by. Southerners have the habit of stopping and talking in the streets, and I heard many conversations not intended to meet my ears. I heard slave-hunters planning how to catch some poor fugitive. Several times I heard allusions to Dr. Flint, myself, and the history of my children, who, perhaps, were playing near the gate. One would say, "I wouldn't move my little finger to catch her, as old Flint's property." Another would say, "I'll catch *any* nigger for the reward. A man ought to have what belongs to him, if he *is* a damned brute." The opinion was often expressed that I was in the Free States. Very rarely did any one suggest that I might be in the vicinity. Had the least suspicion rested on my grandmother's house, it would have been burned to the ground. But it was the last place they thought of. Yet there was no place, where slavery existed, that could have afforded me so good a place of concealment.

Dr. Flint and his family repeatedly tried to coax and bribe my children to tell something they had heard said about me. One day the doctor took them into a shop, and offered them some bright little silver pieces and gay handkerchiefs if they would tell where their mother was. Ellen shrank away from him, and would not speak; but Benny spoke up, and said, "Dr. Flint, I don't know where my mother is. I guess she's in New York; and when you go there again, I wish you'd ask her to come home, for I want to see her; but if you put her in jail, or tell her you'll cut her head off, I'll tell her to go right back."

XXII. Christmas Festivities.

Christmas was approaching. Grandmother brought me materials, and I busied myself making some new garments and little playthings for my children. Were it not that hiring day is near at hand, and many families are fearfully looking forward to the probability of separation in a few days, Christmas might be a happy season for the poor slaves. Even slave mothers try to gladden the hearts of their little ones on that occasion. Benny and Ellen had their Christmas stockings filled. Their imprisoned mother could not have the privilege of witnessing their surprise and joy. But I had the pleasure of peeping at them as they went into the street with their new suits on. I heard Benny ask a little playmate whether Santa Claus brought him any thing. "Yes," replied the boy; "but Santa Claus ain't a real man. It's the children's mothers that put things into the stockings." "No, that can't be," replied Benny, "for Santa Claus brought Ellen and me these new clothes, and my mother has been gone this long time."

How I longed to tell him that his mother made those garments, and that many a tear fell on them while she worked!

Every child rises early on Christmas morning to see the Johnkannaus. Without them, Christmas would be shorn of its greatest attraction. They consist of companies of slaves from the plantations, generally of the lower class. Two athletic men, in calico wrappers, have a net thrown over them, covered with all manner of bright-colored stripes. Cows' tails are fastened to their backs, and their heads are decorated with horns. A box, covered with sheepskin, is called the gumbo box. A dozen beat on this, while others strike triangles and jawbones, to which bands of dancers keep time. For a month previous they are composing songs, which are sung on this occasion. These companies, of a hundred each, turn out early in the morning, and are allowed to go round till twelve o'clock, begging for contributions. Not a door is left unvisited where there is the least chance of obtaining a penny or a glass of rum. They do not drink while they are out, but carry the rum home in jugs, to have a carousal. These Christmas donations frequently amount to twenty or thirty dollars. It is seldom that any white man or child refuses to give them a trifle. If he does, they regale his ears with the following song:—

Poor massa, so dey say;
Down in de heel, so dey say;
Got no money, so dey say;
Not one shillin, so dey say;
God A'mighty bress you, so dey say.

Christmas is a day of feasting, both with white and colored people. Slaves, who are lucky enough to have a few shillings, are sure to spend them for good eating; and many a turkey and pig is captured, without saying, "By your leave, sir." Those who cannot obtain these, cook a 'possum, or a raccoon, from which savory dishes can be made. My grandmother raised poultry and pigs for sale and it was her established custom to have both a turkey and a pig roasted for Christmas dinner.

On this occasion, I was warned to keep extremely quiet, because two guests had been invited. One was the town constable, and the other was a free colored man, who tried to pass himself off for white, and who was always ready to do any mean work for the sake of currying favor with white people. My grandmother had a motive for inviting them. She managed to take them all over the house. All the rooms on the lower floor were thrown open for them to pass in and out; and after dinner, they were invited up stairs to look at a fine mocking bird my uncle had just brought home. There, too, the rooms were all thrown open that they might look in. When I heard them talking on the piazza, my heart almost stood still. I knew this colored man had spent many nights hunting for me. Every body knew he had the blood of a slave father in his veins; but for the sake of passing himself off for white, he was ready to kiss the slaveholders' feet. How I despised him! As for the constable, he wore no false colors. The duties of his office were despicable, but he was superior to his companion, inasmuch as he did not pretend to be what he was not. Any white man, who could raise money enough to buy a slave, would have considered himself degraded by being a constable; but the office enabled its possessor to exercise authority. If he found any slave out after nine o'clock, he could whip him as much as he liked; and that was a privilege to be coveted. When the guests were ready to depart, my grandmother gave each of them some of her nice pudding, as a present for their wives. Through my peep-hole I saw them go out of the gate, and I was glad when it closed after them. So passed the first Christmas in my den.

XXIII. Still In Prison.

When spring returned, and I took in the little patch of green the aperture commanded, I asked myself how many more summers and winters I must be condemned to spend thus. I longed to draw in a plentiful draught of fresh air, to stretch my cramped limbs, to have room to stand erect, to feel the earth under my feet again. My relatives were constantly on the lookout for a chance of escape; but none offered that seemed practicable, and even tolerably safe. The hot summer came again, and made the turpentine drop from the thin roof over my head.

During the long nights I was restless for want of air, and I had no room to toss and turn. There was but one compensation; the atmosphere was so stifled that even mosquitos would not condescend to buzz in it. With all my detestation of Dr. Flint, I could hardly wish him a worse punishment, either in this world or that which is to come, than to suffer what I suffered in one single summer. Yet the laws allowed *him* to be out in the free air, while I, guiltless of crime, was pent up here, as the only means of avoiding the cruelties the laws allowed him to inflict upon me! I don't know what kept life within me. Again and again, I thought I should die before long; but I saw the leaves of another autumn whirl through the air, and felt the touch of another winter. In summer the most terrible thunder storms were acceptable, for the rain came through the roof, and I rolled up my bed that it might cool the hot boards under it. Later in the season, storms sometimes wet my clothes through and through, and that was not comfortable when the air grew chilly. Moderate storms I could keep out by filling the chinks with oakum.

But uncomfortable as my situation was, I had glimpses of things out of doors, which made me thankful for my wretched hiding-place. One day I saw a slave pass our gate, muttering, "It's his own, and he can kill it if he will." My grandmother told me that woman's history. Her mistress had that day seen her baby for the first time, and in the lineaments of its fair face she saw a likeness to her husband. She turned the bondwoman and her child out of doors, and forbade her ever to return. The slave went to her master, and told him what had happened. He promised to talk with her mistress, and make it all right. The next day she and her baby were sold to a Georgia trader.

Another time I saw a woman rush wildly by, pursued by two men. She was a slave, the wet nurse of her mistress's children. For some trifling offence her mistress ordered her to be stripped and whipped. To escape the degradation and the torture, she rushed to the river, jumped in, and ended her wrongs in death.

Senator Brown, of Mississippi, could not be ignorant of many such facts as these, for they are of frequent occurrence in every Southern State. Yet he stood up in the Congress of the United States, and declared that slavery was "a great moral, social, and political blessing; a blessing to the master, and a blessing to the slave!"

I suffered much more during the second winter than I did during the first. My limbs were benumbed by inaction, and the cold filled them with cramp. I had a very painful sensation of coldness in my head; even my face and tongue stiffened, and I lost the power of speech. Of course it was impossible, under the circumstances, to summon any physician. My brother William came and did all he could for me. Uncle Phillip also watched tenderly over me; and poor grandmother crept up and down to inquire whether there were any signs of returning life. I was restored to consciousness by the dashing of cold water in my face, and found myself leaning against my brother's arm, while he bent over me with streaming eyes. He afterwards told me he thought I was dying, for I had been in an unconscious state sixteen hours. I next became delirious, and was in great danger of betraying myself and my friends. To prevent this, they stupefied me with drugs. I remained in bed six weeks, weary in body and sick at heart. How to get medical advice was the question. William finally went to a Thompsonian doctor, and described himself as having all my pains and aches. He returned with herbs, roots, and ointment. He was especially charged to rub on the ointment by a fire; but how could a fire be made in my little den? Charcoal in a furnace was tried, but there was no outlet for the gas, and it nearly cost me my life. Afterwards coals, already kindled, were brought up in an iron pan, and placed on bricks. I was so weak, and it was so long since I had enjoyed the warmth of a fire, that those few coals actually made me weep. I think the medicines did me some good; but my recovery was very slow. Dark thoughts passed through my mind as I lay there day after day. I tried to be thankful for my little cell, dismal as it was, and even to love it, as part of the price I had paid for the redemption of my children. Sometimes I thought God was a compassionate Father, who would forgive my sins for the sake of my sufferings. At other times, it seemed to me there was no justice or mercy in the divine government. I asked why the curse of slavery was permitted to exist, and why I had been so persecuted and wronged from youth upward. These things took the shape of mystery, which is to this day not so clear to my soul as I trust it will be hereafter.

In the midst of my illness, grandmother broke down under the weight and anxiety and toil. The idea of losing her, who had always been my best friend and a mother to my children, was the sorest trial I had yet had. O, how earnestly I prayed that she might recover! How hard it seemed, that I could not tend upon her, who had so long and so tenderly watched over me!

One day the screams of a child nerved me with strength to crawl to my peeping-hole, and I saw my son covered with blood. A fierce dog, usually kept chained, had seized and bitten him. A doctor was sent for, and I heard the groans and screams of my child while the wounds were being sewed up. O, what torture to a mother's heart, to listen to this and be unable to go to him!

But childhood is like a day in spring, alternately shower and sunshine. Before night Benny was bright and lively, threatening the destruction of the dog; and great was his delight when the doctor told him the next day that the dog had bitten another boy and been shot. Benny recovered from his wounds; but it was long before he could walk.

When my grandmother's illness became known, many ladies, who were her customers, called to bring her some little comforts, and to inquire whether she had every thing she wanted. Aunt Nancy one night asked permission to watch with her sick mother, and Mrs. Flint replied, "I don't see any need of your going. I can't spare you." But when she found other ladies in the neighborhood were so attentive, not wishing to be outdone in Christian charity, she also sallied forth, in magnificent condescension, and stood by the bedside of her who had loved her in her infancy, and who had been repaid by such grievous wrongs. She seemed surprised to find her so ill, and scolded uncle Phillip for not sending for Dr. Flint. She herself sent for him immediately, and he came. Secure as I was in my retreat, I should have been terrified if I had known he was so near me. He pronounced my grandmother in a very critical situation, and said if her attending physician wished it, he would visit her. Nobody wished to have him coming to the house at all hours, and we were not disposed to give him a chance to make out a long bill.

As Mrs. Flint went out, Sally told her the reason Benny was lame was, that a dog had bitten him. "I'm glad of it," replied she. "I wish he had killed him. It would be good news to send to his mother. *Her* day will come. The dogs will grab *her* yet." With these Christian words she and her husband departed, and, to my great satisfaction, returned no more.

I learned from uncle Phillip, with feelings of unspeakable joy and gratitude, that the crisis was passed and grandmother would live. I could now say from my heart, "God is merciful. He has spared me the anguish of feeling that I caused her death."

TEN DAYS IN A MAD-HOUSE

Nellie Bly

Nellie Bly (1864–1922) was an American journalist and adventurer. Bly challenged herself to a round-the-world voyage in the style of Jules Verne's *Around The World In Eighty Days* – but she completed the trip in just seventy-two. Bly was an accomplished journalist who undertook an investigation of an asylum by posing as a mental patient in order to report on the deplorable conditions for the patients inside. Her report caused a stir and led to a re-evaluation of mental health practices in American asylums. She is considered an early feminist pioneer and, in 1998, was posthumously inducted into the National Women's Hall of Fame.

Chapter X

My First Supper

This examination over, we heard some one yell, "Go out into the hall." One of the patients kindly explained that this was an invitation to supper. We late comers tried to keep together, so we entered the hall and stood at the door where all the women had crowded. How we shivered as we stood there! The windows were open and the draught went whizzing through the hall. The patients looked blue with cold, and the minutes stretched into a quarter of an hour. At last one of the nurses went forward and unlocked a door, through which we all crowded to a landing of the stairway. Here again came a long halt directly before an open window.

"How very imprudent for the attendants to keep these thinly clad women standing here in the cold," said Miss Neville.

I looked at the poor crazy captives shivering, and added, emphatically, "It's horribly brutal." While they stood there I thought I would not relish supper that night. They looked so lost and hopeless. Some were chattering nonsense to invisible persons, others were laughing or crying aimlessly, and one old, gray-haired woman was nudging me, and, with winks and sage noddings of the head and pitiful uplifting of the eyes and hands, was assuring me that I must not mind the poor creatures, as they were all mad. "Stop at the heater," was then ordered, "and get in line, two by two." "Mary, get a companion." "How many times must I tell you to keep in line?" "Stand still," and, as the orders were issued, a shove and a push were administered, and often a slap on the ears. After this third and final halt, we were marched into a long, narrow dining-room, where a rush was made for the table.

The table reached the length of the room and was uncovered and uninviting. Long benches without backs were put for the patients to sit on, and over these they had to crawl in order to face the table. Placed closed together all along the table were large dressing-bowls filled with a pinkish-looking stuff which the patients called tea. By each bowl was laid a piece of bread, cut thick and buttered. A small saucer containing five prunes accompanied the bread. One fat woman made a rush, and jerking up several saucers from those around her emptied their contents into her own saucer. Then while holding to her own bowl she lifted up another and drained its contents at one gulp. This she did to a second bowl in shorter time than it takes to tell it. Indeed, I was so amused at her successful grabbings that when I looked at my own share the woman opposite, without so much as by your leave, grabbed my bread and left me without any.

Another patient, seeing this, kindly offered me hers, but I declined with thanks and turned to the nurse and asked for more. As she flung a thick piece down on the table she made some remark about the fact that if I forgot where my home was I had not forgotten how to eat. I tried the bread, but the butter was so horrible that one could not eat it. A blue-eyed German girl on the opposite side of the table told me I could have bread unbuttered if I wished, and that very few were able to eat the butter. I turned my attention to the prunes and found that very few of them would be sufficient. A patient near asked me to give them to her. I did so. My bowl of tea was all that was left. I tasted, and one taste was enough. It had no sugar, and it tasted as if it had been made in copper. It was as weak as water. This was also transferred to a hungrier patient, in spite of the protest of Miss Neville.

"You must force the food down," she said, "else you will be sick, and who know but what, with these surroundings, you may go crazy. To have a good brain the stomach must be cared for."

"It is impossible for me to eat that stuff," I replied, and, despite all her urging, I ate nothing that night.

It did not require much time for the patients to consume all that was eatable on the table, and then we got our orders to form in line in the hall. When this was done the doors before us were unlocked and we were ordered to proceed back to the sitting-room. Many of the patients crowded near us, and I was again urged to play, both by them and by the nurses. To please the patients I promised to play and Miss Tillie Mayard was to sing. The first thing she asked me to play was "Rock-a-bye Baby," and I did so. She sang it beautifully.

Chapter XI
In the Bath

A few more songs and we were told to go with Miss Grupe. We were taken into a cold, wet bathroom, and I was ordered to undress. Did I protest? Well, I never grew so earnest in my life as when I tried to beg off. They said if I did not they would use force and that it would not be very gentle. At this I noticed one of the craziest women in the ward standing by the filled bathtub with a large, discolored rag in her hands. She was chattering away to herself and chuckling in a manner which seemed to me fiendish. I knew now what was to be done with me. I shivered. They began to undress me, and one by one they pulled off my clothes. At last everything was gone excepting one garment. "I will not remove it," I said vehemently, but they took it off. I gave one glance at the group of patients gathered at the door watching the scene, and I jumped into the bathtub with more energy than grace.

The water was ice-cold, and I again began to protest. How useless it all was! I begged, at least, that the patients be made to go away, but was ordered to shut up. The crazy woman began to scrub me. I can find no other word that will express it but scrubbing. From a small tin pan she took some soft soap and rubbed it all over me, even all over my face and my pretty hair. I was at last past seeing or speaking, although I had begged that my hair be left untouched. Rub, rub, rub, went the old woman, chattering to herself. My teeth chattered and my limbs were goose-fleshed and blue with cold. Suddenly I got, one after the other, three buckets of water over my head – ice-cold water, too – into my eyes, my ears, my nose and my mouth. I think I experienced some of the sensations of a drowning person as they dragged me, gasping, shivering and quaking, from the tub. For once I did look insane. I caught a glance of the indescribable look on the faces of my companions, who had witnessed my fate and knew theirs was surely following. Unable to control myself at the absurd picture I presented, I burst into roars of laughter. They put me, dripping wet, into a short canton flannel slip, labeled across the extreme end in large black letters, "Lunatic Asylum, B. I., H. 6." The letters meant Blackwell's Island, Hall 6.

By this time Miss Mayard had been undressed, and, much as I hated my recent bath, I would have taken another if by it I could have saved her the experience. Imagine plunging that sick girl into a cold bath when it made me, who have never been ill, shake as if with ague. I heard her explain to Miss

Grupe that her head was still sore from her illness. Her hair was short and had mostly come out, and she asked that the crazy woman be made to rub more gently, but Miss Grupe said:

"There isn't much fear of hurting you. Shut up, or you'll get it worse." Miss Mayard did shut up, and that was my last look at her for the night.

I was hurried into a room where there were six beds, and had been put into bed when some one came along and jerked me out again, saying:

"Nellie Brown has to be put in a room alone to-night, for I suppose she's noisy."

I was taken to room 28 and left to try and make an impression on the bed. It was an impossible task. The bed had been made high in the center and sloping on either side. At the first touch my head flooded the pillow with water, and my wet slip transferred some of its dampness to the sheet. When Miss Grupe came in I asked if I could not have a night-gown.

"We have not such things in this institution," she said.

"I do not like to sleep without," I replied.

"Well, I don't care about that," she said. "You are in a public institution now, and you can't expect to get anything. This is charity, and you should be thankful for what you get."

"But the city pays to keep these places up," I urged, "and pays people to be kind to the unfortunates brought here."

"Well, you don't need to expect any kindness here, for you won't get it," she said, and she went out and closed the door.

A sheet and an oilcloth were under me, and a sheet and black wool blanket above. I never felt anything so annoying as that wool blanket as I tried to keep it around my shoulders to stop the chills from getting underneath. When I pulled it up I left my feet bare, and when I pulled it down my shoulders were exposed. There was absolutely nothing in the room but the bed and myself. As the door had been locked I imagined I should be left alone for the night, but I heard the sound of the heavy tread of two women down the hall. They stopped at every door, unlocked it, and in a few moments I could hear them relock it. This they did without the least attempt at quietness down the whole length of the opposite side of the hall and up to my room. Here they paused. The key was inserted in the lock and turned. I watched those about to enter. In they came, dressed in brown and white striped dresses, fastened by brass buttons, large, white aprons, a heavy green cord about the waist, from which dangled a bunch of large keys, and small, white caps on their heads. Being dressed as were the attendants of the day, I knew they were nurses. The first one carried a lantern, and she flashed its light into my face while she said to her assistant:

"This is Nellie Brown." Looking at her, I asked:

"Who are you?"

"The night nurse, my dear," she replied, and, wishing that I would sleep well, she went out and locked the door after her. Several times during the night they came into my room, and even had I been able to sleep, the unlocking of the heavy door, their loud talking, and heavy tread, would have awakened me.

I could not sleep, so I lay in bed picturing to myself the horrors in case a fire should break out in the asylum. Every door is locked separately and the windows are heavily barred, so that escape is impossible. In the one building alone there are, I think Dr. Ingram told me, some three hundred women. They are locked, one to ten to a room. It is impossible to get out unless these doors are unlocked. A fire is not improbable, but one of the most likely occurrences. Should the building burn, the jailers or nurses would never think of releasing their crazy patients. This I can prove to you later when I come to tell of their cruel treatment of the poor things intrusted to their care. As I say, in case of fire, not a dozen women could escape. All would be left to roast to death. Even if the nurses were kind, which they are not, it would require more presence of mind than women of their class possess to risk the flames and their own lives while they unlocked the hundred doors for the insane prisoners. Unless there is a change there will some day be a tale of horror never equaled.

In this connection is an amusing incident which happened just previous to my release. I was talking with Dr. Ingram about many things, and at last told him what I thought would be the result of a fire.

"The nurses are expected to open the doors," he said.

"But you know positively that they would not wait to do that," I said, "and these women would burn to death."

He sat silent, unable to contradict my assertion.

"Why don't you have it changed?" I asked.

"What can I do?" he replied. "I offer suggestions until my brain is tired, but what good does it do? What would you do?" he asked, turning to me, the proclaimed insane girl.

"Well, I should insist on them having locks put in, as I have seen in some places, that by turning a crank at the end of the hall you can lock or unlock every door on the one side. Then there would be some chance of escape. Now, every door being locked separately, there is absolutely none."

Dr. Ingram turned to me with an anxious look on his kind face as he asked, slowly:

"Nellie Brown, what institution have you been an inmate of before you came here?"

"None. I never was confined in any institution, except boarding-school, in my life."

"Where then did you see the locks you have described?"

I had seen them in the new Western Penitentiary at Pittsburg, Pa., but I did not dare say so. I merely answered:

"Oh, I have seen them in a place I was in – I mean as a visitor."

"There is only one place I know of where they have those locks," he said, sadly, "and that is at Sing Sing."

The inference is conclusive. I laughed very heartily over the implied accusation, and tried to assure him that I had never, up to date, been an inmate of Sing Sing or even ever visited it.

Just as the morning began to dawn I went to sleep. It did not seem many moments until I was rudely awakened and told to get up, the window being opened and the clothing pulled off me. My hair was still wet and I had pains all through me, as if I had the rheumatism. Some clothing was flung on the floor and I was told to put it on. I asked for my own, but was told to take what I got and keep quiet by the apparently head nurse, Miss Grady. I looked at it. One underskirt made of coarse dark cotton goods and a cheap white calico dress with a black spot in it. I tied the strings of the skirt around me and put on the little dress. It was made, as are all those worn by the patients, into a straight tight waist sewed on to a straight skirt. As I buttoned the waist I noticed the underskirt was about six inches longer than the upper, and for a moment I sat down on the bed and laughed at my own appearance. No woman ever longed for a mirror more than I did at that moment.

I saw the other patients hurrying past in the hall, so I decided not to lose anything that might be going on. We numbered forty-five patients in Hall 6, and were sent to the bathroom, where there were two coarse towels. I watched crazy patients who had the most dangerous eruptions all over their faces dry on the towels and then saw women with clean skins turn to use them. I went to the bathtub and washed my face at the running faucet and my underskirt did duty for a towel.

Before I had completed my ablutions a bench was brought into the bathroom. Miss Grupe and Miss McCarten came in with combs in their hands. We were told so sit down on the bench, and the hair of forty-five women was combed with one patient, two nurses, and six combs. As I saw some of the sore heads combed I thought this was another dose I had not bargained for. Miss Tillie Mayard had her own comb, but it was taken from her by Miss Grady. Oh, that combing! I never realized before what the expression "I'll give you a combing" meant, but I knew then. My hair, all matted and wet from the night previous, was pulled and jerked, and, after expostulating to no avail, I set my teeth and endured the pain. They refused to give me my hairpins, and my hair was arranged in one plait and tied with a red cotton rag. My curly bangs refused to stay back, so that at least was left of my former glory.

After this we went to the sitting-room and I looked for my companions. At first I looked vainly, unable to distinguish them from the other patients, but after awhile I recognized Miss Mayard by her short hair.

"How did you sleep after your cold bath?"

"I almost froze, and then the noise kept me awake. It's dreadful! My nerves were so unstrung before I came here, and I fear I shall not be able to stand the strain."

I did the best I could to cheer her. I asked that we be given additional clothing, at least as much as custom says women shall wear, but they told me to shut up; that we had as much as they intended to give us.

We were compelled to get up at 5.30 o'clock, and at 7.15 we were told to collect in the hall, where the experience of waiting, as on the evening previous, was repeated. When we got into the dining-room at last we found a bowl of cold tea, a slice of buttered bread and a saucer of oatmeal, with molasses on it, for each patient. I was hungry, but the food would not down. I asked for unbuttered bread and was given it. I cannot tell you of anything which is the same dirty, black color. It was hard, and in places nothing more than dried dough. I found a spider in my slice, so I did not eat it. I tried the oatmeal and molasses, but it was wretched, and so I endeavored, but without much show of success, to choke down the tea.

After we were back to the sitting-room a number of women were ordered to make the beds, and some of the patients were put to scrubbing and others given different duties which covered all the work in the hall. It is not the attendants who keep the institution so nice for the poor patients, as I had always thought, but the patients, who do it all themselves – even to cleaning the nurses' bedrooms and caring for their clothing.

About 9.30 the new patients, of which I was one, were told to go out to see the doctor. I was taken in and my lungs and my heart were examined by the flirty young doctor who was the first to see us the day we entered. The one who made out the report, if I mistake not, was the assistant superintendent, Ingram. A few questions and I was allowed to return to the sitting-room.

I came in and saw Miss Grady with my note-book and long lead pencil, bought just for the occasion.

"I want my book and pencil," I said, quite truthfully. "It helps me remember things."

I was very anxious to get it to make notes in and was disappointed when she said:

"You can't have it, so shut up."

Some days after I asked Dr. Ingram if I could have it, and he promised to consider the matter. When I again referred to it, he said that Miss Grady said I only brought a book there; and that I had no pencil. I was provoked, and insisted that I had, whereupon I was advised to fight against the imaginations of my brain.

After the housework was completed by the patients, and as the day was fine, but cold, we were told to go out in the hall and get on shawls and hats for a walk. Poor patients! How eager they were for a breath of air; how eager for a slight release from their prison. They went swiftly into the hall and there was a skirmish for hats. Such hats!

Chapter XII
Promenading with Lunatics

I shall never forget my first walk. When all the patients had donned the white straw hats, such as bathers wear at Coney Island, I could not but laugh at their comical appearances. I could not distinguish one woman from another. I lost Miss Neville, and had to take my hat off and search for her. When we met we put our hats on and laughed at one another. Two by two we formed in line, and guarded by the attendants we went out a back way on to the walks.

We had not gone many paces when I saw, proceeding from every walk, long lines of women guarded by nurses. How many there were! Every way I looked I could see them in the queer dresses, comical straw hats and shawls, marching slowly around. I eagerly watched the passing lines and a thrill of horror crept over me at the sight. Vacant eyes and meaningless faces, and their tongues uttered meaningless nonsense. One crowd passed and I noted by nose as well as eyes, that they were fearfully dirty.

"Who are they?" I asked of a patient near me.

"They are considered the most violent on the island," she replied. "They are from the Lodge, the first building with the high steps." Some were yelling, some were cursing, others were singing or praying or preaching, as the fancy struck them, and they made up the most miserable collection of humanity I had ever seen. As the din of their passing faded in the distance there came another sight I can never forget:

A long cable rope fastened to wide leather belts, and these belts locked around the waists of fifty-two women. At the end of the rope was a heavy iron cart, and in it two women – one nursing a sore foot, another screaming at some nurse, saying: "You beat me and I shall not forget it. You want to kill me," and then she would sob and cry. The women "on the rope," as the patients call it, were each busy on their individual freaks. Some were yelling all the while. One who had blue eyes saw me look at her, and she turned as far as she could, talking and smiling, with that terrible, horrifying look of absolute insanity stamped on her. The doctors might safely judge on her case. The horror of that sight to one who had never been near an insane person before, was something unspeakable.

"God help them!" breathed Miss Neville. "It is so dreadful I cannot look."

On they passed, but for their places to be filled by more. Can you imagine the sight? According to one of the physicians there are 1600 insane women on Blackwell's Island.

Mad! what can be half so horrible? My heart thrilled with pity when I looked on old, gray-haired women talking aimlessly to space. One woman had on a straightjacket, and two women had to drag her along. Crippled, blind, old, young, homely, and pretty; one senseless mass of humanity. No fate could be worse.

I looked at the pretty lawns, which I had once thought was such a comfort to the poor creatures confined on the Island, and laughed at my own notions. What enjoyment is it to them? They are not allowed on the grass – it is only to look at. I saw some patients eagerly and caressingly lift a nut or a colored leaf that had fallen on the path. But they were not permitted to keep them. The nurses would always compel them to throw their little bit of God's comfort away.

As I passed a low pavilion, where a crowd of helpless lunatics were confined, I read a motto on the wall, "While I live I hope." The absurdity of it struck me forcibly. I would have liked to put above the gates that open to the asylum, "He who enters here leaveth hope behind."

During the walk I was annoyed a great deal by nurses who had heard my romantic story calling to those in charge of us to ask which one I was. I was pointed out repeatedly.

It was not long until the dinner hour arrived and I was so hungry that I felt I could eat anything. The same old story of standing for a half and three-quarters of an hour in the hall was repeated before we got down to our dinners. The bowls in which we had had our tea were now filled with soup, and on a plate was one cold boiled potato and a chunk of beef, which on investigation, proved to be slightly spoiled. There were no knives or forks, and the patients looked fairly savage as they took the tough beef in their fingers and pulled in opposition to their teeth. Those toothless or with poor teeth could not eat it. One tablespoon was given for the soup, and a piece of bread was the final entree. Butter is never allowed at dinner nor coffee or tea. Miss Mayard could not eat, and I saw many of the sick ones turn away in disgust. I was getting very weak from the want of food and tried to eat a slice of bread. After the first few bites hunger asserted itself, and I was able to eat all but the crusts of the one slice.

Superintendent Dent went through the sitting-room, giving an occasional

"How do you do?" "How are you to-day?" here and there among the patients. His voice was as cold as the hall, and the patients made no movement to tell him of their sufferings. I asked some of them to tell how they were suffering from the cold and insufficiency of clothing, but they replied that the nurse would beat them if they told.

I was never so tired as I grew sitting on those benches. Several of the patients would sit on one foot or sideways to make a change, but they were always reproved and told to sit up straight. If they talked they were scolded and told to shut up; if they wanted to walk around in order to take the stiffness out of them, they were told to sit down and be still. What, excepting torture, would produce insanity quicker than this treatment? Here is a class of women sent to be cured. I would like the expert physicians who are condemning me for my action, which has proven their ability, to take a perfectly sane and healthy woman, shut her up and make her sit from 6 A. M. until 8 P. M. on straight-back benches, do not allow her to talk or move during these hours, give her no reading and let her know nothing of the world or its doings, give her bad food and harsh treatment, and see how long it will take to make her insane. Two months would make her a mental and physical wreck.

I have described my first day in the asylum, and as my other nine were exactly the same in the general run of things it would be tiresome to tell about each. In giving this story I expect to be contradicted by many who are exposed. I merely tell in common words, without exaggeration, of my life in a mad-house for ten days. The eating was one of the most horrible things. Excepting the first two days after I entered the asylum, there was no salt for the food. The hungry and even famishing women made an attempt to eat the horrible messes. Mustard and vinegar were put on meat and in soup to give it a taste, but it only helped to make it worse. Even that was all consumed after two days, and the patients had to try to choke down fresh fish, just boiled in water, without salt, pepper or butter; mutton, beef and potatoes without the faintest seasoning. The most insane refused to swallow the food and were threatened with punishment. In our short walks we passed the kitchen where food was prepared for the nurses and doctors. There we got glimpses of melons and grapes and all kinds of fruits, beautiful white bread and nice meats, and the hungry feeling would be increased tenfold. I spoke to some of the physicians, but it had no effect, and when I was taken away the food was yet unsalted.

My heart ached to see the sick patients grow sicker over the table. I saw Miss Tillie Mayard so suddenly overcome at a bite that she had to rush from the dining-room and then got a scolding for doing so. When the patients complained of the food they were told to shut up; that they would not have as good if they were at home, and that it was too good for charity patients.

A German girl, Louise – I have forgotten her last name – did not eat for several days and at last one morning she was missing. From the conversation of the nurses I found she was suffering from a high fever. Poor thing! she told me she unceasingly prayed for death. I watched the nurses make a patient carry such food as the well ones were refusing up to Louise's room. Think of that stuff for a fever patient! Of course, she refused it. Then I saw a nurse, Miss McCarten, go to test her temperature, and she returned with a report of it being some 150 degrees. I smiled at the report, and Miss Grupe, seeing it, asked me how high my temperature had ever run. I refused to answer. Miss Grady then decided to try her ability. She returned with the report of 99 degrees.

Miss Tillie Mayard suffered more than any of us from the cold, and yet she tried to follow my advice to be cheerful and try to keep up for a short time. Superintendent Dent brought in a man to see me. He felt my pulse and my head and examined my tongue. I told them how cold it was, and assured them that I did not need medical aid, but that Miss Mayard did, and they should transfer their attentions to her. They did not answer me, and I was pleased to see Miss Mayard leave her place and come forward to them. She spoke to the doctors and told them she was ill, but they paid no attention to her. The nurses came and dragged her back to the bench, and after the doctors left they said, "After awhile, when you see that the doctors will not notice you, you will quit running up to them." Before the doctors left me I heard one say – I cannot give it in his exact words – that my pulse and eyes were not that of an insane girl, but Superintendent Dent assured him that in cases such as mine such tests failed. After watching me for awhile he said my face was the brightest he had ever seen for a lunatic. The nurses had on heavy undergarments and coats, but they refused to give us shawls.

Nearly all night long I listened to a woman cry about the cold and beg for God to let her die. Another one yelled "Murder!" at frequent intervals and "Police!" at others until my flesh felt creepy.

The second morning, after we had begun our endless "set" for the day, two of the nurses, assisted by some patients, brought the woman in who had begged the night previous for God to take her home. I was not surprised at her prayer. She appeared easily seventy years old, and she was blind. Although the halls were freezing-cold, that old woman had no more clothing on than the rest of us, which I have described. When she was brought into the sitting-room and placed on the hard bench, she cried:

"Oh, what are you doing with me? I am cold, so cold. Why can't I stay in bed or have a shawl?" and then she would get up and endeavor to feel her way to leave the room. Sometimes the attendants would jerk her back to the bench, and again they would let her walk and heartlessly laugh when she bumped against the table or the edge of the benches. At one time she said the heavy shoes which charity provides hurt her feet, and she took them off. The nurses made two patients put them on her again, and when she did it several times, and fought against having them on, I counted seven people at her at once trying to put the shoes on her. The old woman then tried to lie down on the bench, but they pulled her up again. It sounded so pitiful to hear her cry:

"Oh, give me a pillow and pull the covers over me, I am so cold."

At this I saw Miss Grupe sit down on her and run her cold hands over the old woman's face and down inside the neck of her dress. At the old woman's cries she laughed savagely, as did the other nurses, and repeated her cruel action. That day the old woman was carried away to another ward.

INTO THE WILD

Jon Krakauer

Christopher Johnson McCandless (1968–1992), known by his alias 'Alexander Supertramp', was an American adventurer and outdoor enthusiast. His highly romanticized escape from organized society, and journey to the wilds of Alaska – where he died in the early 1990s – have been the subject of a great deal of speculation. McCandless, rumored to have survived an abusive childhood, was a capable outdoorsman who, in 1992, hitchhiked from South Dakota to Fairbanks, Alaska and lived off the land with only a .22 Calibre Long Rifle and minimal survival equipment. Eventually succumbing to starvation or food poisoning, McCandless was found two weeks following his death by a group of hunters seeking shelter in the abandoned bus in which he had been living. He left behind a number of personal belongings, including a journal and photographs that documented his slow descent into starvation and eventual death.

Jon Krakauer (born 1954) is an American writer and mountaineer. He was a member of the ill-fated expedition to summit Mount Everest in 1996, one of the deadliest disasters in the history of climbing. Both his book about the expedition, *Into Thin Air*, and *Into the Wild*, about John McCandless, have been international bestsellers. *Into the Wild* was adapted into a feature film by Sean Penn in 2007.

He was, at long last, about to be alone in the vast Alaska wilds.

As he trudged expectantly down the trail in a fake-fur parka, his rifle slung over one shoulder, the only food McCandless carried was a ten-pound bag of long-grained rice – and the two sandwiches and bag of corn chips that Gallien had contributed. A year earlier he'd subsisted for more than a month beside the Gulf of California on five pounds of rice and a bounty of fish caught with a cheap rod and reel, an experience that made him confident he could harvest enough food to survive an extended stay in the Alaska wilderness, too.

The heaviest item in McCandless's half-full backpack was his library: nine or ten paperbound books, most of which had been given to him by

Jan Burres in Niland. Among these volumes were titles by Thoreau and Tolstoy and Gogol, but McCandless was no literary snob: He simply carried what he thought he might enjoy reading, including mass-market books by Michael Crichton, Robert Pirsig, and Louis l'Amour. Having neglected to pack writing paper, he began a laconic journal on some blank pages in the back of *Tanaina Plantlore*.

The Healy terminus of the Stampede Trail is traveled by a handful of dog mushers, ski tourers, and snow-machine enthusiasts during the winter months, but only until the frozen rivers begin to break up, in late March or early April. By the time McCandless headed into the bush, there was open water flowing on most of the larger streams, and nobody had been very far down the trail for two or three weeks; only the faint remnants of a packed snow-machine track remained for him to follow.

McCandless reached the Teklanika River his second day out. Although the banks were lined with a jagged shelf of frozen overflow, no ice bridges spanned the channel of open water, so he was forced to wade. There had been a big thaw in early April, and breakup had come early in 1992, but the weather had turned cold again, so the river's volume was quite low when McCandless crossed – probably thigh-deep at most – allowing him to splash to the other side without difficulty. He never suspected that in so doing, he was crossing his Rubicon. To McCandless's inexperienced eye, there was nothing to suggest that two months hence, as the glaciers and snowfields at the Teklanika's headwater thawed in the summer heat, its discharge would multiply nine or ten times in volume, transforming the river into a deep, violent torrent that bore no resemblance to the gentle brook he'd blithely waded across in April.

From his journal we know that on April 29, McCandless fell through the ice somewhere. It probably happened as he traversed a series of melting beaver ponds just beyond the Teklanika's western bank, but there is nothing to indicate that he suffered any harm in the mishap. A day later, as the trail crested a ridge, he got his first glimpse of Mt. McKinley's high, blinding-white bulwarks, and a day after that, May 1, some twenty miles down the trail from where he was dropped by Gallien, he stumbled upon the old bus beside the Sushana River. It was outfitted with a bunk and a barrel stove, and previous visitors had left the improvised shelter stocked with matches, bug dope, and other essentials. "Magic Bus Day," he wrote in his journal. He decided to lay over for a while in the vehicle and take advantage of its crude comforts.

He was elated to be there. Inside the bus, on a sheet of weathered plywood spanning a broken window, McCandless scrawled an exultant declaration of independence:

TWO YEARS HE WALKS THE EARTH. NO PHONE, NO POOL, NO PETS, NO CIGARETTES. ULTIMATE FREEDOM. AN EXTREMIST. AN AESTHETIC VOYAGER WHOSE HOME IS THE ROAD. ESCAPED FROM ATLANTA. THOU SHALT NOT RETURN, 'CAUSE "THE WEST IS THE BEST." AND NOW AFTER TWO RAMBLING YEARS COMES THE FINAL AND GREATEST ADVENTURE. THE CLIMACTIC BATTLE TO KILL THE FALSE BEING WITHIN AND VICTORIOUSLY CONCLUDE THE SPIRITUAL PILGRIMAGE. TEN DAYS AND NIGHTS OF FREIGHT TRAINS AND HITCHHIKING BRING HIM TO THE GREAT WHITE NORTH. NO LONGER TO BE POISONED BY CIVILIZATION HE FLEES, AND WALKS ALONE UPON THE LAND TO BECOME LOST IN THE WILD.

ALEXANDER SUPERTRAMP
MAY 1992

Reality, however, was quick to intrude on McCandless's reverie. He had difficulty killing game, and the daily journal entries during his first week in the bush include "Weakness," "Snowed in," and "Disaster." He saw but did not shoot a grizzly on May 2, shot at but missed some ducks on May 4, and finally killed and ate a spruce grouse on May 5; but he didn't shoot anything else until May 9, when he bagged a single small squirrel, by which point he'd written "4th day famine" in the journal.

But soon thereafter his fortunes took a sharp turn for the better. By mid-May the sun was circling high in the heavens, flooding the taiga with light. The sun dipped below the northern horizon for fewer than four hours out of every twenty-four, and at midnight the sky was still bright enough to read by. Everywhere but on the north-facing slopes and in the shadowy ravines, the snowpack had melted down to bare ground, exposing the previous season's rose hips and lingonberries, which McCandless gathered and ate in great quantity.

He also became much more successful at hunting game and for the next six weeks feasted regularly on squirrel, spruce grouse, duck, goose, and porcupine. On May 22, a crown fell off one of his molars, but the event didn't seem to dampen his spirits much, because the following day he scrambled up the nameless, humplike, three-thousand-foot butte that rises directly north of the bus, giving him a view of the whole icy sweep of the Alaska Range and mile after mile of uninhabited country. His journal entry for the day is characteristically terse but unmistakably joyous: "CLIMB MOUNTAIN!"

McCandless had told Gallien that he intended to remain on the move during his stay in the bush. 'I'm just going to take off and keep walking west," he'd said. "I might walk all the way to the Bering Sea." On May 5, after pausing for four days at the bus, he resumed his perambulation. From the snapshots recovered with his Minolta, it appears that McCandless lost (or intentionally left) the by now indistinct Stampede Trail and headed west and north through the hills above the Sushana River, hunting game as he went.

It was slow going. In order to feed himself, he had to devote a large part of each day to stalking animals. Moreover, as the ground thawed, his route turned into a gauntlet of boggy muskeg and impenetrable alder, and McCandless belatedly came to appreciate one of the fundamental (if counterintuitive) axioms of the North: winter, not summer, is the preferred season for traveling overland through the bush.

Faced with the obvious folly of his original ambition, to walk five hundred miles to tidewater, he reconsidered his plans. On May 19, having traveled no farther west than the Toklat River – less than fifteen miles beyond the bus – he turned around. A week later he was back at the derelict vehicle, apparently without regret. He'd decided that the Sushana drainage was plenty wild to suit his purposes and that Fairbanks bus 142 would make a fine base camp for the remainder of the summer.

Ironically, the wilderness surrounding the bus – the patch of overgrown country where McCandless was determined "to become lost in the wild" – scarcely qualifies as wilderness by Alaska standards. Less than thirty miles to the east is a major thoroughfare, the George Parks Highway. Just sixteen miles to the north, beyond an escarpment of the Outer Range, hundreds of tourists rumble daily into Denali Park over a road patrolled by the National Park Service. And unbeknownst to the Aesthetic Voyager, scattered within a six-mile radius of the bus are four cabins (although none happened to be occupied during the summer of 1992).

But despite the relative proximity of the bus to civilization, for all practical purposes McCandless was cut off from the rest of the world. He spent nearly four months in the bush all told, and during that period he didn't encounter another living soul. In the end the Sushana River site was sufficiently remote to cost him his life.

In the last week of May, after moving his few possessions into the bus, McCandless wrote a list of housekeeping chores on a parchmentlike strip of birch bark: collect and store ice from the river for refrigerating meat, cover the vehicle's missing windows with plastic, lay in a supply of firewood, clean the accumulation of old ash from the stove. And under the heading "LONG TERM" he drew up a list of more ambitious tasks: map the area, improvise a bathtub, collect skins and feathers to sew into clothing, construct a bridge across a nearby creek, repair mess kit, blaze a network of hunting trails.

The diary entries following his return to the bus catalog a bounty of wild meat. May 28: "Gourmet Duck!" June 1: "5 Squirrel." June 2: "Porcupine, Ptarmigan, 4 Squirrel, Grey Bird." June 3: "Another Porcupine! 4 Squirrel, 2 Grey Bird, Ash Bird." June 4: "A THIRD PORCUPINE! Squirrel, Grey Bird." On June 5, he shot a Canada goose as big as a Christmas turkey. Then, on June 9, he bagged the biggest prize of all: "MOOSE!" he recorded in the journal. Overjoyed, the proud hunter took a photograph of himself kneeling over his trophy, rifle thrust triumphantly overhead, his features distorted in a rictus of ecstasy and amazement, like some unemployed janitor who'd gone to Reno and won a million-dollar jackpot.

Although McCandless was enough of a realist to know that hunting game was an unavoidable component of living off the land, he had always been ambivalent about killing animals. That ambivalence turned to remorse soon after he shot the moose. It was relatively small, weighing perhaps six hundred or seven hundred pounds, but it nevertheless amounted to a huge quantity of meat. Believing that it was morally indefensible to waste any part of an animal that has been shot for food, McCandless spent six days toiling to preserve what he had killed before it spoiled. He butchered the carcass under a thick cloud of flies and mosquitoes, boiled the organs into a stew, and then laboriously excavated a burrow in the face of the rocky stream bank directly below the bus, in which he tried to cure, by smoking, the immense slabs of purple flesh.

Alaskan hunters know that the easiest way to preserve meat in the bush is to slice it into thin strips and then air-dry it on a makeshift rack. But McCandless, in his naïveté, relied on the advice of hunters he'd consulted in South Dakota, who advised him to smoke his meat, not an easy task under the circumstances. "Butchering extremely difficult," he wrote in the journal on June 10."Fly and mosquito hordes. Remove intestines, liver, kidneys, one Jung, steaks. Get hindquarters and leg to stream."

June 11: "Remove heart and other lung. Two front legs and head. Get rest to stream. Haul near cave. Try to protect with smoker."

June 12: "Remove half rib-cage and steaks. Can only work nights. Keep smokers going."

June 13: "Get remainder of rib-cage, shoulder and neck to cave. Start smoking."

June 14: "Maggots already! Smoking appears ineffective. Don't know, looks like disaster. I now wish I had never shot the moose. One of the greatest tragedies of my life."

At that point he gave up on preserving the bulk of the meat and abandoned the carcass to the wolves. Although he castigated himself severely for this waste of a life he'd taken; a day later McCandless appeared to regain some perspective, for his journal notes, "henceforth will learn to accept my errors, however great they be."

Shortly after the moose episode McCandless began to read Thoreau's *Walden*. In the chapter titled "Higher Laws," in which Thoreau ruminates on the morality of eating, McCandless highlighted, "when I had caught and cleaned and cooked and eaten my fish, they seemed not to have fed me essentially. It was insignificant and unnecessary, and cost more than it came to."

"THE MOOSE," McCandless wrote in the margin. And in the same passage he marked,

> *The repugnance to animal food is not the effect of experience, but is an instinct. It appeared more beautiful to live low and fare hard in many respects; and though I never did so, I went far enough to please my imagination. I believe that every man who has ever been earnest to preserve his higher or poetic faculties in the best condition has been particularly inclined to abstain from animal food, and from much food of any kind....*
>
> *It is hard to provide and cook so simple and clean a diet as will not offend the imagination; but this, I think, is to be fed when we feed the body; they should both sit down at the same table. Yet perhaps this may be done. The fruits eaten temperately need not make us ashamed of our appetites, nor interrupt the worthiest pursuits. But put an extra condiment into your dish, and it will poison you.*

"YES," wrote McCandless and, two pages later; "Consciousness of food. Eat and cook with concentration....Holy Food." On the back pages of the book that served as his journal, he declared:

> *I am reborn. This is my dawn. Real life has just begun.*
> *Deliberate Living: Conscious attention to the basics of life, and a constant attention to your immediate environment and its concerns, example → A job, a task, a book; anything requiring efficient concentration (Circumstance has no value. It is how one relates to a situation that has value. All true meaning resides in the personal relationship to a phenomenon, what it means to you).*
> *The Great Holiness of **FOOD**, the Vital Heat.*
> *Positivism, the Insurpassable Joy of the Life Aesthetic.*
> *Absolute Truth and Honesty.*
> *Reality.*
> *Independence.*
> *Finality–Stability–Consistency.*

As McCandless gradually stopped rebuking himself for the waste of the moose, the contentment that began in mid-May resumed and seemed to continue through early July. Then, in the midst of this idyll, came the first of two pivotal setbacks.

Satisfied, apparently, with what he had learned during his two months of solitary life in the wild, McCandless decided to return to civilization: It was time to bring his "final and greatest adventure" to a close and get himself back to the world of men and women, where he could chug a beer, talk philosophy, enthrall strangers with tales of what he'd done. He seemed to have moved beyond his need to assert so adamantly his autonomy, his need to separate himself from his parents. Maybe he was prepared to forgive their imperfections; maybe he was even prepared to forgive some of his own. McCandless seemed ready, perhaps, to go home.

Or maybe not; we can do no more than speculate about what he intended to do after he walked out of the bush. There is no question, however, that he intended to walk out.

Writing on a piece of birch bark, he made a list of things to do before he departed "Patch Jeans, Shave!, Organize empty pack...." Shortly thereafter he propped his Minolta on an empty oil drum and took a snapshot of himself brandishing a yellow disposable razor and grinning at the camera, clean-shaven, with new patches cut from an army blanket stitched onto the knees of his filthy jeans. He looks healthy but alarmingly gaunt. Already his cheeks are sunken. The tendons in his neck stand out like taut cables.

On July 2, McCandless finished reading Tolstoy's "Family Happiness," having marked several passaged that moved him:

> *He was right in saying that the only certain happiness in life is to live for others....*
>
> *I have lived through much, and now I think I have found what is needed for happiness. A quiet secluded life in the country, with the possibility of being useful to people to whom it is easy to do good, and who are not accustomed to have it done to them; then work which one hopes may be of some use; then rest, nature, books, music, love for one's neighbor – such is my idea of happiness. And then, on top of all that, you for a mate, and children, perhaps – what more can the heart of a man desire?*

Then, on July 3, he shouldered his backpack and began the twenty-mile hike to the improved road. Two days later, halfway there, he arrived in heavy rain at the beaver ponds that blocked access to the west bank of the Teklanika River. In April they'd been frozen over and hadn't presented an obstacle.

Now he must have been alarmed to find a three-acre lake covering the trail. To avoid having to wade through the murky chest-deep water, he scrambled up a steep hillside, bypassed the ponds on the north, and then dropped back down to the river at the mouth of the gorge.

When he'd first crossed the river, sixty-seven days earlier in the freezing temperatures of April, it had been an icy but gentle knee-deep creek, and he'd simply strolled across it. On July 5, however, the Teklanika was at full flood, swollen with rain and snowmelt from glaciers high in the Alaska Range, running cold and fast.

If he could reach the far shore, the remainder of the hike to the highway would be easy, but to get there he would have to negotiate a channel some one hundred feet wide. The water, opaque with glacial sediment and only a few degrees warmer than the ice it had so recently been, was the color of wet concrete. Too deep to wade, it rumbled like a freight train. The powerful current would quickly knock him off his feet and carry him away.

McCandless was a weak swimmer and had confessed to several people that he was in fact afraid of the water. Attempting to swim the numbingly cold torrent or even to paddle some sort of improvised raft across seemed too risky to consider. Just downstream from where the trail met the river, the Teklanika erupted into a chaos of boiling whitewater as it accelerated through the narrow gorge. Long before he could swim or paddle to the far shore, he'd be pulled into these rapids and drowned.

In his journal he now wrote, "Disaster .... Rained in. River look impossible. Lonely, scared." He concluded, correctly, that he would probably be swept to his death if he attempted to cross the Teklanika at that place, in those conditions. It would be suicidal; it was simply not an option.

If McCandless had walked a mile or so upstream, he would have discovered that the river broadened into a maze of braided channels. If he'd scouted carefully, by trial and error he might have found a place where these braids were only chest-deep. As strong as the current was running, it would have certainly knocked him off his feet, but by dog-paddling and hopping along the bottom as he drifted downstream, he could conceivably have made it across before being carried into the gorge or succumbing to hypothermia.

But it would still have been a very risky proposition, and at that point McCandless had no reason to take such a risk. He'd been fending for himself quite nicely in the country. He probably understood that if he was patient and waited, the river would eventually drop to a level where it could be safely forded. After weighing his options, therefore, he settled on the most prudent course. He turned around and began walking to the west, back toward the bus, back into the fickle heart of the bush.

ALONE ON THE WALL

Alex Honnold

Alex Honnold (born 1985) is an American rock climber and the only person to free solo climb (climbing without harness or ropes) El Capitan, a three-thousand-foot rock face in Yosemite National Park, California. A college dropout who lived for ten years out of his car, Honnold burst onto the professional rock-climbing scene in 2007 after repeating Peter Croft's 1987 feat of scaling two Yosemite cliff faces in a single day. He holds many rock-climbing records and has been the subject of several books and the Oscar-winning documentary *Free Solo*.

I sat on the sloping ledge below the crux and rested for a moment. I felt strong and somewhat impatient, but resting seemed like the smart move. After maybe two minutes of stillness, I began preparing. I peed off the ledge, slightly hunched over myself so that I wasn't in full view of the camera above. I systematically tightened my shoes, pulling on each row of laces until they creaked. I quieted my music and chalked up. There were ten moves of climbing above me. It all came down to ten moves. My life could be distilled down to ten moves.

I pressed up onto the starting ledge and chalked again. My left hand found the sloping crack that marks the beginning of the Boulder Problem. My right hand crimped down on a sloping ripple on the smooth wall above me. Both hands flexed hard, as if I was trying to crack my knuckles. Game on!

I raised my left foot to a tiny chip, a hold so small that it was really just body tension holding my foot in place. My right foot matched next to it on an equally poor foot, taking only enough weight for me to throw my left foot up onto a small tooth below my left hand. My left foot and hand were now in opposition, allowing me to raise my right hand up to a higher spot in the same sloping crack as my left. My thumb spragged across the crack to push off the opposing wall. Right arm and left foot were in opposition and I pushed up so my left hand could grab an undercling above. My right foot stood onto the opening crimp and I was able to rock my weight onto it. Just like that, I was done with the first three moves and able to rest again. My right hand went up to a good crimp, though really it was just a placeholder.

All my weight was on my right foot and I would have been able to stand with no hands, even though the foothold was maybe an inch wide.

I chalked both hands and took a few breaths. I didn't think at all about my position or how I felt, I just carried forward on autopilot. Left hand crimped another side pull. Left foot jammed into the lower crack, somewhat sideways. Right foot came up high by my waist and toed into a tiny scoop. The tension between my left hand and my right foot created stability, and I raised my right hand up to a very small but downpulling crimp. My left hand palmed down on the wall below me and I raised my left foot extra-high onto a very sloping shelf. As soon as the foot was placed, I drove down with it and stood up into the "thumbercling" with my left thumb. The hold was an upside-down ripple, maybe an inch long and a few millimeters wide – so small that it would never qualify as a handhold except for the fact that most of my weight was on my feet. I placed my hand perfectly, my thumb going onto the more sloping part in order to leave the better inch of space for my index and middle fingers to curl over. My right foot sneaked in and matched next to my left, and my left foot kicked out leftward to a slippery black knob.

The whole route came down to the next four moves, the hardest of the whole sequence. My right hand came in to my left and I removed my index and middle fingers, which had been stuck to the wall only by the tension between my right big toe and my left thumb. My right thumb took the fingers' place on the good part of the ripple and my left hand shot out left to grab the sloping loaf. I squeezed it for all I was worth and moved my right hand back to the initial down-pulling crimp. I was in an iron cross, fully spanned between the two holds. Right foot through to a sloping dish, hips open, left foot across to a tiny chip, chalk my right hand, bring it in to the top of the loaf. Now I was set up for the karate kick, though I spent no time thinking about it. My right foot came in to the crucial small edge, perfectly placed to provide counterpressure when I kicked my left foot across. I subtly switched the position of my left hand, making it feel slightly more secure as I squeezed my two hands together and crushed the loaf.

As if on autopilot, my left foot shot out perfectly perpendicular to my body, full extension, three feet to the left. It hit the far wall of the corner exactly where it needed to be. What used to feel like a desperate, falling kick now felt like an easy foot placement. My feet felt welded to the wall. Months of stretching paid off, as I brought my left hand to the crack next to my foot. My right hand switched to palming downward and I felt secure, balanced between my left foot and my right palm. I reached my left hand up to a big edge and I was done. I was through the Boulder Problem.

I felt a flood of elation, or maybe just relief. I was suddenly aware of the world around me again, including the cameras fixed to the wall on either side of me. I said something like, "Oh, yeah!" into the camera in front of me. I started laughing. I romped up the final few easy moves to the faint ledge above. I stood on the ledge for a minute, breathing hard, exultant. I knew that somewhere in the meadow 2,300 feet below, Mikey was watching me through the long shot. I pumped my fists over my head, facing the meadow, wondering whether they saw me. For once, I was somewhat glad that l had an audience. I felt like a hero. I still had ten pitches up to 5.12b above me, but I felt like I'd cleared the final hurdle. Now it was just cruising to the finish line.

I loosened my shoes, letting my toes rest for the next several hundred feet of easier climbing. The Sewer was relatively dry, a clear sign that the seasons were slowly changing. But either way, the conditions didn't really matter to me: I jammed and slithered my way up the wet chimney with ease. That got me to the Block, a big sloping ledge the size of a picnic table where people usually camp. I didn't stop. The pitch above it was one of my favorites on the route, delightful face climbing on good flakes, leading to a perfectly cut crack.

As I traversed the flakes the Enduro Corner came into view above me. I could see my friend Sam Crossley dangling on the lip of the Salathé headwall, looking straight down the barrel of the corner, and Jimmy Chin dangling in space out to the left. They were simultaneously close and far – only a few hundred feet away, but in a totally different world. They were filming everything, though I thought nothing of it. I stopped on Sous le Toit ("Under the Roof," because it's below the Salathé headwall), the ledge below the corner, to tighten my shoes for the next 200 feet. Several backpacks were attached to the anchor on the ledge – all the controls for the cameras down below. I asked if there was any water, but Jimmy said no. I'd have to wait until I got to the Round Table. I rested for a moment, more because I felt like I ought to than anything else. I felt strong and wanted to sprint to the summit. But I was trying to be responsible and pace myself.

The corner represented the last real difficulty on the route, a 5.11+ pitch into a 5.12b corner into a 5.12a face traverse without any great stances in between. I tried to climb at an even, measured pace to control my pump. Not that it really mattered – by this point, I felt indestructible. But I still wanted to do it right. The first half of the corner is predominantly jamming, probably my best strength as a climber. It felt fun and secure. I stopped for a moment at the anchor, balanced with one foot on one side of the corner and my back on the other. I took a few deep breaths and started up the second pitch. It's normally liebacked, but that technique always felt wildly insecure to me. I chose instead to chimney it, my left side pushing inside the faint groove and my right foot on the outside wall. It was a much less elegant way to climb, but it felt safe, and that's all I cared about. Much like the Monster down below, it was hard work to move upward, but it felt secure. And sometimes it's nice to rely on the big pushing muscles instead of tiny fingerholds. I knew that Jimmy and Sam were still filming behind and above me, but because I was facing into the corner, I couldn't see them and thought nothing of their existence. My world consisted entirely of a grainy granite crack and my chalky hands.

The upper corner pinched down at the crux, forcing me to lieback off two fingertip-wide piton scars for a few moves. I did exactly what I was supposed to do: lead with my left hand, right foot up into the corner, left foot up higher onto the face, bring my right hand to a small intermediate crimp in the crack, shuffle my body higher, then reach right hand up to a good fingerlock. It was all executed mechanically, my body doing exactly what it had practiced. The final jamming to the anchor felt the easiest it ever had, probably because I had no rope drag and was up there much earlier in the day.

At the anchor, I rested in a faint stem, feet splayed out on either side of the corner and hands alternating on a good little horn. Jimmy said something to me; it didn't register. I think he started moving upward on his fixed ropes in order to be in position on the summit. I didn't care. I took a few breaths and started the distinctive Freerider traverse.

The main difference between Freerider and the Salathé is that Freerider cuts left below the Salathé headwall and avoids the 200 feet of overhanging crack climbing that make the Salathé one of the most iconic routes in the world. The traverse is the sneaky way to bypass the headwall, climbing four pitches up to 5.12a instead of the wild 5.13c Salathé headwall.

The traverse always felt like sport climbing, specific moves between distinct holds. Because of that, I found it much easier to remember than the cracks, which all look about the same. And because the holds are mostly down-pulling and quite positive, it felt quite safe to me. Seven moves into the traverse, I reached high and left to an enormous hole right on the edge of the corner. As my weight swung around to the left, the entirety of the southwest face of El Cap swept below my feet. There was a full 2,500 feet of exposure below me as I dangled from the hole, yet I felt perfectly safe. The hole was positive and I felt strong. Cheyne Lempe was hanging above me, shooting this side of the corner. I didn't really notice and neither of us said anything. A few moves on big holes led me to a faint stance in a corner. I took a few breaths, chalked up, and executed the final seven-move sequence that guarded the Round Table. I was truly elated. Ecstatic. Effusive, even! I babbled to Cheyne as he jugged up to get onto the ledge with me and give me some water. We high-fived. I felt like giving him a bear hug. I still had 500 feet to go, but it was in the bag. Finishing was just a formality. But still, I popped my shoes off for a minute and ate some energy blocks.

I checked my timer. I'd been on route for about three and a half hours! Outrageous! I could break four hours if I hustled. I'd expected to be climbing for closer to five hours. Four felt amazing. I turned up the volume on my tunes and tied my shoes. Back to work.

Since I'd soloed the Easy Rider a few weeks earlier, this all felt like revisiting an old friend. Except that before, I was breaking some mental barriers and I wasn't well enough warmed up. This time I was on fire. I felt unstoppable. I climbed the corner crack above the Round Table in minutes, not bothering to stem or rest, but just jamming it straight on. I felt like I was swimming upward – long moves between good jams. I was locked into the mountain. There was no one filming, though I could hear Jimmy around the corner from me jugging upward as fast as he could. I occasionally bird-whistled so he could keep track of my progress. I'm not sure whether he could hear, but it was fun to think that we were racing.

As I climbed the Scotty Burke Offwidth, named after a former Yosemite Search and Rescue member who was a prominent climber in his time, I thought, "This is the best this pitch has ever been climbed!" I was moving so quickly and easily that it felt like a different route. But then, just as quickly I thought, "Humility! Stay focused!" I didn't want to fall off just because I got sloppy.

Either way, I was done in minutes. I met Jimmy and Josh Huckabee, the team's rigger, on the ledge above and gave more high fives. Jimmy kept jugging as fast as he could. I teased him to go faster. I gave him a brief head start by chatting with Josh, but I couldn't resist long. I wanted to be done.

The 5:11 finger crack above felt trivially easy. Secure jamming up a vertical crack was a delight. And I was still in the shade! I'd climbed so quickly that even the uppermost parts of the route were still shaded, the first time I'd ever experienced that. It was a pleasant surprise. The 5.10d boulder problem above it felt like a formality, just a few more moves keeping me from the summit. I caught Jimmy again on the ledge above as he switched from one rope to the next. He was winded from his long commute. I gave him another fifteen-second head start, but I couldn't contain myself. All that remained was 100 feet of 5.6, basically walking to the summit. I charged! Moments later, I was on top. I'd free soloed El Cap!

MOBY DICK

Herman Melville

Captain Ahab is the fictional protagonist of **Herman Melville**'s classic work *Moby Dick*. Ahab is a man possessed with a single pathological preoccupation: he wanders the seas in search of the great white whale that claimed his leg many years before. The book is a portrait of fanaticism that is based, in part, on the story of the *Essex,* an American whaling vessel that was sunk by a white whale. The tale of the *Essex* is one of only a few recorded instances of a whale ramming a ship, and provided a dramatic backdrop for Melville's masterful tale.

"An hour," said Ahab, standing rooted in his boat's stern; and he gazed beyond the whale's place, towards the dim blue spaces and wide wooing vacancies to leeward. It was only an instant; for again his eyes seemed whirling round in his head as he swept the watery circle. The breeze now freshened; the sea began to swell.

"The birds!—the birds!" cried Tashtego.

In long Indian file, as when herons take wing, the white birds were now all flying towards Ahab's boat; and when within a few yards began fluttering over the water there, wheeling round and round, with joyous, expectant cries. Their vision was keener than man's; Ahab could discover no sign in the sea. But suddenly as he peered down and down into its depths, he profoundly saw a white living spot no bigger than a white weasel, with wonderful celerity uprising, and magnifying as it rose, till it turned, and then there were plainly revealed two long crooked rows of white, glistening teeth, floating up from the undiscoverable bottom. It was Moby Dick's open mouth and scrolled jaw; his vast, shadowed bulk still half blending with the blue of the sea. The glittering mouth yawned beneath the boat like an open-doored marble tomb; and giving one sidelong sweep with his steering oar, Ahab whirled the craft aside from this tremendous apparition. Then, calling upon Fedallah to change places with him, went forward to the bows, and seizing Perth's harpoon, commanded his crew to grasp their oars and stand by to stern.

Now, by reason of this timely spinning round the boat upon its axis, its bow, by anticipation, was made to face the whale's head while yet under water. But as if perceiving this stratagem, Moby Dick, with that malicious intelligence ascribed to him, sidelingly transplanted himself, as it were, in an instant, shooting his pleated head lengthwise beneath the boat.

Through and through; through every plank and each rib, it thrilled for an instant, the whale obliquely lying on his back, in the manner of a biting shark, slowly and feelingly taking its bows full within his mouth, so that the long, narrow, scrolled lower jaw curled high up into the open air, and one of the teeth caught in a row-lock. The bluish pearl-white of the inside of the jaw was within six inches of Ahab's head, and reached higher than that. In this attitude the White Whale now shook the slight cedar as a mildly cruel cat her mouse. With unastonished eyes Fedallah gazed, and crossed his arms; but the tiger-yellow crew were tumbling over each other's heads to gain the uttermost stern.

And now, while both elastic gunwales were springing in and out, as the whale dallied with the doomed craft in this devilish way; and from his body being submerged beneath the boat, he could not be darted at from the bows, for the bows were almost inside of him, as it were; and while the other boats involuntarily paused, as before a quick crisis impossible to withstand, then it was that monomaniac Ahab, furious with this tantalizing vicinity of his foe, which placed him all alive and helpless in the very jaws he hated; frenzied with all this, he seized the long bone with his naked hands, and wildly strove to wrench it from its gripe. As now he thus vainly strove, the jaw slipped from him; the frail gunwales bent in, collapsed, and snapped, as both jaws, like an enormous shears, sliding further aft, bit the craft completely in twain, and locked themselves fast again in the sea, midway between the two floating wrecks. These floated aside, the broken ends drooping, the crew at the stern-wreck clinging to the gunwales, and striving to hold fast to the oars to lash them across.

At that preluding moment, ere the boat was yet snapped, Ahab, the first to perceive the whale's intent, by the crafty upraising of his head, a movement that loosed his hold for the time; at that moment his hand had made one final effort to push the boat out of the bite. But only slipping further into the whale's mouth, and tilting over sideways as it slipped, the boat had shaken off his hold on the jaw; spilled him out of it, as he leaned to the push; and so he fell flat-faced upon the sea.

Ripplingly withdrawing from his prey, Moby Dick now lay at a little distance, vertically thrusting his oblong white head up and down in the billows; and at the same time slowly revolving his whole spindled body; so that when his vast wrinkled forehead rose—some twenty or more feet out of the water—the now rising swells, with all their confluent waves, dazzlingly broke against it; vindictively tossing their shivered spray still higher into the air.* So, in a gale, the but half baffled Channel billows only recoil from the base of the Eddystone, triumphantly to overleap its summit with their scud.

But soon resuming his horizontal attitude, Moby Dick swam swiftly round and round the wrecked crew; sideways churning the water in his vengeful wake, as if lashing himself up to still another and more deadly assault. The sight of the splintered boat seemed to madden him, as the blood of grapes and mulberries cast before Antiochus's elephants in the book of Maccabees. Meanwhile Ahab half smothered in the foam of the whale's insolent tail, and too much of a cripple to swim,—though he could still keep afloat, even in the heart of such a whirlpool as that; helpless Ahab's head was seen, like a tossed bubble which the least chance shock might burst. From the boat's fragmentary stern, Fedallah incuriously and mildly eyed him; the clinging crew, at the other drifting end, could not succor him; more than enough was it for them to look to themselves. For so revolvingly appalling was the White Whale's aspect, and so planetarily swift the ever-contracting circles he made, that he seemed horizontally swooping upon them. And though the other boats, unharmed, still hovered hard by; still they dared not pull into the eddy to strike, lest that should be the signal for the instant destruction of the jeopardized castaways, Ahab and all; nor in that case could they themselves hope to escape. With straining eyes, then, they remained on the outer edge of the direful zone, whose centre had now become the old man's head.

Meantime, from the beginning all this had been descried from the ship's mast heads; and squaring her yards, she had borne down upon the scene; and was now so nigh, that Ahab in the water hailed her!—"Sail on the"—but that moment a breaking sea dashed on him from Moby Dick, and whelmed him for the time. But struggling out of it again, and chancing to rise on a towering crest, he shouted,—"Sail on the whale!—Drive him off!"

The Pequod's prows were pointed; and breaking up the charmed circle, she effectually parted the white whale from his victim. As he sullenly swam off, the boats flew to the rescue.

Dragged into Stubb's boat with blood-shot, blinded eyes, the white brine caking in his wrinkles; the long tension of Ahab's bodily strength did crack, and helplessly he yielded to his body's doom: for a time, lying all crushed in the bottom of Stubb's boat, like one trodden under foot of herds of elephants. Far inland, nameless wails came from him, as desolate sounds from out ravines.

* This motion is peculiar to the sperm whale. It receives its designation (pitchpoling) from its being likened to that preliminary up-and-down poise of the whale-lance, in the exercise called pitchpoling, previously described. By this motion the whale must best and most comprehensively view whatever objects may be encircling him.

But this intensity of his physical prostration did but so much the more abbreviate it. In an instant's compass, great hearts sometimes condense to one deep pang, the sum total of those shallow pains kindly diffused through feebler men's whole lives. And so, such hearts, though summary in each one suffering; still, if the gods decree it, in their life-time aggregate a whole age of woe, wholly made up of instantaneous intensities; for even in their pointless centres, those noble natures contain the entire circumferences of inferior souls.

"The harpoon," said Ahab, half way rising, and draggingly leaning on one bended arm—"is it safe?"

"Aye, sir, for it was not darted; this is it," said Stubb, showing it.

"Lay it before me;—any missing men?"

"One, two, three, four, five;—there were five oars, sir, and here are five men."

"That's good.—Help me, man; I wish to stand. So, so, I see him! there! there! going to leeward still; what a leaping spout!—Hands off from me! The eternal sap runs up in Ahab's bones again! Set the sail; out oars; the helm!"

It is often the case that when a boat is stove, its crew, being picked up by another boat, help to work that second boat; and the chase is thus continued with what is called double-banked oars. It was thus now. But the added power of the boat did not equal the added power of the whale, for he seemed to have treble-banked his every fin; swimming with a velocity which plainly showed, that if now, under these circumstances, pushed on, the chase would prove an indefinitely prolonged, if not a hopeless one; nor could any crew endure for so long a period, such an unintermitted, intense straining at the oar; a thing barely tolerable only in some one brief vicissitude. The ship itself, then, as it sometimes happens, offered the most promising intermediate means of overtaking the chase. Accordingly, the boats now made for her, and were soon swayed up to their cranes—the two parts of the wrecked boat having been previously secured by her—and then hoisting everything to her side, and stacking her canvas high up, and sideways outstretching it with stun-sails, like the double-jointed wings of an albatross; the Pequod bore down in the leeward wake of Moby-Dick. At the well known, methodic intervals, the whale's glittering spout was regularly announced from the manned mast-heads; and when he would be reported as just gone down, Ahab would take the time, and then pacing the deck, binnacle-watch in hand, so soon as the last second of the allotted hour expired, his voice was heard.—"Whose is the doubloon now? D'ye see him?" and if the reply was, No, sir! straightway he commanded them to lift him to his perch. In this way the day wore on; Ahab, now aloft and motionless; anon, unrestingly pacing the planks.

As he was thus walking, uttering no sound, except to hail the men aloft, or to bid them hoist a sail still higher, or to spread one to a still greater breadth—thus to and fro pacing, beneath his slouched hat, at every turn he passed his own wrecked boat, which had been dropped upon the quarter-deck, and lay there reversed; broken bow to shattered stern. At last he paused before it; and as in an already over-clouded sky fresh troops of clouds will sometimes sail across, so over the old man's face there now stole some such added gloom as this.

Stubb saw him pause; and perhaps intending, not vainly, though, to evince his own unabated fortitude, and thus keep up a valiant place in his Captain's mind, he advanced, and eyeing the wreck exclaimed—"The thistle the ass refused; it pricked his mouth too keenly, sir; ha! ha!"

"What soulless thing is this that laughs before a wreck? Man, man! did I not know thee brave as fearless fire (and as mechanical) I could swear thou wert a poltroon. Groan nor laugh should be heard before a wreck."

"Aye, sir," said Starbuck drawing near, "'tis a solemn sight; an omen, and an ill one."

"Omen? omen?—the dictionary! If the gods think to speak outright to man, they will honorably speak outright; not shake their heads, and give an old wives' darkling hint.—Begone! Ye two are the opposite poles of one thing; Starbuck is Stubb reversed, and Stubb is Starbuck; and ye two are all mankind; and Ahab stands alone among the millions of the peopled earth, nor gods nor men his neighbors! Cold, cold—I shiver!—How now? Aloft there! D'ye see him? Sing out for every spout, though he spout ten times a second!"

The day was nearly done; only the hem of his golden robe was rustling. Soon, it was almost dark, but the look-out men still remained unset.

"Can't see the spout now, sir;—too dark"—cried a voice from the air.

"How heading when last seen?"

"As before, sir,—straight to leeward."

"Good! he will travel slower now 'tis night. Down royals and top-gallant stun-sails, Mr. Starbuck. We must not run over him before morning; he's making a passage now, and may heave-to a while. Helm there! keep her full before the wind!—Aloft! come down!—Mr. Stubb, send a fresh hand to the fore-mast head, and see it manned till morning."—Then advancing towards the doubloon in the main-mast—"Men, this gold is mine, for I earned it; but I shall let it abide here till the White Whale is dead; and then, whosoever of ye first raises him, upon the day he shall be killed, this gold is that man's; and if on that day I shall again raise him, then, ten times its sum shall be divided among all of ye! Away now!—the deck is thine, sir!"

And so saying, he placed himself half way within the scuttle, and slouching his hat, stood there till dawn, except when at intervals rousing himself to see how the night wore on.

RACE AGAINST TIME

Ellen MacArthur

Ellen MacArthur (born 1976) is a retired sailor who, in February 2005, broke the world record for the fastest solo circumnavigation of the globe. Born in Derbyshire, MacArthur's first experience of sailing was with her aunt at the age of eight and she later saved up her school dinner money for three years to buy her first boat. She first made headlines in 2001 when she came second in the Vendée Globe, considered the most gruelling yacht race in the world, at the age of twenty-four. Her famous record came a few years later, in the trimaran *B&Q; The Observer* described Ellen as 'the first true heroine of the twenty-first century'. In 2003, she founded the Ellen MacArthur Cancer Trust, which helps young cancer survivors rebuild their confidence through sailing, and in 2010, after retiring from professional sailing, she established The Ellen MacArthur Foundation, dedicated to creating a circular economy to eliminate waste and pollution.

DAY SIXTY-SIX 1/02/05

3 DAYS 6 HOURS AHEAD
700 MILES WSW OF THE CANARY ISLANDS

It's great to be three days ahead again, but I'm taking each day as it comes. We're having a good run right now; Francis also had a good run at the end. It's swings and roundabouts, and as we get closer the difference is going to matter to a much greater extent. We have good conditions right now, but after tomorrow things will change very quickly, and we're going to end up in the high-pressure system – this will probably be upwind, so our speeds are going to drop incredibly. I think I just have to be realistic and do my best. I've got to sail the boat as best I can, and the result will be visible with time. There is nothing more I can do; there's no point in me sitting here saying I'm going to break the record, because quite frankly there is a very good chance I won't, so I've just got to just do my best and see what happens.

Right now, I'm sailing with blue skies on a heading just to the east of north, roughly about 600 miles south of the Azores. Things have been pretty tough since Cape Horn: we've had no wind or Doldrums or difficult conditions. But on the whole I feel pretty happy to be where we are, happy to be sailing at the speed we're sailing. The worry is still there, though. There's a high-pressure system sitting in front of us with no wind in the centre, and it's moving around all the time.

As we head into the high, we'll get the sun back, but the wind will go light and we might have to gybe around the top of the high, go through all the sail changes possible, and then sail upwind to the finish. It doesn't fill me with joy to think of the sail changes – every muscle and joint is hurting. It's hard to imagine being in much before the record at the moment.

I'm looking forward to the finish, to feeling that I can switch my brain off. It's been so intense and concentrated over the last few months that not having to look at the sea or the wind, or think about the batteries, or look after myself to the same extent that I've had to over the last couple of months will be a relief. It's going to be fantastic to see the team and my friends and those people who have supported me. But, right now, the relief is the one thing I'm looking forward to more than anything else. Life out here is incredibly stressful but also amazing. The priorities are different. And, although I have no control over the weather, I have complete control over the boat – she's my responsibility, and I know that as soon as I cross the finish line, whether I have a record or not, that control will vanish. Life is going to change in a very dramatic way.

DAY SIXTY-SEVEN 2/02/05

3 DAYS 10 HOURS AHEAD
330 MILES SWW OF THE AZORES ISLANDS

I didn't get much sleep last night at all; it was rough, and the breeze was up and down like a yo-yo. The breeze would die to 14 knots, and then within two seconds it would kick up to 28, and then within a minute it would die back down to 14 again. The forecasts are a bit of a waste of time right now; the weather models bear no resemblance to what we have. I just can't get any sleep, because every time I get the sails right and I think I can rest, the alarm suddenly goes off and there're 27 knots of breeze again.

I eventually slept in my oilskins in the cuddy. It was in-and-out-of-consciousness sleep, rather than solid sleep. I lay down on the bunk for an hour again this morning, but I couldn't sleep at all. I was getting thrown against the side of the boat, and the alarms were going off.

I don't know whether to change up to the solent – I'm now on two reefs and a smaller staysail – it's right on the limit in the gusts, but I've got only 18 knots right now. I need to be going faster, but the sail change takes twice as long as it did earlier in the trip because I'm so much more tired now. And I don't want to risk breaking the solent. The problem is that we have to make more gains now, as the weather looks terrible ahead. Current routeing shows me in late Tuesday, and the trend is getting worse. Now is the only time to make gains.

DAY SIXTY-EIGHT 3/02/05

3 DAYS 6 HOURS AHEAD
15 MILES EAST OF TERCEIRA ISLAND, AZORES ARCHIPELAGO

I can't believe what we've been through. We stopped for two hours with two knots of breeze from the wrong direction. The breeze went into the north-west, and we were heading back towards the island, only fourteen miles away. We've moved some, we've stopped, we've talked about tacking, we've taken reefs out, put reefs in, we've had 22 knots, then we've had 4 knots, then 19 knots, everything. Just in the last two hours we've done seven gybes. The wind was going round in circles. It's back now, though. I'm totally drained. I hope the wind doesn't go higher: I don't want to have to change from the genoa to a smaller headsail. Somehow I've got to rest, but that's so hard because now the breeze has fallen out. We were going OK, but we're not any more. The pilot's just gone off twice, never a good time at the beginning of a twelve-hour stretch of no wind. It really is light. Light variable, 4–7 knots, you can sail with 4–7 knots, but I just can't rest, I can't rest with the pilot going off. I can deal with not going anywhere, but I can't deal with the alarm going off all the time.

DAY SIXTY-NINE 4/02/05

2 DAYS 13 HOURS AHEAD
690 MILES WEST OF VIGO, SPAIN

Very stressful last night, very stressful. I had about six hours of absolute hell. We had 4 knots of breeze and were able to sail, and then the breeze would just die for forty-five minutes, spinning round and round and round and round. I furled the genoa and then sat and watched it for forty-five minutes because it was pointless trying to sail. It stabilized for about twenty minutes in one direction at about 3.5 knots, so we could just about get moving, and then it suddenly decided it was going to go through 360 degrees four times, which is obviously quite hard to get sailing in. It finally stabilized after a few hours ESE, and then I could see it coming back up into the north. The bubble of high wasn't as high north as we'd thought but more south-west, and we went right over the top of it, which is why we lost the breeze so early. We're through, though, we're through the high bubble, and we're going to have good light winds for a while. Hopefully there's no second bubble at the same time, so we should be OK. The breeze should come in from the north and then strengthen its arse off, basically! The routeing software wasn't showing that we were going to get very far north, but, as long as the breeze stays where it is, I'm gaining to the north, which is fantastic.

In the end, I duct-taped the alarm up because it was so noisy – I just couldn't bear to hear it any more! But I didn't lose the plot once last night, not even for a second, not in six hours of a complete nightmare. What makes it hard is when you imagine one thing happening and then something very different takes place. You don't know where you are with it, you don't know what's going on, but I never lost it. Yet the night before I couldn't deal with anything, nothing – it's amazing what that sleep difference makes. Three hours yesterday and a few slices during the night. As soon as the winds stabilize, I'll try to sleep some more.

Last night I actually felt hungry, which is a miracle, so around 3.00 a.m. I made my dinner. Your objective is to sleep while the breeze is stable, but you've got to eat as well. I burnt my mouth, as I was shovelling my food down too fast so I would have the chance to sleep even for five minutes! You do what you've got to do to survive. So much of this trip has reflected this basic survival instinct – even down to what you can eat and what you can digest. The only things I remotely want to eat are milk products. I've nearly run out of powdered milk because I've used it all in my tea and on my breakfast cereal. I don't know if it's my body wanting fat because I've had no fat in my diet for the last few months. (Thanks, generator!)

Yesterday a ship passed down my port side – I was on port tack sailing with genoa. There was no blip (on the active echo), no nothing, I didn't see anything. That was quite unnerving, really. Now we're tacking off Finisterre and then heading north – this will take us straight up through the shipping route, so we're going to be on full alert.

Hi, team

I'm sitting here with tears in my eyes, not really knowing what to do with myself. I cannot articulate how I feel; I doubt I shall ever be able to express what this trip has put me through, or continues to put me through ... There have been some incredible moments, but there have also been those moments which are far too painful to bring back ... The hardest part is that I know there is little resilience left ... I'm running so close to empty ...

I'm running so close to empty that I believe it's only the energy from others that's keeping me going. Physically I'm exhausted – not just from the effort of sailing Mobi so hard but also from the constant motion, which makes even standing still impossible. On a scale of 1 to 10 this has been a 9 point something, and I'd stick the Vendée Globe on a 5 max.

To put it briefly, this trip has taken pretty much ALL I have, every last drop and ounce. It has taken everything to get this far – and we're still not there yet. I have never attempted something as hard as this before – I want to tell you now that this will take a long time to recover from ... mentally more than anything else ... though you know that I will be brave and give my return to 'normal' life all I can ... please note that there are NO reserves, and that I'm pretty fragile right now.

I just want you to know how I am inside ... I'm a pretty tough person, but this has taken everything. I chose to do this and I really don't need any sympathy from anyone, quite the reverse, but I do need to know you understand how totally exhausted I really am.

exx

DAY SEVENTY 5/02/05

1 DAY 23 HOURS AHEAD
250 MILES WNW OF CAPE FINISTERRE

It's pretty bad already; it's going to be a whole lot worse. The models where I am now say I should have 18 knots of breeze, but I've got a 28-knot average already, gusting 33, and it's not supposed to get bad for another twelve hours. We're going to be lucky to come through this without breaking something or capsizing, to be frank. The waves are going to be absolutely huge and we're going to be going straight across them, which is the worst thing you could possibly do. I'm really worried. Just got to keep things together for the next twenty-four hours.

At the beginning of the night, I managed to get about an hour and a half of sleep because the breeze died. But then I had hours and hours in the night when I couldn't sleep. I was so cold, it's freezing out here, absolutely freezing. I just couldn't get warm, and there were ships around as well. I tell you something, I'm going to be looking forward to sunrise tomorrow morning.

The last hour has been more stable at least – there were a couple of spikes but it is generally OK. I really don't want to bust anything – when the wind went down to 15 knots it was terrible – everything shakes, you're not even loaded and the boat just falls and that's awful. I can't relax at all because it's not a relaxing situation, and it's not like 'Don't worry, you'll be finished in three days,' because right now we're facing the worst conditions from a boat-break point of view that we've had on the entire trip without a doubt. Yesterday the finish seemed quite close; now it feels a very long way away ...

DAY SEVENTY-ONE 6/02/05

1 DAY 23 HOURS AHEAD
250 MILES WNW OF CAPE FINISTERRE

The breeze is still oscillating the whole time. It's so hard to keep the boat going – boat speed at the moment is averaging at 12.7 knots, which is terrible. We had a few really big waves in the night: I was literally thrown out of the bunk by one that broke right over the boat and filled the cockpit – it was lucky I had the door shut. The cuddy was full, everything was awash, all the ropes were swimming around in the cockpit – there must have been a ton of water in there, and I was a bit worried about the structure.

I spent a few hours in my bunk – it was hard, very rough and cold. But, to be honest, it wasn't as cold as the night before. The night before I suffered terribly from the cold.

I really worked hard last night – I'm so tired, but I just want to get across the line. It's been a massive project so far, and a lot of energy has gone in from a lot of people. I feel this is the last opportunity to tie the knot – I want to get it right, I want to do it as swiftly and as sweetly as possible. But I'm very tired: I had less than an hour's sleep yesterday night, I've been up all day today, and 'm going to be up all night tonight. I'm just trying to keep things together until we cross that line.

I have absolutely no idea when we're going to cross the finish line. It should be some time tomorrow night. We had 18 knots of breeze an hour ago; we've now got an average of 11 knots. There are some clouds to windward, we've got the tide to contend with, and the breeze is going aft, so there are about four different things that could come into play which could change the fact of me crossing the line some time from midnight to sunrise. I have no idea, it really is a mystery right now – I just want to get there as fast as possible.

I've got the radar alarm going off all the time at the moment. There's always a ship in the area – I'm going to be dealing with ships from now until the finish, including the traffic separation scheme – so that's probably another 12–18 hours of getting round ships. The last few days of any race or record attempt are always very hard, and this is absolutely no exception. Right now I'm just concentrating on getting to the line as fast as possible.

I know it's going to be a very long night.

DAY SEVENTY-TWO 7/02/05

1 DAY 8 HOURS AHEAD
100 MILES TO THE FINISH LINE

It's 11.00 a.m., and the last twenty-four hours have been completely dreadful. We've had everything from full-on gusts of 40 knots to huge seas in the tail end of a storm yesterday. We sailed out of that sea state during the night with some very strong gusts, and we had to tack five times to get through a small low-pressure system off the north of Spain, which has proved very, very complicated.

There is definitely still a chance to break the record, as long as I don't hit anything or break anything between where I am now and the finish line. I stand a good chance, but it's going to be, as always, very, very difficult. I was hoping to be in before sunset tonight, but that looks absolutely impossible now. We had a very difficult night last night. I managed only about fifteen minutes of sleep – there were ships everywhere and an exceptionally changeable breeze. It's been a full-on night and I am very, very tired. I've got to get it right. I'm just trying to keep things together until we've crossed that line.

Getting close to land is a strange feeling because it's been a long time since we've even seen any. We didn't see Cape Horn; the islands in the South Atlantic are all that we've seen. There's not been a lot of land sighted from *B&Q*, so the thought of coming back to land is pretty novel in itself. When my brain actually allows itself to relax, then maybe I'll be able to take in what's going on around me.

I can't wait to get in. It's been a very, very long trip and an exceptionally hard one. I'll be glad to be crossing that finish line and finally feeling a little bit of relief.

When I crossed the line, I felt like collapsing on the cockpit floor and just falling asleep. The pure fact that you can actually finally let go, that when you cross the line it's over, it's just over, and you don't have to worry any more, that was the biggest emotion – huge relief. So now I'm elated and at the same time absolutely drained, it's been a very tough trip. I've got a mix of emotions in my mind.

The first real feeling of joy was when the guys scrambled on board. They had been following every single step of the journey, but finally I was home – and they were there in flesh and blood, right with me. I've not looked anyone in the eyes for over two months. Just to be able to see and touch people again felt so special. To hug them and share that moment made the record worth something. Prior to that it had just seemed one long struggle.

You have to believe you have a chance when you start on something like this; otherwise you would never have the motivation and drive to build the boat and prepare the boat and to get everything ready. You have to believe that you can do it. I thought Francis's record was beatable – Francis agreed that it was beatable – but to do it the first time, I really didn't think that was possible. It's been a huge challenge, and it's been just a sleigh ride of ups and downs and five-day leads and losing days. I'm very determined when I decide to do something. I will give everything I have to do it. The drive to do that comes from the fact that I work with an unbelievable team, and am supported by some incredible people. And I'm not just out there doing it for me, I'm doing it for everybody. And when the chips are down and things are going wrong, I don't want to let anyone else down, and that's probably one of the biggest motivations.

And now I'm still smiling and I'm on my way home.

A SENSE OF THE WORLD

Jason Roberts

James Holman (1786–1857) was a British adventurer and amateur sociologist who wrote extensively about his travels. Rendered completely blind in his early adulthood, Holman overcame his lack of sight and debilitating pain to become the first blind man to circumnavigate the globe, a feat he accomplished in 1832. His travels were wholly unprecedented given the sheer extent of the distance and Holman's unique method of using human echolocation to guide himself. The Holman river in Equitorial Guinea is named after him in recognition of his staunch abolitionism and dedication to ending the slave trade in the region.

Jason Roberts is an American author who lives in California. *A Sense of the World* was a national bestseller, a finalist for the National Book Critics Circle Award and longlisted for the *Guardian* First Book Award.

The blind man paused to feel the end of his walking stick. It was scorched and blackened, a few moments shy of bursting into flame. He rolled the tip in the abundant ash to cool it, then continued his progress upward, toward the mouth of the volcano.

His friend, a young Irish surgeon named Robert Madden, looked on with both clinical interest and personal fascination.

> *The great eruption of June, 1821 was witnessed by me. I accompanied to the mount the celebrated blind traveller, Lieutenant Holman, the evening of which the violence was at its greatest height.*

From a distance, thirty-four-year-old James Holman didn't look blind. The prodding of his stick seemed more like a bit of swagger than an act of orientation. Despite his condition he was still a uniformed officer of the British Royal Navy, although that uniform was at the moment irreverently topped with a broad straw hat. He cut an attractive figure: lean, above middle height, clearly accustomed both to command and to life outdoors. His bearing was straight-backed and confident, and his youthful features radiated an intelligent enthusiasm. Jane Austen would have recognized him immediately as a Military Gentleman, dashing yet soulful, suitable for a central role in one of her romances.

> *He insisted on walking over places where we could hear the crackling effects of the fire on the lava beneath our feet, and on a level with the brim of the new crater, which was then pouring forth showers of fire and smoke, and lava, and occasionally masses of rock of amazing dimensions, to an enormous height in the air.*

There was nervous talk of halting the ascent, but Lieutenant Holman remained in firm command of his miniature expedition. He had begun the climb quite willing to proceed alone, and from his two companions he "rather looked for amusement and information than guidance and protection." Signore Salvatori had offered to take them part of the way on muleback., but Holman had insisted on hiking the full distance.

"I see things better with my feet," he explained.

Decades later, when Dr. Madden had abandoned medicine and drifted into hack writing, he recounted that evening in a memoir. It wasn't even his memoir. He was churning out *The Literary Life and Correspondence of the Countess of Blessington* when a chapter set in Italy moved him to slip a personal reverie into the pages:

> *A change of wind must inevitably have buried us, either beneath the ashes or the molten lava. The huge rocks generally fell back into the crater from which they issued. The ground was glowing with heat beneath our feet, which often obliged us to shift our position.*

They arrived at the very edge of the crater. Salvatori, of a family that had served as field guides on Mount Vesuvius for generations, announced that Holman had just made history. Attempts on the volcano were few enough that the king of Naples received a personal report on them all, and Salvatori looked forward to informing His Majesty of the first sightless man to reach the summit. But now it was time to turn back. The air was scarcely breathable, the ground audibly unstable. They were shifting their stances, almost hopping. Too much contact with the fuming ground could burn their feet through their shoes.

Holman lingered as long as he could, savoring the scent of sulphur, the sounds of earth in motion, the very omnipresence of the heat.

When at last they retreated a few yards, Holman shook the ashes from his shoes. Then he calmly pointed out that this was the summit of the *active* volcano—not of the mountain itself. Another, dormant crater awaited higher up the slope, and he intended to climb until there was nothing further to climb. After they cleared the active lava fields it would get quite cold, and he hoped to keep a brisk pace. He had neglected to bring a coat.

It was time for Salvatori to declare an impossibility. Perhaps the lieutenant was forgetting that it was well past midnight. The moon was setting, and outside of the eruption's glow there would soon be nothing but a dangerous darkness. This was, of course, a handicap to Madden and Salvatori, not Holman, who gracefully conceded the point. A quiet settled over them as they began to retrace their steps.

> *The view of the bay of Naples and of the distant city, from the summit of Vesuvius on a beautiful moonlight night without a cloud in the sky, such as we had the good fortune to enjoy, was almost magic in its effect; such serenity and repose and beauty in perfect stillness, formed a striking contrast with the lurid glare of the red hot masses that were emitted from the volcano, and the frightful bellowings of the burning mountain on which we stood.*

It was still dark when they reached the nearest shelter, a hermit's cabin four miles distant. The occupant, something of an entrepreneur-monk, aggressively priced and freely poured cupfuls of a local wine called Lachrymae Christi, Tears of Christ. A group of Austrians slumbered on couches, waiting for sunrise and their shot at the summit. In the warmth of the fire and the wine, Holman allowed elation to win out over exhaustion. The first, the very first.

It was a triumph, not so much of his courage as of his ability to charm a path in a given direction. Vesuvius was at its most violent in living memory, and yet he had managed to deflect the grave concern of his friends, circumvent Salvatori's professional caution, and march to the very precipice, accompanied but at his own pace, under his own power. Such reasserted freedom wasn't just a sop to his dignity. It was, in his opinion, the only thing keeping him alive.

Two years earlier Holman had been a bedridden invalid, slowly retreating from life. Unable to cure either his blindness or the wracking pain that made an agony of motion, his physicians decided that a warmer climate might provide, if not a cure, then at least a comfortable setting for his remaining days. Left to find his own attendants to convey him to the Mediterranean, the patient—acting out of poverty, pride, or sheer self-destructiveness—did nothing of the sort. Instead he hobbled onboard a ferry bound for France, quite alone.

Six months later he emerged at the Mediterranean coast, bruised and wearied by the journey but also reinvigorated. Solitary travel, he'd found, was the collision of chaos and momentum, a constant, welcome assault on his senses and attention. It distracted him from his pain, and sparked new energies within. Instead of lingering at his supposed destination he'd chosen to keep moving, to cling to the road like a lifeline.

It had led him here, to the volcano's edge. And now he was savoring the realization that it could lead him anywhere.

The monk kept a visitors' book, and Salvatori urged him to make an entry. A quill was dipped in ink and placed in his hand, then gently guided to the page. Holman, a poet long before he was an officer, thought for a moment, then made what he hoped was a legible approximation of writing:

Some difficulties meet, full many
I find them not, nor seek for any.

He handed back the book and quill, not realizing he had just composed his lifetime's motto. Years later, at the height of his fame, he would make a habit of adding this scrap of verse to his autograph, as if it were a private pact with the world. *I wish to pass through, not vanquish. Do not vanquish me.*

Holman rose from his chair, not needing to ask the hour. The monk was no longer feeding the fire. The Austrians were stirring more in their sleep. Without touching, he knew precisely where the cabin's windows were—he could tap them with his walking stick if he wished—and now a new warmth was faintly radiating from the panes. The sun was rising. There was sufficient light on the path for his companions.

It was time to go.

MAN'S SEARCH FOR MEANING

Viktor E. Frankl

Viktor E. Frankl (1905–1997) was an Austrian neurologist, psychologist, author and Holocaust survivor. He was the founder of logotherapy, a psychotherapeutic school which considers the innate quest for meaning in life as humanity's most powerful motivating force. Whilst living in the horrific conditions of the Nazi concentration camps of Auschwitz and Theresienstadt, Frankl arrived at the conclusion that people who have an inner sense of meaning – an ideology, faith or person that gives their existence purpose – can endure virtually any torment. Conversely, he believed those with no interior sense of meaning can suffer and perish in even the most ideal conditions. For Frankl, the desire for the love of his wife allowed him to endure starvation, beatings and endless psychological torment. When his camp was liberated by American forces in 1945, he learned that his wife had been killed in another camp. Overcome by grief, he threw himself into his psychological research, which provided him with a new sense of purpose. *Man's Search For Meaning*, published in 1946, is based on his experiences within the camps.

In attempting this psychological presentation and a psychopathological explanation of the typical characteristics of a concentration camp inmate, I may give the impression that the human being is completely and unavoidably influenced by his surroundings. (In this case the surroundings being the unique structure of camp life, which forced the prisoner to conform his conduct to a certain set pattern.) But what about human liberty? Is there no spiritual freedom in regard to behavior and reaction to any given surroundings? Is that theory true which would have us believe that man is no more than a product of many conditional and environmental factors – be they of a biological, psychological or sociological nature? Is man but an accidental product of these? Most important, do the prisoners' reactions to the singular world of the concentration camp prove that man cannot escape the influences of his surroundings? Does man have no choice of action in the face of such circumstances?

We can answer these questions from experience as well as on principle. The experiences of camp life show that man does have a choice of action. There were enough examples, often of a heroic nature, which proved that apathy could be overcome, irritability suppressed. Man *can* preserve a vestige of spiritual freedom, of independence of mind, even in such terrible conditions of psychic and physical stress.

We who lived in concentration camps can remember the men who walked through the huts comforting others, giving away their last piece of bread. They may have been few in number, but they offer sufficient proof that everything can be taken from a man but one thing: the last of the human freedoms – to choose one's attitude in any given set of circumstances, to choose one's own way.

And there were always choices to make. Every day, every hour, offered the opportunity to make a decision, a decision which determined whether you would. or would not submit to those powers which threatened to rob you of your very self, your inner freedom; which determined whether or not you would become the plaything of circumstance, renouncing freedom and dignity to become molded into the form of the typical inmate.

Seen from this point of view, the mental reactions of the inmates of a concentration camp must seem more to us than the mere expression of certain physical and sociological conditions. Even though conditions such as lack of sleep, insufficient food and various mental stresses may suggest that the inmates were bound to react in certain ways, in the final analysis it becomes clear that the sort of person the prisoner became was the result of an inner decision, and not the result of camp influences alone. Fundamentally, therefore, any man can, even under such circumstances, decide what shall become of him – mentally and spiritually. He may retain his human dignity even in a concentration camp. Dostoevski said once, "There is only one thing that I dread: not to be worthy of my sufferings." These words frequently came to my mind after I became acquainted with those martyrs whose behavior in camp, whose suffering and death, bore witness to the fact that the last inner freedom cannot be lost. It can be said that they were worthy of their sufferings; the way they bore their suffering was a genuine inner achievement. It is this spiritual freedom – which cannot be taken away – that makes life meaningful and purposeful.

An active life serves the purpose of giving man the opportunity to realize values in creative work, while a passive life of enjoyment affords him the opportunity to obtain fulfillment in experiencing beauty, art, or nature. But there is also purpose in that life which is almost barren of both creation and enjoyment and which admits of but one possibility of high moral behavior: namely, in man's attitude to existence, an existence restricted by external forces. A creative life and a life of enjoyment are banned to him. But not only creativeness and enjoyment are meaningful. If there is a meaning in life at all, then there must be a meaning in suffering. Suffering is an ineradicable part of life, even as fate and death. Without suffering and death human life cannot be complete.

The way in which a man accepts his fate and all the suffering it entails, the way in which he takes up his cross, gives him ample opportunity – even under the most difficult circumstances – to add a deeper meaning to his life. It may remain brave, dignified and unselfish. Or in the bitter fight for self-preservation he may forget his human dignity and become no more than an animal. Here lies the chance for a man either to make use of or to forgo the opportunities of attaining the moral values that a difficult situation may afford him. And this decides whether he is worthy of his sufferings or not.

Do not think that these considerations are unworldly and too far removed from real life. It is true that only a few people are capable of reaching such high moral standards. Of the prisoners only a few kept their full inner liberty and obtained those values which their suffering afforded, but even one such example is sufficient proof that man's inner strength may raise him above his outward fate. Such men are not only in concentration camps. Everywhere man is confronted with fate, with the chance of achieving something through his own suffering.

Take the fate of the sick – especially those who are incurable. I once read a letter written by a young invalid, in which he told a friend that he had just found out he would not live for long, that even an operation would be of no help. He wrote further that he remembered a film he had seen in which a man was portrayed who waited for death in a courageous and dignified way. The boy had thought it a great accomplishment to meet death so well. Now – he wrote – fate was offering him a similar chance.

Those of us who saw the film called *Resurrection* – taken from a book by Tolstoy – years ago, may have had similar thoughts. Here were great destinies and great men. For us, at that time, there was no great fate; there was no chance to achieve such greatness. After the picture we went to the nearest café, and over a cup of coffee and a sandwich we forgot the strange metaphysical thoughts which for one moment had crossed our minds. But when we ourselves were confronted with a great destiny and faced with the decision of meeting it with equal spiritual greatness, by then we had forgotten our youthful resolutions of long ago, and we failed.

Perhaps there came a day for some of us when we saw the same film again, or a similar one. But by then other pictures may have simultaneously unrolled before one's inner eye; pictures of people who attained much more in their lives than a sentimental film could show. Some details of a particular man's inner greatness may have come to one's mind, like the story of the young woman whose death I witnessed in a concentration camp. It is a simple story. There is little to tell and it may sound as if I had invented it; but to me it seems like a poem.

This young woman knew that she would die in the next few days. But when I talked to her she was cheerful in spite of this knowledge. "I am grateful that fate has hit me so hard," she told me. "In my former life I was spoiled and did not take spiritual accomplishments seriously." Pointing through the window of the hut, she said, "This tree here is the only friend I have in my loneliness." Through that window she could see just one branch of a chestnut tree, and on the branch were two blossoms. "I often talk to this tree," she said to me. I was startled and didn't quite know how to take her words. Was she delirious? Did she have occasional hallucinations? Anxiously I asked her if the tree replied. "Yes." What did it say to her? She answered, "It said to me, 'I am here – I am here – I am life, eternal life.'"

APPENDIX

ACKNOWLEDGEMENTS

This book is my nod to those who have shown courage: the mavericks, the survivors, the explorers, the daring, the wanderers and the oppressed. This is a compilation for the brave, for those who stayed the course, come what may. Their stories ought to be remembered.

There are many people that I must thank for their assistance in compiling this anthology.

I must give special mention to Charles McBryde for his hard work in creating a long list of contenders and for his research into the biographies.

Evangeline Modell was instrumental in selecting the right extracts, making sure there was balance and ensuring the book found its flow. Her help has been invaluable throughout.

As ever, I owe the book to my fabulous agent Jo Cantello, and Georgina Blackwell at my publisher, Head of Zeus, as well as all of the team involved in the edit.

I am, of course, eternally indebted to all those authors who have contributed to this book. Their extraordinary feats, bravery and determination as well as generosity in sharing their stories has inspired countless people around the world and will continue to do so for generations to come. I thank them for inspiring me too, and salute them all.

EXTENDED COPYRIGHT

ROBERT CALDER: *A Hero for the Americas: The Legend of Gonzalo Guerrero*, University of Regina Press, copyright © Robert Calder, 2017. Reprinted by permission of University of Regina Press. All rights reserved;

JON T. COLEMAN: "Just Another Word for Nothing Left to Chew" from *Here Lies Hugh Glass*, copyright © Jon T. Coleman, 2012. Reproduced by permission of Hill and Wang, a division of Farrar, Straus and Giroux. All Rights Reserved;

MARIE COLVIN: *On the Frontline*, HarperPress, 2012, copyright © *Sunday Times*, 2012. Reproduced by permission of HarperCollins Publishers Ltd;

LAURA DEKKER: *One Girl, One Dream*, HarperCollins Publishers (New Zealand) Limited, 2017. Reproduced by permission of the publisher;

ELISABETH ELLIOT: *Through Gates of Splendor*, OM / Send the Light, pp. 183–189, copyright © 1956, 1957. Reproduced with thanks from The Elisabeth Elliot Foundation, ElisabethElliot.org, and Tyndale Publishing;

SHŪSAKU ENDŌ: *Silence*, translated by William Johnston, Picador, 2015, copyright © The Heirs of Shūsaku Endō, 1966, translation copyright © William Johnston, 1969. Reproduced by permission of the Licensor through PLSClear;

RANUPLH FIENNES: *Living Dangerously*, Macmillan, copyright © Sir Ranulph Fiennes, 1987. Reproduced by permission of Curtis Brown Group Ltd, London on behalf of Ranulph Fiennes;

VIKTOR E. FRANKL: *Man's Search for Meaning*, Rider, copyright © Viktor E Frankl 1959, 1962, 1984, 1992, 2004. Reprinted by permission of The Random House Group Limited;

PETER FREUCHEN: *Vagrant Viking: My life and Adventures*, copyright © Peter Freuchen, 1953, renewed 1981 by Dagmar Gale Freuchen. Reproduced by permission of Don Congdon Associates, Inc.;

MAHATMA GANDHI: *Selected Political Writings*, Hackett Publishing, 1996. Reproduced by permission of Navajivan Trust;

YOSSI GHINSBERG: *Lost in the Jungle*, Summersdale, copyright © 2006, pp. 241–251. Reproduced by kind permission of the author;

ROBERT GOODWIN: *Crossing the Continent, 1527-1540*, HarperCollins, 2008, copyright © Robert Goodwin, 2008. Reproduced by permission of HarperCollins Publishers and the author;

DAVID GRANN: *The Lost City of Z*, Simon & Schuster UK Limited, 2017, copyright © David Grann, 2005, 2009. Reproduced by permission of the publisher;

DAVID GRANN: *The White Darkness*, Simon & Schuster UK Limited, 2018, copyright © David Grann, Inc, 2018. Reproduced by permission of the publisher;

GRAHAM GREENE: *The Power and the Glory*, Vintage, 2001, copyright © Graham Greene, 1940. Reproduced by permission of David Higham Associates;

HEINRICH HARRER: *Seven Years in Tibet*, translated by Rupert Hart-Davis, Harper Perennial, 2005, copyright © Heinrich Harrer, 1952, translation copyright © Rupert Hart-Davis, 1953. Reproduced by permission of HarperCollins Publishers Ltd; and the Estate of the translator;

SVEN HEDIN: *My Life as an Explorer*, translated by Alfhild Heubsch, copyright © 1925. Permission granted by the Sven Hedin Foundation, Royal Swedish Academy of Sciences;

THOR HEYERDAHL: *The Kon-Tiki Expedition: By Raft Across the South Seas*, Flamingo, 1996, copyright © George Allen & Unwin (Publishers) Ltd, 1950, 1982. Reproduced by permission of HarperCollins Publishers Ltd; and Gyldendal Norsk Forlag;

EDMUND HILLARY, Sir: *Nothing Venture, Nothing Win*, Hodder & Stoughton, 1975, pp. 184–191. Reproduced by permission of Ed Hillary IP Ltd;

ALEX HONNOLD: *Alone on the Wall*, Macmillan, 2019, copyright © Alex Honnold and David Roberts, 2015, 2018. Reproduced by permission of the Licensor through PLSClear;

IMMACULÉE ILIBAGIZA: *Left to Tell: Discovering God Amidst the Rwandan Holocaust*, Hay House, 2006, pp. 102–106. Reproduced by permission of the publisher;

DONALD JOHANSON and MAITLAND EDEY: *Lucy: The Beginnings of Humankind*, Touchstone, 1990, copyright © Donald Johanson and Maitland Edey, 1981. Reproduced by permission of SLL/Sterling Lord Literistic, Inc.;

SCOTT KELLY: *Endurance: A Year in Space, A Lifetime of Discovery*, Doubleday, Black Swan 2018, copyright © Scott Kelly, 2017. Reproduced by permission of The Random House Group Limited;

JULIANE KOEPCKE: *When I Fell from The Sky: The True Story of One Woman's Miraculous Survival*, translated by Ross Benjamin, Nicholas Brealey Publishing, an imprint of John Murray Press, 2012, copyright © Juliane Koepcke and Piper Verlag, Munich 2012; translation copyright © Ross Benjamin, 2011. Reproduced by permission of the Licensor through PLSClear;

JOHN KRAKAUER: *Into The Wild*, Pan Macmillan 2007, copyright © John Krakauer, 1996. Reproduced by permission of the Licensor through PLSClear;

DALAI LAMA: *My Land and My People*, copyright © 1997. Reprinted by permission of Grand Central Publishing, an imprint of Hachette Book Group, Inc.;

ALFRED LANSING: *Endurance: Shackleton's Incredible Voyage to the Antarctic*, Phoenix, 2000, copyright © Alfred Lansing, 1959. Reproduced by permission of the Licensor through PLSClear;

MARTIN LUTHER KING, JR,. DR.: 'Letter from a Birmingham Jail', copyright © 1963 by Dr. Martin Luther King, Jr. Renewed © 1991 by Coretta Scott King. Reprinted by arrangement with The Heirs to the Estate of Martin Luther King Jr., c/o Writers House as agent for the proprietor New York, NY;

ELLEN MACARTHUR: *Race Against Time*, Michael Joseph, copyright © Ellen MacArthur, 2005. Reproduced by permission of Penguin Books Limited;

ELLA MAILLART: *The Cruel Way*, John Murray Publishers, 2021, copyright © Ella Maillart, 1947. Reproduced by permission of the Licensor through PLSClear;

NELSON MANDELA: *Long Walk to Freedom: The Autobiography of Nelson Mandela*, excerpt, Abacus, copyright © 1994, pp. 747–751. Reproduced by permission of the Nelson Mandela Foundation;

DAVID MASSOLA: *The Great Cave Rescue*, Duckworth Books, 2019, pp. 53–57. Reproduced by permission of Duckworth Books and Allen & Unwin Publishers, https://www.allenandunwin.com;

ROBERT D. MATZEN: *Mission: Jimmy Stewart and the Fight for Europe*, GoodKnight Books, copyright © 2016, pp. 239–250. Reproduced by kind permission of the author;

STEPHEN MCGINTY: *The Dive*, copyright © Stephen McGinty, 2021. Reproduced by permission of HarperCollins Publishers Ltd;

STEPHEN MCGOWN: *Six Years a Hostage*, Robinson, 2021, copyright © Stephen McGown, 2020. Reproduced by permission of the Licensor through PLSClear; and Maverick 451, a division of Daily Maverick;

NICHOLAS MONSARRAT: *The Cruel Sea*, Orion, 2009, copyright © Nicholas Monsarrat, 1951. Reproduced by permission of The Marsh Agency Ltd on behalf of The Estate of Nicholas Monsarrat;

MARÍA ELENA MOYANO: *The Autobiography of María Elena Moyano: The Life and Death of a Peruvian Activist*, translated by Patricia Taylor Edmisten, edited by Diana Miloslavich Tupac, University Press of Florida, 2000, pp. 62–72. Reprinted by permission of the University Press of Florida;

ERIC NEWBY: *Love and War in the Appenine Mountains,* HarperPress, 2010, copyright © Eric Newby, 1971. Reproduced by permission of HarperCollins Publishers Ltd;

JENNIFER NIVEN: *Ada Blackjack: A True Story of Survival in the Arctic*, Hyperion, 2004, copyright © Jennifer Niven, 2003, pp. 215–222. Reproduced by permission of Levin, Greenberg, Rostan Literary Agency;

DAVID NOTT: *War Doctor*, Picador, pp. 14–16, 19–24, copyright © 2019. Reproduced by permission of David Higham Associates;

HIROO ONODA: *No Surrender: My Thirty-Year War*, translated by Charles S. Terry, Bluejacket Books, 1999, copyright © Kodansha International Ltd, 1974. Reproduced by permission of Naval Institute Press;

NANDO PARRADO: *Miracle in the Andes: 72 Days on the Mountain and My Long Trek Home*, Orion, 2006, copyright © Nando Parrado, 2006. Reproduced by permission of the Licensor through PLSClear;

NIMSDAI PURJA: *Beyond Possible*, Hodder & Stoughton, 2020, copyright © Nirmal Purja, 2020. Reproduced by permission of the Licensor through PLSClear;

ARON RALSTON: *Between a Rock and a Hard Place*, Simon & Schuster UK Limited, 2004, copyright © Aron Ralston, 2004. Reproduced by permission of the publisher;

JASON ROBERTS: *A Sense of the World*, Simon & Schuster UK Limited, 2006, copyright © Jason Roberts, 2006. Reproduced by permission of the publisher and the author;

DAVID SHEFF: *The Buddhist on Death Row*, HQ, 2020, copyright © David Sheff, 2020. Reproduced by permission of HarperCollins Publishers Ltd;

JOE SIMPSON: *Touching the Void: The True Story of One Man's Miraculous Survival,* Jonathan Cape, 1988, Vintage, 1997, copyright © Joe Simpson, 1988. Reproduced by permission of The Random House Group Limited;

ALEKSANDR SOLZHENITSYN: *The Gulag Archipelago*, copyright © The Russian Social Fund, 1985. Reprinted by permission of The Random House Group Limited;

ED STAFFORD: *Walking the Amazon*, Virgin Books, an imprint of Ebury Publishing, copyright © Ed Stafford 2011. Reproduced by permission of The Random House Group Limited;

FREYA STARK: *The Valley of the Assassins*, John Murray Publishers, 2021, copyright © Freya Stark, 1934. Reproduced by permission of the Licensor through PLSClear;

WILFRED THESIGER: *Arabian Sands*, Penguin Books, 2007, copyright © Wilfred Thesiger, 1959. Reproduced by permission of Curtis Brown Group Ltd, London on behalf of The Estate of Wilfred Thesiger;

LOUNG UNG: *First They Killed My Father,* 2001, first published in the USA by HarperCollins, 2000; published in the UK by Mainstream, 2017, copyright © Loung Ung, 2000. Reproduced by permission of CreativeWell, Inc. (USA). All rights reserved;

TERRY WAITE: *Taken on Trust*, Hodder & Stoughton, 1993, copyright © Terry Waite, 1993. Reproduced by permission of the Licensor through PLSClear;

SARA WHEELER: *Terra Incognita*, Vintage, copyright © Sara Wheeler, 1996. Reproduced by permission of The Random House Group Limited;

MALALA YOUSAFZAI: *I am Malala: The Girl Who Stood Up for Education and Was Shot by the Taliban*, with Christina Lamb, Weidenfeld & Nicholson, an imprint of Orion Publishing Group Ltd, 2013, copyright © Salazai Limited, 2013. Reproduced by permission of the Licensor through PLSClear.